Wave Motion and Applied Science

Wave Motion and Applied Science

Editor:

Olga Moreira, Ph.D.

ArclerPress

www.arclerpress.com

Wave Motion and Applied Science

Editor: Olga Moreira, Ph.D.

© 2017 Arcler Press LLC
708 3rd Avenue, 6th Floor
New York
NY 10017
United States of America
www.arclerpress.com

ISBN: 978-1-68094-369-6

Library of Congress Control Number: 2016950947

About the Editor

Olga Moreira, Ph.D.

Olga Moreira obtained her Ph.D. in Astrophysics from the University of Liege (Belgium) in 2010, her BSc. in Physics and Applied Mathematics from the University of Porto (Portugal). Her post-graduate travels and international collaborations with the European Space Agency (ESA) and European Southern Observatory (ESO) led to great personal and professional growth as a scientist. Currently, she is working as an independent researcher, technical writer, and editor in the fields of Mathematics, Physics, Astronomy and Astrophysics.

Contents

List of Contributors

Paulo Roberto de Freitas Teixeira
Engineering School, Federal University of Rio Grande - FURG, Rio Grande, Brazil

Alexey Androsov
Alfred Wegener Institute for Polar and Marine Research, POB 120161, 27515 Bremerhaven, Germany

Sven Harig
Alfred Wegener Institute for Polar and Marine Research, POB 120161, 27515 Bremerhaven, Germany

Annika Fuchs
Alfred Wegener Institute for Polar and Marine Research, POB 120161, 27515 Bremerhaven, Germany

Antonia Immerz
Alfred Wegener Institute for Polar and Marine Research, POB 120161, 27515 Bremerhaven, Germany

Natalja Rakowsky
Alfred Wegener Institute for Polar and Marine Research, POB 120161, 27515 Bremerhaven, Germany

Wolfgang Hiller
Alfred Wegener Institute for Polar and Marine Research, POB 120161, 27515 Bremerhaven, Germany

Sergey Danilov
Alfred Wegener Institute for Polar and Marine Research, POB 120161, 27515 Bremerhaven, Germany

Yoko Tono
Department of Deep Earth Structure and Dynamics ResearchJapan Agency for Marine-Earth Science and Technology

Kiwamu Nishida
Earthquake Research InstituteUniversity of Tokyo

Yoshio Fukao
Research and Development Center for Earthquake and TsunamiJapan Agency for Marine-Earth Science and Technology

Akiko To
Research and Development Center for Earthquake and TsunamiJapan Agency for Marine-Earth Science and Technology

Narumi Takahashi
Research and Development Center for Earthquake and TsunamiJapan Agency for Marine-Earth Science and Technology

Oleg A Godin
Cooperative Institute for Research in Environmental SciencesUniversity of Colorado
Physical Sciences DivisionNOAA Earth System Research Laboratory

Nikolay A Zabotin
Cooperative Institute for Research in Environmental SciencesUniversity of Colorado
Department of Electrical, Computer, and Energy EngineeringUniversity of Colorado

Terence W Bullett
Cooperative Institute for Research in Environmental SciencesUniversity of Colorado
NOAA National Geophysical Data Center

Takuji Waseda

Department of Ocean Technology Policy and Environment, Graduate School of Frontier ScienceThe University of Tokyo
Application Laboratory, Japan Agency for Marine-Earth Science and Technology

Masato Sinchi
Department of Ocean Technology Policy and Environment, Graduate School of Frontier ScienceThe University of Tokyo

Keiji Kiyomatsu
Department of Ocean Technology Policy and Environment, Graduate School of Frontier ScienceThe University of Tokyo

Tomoya Nishida
Department of Ocean Technology Policy and Environment, Graduate School of Frontier ScienceThe University of Tokyo

Shunsuke Takahashi
Department of Ocean Technology Policy and Environment, Graduate School of Frontier ScienceThe University of Tokyo

Sho Asaumi
Department of Ocean Technology Policy and Environment, Graduate School of Frontier ScienceThe University of Tokyo

Yoshimi Kawai
Research and Development Center for Global Change, Japan Agency for Marine-Earth Science and Technology

Hitoshi Tamura
Application Laboratory, Japan Agency for Marine-Earth Science and Technology

Yasumasa Miyazawa
Application Laboratory, Japan Agency for Marine-Earth Science and Technology

Benoit Frisquet
Laboratoire Interdisciplinaire Carnot de Bourgogne (ICB), UMR 6303 CNRS - Université de Bourgogne Franche-Comté, F-21078 Dijon, France

Bertrand Kibler
Laboratoire Interdisciplinaire Carnot de Bourgogne (ICB), UMR 6303 CNRS - Université de Bourgogne Franche-Comté, F-21078 Dijon, France

Philippe Morin
Laboratoire Interdisciplinaire Carnot de Bourgogne (ICB), UMR 6303 CNRS - Université de Bourgogne Franche-Comté, F-21078 Dijon, France

Fabio Baronio
Department of Information Engineering, Università di Brescia, and INO-CNR, Brescia, Italy

Matteo Conforti
Univ. Lille, CNRS, UMR8523 - PhLAM - Physique des Lasers Atomes et Molécules, F-59000 Lille, France

Guy Millot
Laboratoire Interdisciplinaire Carnot de Bourgogne (ICB), UMR 6303 CNRS - Université de Bourgogne Franche-Comté, F-21078 Dijon, France

Stefan Wabnitz
Department of Information Engineering, Università di Brescia, and INO-CNR, Brescia, Italy

Shunsuke Takemura
Department of Material System Science, Graduate School of NanobioscienceYokohama City University

Mai Akatsu
Department of Material System Science, Graduate School of NanobioscienceYokohama City University

Kei Masuda
Department of Material System Science, Graduate School of NanobioscienceYokohama City University

Kengo Kajikawa
Department of Material System Science, Graduate School of NanobioscienceYokohama City University

Kazuo Yoshimoto
Department of Material System Science, Graduate School of NanobioscienceYokohama City University

Arben Pitarka
Atmospheric, Earth and Energy Division, Lawrence Livermore National Laboratory, 7000 East Avenue

Abdullah Al-Amri
Department of Geology and Geophysics, King Saud University, Riyadh 11451, Saudi Arabia.

Michael E. Pasyanos
Atmospheric, Earth and Energy Division, Lawrence Livermore National Laboratory, 7000 East Avenue

Arthur J. Rodgers
Atmospheric, Earth and Energy Division, Lawrence Livermore National Laboratory, 7000 East Avenue

Robert J. Mellors
Atmospheric, Earth and Energy Division, Lawrence Livermore National Laboratory, 7000 East Avenue

Petra Koucká Knížová
Institute of Atmospheric Physic, Czech Academy of Sciences Czech Republic

Zbyšek Mošna
Institute of Atmospheric Physic, Czech Academy of Sciences Czech Republic

J. Kraskiewicz
Group of Astrophysics, Institute of Physics, UMCS, ul. Radziszewskiego 10, 20-031 Lublin, Poland

K. Murawski
Group of Astrophysics, Institute of Physics, UMCS, ul. Radziszewskiego 10, 20-031 Lublin, Poland

A. Solov'ev
Central (Pulkovo) Astronomical Observatory of the Russian Academy of Science, Pulkovskoe Shausse, 65/1, 196140, Saint-Petersburg, Russia
Kalmyk State University, Elista, 358000, Russia

A.K. Srivastava
Department of Physics, Indian Institute of Technology (BHU), Varanasi 221005, India

R.J. Morton
Department of Mathematics and Information Sciences, Northumbria University, Newcastle Upon Tyne NE1 8ST, UK.
High Altitude Observatory, National Center for Atmospheric Research, Boulder, Colorado 80307-3000, USA.

S. Tomczyk
High Altitude Observatory, National Center for Atmospheric Research, Boulder, Colorado 80307-3000, USA.

R. Pinto
UPS-OMP, IRAP, Universite´ de Toulouse, 14 Avenue Edouard Belin -314000 Toulouse, France.
CNRS, IRAP, 9 Avenue colonel Roche, BP 44346, F-31028 Toulouse, France. Correspondence and requests for materials should be addressed to R.J.M.

Ivan Fernandez-Corbaton
Department of Physics & Astronomy, Macquarie University, Australia
ARC Center for Engineered Quantum Systems, Macquarie University, North Ryde, New South Wales 2109, Australia
Institute of Nanotechnology, Karlsruhe Institute of Technology, 76021 Karlsruhe, Germany

Mauro Cirio
Department of Physics & Astronomy, Macquarie University, Australia
ARC Center for Engineered Quantum Systems, Macquarie University, North Ryde, New South Wales 2109, Australia

Alexander B¨use
Department of Physics & Astronomy, Macquarie University, Australia
ARC Center for Engineered Quantum Systems, Macquarie University, North Ryde, New South Wales 2109, Australia

Lucas Lamata
Department of Physical Chemistry, University of the Basque Country UPV/EHU, Apartado 644, E-48080 Bilbao, Spain

Enrique Solano
Department of Physical Chemistry, University of the Basque Country UPV/EHU, Apartado 644, E-48080 Bilbao, Spain
IKERBASQUE, Basque Foundation for Science, Maria Diaz de Haro 3, 48013 Bilbao, Spain

Gabriel Molina-Terriza
Department of Physics & Astronomy, Macquarie University, Australia
ARC Center for Engineered Quantum Systems, Macquarie University, North Ryde, New South Wales 2109, Australia

Hitendra K. Malik
Department of Physics, Indian Institute of Technology Delhi, New Delhi, India

Dustin J. Kreft
University of Wisconsin – Madison Department of Electrical and Computer Engineering U.S.A.

Robert H. Blick
University of Wisconsin – Madison Department of Electrical and Computer Engineering U.S.A.

Somsak Akatimagool
Department of Teacher Training in Electrical Engineering, King Mongkut's University of Technology North Bangkok, Thailand

Saran Choocadee
Department of Teacher Training in Electrical Engineering, King Mongkut's University of Technology North Bangkok, Thailand

PhilippWessels
The Hamburg Centre for Ultrafast Imaging (CUI), Luruper Chaussee 149, 22761 Hamburg, Germany.
Center for Optical Quantum Technologies (ZOQ), University of Hamburg, Luruper Chaussee 149, 22761 Hamburg, Germany.

AndreasVogel
Institut für Nanostruktur- und Festkörperphysik (INF), University of Hamburg, Jungiusstraße 11, 20355 Hamburg, Germany

Jan-NiklasTödt
Institut für Nanostruktur- und Festkörperphysik (INF), University of Hamburg, Jungiusstraße 11, 20355 Hamburg, Germany

MarekWieland
The Hamburg Centre for Ultrafast Imaging (CUI), Luruper Chaussee 149, 22761 Hamburg, Germany.
Institut für Experimentalphysik, University of Hamburg, Luruper Chaussee 149, 22761 Hamburg,Germany.

Guido Meier
The Hamburg Centre for Ultrafast Imaging (CUI), Luruper Chaussee 149, 22761 Hamburg, Germany.
Institut für Nanostruktur- und Festkörperphysik (INF), University of Hamburg, Jungiusstraße 11, 20355 Hamburg, Germany.
Max-Planck-Institute for the Structure and Dynamics of Matter, Luruper Chaussee 149, 22761 Hamburg,Germany. Correspondence and requests for materials should be addressed to P.W

Markus Drescher
The Hamburg Centre for Ultrafast Imaging (CUI), Luruper Chaussee 149, 22761 Hamburg, Germany.
Center for Optical Quantum Technologies (ZOQ), University of Hamburg, Luruper Chaussee 149, 22761 Hamburg, Germany.
Institut für Experimentalphysik, University of Hamburg, Luruper Chaussee 149, 22761 Hamburg,Germany.

A L. Marchant
Joint Quantum Centre (JQC), Durham - Newcastle, Department of Physics, Durham University, Durham DH1 3LE, United Kingdom

T. P. Billam
Jack Dodd Centre for Quantum Technology, Department of Physics, University of Otago, Dunedin 9016, New Zealand

T. P. Wiles
Joint Quantum Centre (JQC), Durham - Newcastle, Department of Physics, Durham University, Durham DH1 3LE, United Kingdom

M. M. H. Yu
Joint Quantum Centre (JQC), Durham - Newcastle, Department of Physics, Durham University, Durham DH1 3LE, United Kingdom

S. A. Gardiner
Joint Quantum Centre (JQC), Durham - Newcastle, Department of Physics, Durham University, Durham DH1 3LE, United Kingdom

S. L. Cornish
Joint Quantum Centre (JQC), Durham - Newcastle, Department of Physics, Durham University, Durham DH1 3LE, United Kingdom

Patrick Winter
Julius-Maximilians-Universität Würzburg, Lehrstuhl für Experimentelle Physik 5, Am Hubland, 97074, Würzburg, Germany

Thomas Kampf
Julius-Maximilians-Universität Würzburg, Lehrstuhl für Experimentelle Physik 5, Am Hubland, 97074, Würzburg, Germany

Xavier Helluy
Julius-Maximilians-Universität Würzburg, Lehrstuhl für Experimentelle Physik 5, Am Hubland, 97074, Würzburg, Germany

Fabian T Gutjahr
Julius-Maximilians-Universität Würzburg, Lehrstuhl für Experimentelle Physik 5, Am Hubland, 97074, Würzburg, Germany

Cord B Meyer
Julius-Maximilians-Universität Würzburg, Lehrstuhl für Experimentelle Physik 5, Am Hubland, 97074, Würzburg, Germany

Eberhard Rommel
Julius-Maximilians-Universität Würzburg, Lehrstuhl für Experimentelle Physik 5, Am Hubland, 97074, Würzburg, Germany

Wolfgang R Bauer
Julius-Maximilians-Universität Würzburg, Medizinische Universitätsklinik, Josef-Schneider Straße 4, 97080, Würzburg, Germany

Peter M Jakob
Julius-Maximilians-Universität Würzburg, Lehrstuhl für Experimentelle Physik 5, Am Hubland, 97074, Würzburg, Germany

Volker Herold
Julius-Maximilians-Universität Würzburg, Lehrstuhl für Experimentelle Physik 5, Am Hubland, 97074, Würzburg, Germany

Marcel Van Der Heijden
Department of Neuroscience, Erasmus MC, Room Ee 1285, Rotterdam, The Netherlands

Corstiaen P. C. Versteegh
Department of Neuroscience, Erasmus MC, Room Ee 1285, Rotterdam, The Netherlands

J. Rankin
Neuromathcomp Team, Inria Sophia Antipolis, 2004 Route des Lucioles-BP 93, Alpes-Maritimes, 06902, France

A. I. Meso
Institut de Neurosciences de la Timone, CNRS et Aix-Marseille Universit´e, Campus Sant´e Timone, 27 Bd Jean Moulin, Marseille 13385, France

G. S. Masson
Institut de Neurosciences de la Timone, CNRS et Aix-Marseille Universit´e, Campus Sant´e Timone, 27 Bd Jean Moulin, Marseille 13385, France

O. Faugeras
Neuromathcomp Team, Inria Sophia Antipolis, 2004 Route des Lucioles-BP 93, Alpes-Maritimes, 06902, France

P. Kornprobst
Neuromathcomp Team, Inria Sophia Antipolis, 2004 Route des Lucioles-BP 93, Alpes-Maritimes, 06902, France

Andre´s Concha
School of Engineering and Sciences, Adolfo Iban˜ez University, Diagonal las Torres 2640, Pen˜alolen, Santiago 7941169, Chile.

Paula Mellado
School of Engineering and Sciences, Adolfo Iban˜ez University, Diagonal las Torres 2640, Pen˜alolen, Santiago 7941169, Chile.

Bernal Morera-Brenes
Laboratorio de Gene´tica Evolutiva, Escuela de Ciencias Biolo´gicas, Universidad Nacional de Costa Rica, 86-3000 Heredia, Costa Rica.

Cristiano Sampaio Costa
Department of Zoology, Institute of Bioscience, Universidade de Sao Paulo, Sao Paulo, Sao Paulo 11461, Brazil.

L. Mahadevan
School of Engineering and Applied Sciences, Harvard University, Cambridge, Massachusetts 02138, USA.
Department of Physics, Harvard University, Cambridge, Massachusetts 02138, USA.

Julia´n Monge-Na´jera
Tropical Biology, Universidad de Costa Rica, 2060 San Jose´, Costa Rica. Correspondence and requests for materials should be addressed to A.C.

Kastberger G
Department of Zoology, University Graz, 8010 Graz, Austria. gerald

Weihmann F
Department of Zoology, University Graz, 8010 Graz, Austria. gerald

Hoetzl T
Department of Zoology, University Graz, 8010 Graz, Austria. gerald

Preface

Waves can be divided into two main groups, mechanical and electromagnetic waves. Mechanical waves require a medium to propagate (i.e. these cannot travel through a vacuum). These can be both longitudinal (particles displacement is parallel to the direction of wave propagation) transverse (particle displacement is perpendicular to the direction of wave propagation) waves which properties and characteristics depend on the restoring forces between particles combined with the inertia of the particles. Examples of mechanical waves are acoustic waves, seismic waves and water waves(wind waves). Electromagnetic waves do not require a medium to propagate they consist of periodic oscillations of electrical and magnetic fields. These are transverse waves in free space (vacuum) and in isotropic, homogeneous media. They are generated whenever charged particles are accelerated and can be characterized by either the frequency or wavelength. Examples are radio waves, microwaves, X-rays, light and gamma waves. There are also gravitational waves which are similar to electromagnetic waves in that they also travel through free space as the result of oscillations or movement in gravitational fields. The first part of this book is centered on the propagation of sea waves and the coupling between wave processes in the ocean, atmosphere, and the solid Earth [1-9]. It starts with a numerical simulation of water waves and interaction between these waves and coastal structures (breakwaters). Next, it overviews transoceanic propagation tsunami waves and ocean infragravity as well as challenge behind detection and the understanding of high amplitude sea waves generation mechanism. It ends with the analysis of long-period ground motions, earthquakes in Arabian Gulf and Kanto Basin. The second part of the book focuses on magnetoacoustic waves in the sun's atmosphere and the gravitational waves in space [10-12]. The third part returns to Earth by focusing on the understanding of electromagnetic waves and waveguide propagation for applications in microcircuits design, electro-mechanical systems, spintronics and many-body quantum systems [13-17]. The last part includes the different applications of wave motion theory in the biomedical engineering, neurosciences, biomechanics and wave-like motions in nature [18-22].

Editor
Olga Moreira, Ph.D.

INTRODUCTION

Motion refers to a change in location. Physics is the scientific study of matter, energy, and the interactions between them. It includes fields such as acoustics, cryogenics, electromagnetism, optics, and mechanics, which refers to the study of how force act on matter or material systems and includes motion physics, the scientific study of movement.

In motion physics, it is typical to speak of the change in location of a "body." Applied force is the initiator of a change in motion, which can mean starting motion, stopping motion, or changing direction. Without an applied force, bodies tend to resist acceleration and to stay at rest if they are at rest or, when moving in a straight line, to continue moving in a straight line.

Motion is usually described using several typical parameters, including velocity. Velocity is the rate of change in a body's position, so it refers to both speed — distance covered in a certain amount of time — and direction, and so it is a vector. Despite this, it is often represented by the equation $v = d/t$, where v represents velocity, d represents distance, and t represents time. It is usually reported in meters per second.

The second parameter is acceleration, which is the change in velocity over time. Like velocity, acceleration is a vector. It is caused by a force applied to the body.

General features

A single, all-encompassing definition for the term wave is not straightforward. A vibration can be defined as a back-and-forth motion around a reference value. However, a vibration is not necessarily a wave. An attempt to define the necessary and sufficient characteristics that qualify a phenomenon to be called a wave results in a fuzzy border line.

The term wave is often intuitively understood as referring to a transport of spatial disturbances that are generally not accompanied by a motion of the medium occupying this space as a whole. In a wave, the energy of a vibration is moving away from the source in the form of a disturbance within the surrounding medium. However, this motion is problematic for a standing wave (for example, a wave on a string), where energy is moving in both directions equally, or for electromagnetic (e.g., light) waves in a vacuum, where the concept of medium does not apply and interaction with a target is the key to wave detection and practical applications. There are water waves on the ocean surface; gamma waves and light waves emitted by the Sun; microwaves used in microwave ovens and in radar equipment; radio waves broadcast by radio stations; and sound waves generated by radio receivers, telephone handsets and living creatures (as voices), to mention only a few wave phenomena.

It may appear that the description of waves is closely related to their physical origin for each specific instance of a wave process. For example, acoustics is distinguished from optics in that sound waves are related to a mechanical rather than an electromagnetic wave transfer caused by vibration. Concepts such as mass, momentum, inertia, or elasticity, become therefore crucial in describing acoustic (as distinct from optic) wave processes. This difference in origin introduces certain wave characteristics particular to the properties of the medium involved. For example, in the case of air: vortices, radiation pressure, shock waves etc.; in the case of solids: Rayleigh waves, dispersion; and so on....

Other properties, however, although usually described in terms of origin, may be generalized to all waves. For such reasons, wave theory represents a particular branch of physics that is concerned with the properties of wave processes independently of their physical origin. For example, based on the mechanical origin of acoustic waves, a moving disturbance in space–time can exist if and only if the medium involved is neither infinitely stiff nor infinitely pliable. If all the parts making up a medium were rigidly bound, then they would all vibrate as one, with no delay in the transmission of the vibration and therefore no wave motion. On the other hand, if all the parts were independent, then there would not be any transmission of the vibration and again, no wave motion. Although the above statements are meaningless in the case of waves that do not require a medium, they reveal a characteristic that is relevant to all waves regardless of origin: within a wave, the phase of a vibration (that is, its position within the vibration cycle) is different for adjacent points in space because the vibration reaches these points at different times.

Wave motion

Wave motion, propagation of disturbances—that is, deviations from a state of rest or equilibrium—from place to place in a regular and organized way. Most familiar are surface waves on water, but both sound and light travel as wavelike disturbances, and the motion of all subatomic particles exhibits wavelike properties. The study of waves therefore forms a topic of central importance in all physical science and engineering.

The simplest types of wave motion are vibrations of elastic media, such as air, crystalline solids, or stretched strings. If, for example, the surface of a metal block is struck a sharp blow, the deformation of the surface material compresses the metal in the vicinity of the surface, and this transmits the disturbance to the layers beneath. The surface relaxes back to its initial configuration, and the compression propagates on into the body of the material at a speed determined by the stiffness of the material. This is an example of a compression wave. The steady transmission of a localized disturbance through an elastic medium is common to many forms of wave motion.

In most systems of interest, two or more disturbances of small amplitude may be superimposed without modifying one another. Conversely, a complicated disturbance may be analyzed into several simple components. In radio transmission, for example, a high-frequency signal can be superimposed on a low-frequency carrier wave and then filtered out intact on reception. In the simplest waves, the disturbance oscillates periodically with a fixed frequency and wavelength. These sinusoidal oscillations form the basis for the study of almost all forms of linear wave motion. In sound, for instance, a single sine wave produces a pure tone, and the distinctive timbre of different musical instruments playing the same note results from the admixture of sine waves of different frequencies. In electronics, the natural rhythmic oscillations of electric currents in tuned circuits are used to produce sinusoidal radio waves.

Although the mathematical properties of all linear waves are common, the waves exhibit various physical manifestations. One important class—electromagnetic waves—represents oscillations of the electromagnetic field. These include infrared radiation, visible light, radio and television, microwave, ultraviolet, X-rays, and gamma rays. Electromagnetic waves are produced by moving electric charges and varying currents, and they can travel through a vacuum. Unlike sound waves, they are not, therefore, disturbances in any medium. Another difference between electromagnetic and sound waves is that the former are transverse, that is, the disturbance occurs in a direction perpendicular to that in which the wave is propagating. Sound waves are longitudinal: they vibrate along the path of their propagation. The propagation of a wave through a medium will depend on the properties of the medium. For example, waves of different frequencies may travel at different speeds, an effect known as dispersion. In the case of light, dispersion leads to the unscrambling of colors and is the mechanism whereby a prism of glass can produce a spectrum. In geophysics, the dispersive propagation of seismic waves can provide information about the constitution of the Earth's interior. Two important characteristics of all waves are the phenomena of diffraction and interference. When a wave disturbance is directed toward a small aperture in a screen or other obstacle, it emerges traveling in a range of directions. Thus, light rays, which normally follow straight paths, can bend upon passing through a small hole: this is the phenomenon known as diffraction.

Interference occurs when two waves are combined and the disturbances overlap. If the waves arrive at a point in phase, enhancement occurs and the disturbance is large. Where the waves are out of phase, their opposing motions cancel and the disturbance is small or nonexistent. The net effect is therefore a distinctive interference pattern of large and small disturbances. Mathematically less tractable is the study of nonlinear waves, which can be very important in many applications. These usually display a more complicated structure and behavior; for example, water waves in a shallow channel can develop a humplike formation known as a soliton, which propagates as a coherent entity. Nonlinear waves are important in systems as diverse as nerve networks and the spiral arms of galaxies.

Mechanical Wave

A mechanical wave is a disturbance that pulsates through some form of medium: a solid, a liquid, or a gas, carrying the original, initiating energy from one place to another. There are three types of mechanical waves: transverse, longitudinal, and boundary.

Simple examples of waves and the media through which they move are sound in air, seismic waves in rock, and the vibrations of a guitar string. One example of mechanical waves are ripples in a pond that move outward from where a stone is dropped. The wave moves through the medium even though the atoms and molecules of the medium itself generally return to their original position after they have been squeezed together, stretched apart or rotated by the wave passing through them. The mechanical wave may move in straight-line motion, rotational motion, or a combination of the two. Waves possess the characteristics of frequency, period, wavelength, and amplitude. The distance between one wave pulse and the next is its wavelength and the time between pulses is the period. For a transverse waves, the amplitude is the distance that the wave displaces the medium. The number of waves passing through the medium over time is the wave's frequency. The higher the frequency of a mechanical wave, the greater the energy it transmits.

Transverse waves transmit their energy in an up and down or a side-to-side fashion, forcing the medium to take the form of a crest and trough — as in a sine wave. A good example of this kind of wave is a seismic S-wave in which the Earth's crust moves up and down, or side to side, perpendicular to the direction of wave propagation. The mechanical wave model of a transverse wave is given by the wave equation, an energy equation, the solution of which is a sine wave possessing the aforementioned characteristics of frequency, period, wavelength, and amplitude. Longitudinal waves are density waves that move through the expansion and compression of particles of the medium. One expansion and compression is one wavelength. Good examples of these waves are sound waves in air and seismic P-waves in the ground. The greater the density of the material, the faster the waves move through it. Boundary waves are mechanical waves that resonate between two dissimilar media. Air cavities and constrained strings in musical instruments, the vibration of a tuning fork, and the crown-like shape of a water droplet splash are good examples of boundary mechanical waves. The resonant sound of the tuning fork is limited by the shape and size of its boundary and the elasticity of the material composing it.

How It Works

In wave motion, energy—the ability to perform work, or to exert force over distance—is transmitted from one place to another without actually moving any matter along the wave. In some types of waves, such as those on the ocean, it might seem as though matter itself has been displaced; that is, it appears that the water has actually moved from its original position. In fact, this is not the case: molecules of water in an ocean wave move up and down, but they do not actually travel with the wave itself. Only the energy is moved.

A wave is an example of a larger class of regular, repeated, and/or back-and-forth types of motion. As with wave motion, these varieties of movement may or may not involve matter, but, in any case, the key component is not matter, but energy. Broadest among these is periodic motion, or motion that is repeated at regular intervals called periods. A period might be the amount of time that it takes an object orbiting another (as, for instance, a satellite going around Earth) to complete one cycle of orbit. With wave motion, a period is the amount of time required to complete one full cycle of the wave, from trough to crest and back to trough.

Harmonic Motion

Harmonic motion is the repeated movement of a particle about a position of equilibrium, or balance. In harmonic motion—or, more specifically, simple harmonic motion—the object moves back and forth under the influence of a force directed toward the position of equilibrium, or the place where the object stops if it ceases to be in motion. A familiar example of harmonic motion, to anyone who has seen an old movie with a clichéd depiction of a hypnotist, is the back-and-forth movement of the hypnotist's watch, as he tries to control the mind of his patient.

One variety of harmonic motion is vibration, which wave motion resembles in some respects. Both wave motion and vibration are periodic, involving the regular repetition of a certain form of movement. In both, there is a continual conversion and reconversion between potential energy (the energy of an object due to its position, as for instance with a sled at the top of a hill) and kinetic energy (the energy of an object due to its motion, as with the sled when sliding down the hill.) The principal difference between vibration and wave motion is that, in the first instance, the energy remains in place, whereas waves actually transport energy from one place to another.

Oscillation

Oscillation is a type of harmonic motion, typically periodic, in one or more dimensions. Suppose a spring is fixed in place to a ceiling, such that it hangs downward. At this point, the spring is in a position of equilibrium. Now, consider what happens if the spring is grasped at a certain point and lifted, then let go. It will, of course, fall downward with the force of gravity until it comes to a stop—but it will not stop at the earlier position of equilibrium. Instead, it will continue downward to a point of maximum tension, where it possesses maximum potential energy as well. Then, it will spring upward again, and as it moves, its kinetic energy increases, while potential energy decreases. At the high point of this period of oscillation, the spring will not be as high as it was before it was originally released, but it will be higher than the position of equilibrium.

Once it falls, the spring will again go lower than the position of equilibrium, but not as low as before—and so on. This is an example of oscillation. Now, imagine what happens if another spring is placed beside the first one, and they are connected by a rubber band. If just the first spring is disturbed, as before, the second spring will still move, because the energy created by the movement of the first spring will be transmitted to the second one via the rubber band. The same will happen if a row of springs, all side-by-side, are attached by multiple rubber bands, and the first spring is once again disturbed: the energy will pass through the rubber bands, from spring to spring, causing the entire row to oscillate. This is similar to what happens in the motion of a wave.

Types and Properties of Waves

There are some types of waves that do not follow regular, repeated patterns; these are discussed below, in the illustration concerning a string, in which a pulse is created and reflected. Of principal concern here, however, is the periodic wave, a series of wave motions, following one after the other in regular succession. Examples of periodic waves include waves on the ocean, sound waves, and electromagnetic waves. The last of these include visible light and radio, among others.

Electromagnetic waves involve only energy; on the other hand, a mechanical wave involves matter as well. Ocean waves are mechanical waves; so, too, are sound waves, as well as the waves produced by pulling a string. It is important to note, again, that the matter itself is not moved from place to place, though it may move in place without leaving its position. For example, water molecules in the crest of an ocean wave rotate in the same direction as the wave, while those in the trough of the wave rotate in a direction opposite to that of the wave, yet there is no net motion of the water: only energy is transmitted along the wave.

Five Properties of Waves

There are three notable interrelated characteristics of periodic waves. One of these is wave speed, symbolized by v and typically calculated in meters per second. Another is wavelength, represented as λ (the Greek letter lambda), which is the distance between a crest and the adjacent crest, or a trough and the adjacent trough. The third is frequency, abbreviated as f, which is the number of waves passing through a given point during the interval of 1 second.

Frequency is measured in terms of cycles per second, or Hertz (Hz), named in honor of nineteenth-century German physicist Heinrich Rudolf Hertz (1857-1894). If a wave has a frequency of 100 Hz, this means that 100 waves are passing through a given point during the interval of 1 second. Higher frequencies are expressed in terms of kilohertz (kHz; 10^3 or 1,000 cycles per second) or megahertz (MHz; 10^6 or 1 million cycles per second.)

Frequency is clearly related to wave speed, and there is also a relationship—though it is not so immediately grasped—between wavelength and speed. Over the interval of 1 second, a given number of waves pass a certain point (frequency), and each wave occupies a certain distance (wavelength). Multiplied by one another, these two properties equal the speed of the wave. This can be stated as a formula: $v = f\lambda$.

Earlier, the term «period» was defined in terms of wave motion as the amount of time required to complete one full cycle of the wave. Period, symbolized by T, can be expressed in terms of frequency, and, thus, can also be related to the other two properties identified above. It is the inverse of frequency, meaning that $T = 1/f$. Furthermore, period is equal to the ratio of wavelength to wave speed; in other words, $T = \lambda/v$.

A fifth property of waves—one not mathematically related to wavelength, wave speed, frequency, or period, is amplitude. Amplitude can be defined as the maximum displacement of oscillating particles from their normal position. For an ocean wave, amplitude is the distance from either the crest or the trough to the level that the ocean would maintain if it were perfectly still.

Wave Shapes

When most people think of waves, naturally, one of the first images that comes to mind is that of waves on the ocean. These are an example of a transverse wave, or one in which the vibration or motion is perpendicular to the direction the wave is moving. (Actually, ocean waves are simply perceived as transverse waves; in fact, as discussed below, their behavior is rather more complicated.) In a longitudinal wave, on the other hand, the movement of vibration is in the same direction as the wave itself.

Transverse waves are easier to visualize, particularly with regard to the aspects of wave motion—for example, frequency and amplitude—discussed above. Yet, longitudinal waves can be understood in terms of a common example. Sound waves, for instance, are longitudinal: thus, when a stereo is turned up to a high volume, the speakers vibrate in the same direction as the sound itself.

A longitudinal wave may be understood as a series of fluctuations in density. If one were to take a coiled spring (such as the toy known as the "Slinky") and release one end while holding the other, the motion of the springs would produce longitudinal waves. As these waves pass through the spring, they cause some portions of it to be compressed and others extended. The distance between each point of compression is the wavelength.

Now, to return to the qualified statement made above: that ocean waves are an example of transverse waves. We perceive them as transverse waves, but, in fact, they are also longitudinal. In fact, all types of waves on the surface of a liquid are a combination of longitudinal and transverse, and are known as surface waves. Thus, if one drops a stone into a body of still water, waves radiate outward (longitudinal), but these waves also have a component that is perpendicular to the surface of the water, meaning that they are also transverse.

Sound Waves

Production

The experience of sound involves production, or the generation of sound waves; transmission, or the movement of those waves from their source; and reception, the principal example of which is hearing. Sound itself is discussed in detail elsewhere. Of primary concern here is the transmission, and to a lesser extent, the production of sound waves.

In terms of production, sound waves are, as noted, longitudinal waves: changes in pressure, or alternations between condensation and rarefaction. Vibration is integral to the generation of sound. When the diaphragm of a loudspeaker pushes outward, it forces nearby air molecules closer together, creating a high-pressure region all around the loudspeaker. The loudspeaker's diaphragm is pushed backward in response, thus freeing up a volume of space for the air molecules. These, then, rush toward the diaphragm, creating a low-pressure region behind the high-pressure one. As a result, the loudspeaker sends out alternating waves of high pressure (condensation) and low pressure (rarefaction).

Frequency and Wavelength

As sound waves pass through a medium such as air, they create fluctuations between condensation and rarefaction. These result in pressure changes that cause the listener's eardrum to vibrate with the same frequency as the sound wave, a vibration that the ear's inner mechanisms translate and pass on to the brain. The range of audibility for the human ear is from 20 Hz to 20 kHz. The lowest note of the eighty-eight keys on a piano is 27 Hz and the highest 4.186 kHz. This places the middle and upper register of the piano well within the optimal range for audibility, which is between 3 and 4 kHz.

Sound travels at a speed of about 1,088 ft (331 m) per second through air at sea level, and the range of sound audible to human ears includes wavelengths as large as 11 ft (3.3 m) and as small as 1.3 in (3.3 cm). Unlike light waves, which are very small, the wavelengths of audible sound are comparable to the sizes of ordinary objects. This creates an interesting contrast between the behaviors of sound and light when confronted with an obstacle to their transmission.

It is fairly easy to block out light by simply holding up a hand in front of one's eyes. When this happens, the Sun casts a shadow on the other side of one's hand. The same action does not work with one's ears and the source of a sound, however, because the wavelengths of sound are large enough to go right past a relatively small object such as a hand. However, if one were to put up a tall, wide cement wall between oneself and the source of a sound—as is often done in areas where an interstate highway passes right by a residential community—the object would be sufficiently large to block out much of the sound.

Radio Waves

Radio waves, like visible light waves, are part of the electromagnetic spectrum. They are characterized by relatively long wavelengths and low frequencies—low, that is, in contrast to the much higher frequencies of both visible and invisible light waves. The frequency range of radio is between 10 KHz and about 2,000 MHz—in other words, from 10,000 Hz to as much as 2 billion Hz—an impressively wide range.

AM radio broadcasts are found between 0.6 and 1.6 MHz, and FM broadcasts between 88 and 108 MHz. Thus, FM is at a much, much higher frequency than AM, with the lowest frequency on the FM dial 55 times as great as the highest on the AM dial. There are other ranges of frequency assigned by the FCC (Federal Communications Commission) to other varieties of radio transmission: for instance, citizens' band (CB) radios are in a region between AM and FM, ranging from 26.985 MHz to 27.405 MHz. Frequency does not indicate power. The power of a radio station is a function of the wattage available to its transmitter: hence, radio stations often promote themselves with announcements such as "operating with 100,000 watts of power…." Thus, an AM station, though it has a much lower frequency than an FM station, may possess more power, depending on the wattage of the transmitter. Indeed, as we shall see, it is precisely because of its high frequency that an FM station lacks the broadcast range of an AM station.

Amplitude and Frequency Modulations

What is the difference between AM and FM? Or to put it another way, why is it that an AM station may be heard halfway across the country, yet its sound on a car radio fades out when the car goes under an over-pass? The difference relates to how the various radio signals are modulated.

A radio signal is simply a carrier: it may carry Morse code, or it may carry complex sounds, but in order to transmit voices and music, its signal must be modulated. This can be done, for instance, by varying the instantaneous amplitude of the radio wave, which is a function of the radio station's power. These variations in amplitude are called amplitude modulation, or AM, and this was the first type of commercial radio to appear. Developed in the period before World War I, AM made its debut as a popular phenomenon shortly after the war.

Ironically, FM (frequency modulation) was developed not long after AM, but it did not become commercially viable until well after World War II. As its name suggests, frequency modulation involves variation in the signal's frequency. The amplitude stays the same, and this—combined with the high frequency—produces a nice, even sound for FM radio.

CHAPTER 1

A NUMERICAL MODEL BASED ON NAVIER-STOKES EQUATIONS TO SIMULATE WATER WAVE PROPAGATION WITH WAVE-STRUCTURE INTERACTION

Paulo Roberto de Freitas Teixeira

INTRODUCTION

The interaction between waves and structures is a very important subject of the coastal engineering. Many numerical models have been developed over recent decades to analyze these types of cases, which involve phenomena that combine reflection, shoaling, refraction, diffraction, breaking and wave-wave interaction. These non-linear effects provide harmonic generation, including energy transferences with high complexity.

Models based on the Laplace equation assume the potential flow, in which the movement is irrotational and the flow is incompressible. Models based on the boundary element technique [1,2,3] and spectral methods [4,5] are some examples. This theory applies neither to viscous flows nor to situations in which there are flow separations, vortex generations and turbulences.

Other models, called depth-integrated models [6], based on a Boussinesq-type equation for variable depth, consider polynomial approximations for the vertical velocity distribution and vertical integration in the resulting equations at a certain depth. The simplified hypotheses, that include slight non linearity and dispersion, limit the applicability of this type of models to shallow and intermediate waters. Several researches have been developed to extend the applicability of these equations, including high order terms, to deeper water and strong non linearity cases in the last decades. Wave propagation phenomena, such as breaking, bottom friction and run-up, have also been included in these extended Boussinesq equation [7,8,9,10,11,12,13,14]. The accuracy of these type of models has recently been improved by the implementation of the multi-layer concept, in which the water column is divided into layers and a velocity profile is adopted to each one [15,16]. Although the accuracy has been improved significantly, the simplified hypotheses, related to the vertical integration along each layer, limit the use of these models to depths without strong variations.

Many efforts have been carried out to develop non hydrostatic models to prevent the difficulties of the Boussinesq models [17]. The non hydrostatic models capture the free surface movement using a function on the horizontal plane, requiring lower vertical discretization by comparison with those that use classic methods to describe the free surface. In some of these models, the pressure and velocity fields are decomposed by hydrostatic and non-hydrostatic pressure to improve their efficiency.

The numerical solution of the fully Navier-Stokes equations to determine the tridimensional velocity and pressure fields and the free surface position demands high computational cost, due to the large horizontal scale of many

Citation: Paulo Roberto de Freitas Teixeira (2013). A Numerical Model Based on Navier-Stokes Equations to Simulate Water Wave Propagation with Wave-Structure Interaction, Wave Propagation Theories and Applications, Dr. Yi Zheng (Ed.), InTech, DOI: 10.5772/51033.

coastal engineering problems. However, in cases in which there are flow separation, vortex shedding and turbulence, these models provide more real results. There are several methods to capture the free surface movements, such as the arbitrary lagrangian eulerian (ALE) [22,46], the marked and cell [25], the volume of fluid [26,27] and the level-set methods [28].

This text describes a code (in Fortran 90 language) that integrates the Navier-Stokes equations using a fractional method to simulate 3D incompressible flow problems with free surface, named FLUINCO [29]. The model employs a semi implicit two-step Taylor Galerkin method to discretize the Navier-Stokes equations in time and space; uses the ALE method and a mesh velocity distribution technique to deal with free surface movement.

To show the applicability of the code, two study cases are analyzed: the wave propagation over a submerged horizontal cylinder and submerged trapezoidal breakwaters. In Section 2, the numerical model is described. Section 3 presents study cases, their results and discussion. Finally, Section 4 concludes the analysis.

NUMERICAL MODEL

Governing Equations

The algorithm is based on the continuity equation, given by:

$$\frac{\partial \rho}{\partial t} = \frac{1}{c^2}\frac{\partial p}{\partial t} = -\frac{\partial U_i}{\partial x_i}\ (i = 1,2,3), \tag{1}$$

and momentum equations, that are represented by the following equations according to the ALE formulation:

$$\frac{\partial U_i}{\partial t} = -\frac{\partial f_{ij}}{\partial x_j} + \frac{\partial \tau_{ij}}{\partial x_j} - \frac{\partial p}{\partial x_i} + \mathrm{w}_j\frac{\partial U_i}{\partial x_j}\ (i,j = 1,2,3), \tag{2}$$

where ρ is the specific mass, p is the pressure, $U_i = \rho \mathrm{v}_i$, $f_{ij} = \mathrm{v}_j(\rho \mathrm{v}_i) = \mathrm{v}_j U_i$, v_i are the velocity components, wi the reference system velocity components and τij is the viscous stress tensor (i,j.=1,2,3).

Semi-Implicit Two-Step Taylor-Galerkin Method

Basically, the algorithm consists of the following steps [30]:

- Calculate non-corrected velocity at Δt/2, obtained by time discretization of Eq. (2), where the pressure term is at t instant, according to Eq. (3).

$$\tilde{U}_i^{n+1/2} = U_i^n - \frac{\Delta t}{2}\left(\frac{\partial f_{ij}^n}{\partial x_j} - \frac{\partial \tau_{ij}^n}{\partial x_j} + \frac{\partial p^n}{\partial x_i} - \rho g_i - \mathrm{w}_j^n\frac{\partial U_i^n}{\partial x_i}\right)\quad (i,j = 1,2,3), \tag{3}$$

where gi are the gravity acceleration components.

- Update the pressure p at t+Δt, obtained by time discretization of Eq. (1), given by the Poisson equation:

$$\frac{1}{c^2}\Delta p = -\Delta t\left[\frac{\partial \tilde{U}_i^{n+1/2}}{\partial x_i} - \frac{\Delta t}{4}\frac{\partial}{\partial x_i}\frac{\partial \Delta p}{\partial x_i}\right], \tag{4}$$

where $\Delta p = p^{n+1} - p^n$ and $i = 1,2,3$.

- Correct the velocity at t+Δt/2, adding the pressure variation term from t to t+Δt/2, according to the equation:

$$U_i^{n+1/2} = \tilde{U}_i^{n+1/2} - \frac{\Delta t}{4}\frac{\partial \Delta p}{\partial x_i}\quad (i = 1,2,3),$$

- Calculate the velocity at t+Δt using variables updated in the previous steps as follows:

$$U_i^{n+1} = U_i^n - \Delta t\left(\frac{\partial f_{ij}^{n+1/2}}{\partial x_j} - \frac{\partial \tau_{ij}^{n+1/2}}{\partial x_j} + \frac{\partial p^{n+1/2}}{\partial x_i} - \mathrm{w}_j^{n+1/2}\frac{\partial U_i^{n+1/2}}{\partial x_i} - \rho g_i\right)\quad (i,j = 1,2,3),$$

Space Discretization

The classical Galerkin weighted residual method is applied to the space discretization by using a tetrahedron element. In the variables at t+Δt/2 instant, a constant shape function **PE** is used, and in the variables at t and t+Δt, a linear shape function **N** is employed. By applying this procedure to Eq. (3), (4), (5) and (6), the following expressions in the matrix form are obtained [29]:

$$\Omega_E^{n+1/2}\bar{\tilde{\mathbf{U}}}_i^{n+1/2} = \mathbf{C}\,\bar{\mathbf{U}}_i^n - \frac{\Delta t}{2}\left[\mathbf{L}_j\left(\bar{\mathbf{f}}_{ij}^n - \bar{\boldsymbol{\tau}}_{ij}^n\right) + \mathbf{L}_i\bar{\mathbf{p}}^n - \mathbf{T}\,\bar{\mathbf{U}}_i^n - \Omega_E^{n+1/2}\overline{\rho g}_i\right] \tag{5}$$

$$\left(\tilde{\mathbf{M}} + \frac{\Delta t^2}{4}\mathbf{H}\right)\Delta\bar{\mathbf{p}} = \Delta t\left(\mathbf{L}_i^T\bar{\tilde{\mathbf{U}}}_i^{n+1/2} + \mathbf{f}_a\right) \tag{6}$$

$$\bar{\mathbf{U}}_i^{n+1/2} = \bar{\tilde{\mathbf{U}}}_i^{n+1/2} - \frac{\Delta t}{4\Omega_E}\mathbf{L}_i\Delta\bar{\mathbf{p}} \tag{7}$$

$$\mathbf{M}^{n+1}\overline{\mathbf{U}}_i^{n+1} = \mathbf{M}^n\overline{\mathbf{U}}_i^n + \Delta t\mathbf{L}_j^T\left(\bar{\mathbf{f}}_{ij}^{n+1/2} - \overline{\mathbf{w}}_j^{n+1/2}\overline{\mathbf{U}}_i^{n+1/2}\right) \tag{8}$$

$$-\Delta t\mathbf{Q}_j\bar{\boldsymbol{\tau}}_{ij}^n + \Delta t\mathbf{Q}_i\left(\bar{\mathbf{p}}^n + \Delta\bar{\mathbf{p}}/2\right) + \Delta t\mathbf{S}_{bi} - \Delta t\mathbf{C}^T\bar{\mathbf{g}}_i \tag{9}$$

where variables with upper bars at n and n+1 instants indicate nodal values, while those at n+1/2 instant represent constant values in the element. The matrices and vectors from Eq. (7) to (10) are volume and surface integrals that can be seen in detail in [30].

Equation (8) is solved using the conjugated gradient method with diagonal pre-conditioning [31]. In Eq. (10), the consistent mass matrix is substituted by the lumped mass matrix, and then this equation is solved iteratively.

The scheme is conditionally stable and the local stability condition for the element E is given by

$$\Delta t_E \leq \beta h_E / |\mathbf{u}| \tag{10}$$

where hE is the characteristic element size, β is the safety factor and u is the fluid velocity.

Mesh Movement

The free surface is the interface between two fluids, water and air, where atmospheric pressure is considered constant (generally the reference value is null). In this interface, the kinematic free surface boundary condition (KFSBC) is imposed. By using the ALE formulation, it is expressed as:

$$\frac{\partial \eta}{\partial t} + {}^{(s)}v_i \frac{\partial \eta}{\partial x_i} = {}^{(s)}v_3 \tag{11}$$

where η is the free surface elevation, ${}^{s}v_3$ is the vertical fluid velocity component and ${}^{s}v_i$ (i=1,2) are the horizontal fluid velocity components in the free surface. The eulerian formulation is used in the x_1 and x_2 directions (horizontal plane) while the ALE formulation is employed in the x_3 or vertical direction.

The time discretization of KFSBC is carried out in the same way as the one for the momentum equations as presented before. After applying expansion in Taylor series, the expressions for η at n+1/2 (first step) and n+1 (second step) instants are obtained:

$$\eta^{n+1/2} = \eta^n + \frac{\Delta t}{2}\left({}^{(s)}v_3 - {}^{(s)}v_1 \frac{\partial \eta}{\partial x_1} - {}^{(s)}v_2 \frac{\partial \eta}{\partial x_2} \right)^n \tag{12}$$

$$\eta^{n+1} = \eta^n + \Delta t\left({}^{(s)}v_3 - {}^{(s)}v_1 \frac{\partial \eta}{\partial x_1} - {}^{(s)}v_2 \frac{\partial \eta}{\partial x_2} \right)^{n+1/2} \tag{13}$$

Linear triangular elements coincident with the face of the tetrahedral elements on the free surface are used to the space discretization by applying the Galerkin method.

The mesh velocity vertical component w_3 is computed to diminish element distortions, keeping prescribed velocities on moving (free surface) and stationary (bottom) boundary surfaces. The mesh movement algorithm adopted in this paper uses a smoothing procedure for the velocities based on these boundary surfaces. The updating of the mesh velocity at point i of the finite element domain is based on the mesh velocity of the points j that belong to the boundary surfaces, and is expressed in the following way [32]:

$$w_3^i = \frac{\sum_{j=1}^{ns} a_{ij} w_3^j}{\sum_{j=1}^{ns} a_{ij}} \tag{14}$$

where ns is the total number of points belonging to the boundary surfaces and aij are the influence coefficients between the point i inside the domain and the point j on the boundary surface given by the following expression:

$$a_{ij} = \frac{1}{d_{ij}^4} \tag{15}$$

with dij being the distance between points i and j. In other words, aij represents the weight that every point j on the boundary surface has on the value of the mesh velocity at points i inside the domain. When dij is low, aij has a high value, favouring the influence of points i, located closer to the boundary surface containing point j.

The free surface elevation, the mesh velocity and the vertical coordinate are updated according to the following steps:

1. Calculate $\eta^{n+1/2}$ and $\tilde{U}_i^{n+1/2}$, Eq. (13) and Eq. (3), respectively.
2. Calculate Δp, Eq. (4).
3. Calculate $U_i^{n+1/2}$, Eq. (3).
4. Calculate U_i^{n+1}, Eq. (6).
5. Calculate η^{n+1}, Eq. (14).
6. Update the mesh velocity w_3 and the vertical coordinate x_3:
   - Calculate the mesh velocity in the free surface at $t + \Delta t$: ${}^{(S)}w_3^{n+1} = \left(\eta^{n+1} - \eta^n\right)/\Delta t$.
   - Calculate the mesh velocity in the interior of the domain at n+1 e n+1/2 by using Eq. (13) and $w_3^{n+1/2} = \frac{(w_3^{n+1} + w_3^n)}{2}$, respectively.

- Update the vertical coordinates in the interior of the domain: $x_3^{n+1/2} = x_3^n + w_3^n \frac{\Delta t}{2}$, $x_3^{n+1} = x_3^n + w_3^{n+1/2} \Delta t$.

Wave Generation And Radiation Conditions

The wave generation is considered imposing the free surface elevation and the fluid velocity components to each time step directly, considering the linear wave equations [33].

The Flather's radiation condition [34] is used to deal with open boundaries. In this method, the Sommerfeld condition to free surface elevation is combined with one-dimensional version of the continuity equation. Then, the normal velocity of the boundary can be expressed by:

$$u = \eta\sqrt{\frac{g}{h}} \tag{17}$$

where g is the gravity acceleration and h is the depth.

STUDY CASES

Submerged Cylinder

The interaction among regular waves and submerged circular cylinders, with their axes parallel to the crests of the incident waves, has been studied analytically, experimentally and numerically by many authors. The presence of an obstacle near the free surface may cause reflected and modified transmitted waves. These phenomena depend on the characteristics of the incident wave, the obstacle geometry and the depth. Many studies of this interaction are available to provide a good example to validate numerical codes.

The first study was developed by [35] and, after that, by [36]. Considering a linear behavior, these authors showed that (a) the cylinder does not reflect any energy, regardless of its ray, depth or wave frequency; (b) the transmitted waves are out of phase, but their amplitudes are the same. Chaplin [37] studied the nonlinear forces and characteristics of the reflected and transmitted waves experimentally. He showed that the reflection is negligible up to the third order. This author and Schonberg and Chaplin [38] presented many experimental and numerical studies concerning the nonlinear interaction among waves and submerged cylinders. A detailed review of analyses for this case can be found in Paixão Conde et al. [39].

This case considers a 5.2 m long and 0.425 m deep channel with a submerged cylinder of r = 0.025m positioned 1.60 m from the wave generator (Figure 1). The cylinder center is 0.075 m (3r) from the free surface. The frequency wave is f = 1.4Hz; its amplitude is a = 0.0119 m and its wavelength is L=0.796 m, characterizing a deep water case.

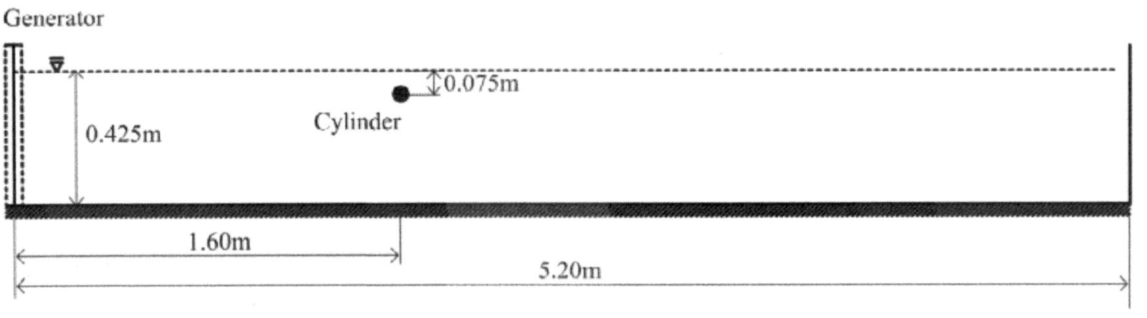

Figure 1. Geometry of the horizontal cylinder case.

Table 1 shows the period, the frequency and the wavelength for the fundamental frequency and its 2nd, 3rd and 4th harmonics, according to the linear wave theory.

Table 1. Period, frequency and wavelength of the fundamental frequency and its 2nd, 3rd and 4th harmonics in the horizontal cylinder case.

	Fundamental	2ndharmonic	3rdharmonic	4thharmonic
T (s)	0.7143	0.3571	0.2381	0.1786
f(Hz)	1.4	2.8	4.2	5.6
L(m)	0.796	0.199	0.0885	0.0498

The mesh, with 173900 nodes and 515623 elements, has one layer of elements in the transversal direction. The average element size on the cylinder boundary is 0.0015.m (105 divisions in the circumference). The element size diminishes from the ends to the region near the cylinder and from the bottom to the free surface. The element sizes on the end where the wave generator is located and on the opposite end are 0.015 m (53 points per fundamental wavelength) and 0.02 m (40 points per fundamental wavelength), respectively. On the bottom, 0.0015m is also used.

The initial conditions are: null velocity components in all domain and hydrostatic pressure (null on the free surface). The wave is generated by imposing the surface elevation and the velocity components. The non-slip condition is imposed to the bottom and to the cylinder wall. The time step is 0.0002s, which satisfies the Courant condition.

Figure 2 shows the free surface elevation obtained by the code and experimental tests, where xc is the horizontal coordinate of the cylinder center. In general, there is agreement between numerical and experimental results [39]. We can notice the free surface disturbance downstream the cylinder. When (x-xc)/L is above 1.7, the numerical results are smoother than the experimental ones, showing the necessity of a refinement in this region.

Figure 3 shows a comparison among numerical and experimental results in terms of free surface elevation on four gauges located at (x-xc)/L equal to -0.503, 0.0692, 0.509 and 1.264 (there is only a numerical result on the first gauge). We can observe the similarity among numerical and experimental results.

Figure 4 shows the streamlines and the velocity modulus distribution at the same instant used in Figure 2. Recirculation and separation cannot be observed at downstream. Due to the oscillatory flow behavior, there is no time for recirculation productions. We can notice the flow acceleration near the cylinder due to the boundary layer effect. The viscous effects have only local influence, without modifying the velocity field far from it.

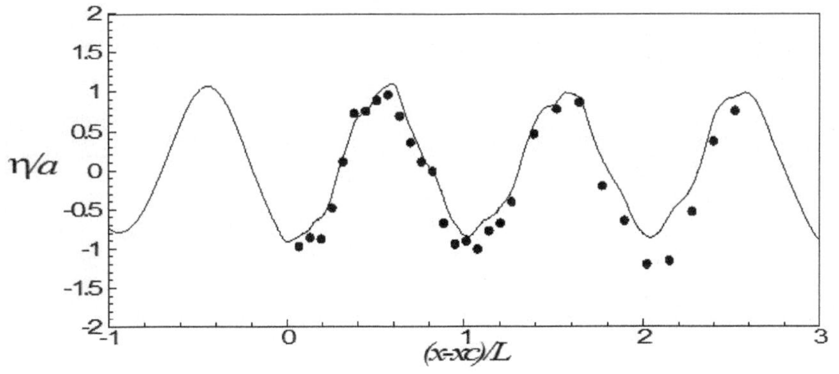

Figure 2. Free surface elevation in the submerged cylinder case (Numerical ¾ ; Experimental ■).

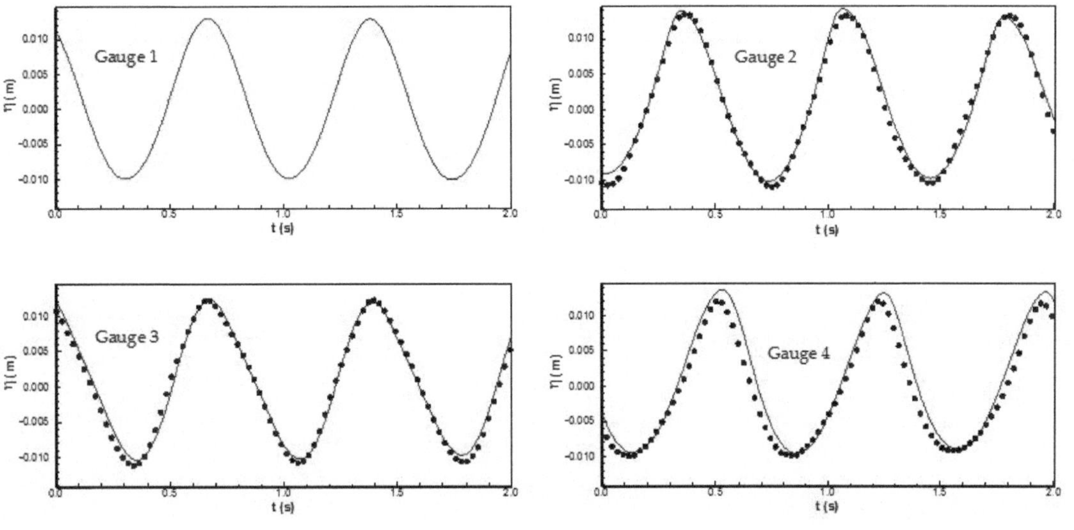

Figure 3. Free surface elevations on the gauges located at (x-xc)/L equal to -0.503, 0.0692, 0.509 and 1.264 in the horizontal cylinder case (Numerical ¾ ; Experimental •).

In Figures 5 and 6, velocity component profiles, u and v, on the same gauge positions are presented. These profiles were constructed at the same instant as that used in Figure 2. According to the linear theory, the maximum value for both horizontal and vertical components is equal to 0.105 m/s. For horizontal components, these values occur on the crest and the trough, while for vertical ones, these values occur on upward and downward zero-crossings. When one component is the maximum, another is null, because the phase difference is 90 degrees.

Gauge 1 ((x-xc)/L = -0.503) is located upstream, near the wave crest; no significant disturbance in u and v profiles is observed. The horizontal velocity component is positive and its maximum value is similar to the theoretical value in the crest. The wave trough passes by gauge 2; the vertical velocity component presents low values and the horizontal velocity component has negative values, reaching the maximum absolute value close to the theoretical ones (0.105 m/s). Gauge 3 is located near the first crest upstream, resulting in high horizontal component values. Finally, gauge 4 is on a region between the trough and upward zero-crossing. Both component profiles are negative and the vertical component magnitude shows how close the gauge is to upward zero-crossing.

The non-slip boundary condition on the bottom does not change the general behavior of the wave propagation, because this case is considered a deep water one.

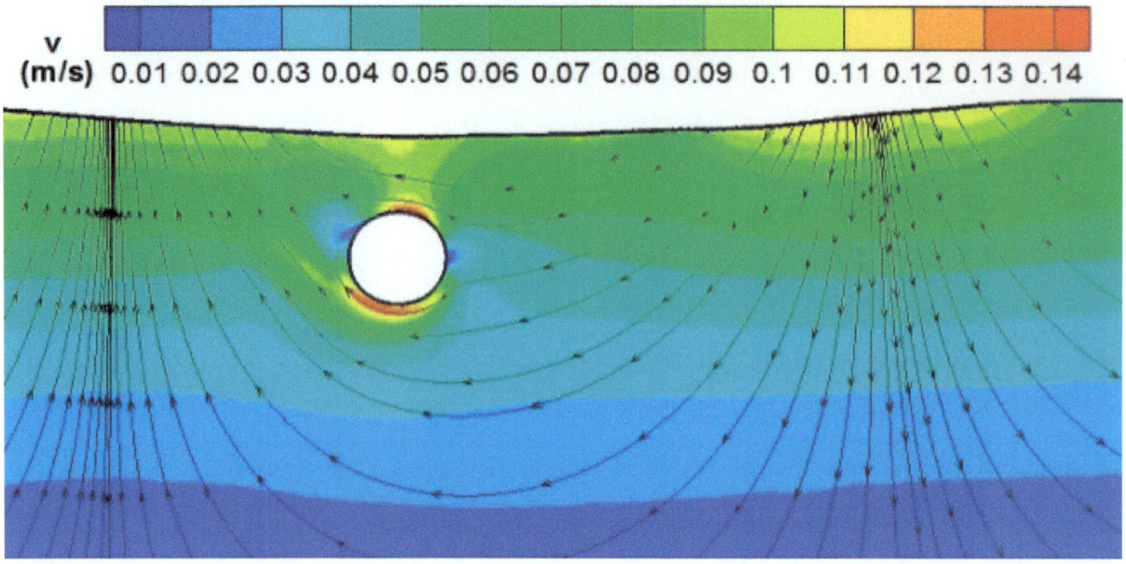

Figure 4. Streamlines and velocity modulus at the instant in which the free-surface elevation was captured (Figure 2) in the horizontal cylinder case.

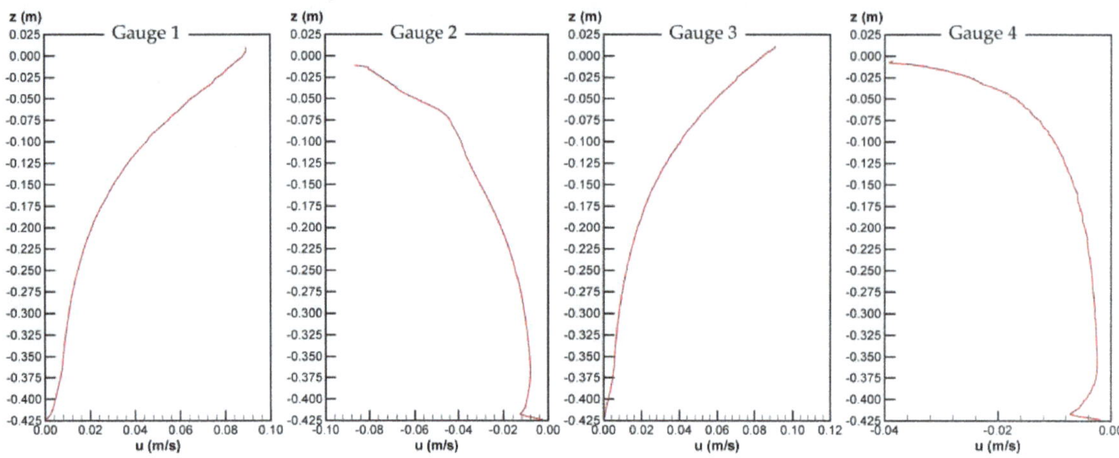

Figure 5. Horizontal velocity components at the same instant used in Figure 2 along the depth on gauges located at (x-xc)/L equal to -0.503, 0.0692, 0.509 and 1.264 in the horizontal cylinder case.

Figure 7 shows the frequency spectra for these four gauges distributed along the channel. In all cases, the energy is concentrated on the fundamental frequency and its harmonic waves. On gauge 1, the fundamental frequency presents most energy and the second harmonic shows a little value. On gauges 2 and 4, located upstream, significant energy up to the third harmonic appears, similar to the experimental results.

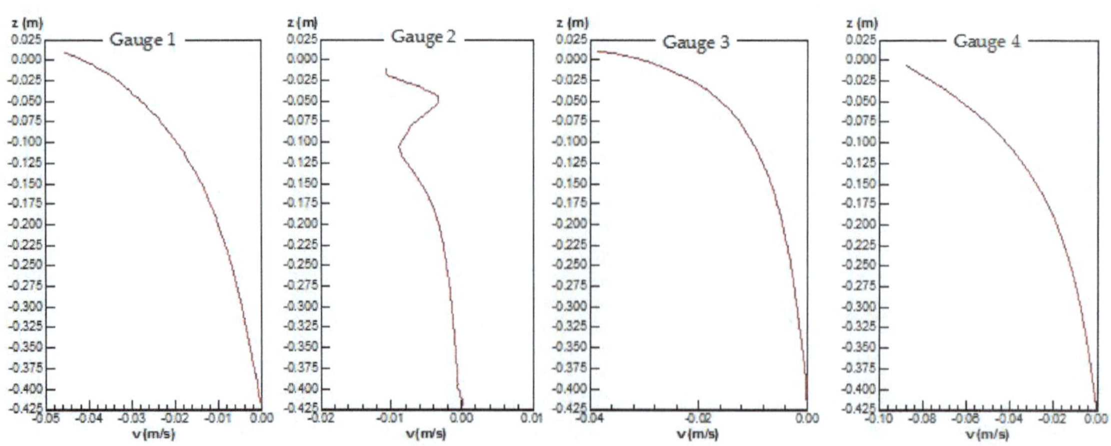

Figure 6. Vertical velocity components at the same instant used in Figure 2 along the depth on gauges located at (x-xc)/L equal to -0.503, 0.0692, 0.509 and 1.264 in the horizontal cylinder case.

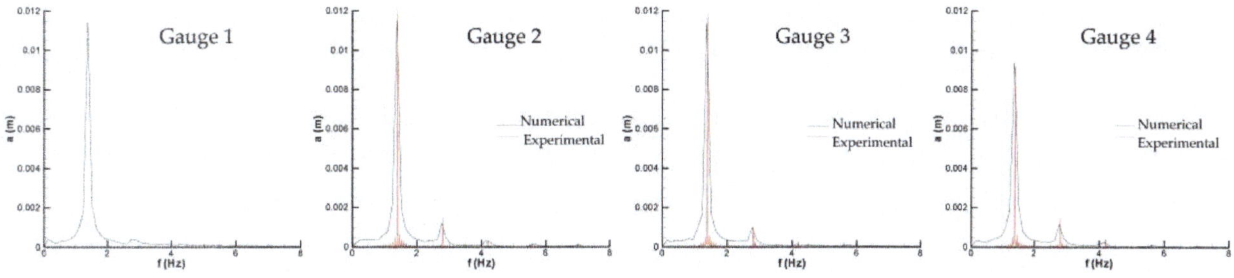

Figure 7. Frequency spectra on gauges located at (x-xc)/L equal to -0.503, 0.0692, 0.509 e 1.264 in the horizontal cylinder case.

Submerged Trapezoidal Breakwaters

Whatever the numerical model characteristics, the simulation of wave propagation over submerged breakwaters are important tests to validate wave propagation models. In these cases, the harmonic generation [40,41] and the vortex formation, depending on the geometry [42], also occur. When waves propagate in deep waters over a submerged obstacle, part of the wave energy is transferred from the primary wave component to their harmonics, contributing to increase non linearity. Harmonic generation phenomena that occur when waves propagate over obstacles, such as natural reefs, were studied theoretically [6], experimentally [43,44,45] and numerically [46,23,17,47,48,45,49,50]. In some situations, the correct simulation of the flow can only be figured out considering the viscosity effects [51]. Huang and Dong [42] studied the interaction between solitary waves and rectangular submerged breakwaters using a model based on 2D Navier-Stokes equations and concluded that the flow around the breakwater is laminar, without turbulence. The experimental studies carried out by Ting and Kim [51] and Zhuang and Lee [52] show that velocity fluctuations do not exist around the breakwater.

Two different configurations of the trapezoidal breakwaters, with different level of non-linearity, are used to test the behaviour of the numerical model. In the first case, the downstream and upstream slopes are 1:20 and 1:10, respectively [44]. In the second one, both slopes are 1:2 [45], where the non-linear effects are more significant.

Breakwater with Slopes 1:20 and 1:10

Figure 8 shows the channel and the submerged breakwater geometries, and the position of the gauges. The channel is 23m in length, 0.4m and 0.1m are the maximum and the minimum depths, respectively. In the channel entrance, a monochromatic wave is generated with a period of 2.02s and an amplitude of 0.01m.

Table 2 presents some parameters for this case study, in which H is the wave length, h is the depth, $k=2\pi/L$ is the wave number and $Ur.=.gHT^2/h^2$ is the Ursell number, where T is the wave period. H/h, even on the platform, has small values in comparison with breaking limit of approximately 0.8 [33]. The case involves intermediate water for the channel $(0.314.<.kh.<.3.142)$ and shallow water for the platform $(kh < 0.314)$. Ursell numbers show that the non-linear effects on the platform are more intensive.

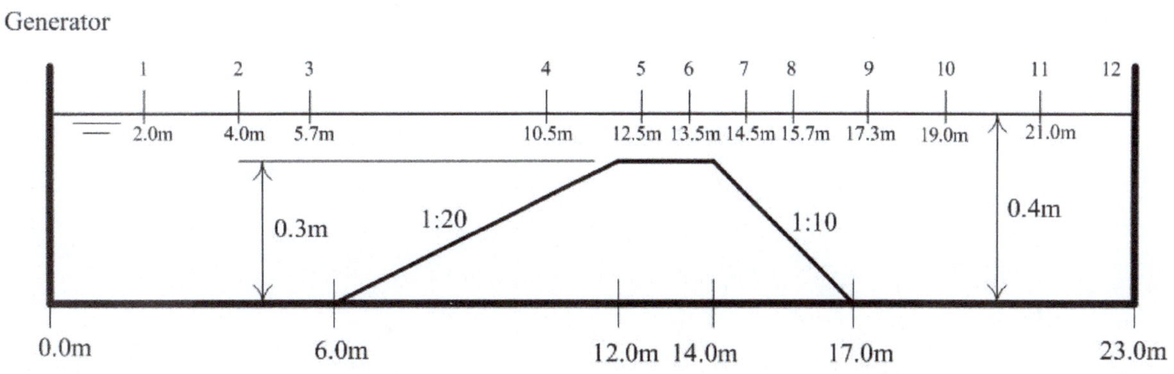

Figure 8. Channel geometry for the 1:20 and 1:10 breakwater

Table 2. Wave parameters for the 1:20 and 1:10 breakwater

	H/h	kh	Ur
Channel (h = 0.4m)	0.050	0.674	5.0
Platform (h = 0.1m)	0.259	0.318	103.6

Table 3 presents periods, frequencies and wavelengths concerning the fundamental frequency and the harmonic components that occur along the wave propagation. The wavelength was estimated according to the dispersion equation of the linear theory. These values are references to determine discretizations in time and space to be used in the modeling.

Table 3. Period, frequency and wavelength concerning the fundamental frequency, and 2nd, 3rd and 4thharmonics for the 1:20 and 1:10 breakwater

	Fundamental	**2ndharmonic**	**3rdharmonic**	**4thharmonic**
Period (s)	2.02	1.01	0.67	0.50
Frequency (Hz)	0.50	1.00	1.50	2.00
Wavelength (m)	3.73	1.46	0.70	0.39

FLUINCO used a mesh with 88700 elements and 37296 nodes. Twenty layers of elements were used in vertical direction, where small elements are located near the bottom and the free surface. Along the channel, the element sizes vary from $\Delta x.=.0.08$m in the boundary to $\Delta x.=.0.025$m around the platform. In the transversal direction, only one layer of elements is used, because the behavior of the flow is bi-dimensional. In the entrance of the domain, the wave generation condition is imposed while at the end the radiation condition is imposed. The velocity components are null on the bottom and the KFSBC is imposed in the free surface. The velocity component perpendicular to the surface is null for lateral walls (symmetry condition). As an initial condition, the velocity field is null and the pressure one is hydrostatic. The time step is 0.003s, a fact that satisfies the Courant stability condition.

Figure 9 shows the free surface elevations in gauge 3, located downstream the breakwater (x=5.7m); in gauge 6, on the platform (x=13.5m); in gauge 8, in the middle of the upstream slope (x=15.7m); and in gauge 11, on the upstream and far from the breakwater (x=23m). Results obtained by numerical model are compared with the experimental ones presented by Dingemans [44].

In general, there is good agreement between numerical results and experimental ones in gauges 3 and 6. In gauge 6, FLUINCO presents slightly smooth surface deformation. In gauges 8 and 11, corresponding to downstream, the nonlinear effects are more significant. The deformations in gauge 8 are well represented by FLUINCO; although the results get closer to the experimental ones in some regions, there are difficulties in representing the deformations related to higher harmonics, possibly due to the lack of an appropriate discretization to capture the nonlinear phenomena.

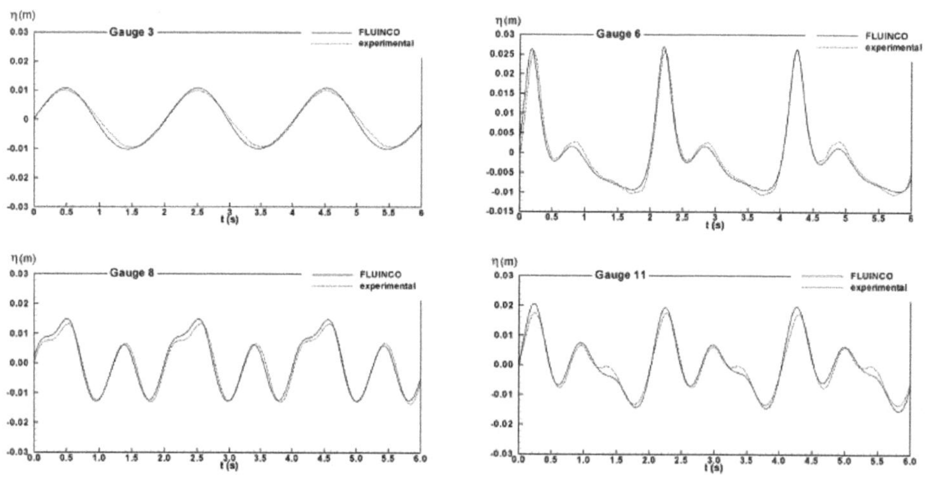

Figure 9. Free surface elevation of the 1:20 and 1:10 breakwater

Figure 10 shows the frequency spectra obtained by the model in the gauges and a comparison with the experimental results. The differences found in the free surface elevation are confirmed in Figure 10, which shows differences in the intensity of harmonic components, mainly in gauges located at the end of the channel. The numerical model adequately simulate the position of the peaks of the fundamental frequency and the harmonic components throughout the domain. However, there are some differences in the amplitude of these peaks, especially in gauges 8 and 11.

Figure 11 presents the streamlines around the upstream slope of the breakwater in eleven instants completing one wave period obtained by FLUINCO. We can observe that the flow separation and the vortex do not exist at all instants, due to the mild inclination of the upstream slope.

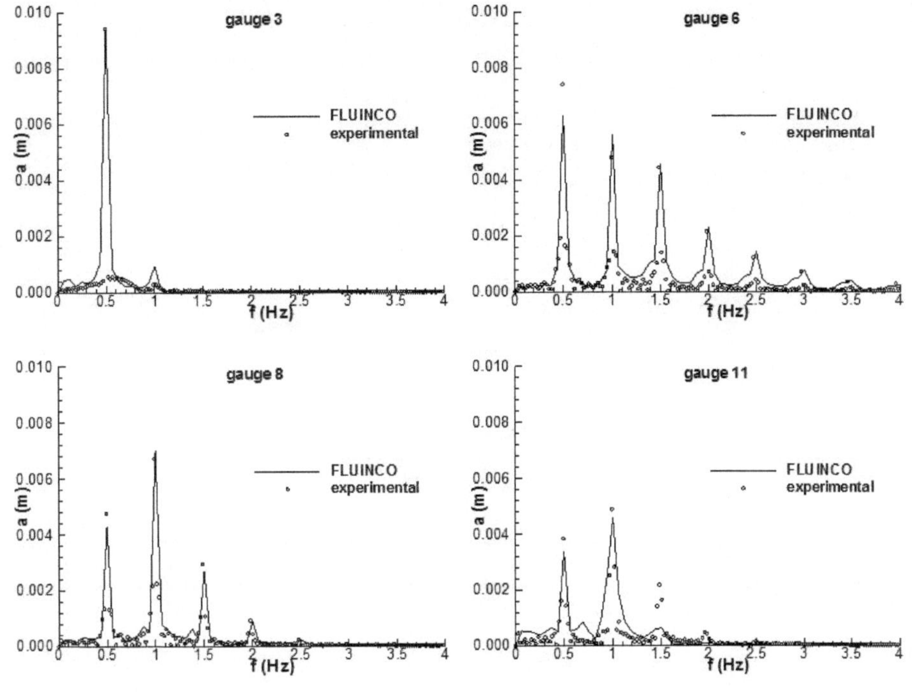

Figure 10. Numerical and experimental frequency spectrum in the gauges of the breakwater 1:20 and 1:10

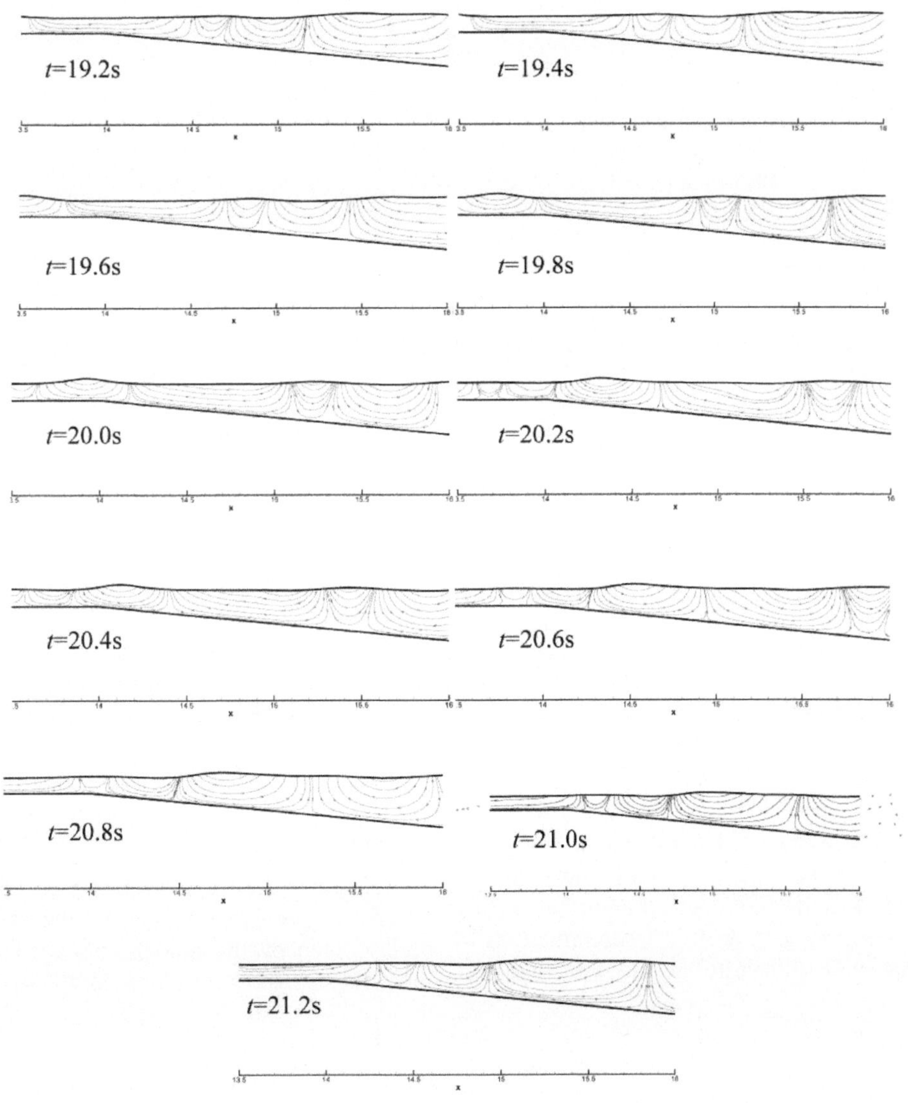

Figure 11. Streamlines of the 1:20 and 1:10 breakwater

Breakwater with Slopes 1:2

In this case, the length of the channel is 35m and the maximum and the minimum depths are 0.5m and 0.15m, respectively (See Figure 13). In the entrance of the channel, a monochromatic wave is generated with a period of 2.68s, related to a wavelength of 5.66m in the channel, and an amplitude of 0.025m. This problem is case 6 studied by Ohyama et al. [45] who analyzed six different types of waves experimentally. Table 4 shows some parameters that characterize the problem, calculated according to the linear theory. The Ursell number on the platform is 210, indicating the strong non-linearity in this region. Parameter H/h shows that breaking does not even occur on the platform.

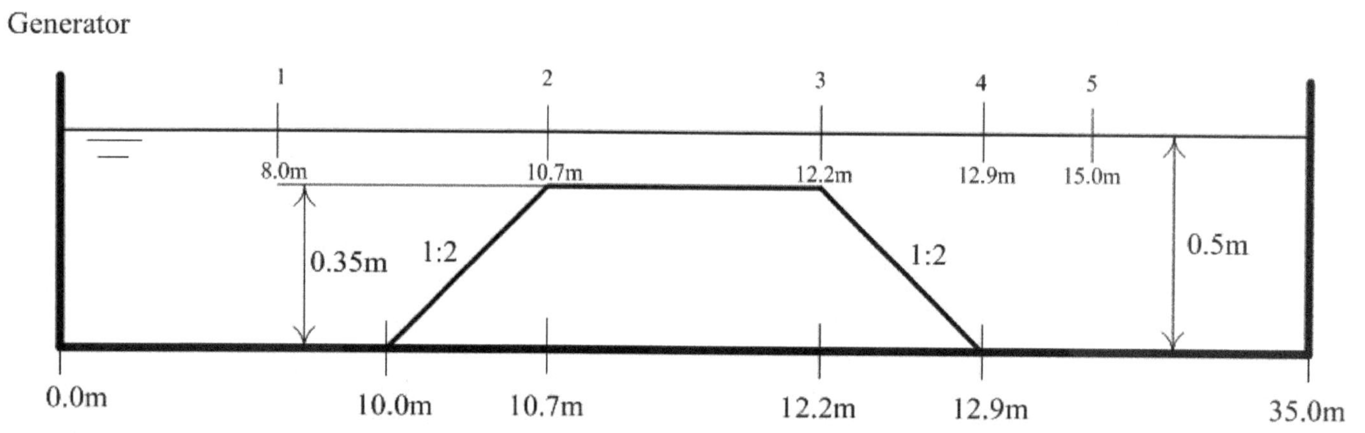

Figure 12. Geometry of the channel for the 1:2 breakwater

Table 4. Wave parameters for the 1:2 breakwater

	H/h	kh	Ur
Channel (h = 0.5m)	0.100	0.555	14.1
Platform (h = 0.15m)	0.355	0.294	210.0

Table 5 shows periods, frequencies and wavelengths concerning the fundamental frequency and the harmonic components that occur along the wave propagation.

A mesh with 120200 elements and 50526 nodes was used for FLUINCO in this simulation. The element sizes along the channel vary between dx=0.08m at the ends and dx=0.01m on the platform. The boundary and the initial conditions are similar to the ones in the previous case, and 0.002s was the time step.

Table 5. Period, frequency and wavelength related to the fundamental frequency, and 2nd, 3rd, and 4th harmonics for the 1:2 breakwaters

	Fundamental	2nd harmonic	3rd harmonic	4th harmonic
Period (s)	2.68	1.34	0.89	0.67
Frequency (Hz)	0.373	0.746	1.124	1.493
Wavelength (m)	5.66	2.42	1.22	0.70

Figure 13 shows the free surface elevations in gauges 3 and 5 (gauge positions are indicated in Figure 5). Numerical results are compared with the experimental ones presented by Ohyama et al. [45]. The FLUINCO model represents the surface deformation recorded in gauge 3 well. The deformations of gauge 5 indicate that the nonlinearity increases. In this case, FLUINCO captures the variation of the surface elevation more accurately.

Figure 14 shows frequency spectra obtained in gauges 3 and 5. The fundamental and the harmonic waves are well represented by the models, but their amplitudes differ. The FLUINCO results are closer for the two gauges.

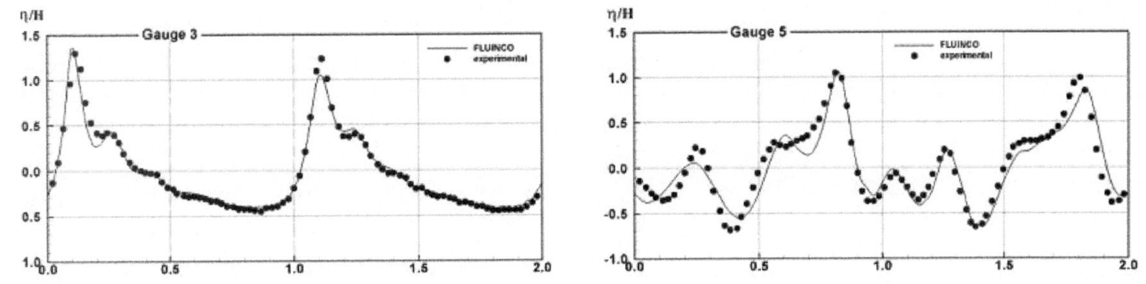

Figure 13. Free surface elevation for the 1:2 breakwater in gauges 3 and 5.

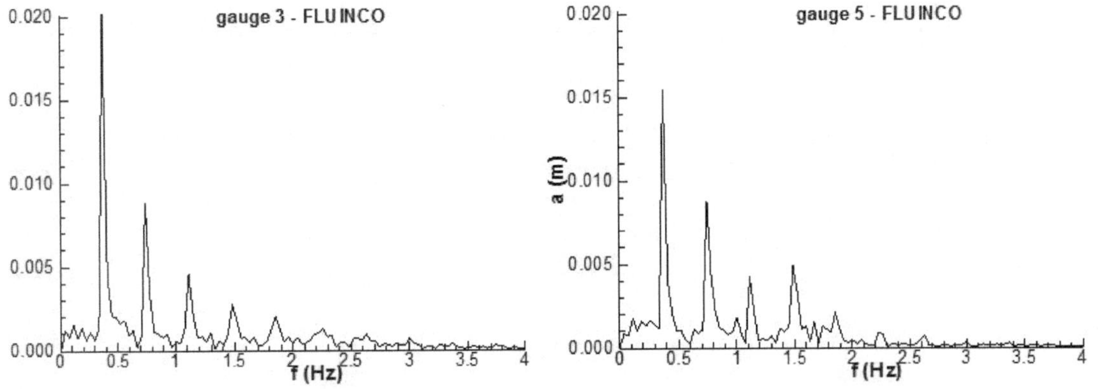

Figure 14. Breakwater case. Frequency spectra in gauges 3 and 5.

Streamlines during one wave period obtained by FLUINCO are presented in Figure 15. Unlike the previous case, a vortex, located between the upstream slope and the bottom, occurred during part of the wave period.

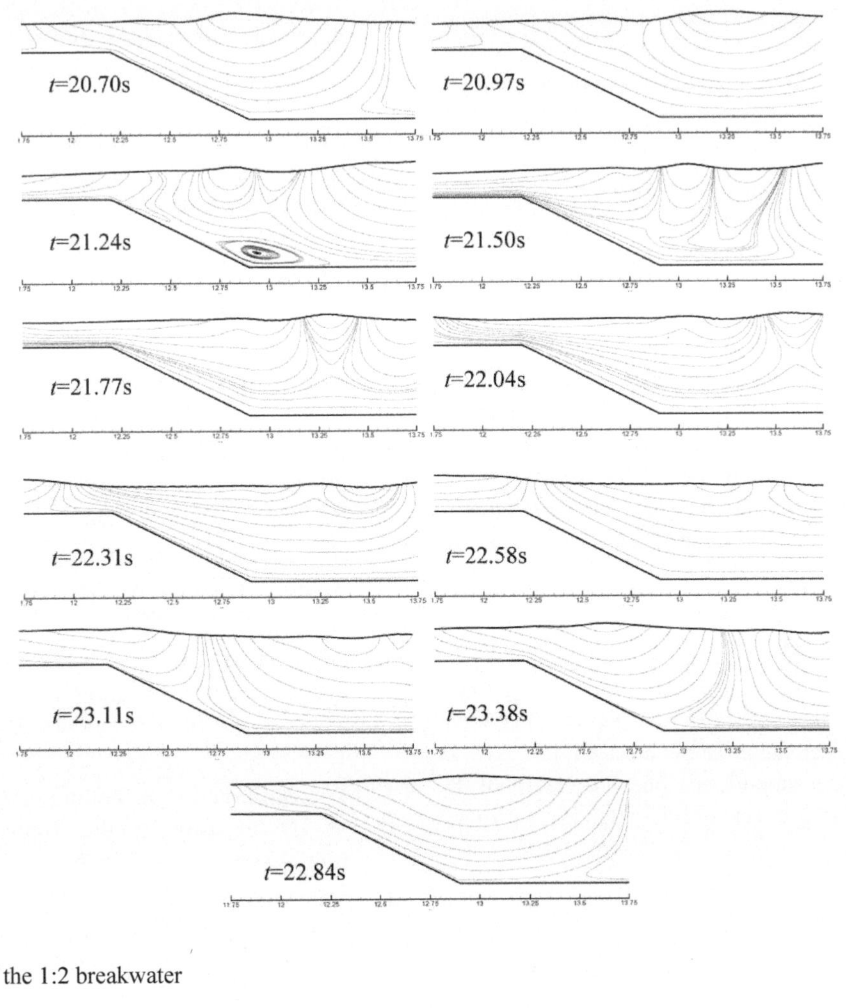

Figure 15. Streamlines of the 1:2 breakwater

CONCLUSIONS

In this text, we showed a model, named FLUINCO, capable of simulating flows on free surface. It is based on the semi-implicit two-step Taylor- Galerkin method to integrate Navier-Stokes equations in time and space. An ALE formulation is employed to describe the free surface movement. The methodology was validated in two study cases: the wave propagation over a submerged horizontal cylinder and submerged trapezoidal breakwaters. Both study cases showed the application of a Navier-Stokes based code, which considers accurately vertical flow effects, in the wave-submerged structure interaction problems.

In the case of the submerged horizontal cylinder, the free surface elevations and the velocity profiles obtained by the model were similar to experimental ones [39]. The numerical results presented a slight free surface deformation downstream, possibly because of the lack of refinement that caused numerical diffusion. In this case, the viscous effects influenced the flow behavior locally whereas the viscosity was not important far from the cylinder. The non-slip condition on the bottom did not modify the wave propagation significantly because it is a deep water case.

In the case of trapezoidal breakwater, two analyses were carried out for different upstream and downstream slopes. The first analysis deals with upstream and downstream slopes of 1:20 and 1:10, respectively. The results obtained by the model were compared with Dingemans' experimental data [44]. A comparison of the surface elevations and the energy spectrum for some gauges along the channel showed that the model provided good results. Although the FLUINCO results have been somewhat smoothed, they were closer to the experimental ones, including the ones in the gauges placed downstream, where nonlinear effects are more significant. Streamlines over a wave period showed that there was no flow separation in this case.

The second analysis consists of two 1:2 breakwater slopes; it showed a strong influence of nonlinear effects on the results of the surface elevation and the energy spectrum. The numerical results were compared with experimental ones presented by Ohyama et al [45]. The vertical velocity field obtained by FLUINCO showed that a vortex of non-turbulent origin was formed in the flow. The model obtained results closer to the experimental ones, including the ones downstream of the breakwater, where the nonlinearity effects are more significant. Both breakwater analyses showed that FLUINCO captures the nonlinear effects of the flow accurately, due to the fact that this model considers the influence of the vertical circulation in the flow.

ACKNOWLEDGEMENT

The author acknowledges Conceição Juana Fortes, Eric Didier and José Manuel Paixão Conde and the partnership between FURG and Laboratório Nacional de Engenharia Civil (LNEC). The author also thanks the support of Conselho Nacional de Desenvolvimento Científico e Tecnológico (CNPq - project 303308/2009-5).

REFERENCES

1. M S Longuet-higgins, and E D Cokelet, 1976The deformation of steep surface waves on water. I. A numerical method of computation, Proc. R. Soc. London A 95. 126
2. M Isaacson, 1982Non-linear wave effects on fixed and floating bodies. J. Fluid. Mech., j. 120267281
3. S. T Grilli, P Guyenne, F Dias, 2001A fully non-linear model for three-dimensional overturning waves over an arbitrary bottom. Int. J. Numer. Meth. Fluids. j. 357829867
4. D G Dommermuth, D Yue, K P (1987A High-order, spectral method for the study of nonlinear gravity waves. J. Fluid Mech. j. 184267288
5. W Bateman, J D Swan, C Taylor, P H (2001On the efficient numerical simulation of directionally spread surface water waves. J. Comput. Phys. j. 1741277305
6. D. H Peregrine, 1967Long waves on a beach. J. Fluid Mech. j. 27815882
7. P A Madsen, R Murray, O R Sorensen, 1991A new form of Boussinesq equations with improved linear dispersion characteristics. Coast Engrg. j. 15371388
8. P A Madsen, H B Bingham, P L-F Liu, 2002A new Boussinesq method for fully nonlinear waves from shallow to deep water. J. Fluid Mech. j. 462130
9. O Nwogu, 1993An alternative form of Boussinesq equations for nearshore wave propagation. J. Waterway, Port, Coastal Engineering. j. 119618638
10. G Wei, J T Kirby, S T Grilli, R Subramanya, 1995A fully nonlinear Boussinesq model for surface waves. Part 1. Highly nonlinear unsteady Wave. J. Fluid Mech. j. 2947192
11. M F Gobbi, and J T Kirby, 1999Wave evolution over submerged sills: Tests of a high-order Boussinesq model. Coastal Engineering. j. 375796
12. H A Schäffer, and P A Madsen, 1995Further enhencements of Boussinesq-type equations. Coastal Engineering. j. 26114
13. Y Agnon, P A Madsen, H Schaffer, 1999A new approach to high order Boussinesq models. Journal of Fluid Mechanics. j. 399319333
14. A B Kennedy, J T Kirby, Q Chen, R A Dalrymple, 2001Boussinesq-type equations with improved nonlinear behaviour. Wave Motion. j. 33225243
15. P Lynett, and P. L-F Liu, 2004A Two-layer, approach to water wave modelling. Proceedings of the Royal Society of London A. j. 46026372669
16. S C Hsiao, P Lynett, H H Hwung, P L-F Liu, 2005Numerical simulations of nonlinear short waves using a multilayer model. ASCE. J. Eng. Mech. j. 131231243
17. V Casulli, and G S Stelling, 1998Numerical simulation of 3D quasi-hydrostatic, free-surface flows. J. Hydr. Eng. ASCE. j. 124678686
18. P K Stansby, and J G Zhou, 1998Shallow-water flow solver with non-hydrostatic pressure: 2D vertical plane problems. Int. J. Numer. Meth. Fluids. j. 28514563

19. M Zijlema, and G S Stelling, 2008Efficient computation of surf zone waves using the nonlinear shallow water equations with non-hydrostatic pressure. Coastal Eng. j. 55780790
20. A Mahadevan, J Oliger, R Street, 1996A Non-hydrostatic, mesoscale ocean model. Part 1: Well-posedness and scaling. J. Phys. Oceanogr. j. 2618681880
21. J Marshall, A Adcroft, C Hill, L Perelman, C Heisey, 1997A Finite-volume, incompressible Navier-Stokes model for studies of the ocean on parallel computers. J. Geophys. Res. j. 10257535766
22. B R Hodges, and R L Street, 1999On simulation of turbulent nonlinear free-surface flows. J. Comput. Phys. j. 151425457
23. P Lin, and C W Li, 2002A σ-coordinate three-dimensional numerical model for surface wave propagation. Int. J. Numer. Meth. Fluids. j. 3810451068
24. J G Zhou, P K Stansby, 1999An arbitrary Lagrangian-Eulerian σ (ALES) model with non-hydrostatic pressure for shallow water flows. Comput. Meth. Appl. Mech. Eng. j. 178199214
25. F H Harlow, and J E Welch, 1965Numerical calculation of time-dependent viscous incompressible flow. Phys. Fluids. j. 821822189
26. P Lin, and P Liu, F L (1998A numerical study of breaking waves in the surf zone. J. Fluid Mech. j. 359239264
27. P D Hieu, T Katsutohi, V T Ca, 2004Numerical simulation of breaking waves using a two-phase flow model. Appl. Math. Model. j. 289831005
28. A Iafrati, Di Mascio A, Campana E F (2001A level set technique applied to unsteady free surface flows. Int. J. Numer. Meth. Fluids. j. 35281297
29. P Teixeira, R F (2011Simulação numérica da interação de escoamentos tridimensionais de fluidos compressíveis e incompressíveis e estruturas deformáveis usando o método de elementos finitos. Doctoral thesis, PPGEC-UFRGS, Porto Alegre, RS (in Portuguese).
30. P Teixeira, R F Awruch, A M (2000Numerical simulation of three dimensional incompressible flows using the finite element method. 8th Brazilian Congress of Thermal Engineering and Sciences- ENCIT, Porto Alegre.
31. J Argyris, St. Doltsinis J, Wuestenberg H, Pimenta P M (1985Finite element solution of viscous flow problems. In: Finite Elements in Fluids.Wiley, New York. 6
32. P Teixeira, R F Awruch, A M (2005Numerical simulation of fluid-structure interaction using the finite element method. Computers & Fluids. j. 34249273
33. R G Dean, and R A Dalrymple, 1994Water Wave Mechanics for Engineers and Scientists. In: Advanced Series on Ocean Engineering 2, World Scientific Publishing Co. Ltd. 470 p.
34. R A Flather, 1976A tidal model of the northwest European continental shelf. Mem. Soc. R. Liege, Ser. j. 106141164
35. W R Dean, 1948On the Reflection of Surface Waves by a Submerged Circular Cylinder. Proc. Camb. Phil. Soc. j. 44483491
36. F Ursell, 1950Surface Waves on Deep Water in presence of a Submerged Circular Cylinder, Proc. Camb. Phil. Soc. j. 46141158
37. J R Chaplin, 2001Nonlinear Wave Interactions with a Submerged Horizontal Cylinder. Proc. 11th Int. Offshore and Polar Eng. Conf., Stavanger. j. 3272279
38. T Schonberg, and J R Chaplin, 2003Computation of Nonlinear Wave Reflections and Transmissions from Submerged Horizontal Cylinder, Int. J. Offshore and Polar Eng. j. 132937
39. Paixão Conde J MDidier E, Lopes M F P, Gato L M C (2007Nonlinear wave diffraction by a submerged horizontal circular cylinder. Proc. 17th Int. Offshore and Polar Eng. Conf. (ISOPE 2007), Lisbon.
40. J W Johnson, R A Fuchs, J R Morison, 1951The damping action of submerged breakwaters. Trans. Am. Geophys. Union. j. 32704718
41. P Jolas, 1960Passage de la houle sur un seuil. Houille Blanche. j. 2148152
42. C J Huang, C M Dong, 1999Wave deformation and vortex generation in water waves propagating over a submerged dike. Coastal Engrg. j. 37123148
43. S Beji, J A Battjes, 1993Experimental investigations of wave propagation over a bar. Coastal Engineering. j. 19151162
44. M W Dingemans, 1994Comparison of computations with Boussinesq-like models and laboratory measurements. Report H-1684.12 Delft Hydraulics. 32 p.
45. T Ohyama, W Kioka, A Tada, 1995Applicability of numerical models to nonlinear dispersive waves. Coastal Engrg. j. 24297313
46. J G Zhou, P K Stansby, 1999An arbitrary Lagrangian-Eulerian σ (ALES) model with non-hydrostatic pressure for shallow water flows. Comput. Meth. Appl. Mech. Eng. j. 178199214
47. H L Yuan, C H Wu, 2004A Two-dimensional, vertical non-hydrostatic σ model with an implicit method for free-surface flows. Int. J. Numer. Meth. Fluids. j. 44811835
48. G S Stelling, M Zijlema, 2005Further experience in computing non-hydrostatic free-surface flow involving with water waves. Int. J. Numer. Meth. Fluids. j. 48169197
49. S Beji, and J A Battjes, 1994Numerical simulation of nonlinear wave propagation over a bar. Coast. Eng. j. 23116
50. Y M Shen, C O Ng, Y H Zheng, 2004Simulation of wave propagation over a submerged bar using the VOF method with a two-equation k-epsilon turbulence modeling. Ocean Eng. j. 318795
51. F Ting, C K Kim, Y.K (1994Vortex generation in water-waves propagating over a submerged obstacle. Coast. Eng. j. 242349
52. F Zhuang, J J Lee, 1996A viscous rotational model for wave overtopping over marine structure. Proc. 25th Int. Conf. Coastal Eng. ASCE. 21782191

CHAPTER 2

TSUNAMI WAVE PROPAGATION

Alexey Androsov, Sven Harig, Annika Fuchs, Antonia Immerz, Natalja Rakowsky, Wolfgang Hiller and Sergey Danilov

INTRODUCTION

Tsunami hazard is connected with loss of human life, flooding of coastal structures, destruction of berthing designs, infrastructures of coastal water areas. In support of the Tsunami Early Warning System for the Indian Ocean, a finite element model TsunAWI for simulations of wave propagation has been developed. It is part of German Indonesian Tsunami Early Warning System (GITEWS) (www.gitews.de) serving to predict arrival times and expected wave heights. TsunAWI is based on an unstructured grid approach employing finite elements to solve the governing equations. Finite-element methods are widely used in studies of wave generation and propagation in different fields of fluid dynamics. They are often employed to simulate propagation of long waves such as ocean tides and tsunamis in the ocean in the framework of shallow-water equations [1–3]. The main reason to prefer FE modelling is that the solution is computed over a mesh that can be adapted to cover basins with complex geometries characterized by irregular bottom topography and coastlines. The purpose of this work is to describe a complex system of tsunami warning, based on the numerical modelling of tsunami events in Indian Ocean. The complexity of the problem stems from insufficient information about the sources generating tsunami, real bottom morphometry and extremely short warning time. In addition, some physical processes, such as interaction of tsunami waves with long tidal waves or nonhydrostatic effects, commonly neglected in models of tsunami wave propagation, may lead to substantial corrections. Our work, therefore, addresses the influence of tidal dynamics on tsunami wave propagation in coastal areas. A tsunami wave is much shorter than tidal waves which explain why tidal waves are usually ignored in tsunami modelling. There are three approaches to account for the interaction between tsunami waves and tides. The first one presumes that the wave propagation is linear, so that tides reduce or augment (depending on their phase) the amplitude of arriving waves [4]. In this case, having simulated tidal patterns in advance, it would be possible to predict arriving waves as a simple superposition of signals. The second approach assumes that the interaction has a nonlinear character caused by changes in the fluid layer thickness in the shallow area [5]. Finally, the third approach suggests that the main effect is due to nonlinear interactions between tidal and tsunami velocities [6]. The studies performed thus far are of theoretical character and do not involve practical examples. As a tsunami wave arrives in coastal regions with rough bathymetry, the nonhydrostatic part of pressure, neglected in the standard hydrostatic configuration of TsunAWI, gains in importance. The account for nonhydrostatic effects corrects the wave propagation speed. The impact of nonhydrostatic effects is investigated using the standard benchmark problem of a solitary wave runup on a plane beach [7]. The organization of present work is as follows. Section 2 introduces the basic equations, the numerical implementation of the model and the mesh generation algorithm. Its final part presents the nonhydrostatic pressure correction algorithm. Section 3 describes the architecture of the Tsunami Early Warning System. In section 4, the system is applied to simulate some realistic scenario of tsunami wave propagation in the Indian Ocean. Section 5 deals with tidal-tsunami interactions whereas section 6 illustrates manifestations of nonhydrostatic effects. Section 7 contains the conclusions of the work.

Citation: Alexey Androsov, Sven Harig, Annika Fuchs, Antonia Immerz, Natalja Rakowsky, Wolfgang Hiller and Sergey Danilov (2013). Tsunami Wave Propagation, Wave Propagation, Dr. Yi Zheng (Ed.), InTech, DOI: 10.5772/51340.

DESCRIPTION OF THE MODEL

Boundary-value problem in Cartesian coordinates

In the domain $\Omega = \{(x, y) \in \Omega, 0 \le t \le T\}$, where Ω is a plane domain with boundary $\partial\Omega$, we consider the vertically averaged equations of motion and continuity

$$\mathbf{v}_t + (\mathbf{v}\cdot\nabla)\mathbf{v} + g\nabla\zeta = \Phi \equiv f\mathbf{k}\times\mathbf{v} - rH^{-1}\mathbf{v}|\mathbf{v}| + H^{-1}\nabla(K_h H\nabla\mathbf{v}), \tag{1}$$

$$\zeta_t + \nabla\cdot(H\mathbf{v}) = 0, \tag{2}$$

where v = (u, v) is the horizontal velocity vector, $H = h + \zeta$ is the total water depth, $H > 0$, h is the unperturbed water depth, and ζ is the surface elevation, $\nabla = (\partial/\partial x, \partial/\partial y)$ is the gradient operator, f the Coriolis parameter, k is the unit vector in the vertical direction, r is the bottom friction coefficient, and K_h is the eddy viscosity coefficient. The set of (1) and (2) is known as the rotating shallow water equations. On the solid part of the boundary, $\partial\Omega_1$, and on its open part, $\partial\Omega_2$, we impose the following boundary conditions

$$v_n|_{\partial\Omega_1} = 0, \quad \Gamma(\mathbf{v},\zeta)|_{\partial\Omega_2} = \Theta_1 \tag{3}$$

where v_n is the velocity normal to $\partial\Omega_1$, Γ is the operator of the boundary conditions and Θ_1 is a vector-function determined by the boundary regime and different for inflow and outflow [8]. In practice, when the full information on the open boundary is unavailable, in place of the second condition (3) one commonly imposes the boundary condition on the elevation $\zeta|\partial\Omega_2 = \psi(x, y, t)$ or the radiation boundary condition $vn = v \cdot n = g/H\zeta$. The latter provides free linear wave passage through the open boundary (when the Coriolis acceleration plays a small role). Here n is the outer unit normal to $\partial\Omega_2$. The accuracy of the reduced boundary-value formulation with only the sea level assigned at the open boundary, was analyzed in [9]. The problem (1)-(3) for the combination $u = (v, \zeta)$ is solved for given initial conditions: $u|_{t=0} = u^0$. The equation of energy for set (1) and (2) has the form

$$\frac{\partial E}{\partial t} + \nabla\cdot\left[H\left(g\zeta + \frac{1}{2}|\mathbf{v}|^2\right)\mathbf{v}\right] = -r|\mathbf{v}|^{3/2} + \mathbf{v}\cdot\nabla(K_h H\nabla\mathbf{v}) \tag{4}$$

where

$$E = \frac{1}{2}\left(H|\mathbf{v}|^2 + g\zeta^2\right) \tag{5}$$

is the total energy per unit area.

Method

The finite element spatial discretization is based on the approach by Hanert et al. [10] with some modifications like added viscous and bottom friction terms, corrected momentum advection terms, radiation boundary condition and nodal lumping of mass matrix in the continuity equation. The basic principles of discretization follow the paper of [10] and are not repeated here. Simulation of tsunami wave propagation benefits from using an explicit time discretization. Indeed, numerical accuracy requires relatively small time steps, which reduces the main advantage of implicit schemes. Furthermore, modelling the inundation processes usually requires very high spatial resolution in coastal regions (up to some tens of meters) and consequently large number of nodes, drastically increasing necessary computational resources in case of implicit temporal discretization. The leap-frog discretization was chosen as a simple and easy to implement method. We rewrite eqs. (1) and (2) in time discrete form,

$$\frac{\mathbf{v}^{n+1} - \mathbf{v}^{n-1}}{2\Delta t} + fk\times\mathbf{v}^n + g\nabla\zeta^n + \frac{r}{H^n}|\mathbf{v}^n|\mathbf{v}^{n+1} - \nabla K_h\nabla\mathbf{v}^{n-1} + (\mathbf{v}^n\nabla)\mathbf{v}^n = 0, \tag{6}$$

$$\frac{\zeta^{n+1} - \zeta^{n-1}}{2\Delta t} + \nabla\cdot(H^n\mathbf{v}^n) = 0. \tag{7}$$

Here Δt is the time step length and n the time index. The leap-frog three-time-level scheme provides second-order accuracy and is neutral within the stability range. This scheme however has a numerical mode which is removed by the standard filtering procedure. Notice that friction and viscosity contributions deviate from the usual leap-frog method.

Momentum advection scheme

Because of the discontinued character of velocity representation, special care is required with respect to the implementation of momentum advection. Earlier experiments with $P^{NC}_1 - P_1$ code revealed problems with spatial noise and instability of momentum advection when the discretization is used in the form described in [10]. A modified implementation without upwinding terms was found to work well, yet when paired with rather high viscous dissipation for removing small-scale noise in the velocity field. In addition, the implementation of the momentum advection for P^{NC}_1 velocities involves cycling over edges in the numerical code, in addition to cycling over elements to assemble the elemental (regular) contributions. This reduces numerical efficiency. This lead us to a simpler approach, which provides some smoothing of the velocity field while removing edge contributions. According to this approach, prior to calculating the advection term in the momentum equation we project the velocity from the P^{NC}_1 to the P_1 space in order to smooth it. To make this projection numerically efficient, nodal quadrature (lumped mass matrix) is employed. The projected velocity is then used to estimate the advection term. Finally we proceed as usual by multiplying the result with a P^{NC}_1 basis function and integrating over the domain [11]. This results in a very stable behaviour

Other implementation details

Wetting and drying, viscosity

For modelling wetting and drying we use a moving boundary technique which utilizes linear least square extrapolation through the wet-dry boundary and into the dry region. We apply "dry node concept" developed in [12] The idea of this concept is to exclude dry nodes from the solution and then to extrapolate elevation to the dry nodes from their wet neighbours. Because the scheme is neutrally stable it demands horizontal viscosity in places of large gradients of the solution. The coefficient of horizontal viscosity is determined by a Smagorinsky parameterization [13].

Code parallelization

Since a large number of scenarios has to be calculated, code optimization and parallelization is crucial. The operational version of TsunAWI is parallelized employing OpenMP. It is therefore limited to shared memory platforms. The parallelization is implemented by defining parallel regions in the numerically most demanding parts of the code and splitting up the corresponding loops, thus sharing the load to the CPUs involved. The remaining part of the code stays serial. The implementation in this ideology needs therefore smaller changes than a full parallel implementation based on MPI, and proves to be numerically efficient.

Mesh generation

The quality of the triangulation of the model domain is crucial for the model results. The meshes used in the following studies were generated by a mesh generator based on the freely available software Triangle by Jonathan Shewchuk [14]. Starting from a model domain defined within a topography/bathymetry data set (in our case GEBCO 30") Triangle is used to generate a mesh based on a refinement rule depending on the water depth and prescribed by the corresponding wave phase velocity and the CFL criterion. The triangulation will be refined until the edges in all triangles fulfill the criterion8

$$\Delta x \le \min\{c\sqrt{gh}, c_g \frac{h}{\nabla h}\} \tag{8}$$

Since the Triangle output is not yet smooth enough for numerical experiments several iterations of smoothing are applied. Smoothing steps consist of edge swapping, torsion smoothing and linear smoothing. Torsion smoothing tries to equal out angles around each node, linear smoothing acts on the distance between nodes. These strategies are described in [15].

Nonhydrostatic approach

Limitations of the shallow water model

The shallow water equations are derived on the assumption that the horizontal motion of the fluid dominates over the vertical one. Using the wave length λ and the reference water depth h as characteristical values of horizontal and vertical motion the ratio

$$\delta := \frac{h}{\lambda} \ll 1, \tag{9}$$

must be fulfilled. In this case, terms containing the vertical component of velocity are very small compared with others and a model reduced to the two horizontal dimensions still provides a good approximation of the three-dimensional flow transport. In the course of neglecting the impact of vertical motion, the hydrostatic approximation is approved: The pressure term is limited to static pressure $p0 = \rho g(\zeta - z)$, as dynamic pressure forces are mainly induced by vertical elevation. With regard to tsunami propagation, condition (9) is satisfied in deep ocean, as the wave length accounts for hundreds of kilometers. When the tsunami reaches coastal regions, the wave length decreases rapidly and the ratio δ as defined above may become less strict, especially in the presence of horizontal inhomogeneities. Waves become dispersive in this limit. They cannot be represented by the standard shallow water equations since in a hydrostatic model the phase velocity $c \approx gh$ of a wave packet is not affected by the wave length. A more accurate model may be required near the shore. We are seeking how to improve TsunAWI so that the nonhydrostatic effects can be taken into account if required.

The nonhydrostatic approach

In search of approach that corrects the given one, it is useful to look at terms neglected so far. The deviation from the hydrostatic approximation to the real pressure

$$p' = p - p_0 \tag{10}$$

is of particular interest and serves as a starting point. Returning to the momentum equation in vertical direction and considering the missing nonhydrostatic pressure term in (1), the depth-averaged momentum equations result in

$$\partial_t \mathbf{v} + (\mathbf{v}\cdot\nabla)\mathbf{v} + g\nabla\zeta + \frac{1}{\rho H}\int_{-h}^{\zeta} \nabla p' \mathrm{d}z = \Phi \tag{11}$$

$$\partial_t w + (\mathbf{v}\cdot\nabla)w \qquad + \frac{1}{\rho H}\int_{-h}^{\zeta} \partial_z p' \mathrm{d}z = 0 \tag{12}$$

in which w describes the depth-averaged vertical velocity component. Both additional unknowns w and pare simplified by assuming linear behavior of non-averaged fields in vertical direction. The values at surface and bottom are partly given by the boundary conditions: the kinematic boundary conditions determine the vertical velocities w_ζ and w_{-h} as

$$w_\zeta = \partial_t\zeta + \mathbf{v}\cdot\nabla\zeta \tag{13}$$

$$w_{-h} = \quad -\mathbf{v}\cdot\nabla h. \tag{14}$$

At the surface, the boundary condition for pressure enforces the nonhydrostatic part of pressure to vanish, just like the hydrostatic counterpart. Hence, p' depends only on nonhydrostatic bottom pressure $q := p'_{-h}$.

Discretization

For solving equations (11) and (12) a two-step procedure is applied, as suggested in [16] in the framework of finite-difference model, used in [17] for wave breaking and run-up issues and realized in a Finite Element/Finite Volume context by [18]. Firstly, the hydrostatic shallow water equations (1), (2) are stepped forward as before, just as the hydrostatic variant of equation (12) with $p' \equiv 0$. The additional unknowns q and w are introduced as vectors containing the values at the nodes of the triangulation in the same way as the sea surface elevation ζ. In a second step the resulting velocity vector ($\tilde{v}$, $\tilde{w}$) is corrected by

$$\mathbf{v}^{n+1} = \tilde{\mathbf{v}}^{n+1} + \frac{2\Delta t}{\rho}\left(\nabla\frac{q^{n+1}}{2} - \frac{q^{n+1}}{2}\frac{\nabla(\zeta^n - h)}{H^n}\right), \tag{15}$$

$$w^{n+1} = \tilde{w}^{n+1} + \frac{2\Delta t}{\rho}\frac{q^{n+1}}{H^n}, \tag{16}$$

in which the correction terms depending on q^{n+1} are nothing else but the calculated integral terms of (11) and (12). The factor 2Δt arises from using the leapfrog time-stepping scheme. This correction is performed at the end of each time step. However, the nonhydrostatic part of bottom pressure q^{n+1} must be estimated before. While p_0 can be calculated explicitly, the computation of q is performed in an implicit manner. By including the equations (15) and (16) in the depth-integrated continuity equation

$$\int_{-h}^{\zeta} \nabla\cdot\mathbf{v} + \partial_z w \ \mathrm{d}z = 0, \tag{17}$$

the nonhydrostatic bottom pressure can be determined. Because of finite-element discretization, equation (17) reduces to a system of linear equations with a large sparse matrix whose entries vary with time step. As concerns the CPU time, computing q proves to be rather expensive compared to the time required by the shallow water model.

THE GERMAN-INDONESIAN TSUNAMI EARLY WARNING SYSTEM GITEWS

The German-Indonesian Tsunami Early Warning System GITEWS was founded after the devastating Indian Ocean Tsunami 2004 as a joint project of German research institutes leaded by GFZ and Indonesian institutes and authorities. In 2005, Indonesia and Germany agreed in a joint declaration to develop a warning system coordinated by the UNESCO Intergovernmental Oceanographic Commission. The system was inaugurated in 2008 and evaluated in 2010 by an international commission including the heads of the four operating worldwide tsunami warning centers.

Architecture of the Indonesian TEWS

The Indonesian warning systems meets the challenge of near field warning with extremely short warning time. As tsunamigenic earthquakes originate close to the shore, the time between a seismic event and the issue of a warning is limited to just a few minutes. The TEWS therefore relies on a repository of tsunami scenarios TSR precomputed with the tsunami model TsunAWI, see section 3.3. Figure 1 shows a schematic overview of the components of the TEWS. Data of various sensors, such as the seismic data analyzed by SeisComp, GPS sensors as well as data from tide gauges and buoys is collected via the Tsunami Service Bus TSB [19], and distributed to the Decision Support System DSS [20], which triggers the Simulation System SIM (see section 3.2) with the available sensor data. Based on this data, the SIM delivers a set of best matching scenarios. The DSS performs a worst case aggregation over these scenarios and visualizes expected wave heights and arrival times for the Indonesian coasts. The system is installed at the warning center in Jakarta/Indonesia and assists the Chief Officer on Duty in assessing the potential tsunami risk. He then disseminates warnings to governmental institutions, local disaster management, action forces and media.

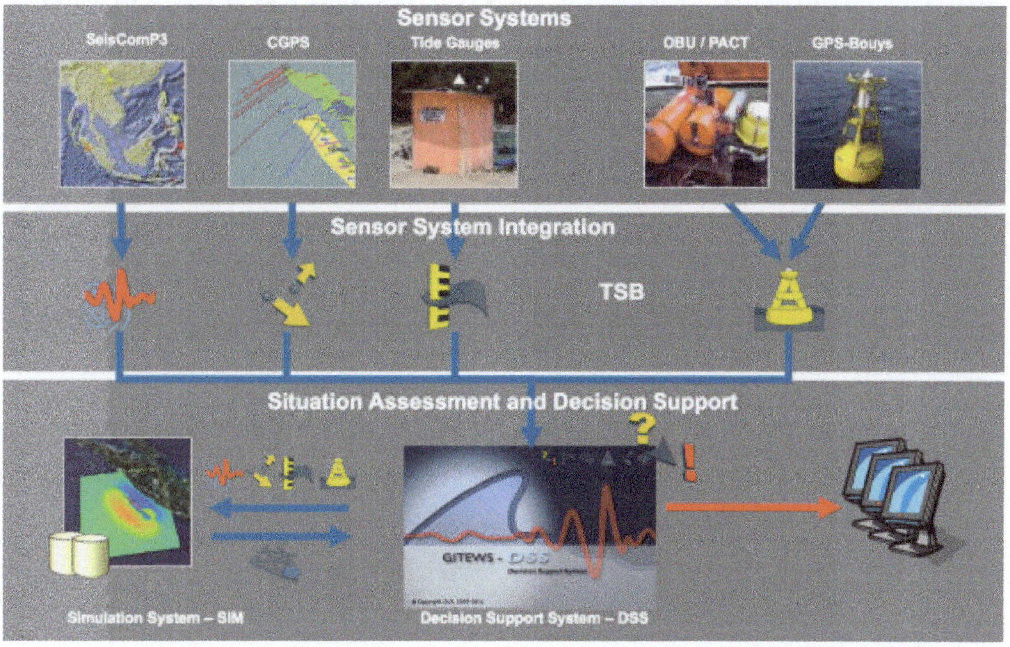

Figure 1. Schematic overview of the early warning process. In case of an earthquake, all available sensor data are gathered by the Decision Support System DSS via the Tsunami Service Bus TSB and sent to the Simulation System SIM. A set of scenarios fitting to the sensor data are delivered and aggregated to an overall perspective.

Simulation System (SIM)

In the following the SIM with its components and the algorithm for selecting the best fitting scenarios to an earthquake event shall be introduced in a nutshell. A more detailed overview of the SIM is given in [21], though referring to a prior version of the selection algorithm. The SIM is written as a Java web application offering web processing WPS and web notification services WNS conforming to the open GIS consortium OGC standard. It is accessible via HTTP-Requests where request and response details are transferred to and from the SIM in XML format.

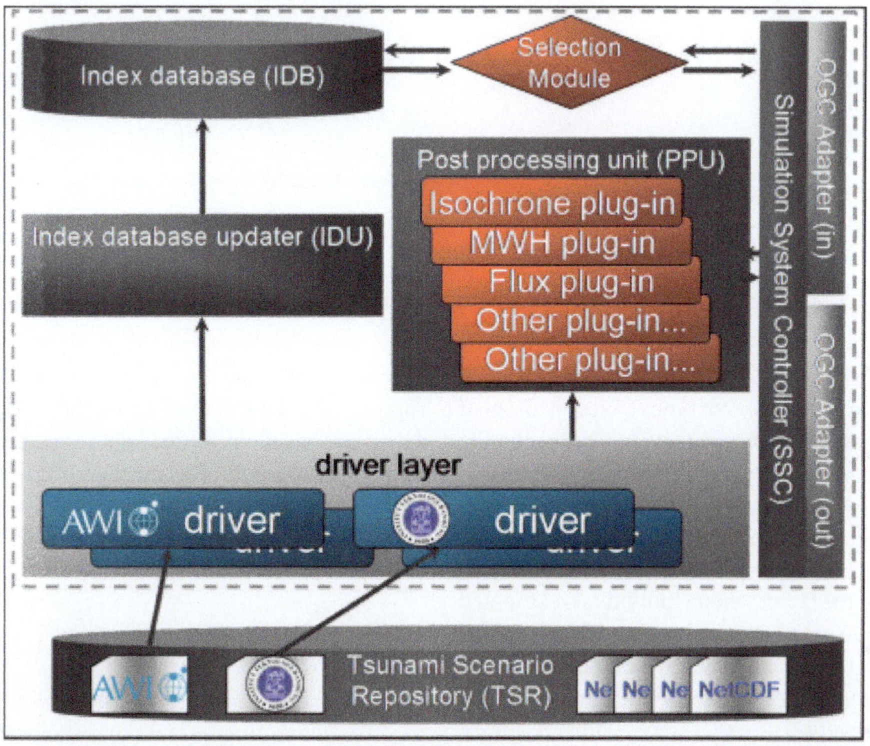

Figure 2. Overview of the SIM software components. Main components are the controller, responsible for overall coordination of processes, the Index Database IDB with data products from the TSR for fast access in case of an event, Postprocessing Unit for extraction of these data products from scenarios, the selection module determining best fitting scenarios for given sensor data, and the OGC-adapter as communication interface.

SIM components

A graphical overview of the software components is given in figure 2.

- Index Database IDB: As mentioned in section 3.1, warning time is crucial for the region of the Sunda Arc. To ensure a response time of the SIM of less than 10 s, significant data products are extracted in advance from each scenario contained in the pre-calculated scenario repository TSR. The IDB contains maximum sea surface height (mwh) and arrivaltimes at coastal forecast zones, isolines of mwh, isochrones, sea surface height time series at tide gauge and buoy locations, and GPS displacement values at GPS station locations.
- Index Database Updater: This component is used to store extracted data products in the index database.
- Postprocessing Unit PPU: The data products are extracted by the post processing plugins compounded in the PPU. The plugins are written in C to employ fast mathematical calculation routines. There task is to extract the information stored in the IDB and to generate SHP-files for visualizing the wave propagation in the DSS. Furthermore, the simulated inundation can be extracted as SHP-files, which are a basis for a priori risk assessment and hazard maps.
- Driver Layer: The driver layer separates the scenario data format from the internal representation of data in the SIM and thus allows to integrate scenarios from different origins and in different formats. One driver per scenario type acts as a translator between scenario data format and the data structure used in the SIM. At the time being, only scenarios calculated with TsunAWI are addressed.
- Scenario Selection Module: The component responsible for the identification of the best fitting scenarios to an earthquake event is described in section 3.2.2.
- Tsunami Scenario Repository TSR: It contains all pre calculated scenarios and is described in more detail in section 3.3

Selection algorithm

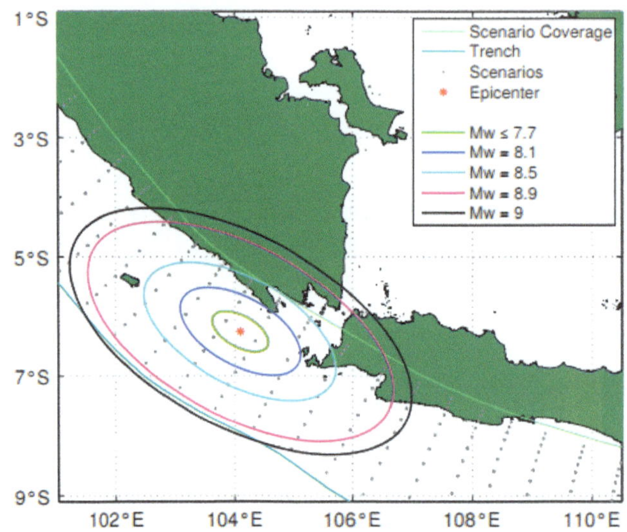

Figure 3. Visualization of the seismic uncertainty ellipse with different size depending on the observed magnitude

The selection algorithm uses a multi sensor approach combining the different available sensor types to acquire a set of best matching scenarios to an earthquake event. By basing the selection on different sensor data types, uncertainties resulting from inaccurate measurements and errors in the tsunami model are reduced. In the selection algorithm as described in [21], the different sensor types and even individual sensor stations could be weighted so to factor their individual estimated uncertainty. Initially, the so called matching values generated for each sensor type for a scenario were accumulated in an overall weighted sum of matching values defining a measure of suitability to the current event. The experiences with real sensor data in the GITEWS project showed that each sensor group has to be regarded separately with its characteristics in mind. The weighted sum over all matching values is therefore replaced by a stepped approach starting with seismic data delivering the most robust values, followed by CGPS data. The algorithm is specialized for the Sunda Arc taking into advantage that the rupture is often oriented along the trench. The preselection of scenarios based on seismic data is performed by calculating an elliptic area around the measured epicenter within which the selected scenarios lie (figure 3). The dimension of the ellipse depends on the measured magnitude MW, the long ellipse-axis is given by

$$r_L = 10^{0.5-[M_W+0.3]-1.8}\text{km}.$$

To ensure that at least one scenario is covered, for small magnitudes r˜L = 180km is chosen. The orientation of the ellipse is derived from the orientation of the trench between the two coordinates found by going rL/2 up an down the trench from the nearest point to the epicenter on the trench. The second important sensor data class in GITEWS are

CGPS dislocation vectors. The sensor data is reliable, and the measurement arrives at the DSS fast, thus allowing for a better estimate of the tsunami hazard in the first few minutes after the earthquake. For each sensor and each scenario (that remains after preselection by the seismic sensor data) the length of the measured and the corresponding scenario dislocation vector are compared. Scenarios for which a defined minimum number of similar GPS values is reached are taken into account for the final set of best fitting scenarios. If several scenarios are chosen for one epicenter, only those with the largest magnitude within the uncertainty range [M_w − 0.5, M_w + 0.3] is kept in the list, because in DSS processing, a worst case aggregation over all scenarios is performed. For each coastal forecast zone, the shortest arrival time is taken from all scenarios in the list, and the highest maximum wave height. The scenarios with lower magnitudes will not account for this aggregation, hence skipping them reduces the amount of data to be processed without changing the result.

The Tsunami Scenario Repository

As of March 2012, the GITEWS Tsunami Scenario Repository TSR contains 3470 scenarios for prototypic ruptures (RuptGen2.1 [22]) with magnitudes in the range of Mw=7.2,7.4,. . . ,9.0 on 528 different epicenters, see figure 4. These scenarios for prototypic ruptures in the Sunda

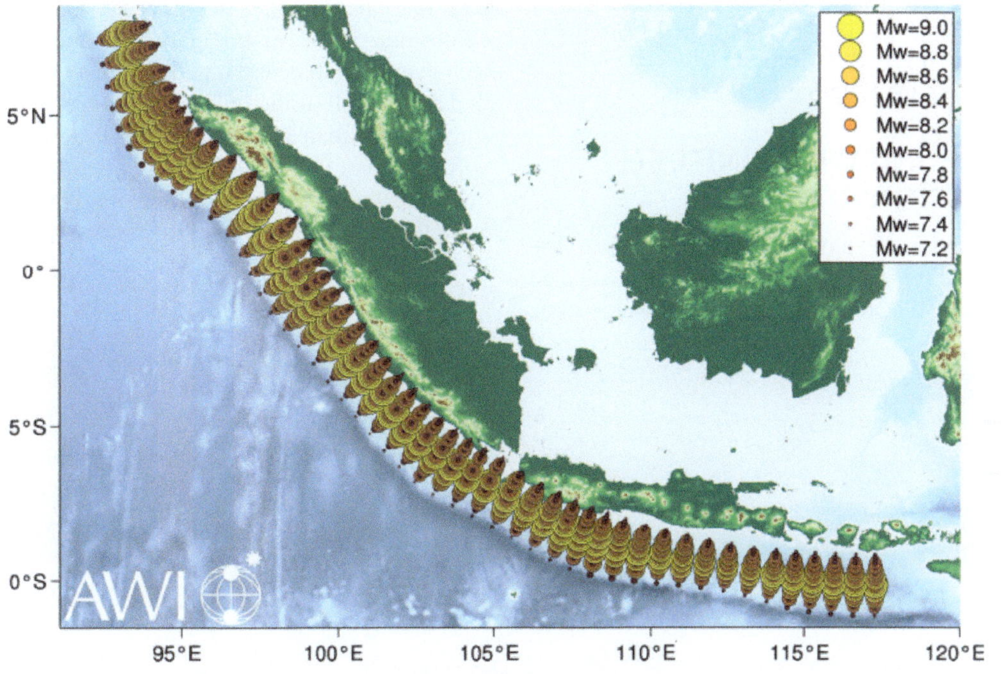

Figure 4. Tsunami scenario repository for GITEWS. Each circle symbolises a scenario with center at the given position. Scenarios for up to ten different magnitudes Mw= 7.2, 7.4, ..., 9.0 are calculated at each of the 528 positions, resulting in 3470 scenarios in total.

Arc are used in the warning situation as well as for a priori risk analysis and hazard maps. The quality of the scenarios, being crucial for both tasks, relies strongly on the numerical implementation of the governing physical equations and on an accurate representation of the bathymetry and topography especially in coastal regions. Therefore, GITEWS included the development and concise validation of the tsunami model TsunAWI, described in detail in section 4. For the TSR, TsunAWI operates on a grid with the resolution changing seamlessly between 14 km in deep sea, 150 m at the coast and down to 50 m in regions of special interest, e.g., densely populated areas and at tide gauges. The run-up scheme provides a realistic approximation of the inundation hazard. The dense net of scenarios allows to evaluate the

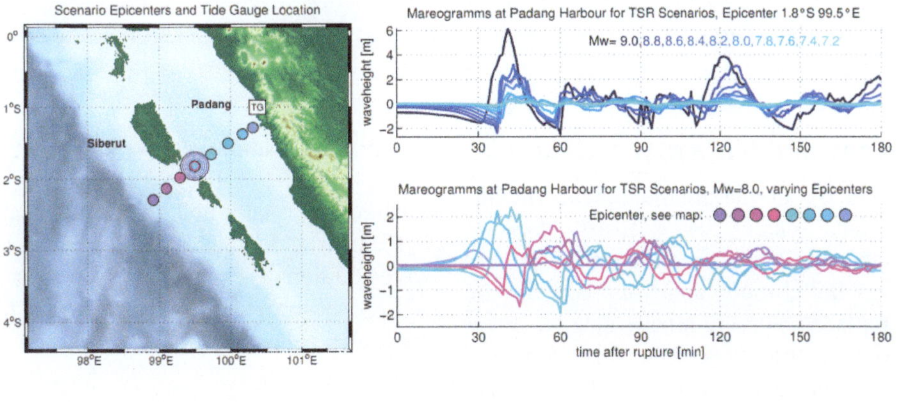

Figure 5. Influence of the magnitude Mw and of the source location on the strength of a tsunami: Mareograms at Padang Harbor for TSR-scenarios originating at the same epicenter with different magnitude (above) and with equal magnitude, but different origin (below).

effect of small changes in epicenter and magnitude as illustrated in figure 5 for artificial mareograms for the harbour of Padang, Sumatra. The logarithmic energy scale of the magnitude is clearly illustrated by the upper mareogram. The lower graph shows more complex features due to varying epicenters on a line perpendicular to the trench. While the magnitude is fixed to Mw=8.0 in this example, the tsunamis originating closest to the coast have the lowest impact. On the one hand, this is due to the initial conditions generated by RuptGen which assumes a source depth of 0 km at the trench increasing up to 100 km under the main islands. A second factor determining the strength of the tsunami is the height of the water column at the origin, and this of course is small in coastal regions. Epicenters southwest off the coast yield increasing wave heights in Padang, until the Mentawai islands are reached. The pink epicenter marks a turning point where the corresponding tsunami is no longer trapped in reflections between Sumatra and the islands, relieving the situation for Padang.

TSUNAMI MODELING WITH FINITE ELEMENTS: APPLICATIONS

TsunAWI is still under development and therefore the performance and model results must be constantly evaluated to ensure a consistent and stable code. Validation needs to be performed on several levels, among these are:

- Consistency / convergence,
- Benchmarks in idealized settings with well defined results,
- Comparisons in real cases with measured data. This section deals mostly with the last case, section 6 includes some cases with idealized bathymetry and compares hydrostatic and non hydrostatic results.

Indian Ocean tsunami in December 2004

For the tsunami generated by the great Sumatra-Andaman earthquake on 26 December 2004, a large amount of data is available. In this section model comparisons with observations from satellite altimetry, tide gauge records, and field surveys are discussed. The mesh (figure 6) employed in this study has been generated to deal with all the stages in the propagation of the tsunami. All of the Indian Ocean is covered, the resolution in the deep ocean is about 15 km. In the Aceh region in the Northern tip of Sumatra, where inundation results of the model are compared to field measurements, the mesh size reaches down to 40 m. The results in this section are closely related to the studies published in [23] where additional information can be found.

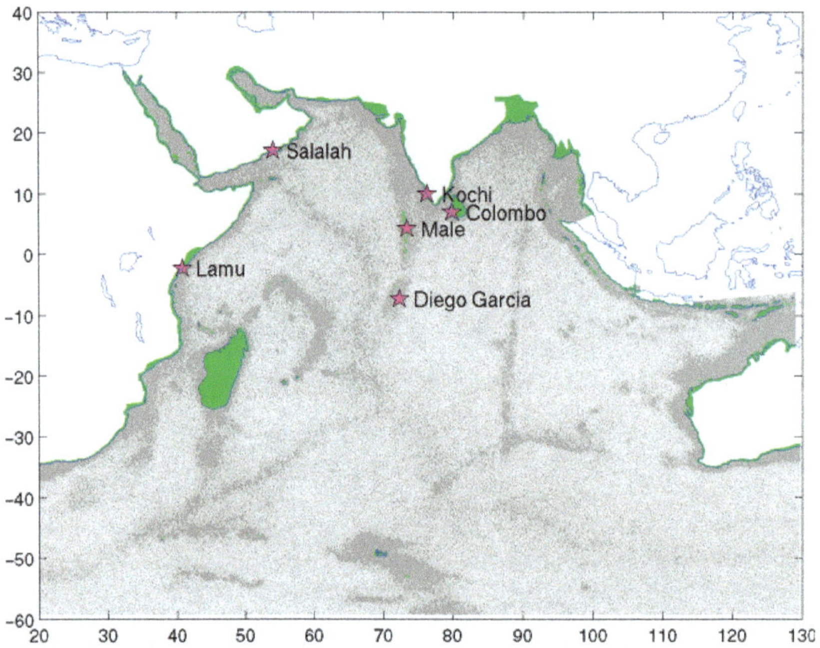

Figure 6. Mesh density in the model domain. The green areas are land nodes contained in the mesh which are initially dry. The resolution ranges from 15 km in the deep ocean to 40 m in the Northern tip of Sumatra.

Available topography and bathymetry data

Topography and bathymetry in the following experiments are based on several data sources. The GEBCO data set [24] is globally available at 30 arc seconds resolution and is used in all meshes as initial topographic and bathymetric information which is replaced by more precise data wherever they are available. Bathymetric data is locally improved by ship measurements and nautical charts. Topographic information is improved by the SRTM data set [25] which is freely available at a resolution of 90 m, however the vertical accuracy is usually not sufficient for model studies of runup in the coastal domain. In the area of Indonesia the SRTM data were additionally processed by the German Aerospace Agency (DLR) and provided at a resolution of 30m. In Aceh region additional bathymetric and topographic data were provided by BPPT Jogyakarta.

Source model

In tsunami modeling it turns out that the exact knowledge of the source, i.e. the initial conditions of the model is of crucial importance for comparisons with data. Usually the source parameters are optimized with respect to certain measurements and normally it's not possible to match model results with different measurements like tide gauges and inundation. The source model used in the following studies is based on the results presented in [26]. The objective of that paper is to optimize the subfaults of the earthquake such that an optimal agreement with certain tide gauge records is obtained. The resulting subfaults are shown in figure 7. In order to demonstrate the impact of source parameters with respect to the matching with data the orientation of the Southern faults is modified as indicated in that figure. The strike angle of subfaults A/C is changed from 340/340 degrees as proposed in [26] to 290/320 degrees. All other parameters can be found in in [23, 26].

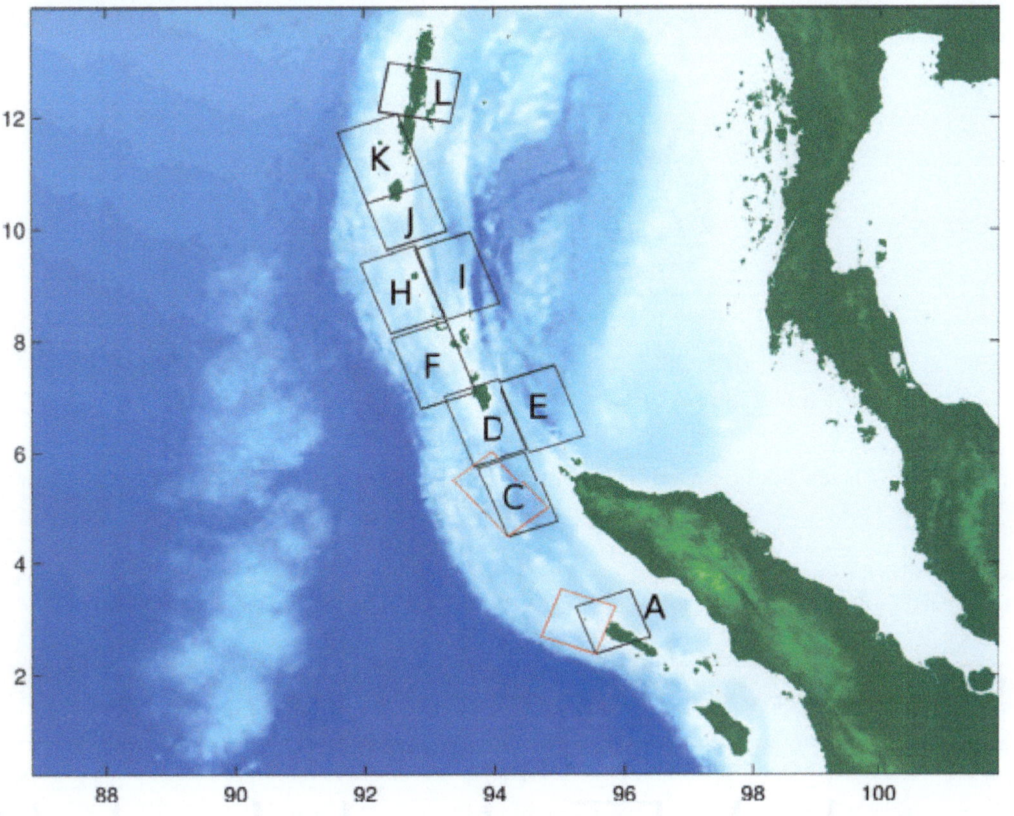

Figure 7. Sub-faults as proposed in [26]. The rupture area has been decomposed into 12 sub-regions. With a rupture speed of 1.7 km/s, the whole rupture process takes 12 min. Faults B and G have zero slip in our experiments and are not displayed. The red faults have different strike angle and show better agreement with satellite altimetry

Model setup

The mesh used in these experiments consists of 5 million nodes and 10 million triangles. The finest resolution enforces a time step of half a second. The model is integrated for 10 hours. In the positions along the satellite tracks the model state is written to a file every second.

This enables a careful comparison between model and satellite altimetry data. Additional information on model setup and initialization is contained in [23].

Satellite altimetry

For the first time the Indian Ocean tsunami was observed by several Satellite missions. Jason 1 (J1) and TOPEX/POSEIDON (T/P) were above the bay of Bengal about two hours after the earthquake, whereas ENVISAT observed the tsunami about 3h20' after the event. In all cases the tsunami signal was extracted from the altimeter measurements by subtracting the data from consecutive cycles. Table 1 summarizes the cycles that were taken into account in the three missions. Figure 8 contains the groundtracks of J1 and T/P whereas figure 9 displays the extracted tsunami signal for J1 and T/P. The model results are interpolated in time and space and extracted from a Hovmöller diagram as shown in figure 10. The J1 signal in figure 9 shows clearly a double peak in the position of the leading wave crest which is due to the partial waves generated by the southern subfaults, whereas in the T/P measurements these partial waves overlap. This behavior is reproduced by the model however as it turns out the matching depends strongly on the fault parameters. Changing the strike angles in fault A from 340 to 290 and in fault C from 340 to 320 improves the matching and lowers the RMS errors as indicated in table 1.

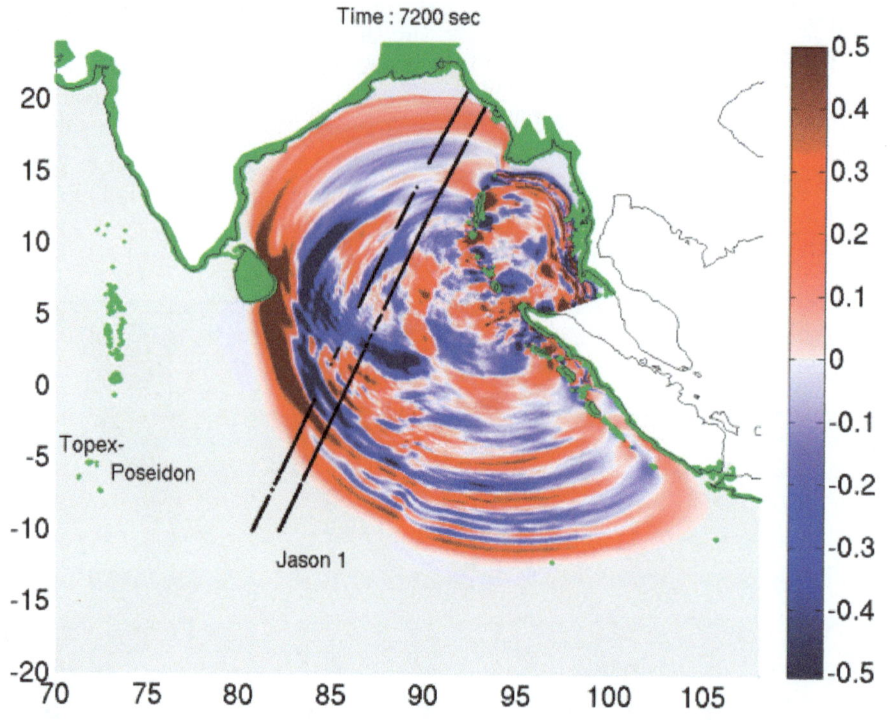

Figure 8. Model snapshot after two hours. Sea surface elevation together with the positions of the satellite tracks Jason 1 and TOPEX/POSEIDON

Table 1. Satellite Missions used in this study for data comparison. The last columns quantify the RMS error obtained in scenarios with different strike angles in subfaults A and C (figure 7).

Mission	Pass	Cycle	Equator time	rms (340/340)	rms (290/320)
Jason 1	129	109	02:55 UTC	0.243 m	0.238 m
Topex/Poseidon	129	452	03:01 UTC	0.223 m	0.164 m

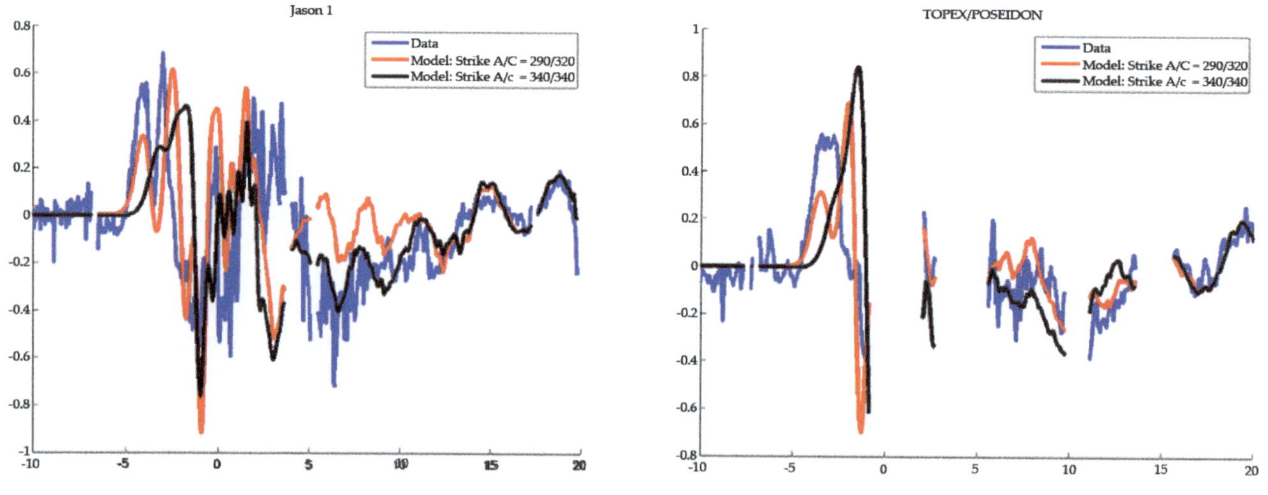

Figure 9. Tsunami signal extracted from satellite tracks Jason 1 and TOPEX/POSEIDON (blue lines) together with model results along the tracks interpolated in space and time (red and black lines).

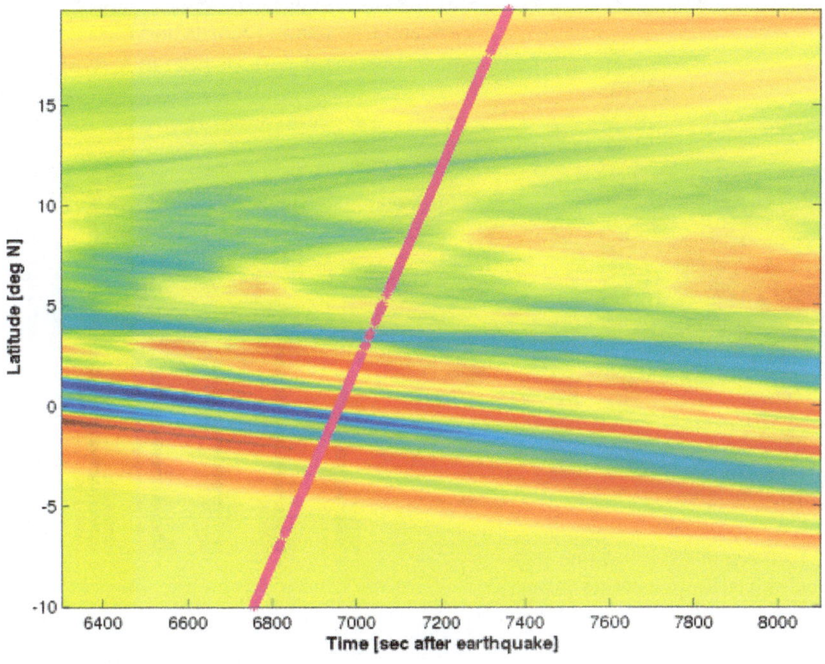

Figure 10. In the location of the satellite tracks (figure 8) the sea surface elevation is written to file every second. From the resulting Hovmöller diagram the model results corresponding to the satellite observations as shown in figure 9 are interpolated in space and time.

Inundation

After the event in December 2004 several field surveys examined the runup and flow depth in the affected regions. In the area of Banda Aceh, which was most heavily hit by the tsunami eight locations with well documented field measurements were chosen to compare model runup to these results. Runup and flow depth depend heavily on the prescribed roughness parameterization. Several experiments with identical initial conditions and varying Manning parameter were conducted. In all cases constant manning number was applied in the whole model domain. Figure 11 shows the locations as well as a comparison of measurements and model results for different roughness parameters in a bar diagram. These results are obtained with strike angles 290/320 in subfaults A/C. From the corresponding rms errors the best fitting Manning number for experiments with constant parameters can be chosen. Figure 12 shows the flow depth in Aceh region together with a line depicting the boundary of inundation as it was obtained from Satellite images (provided by DLR).

Figure 13 displays the flow depth comparison for different strike angles in the Southern subfaults. Also in this respect the modified orientation improves the results considerably.

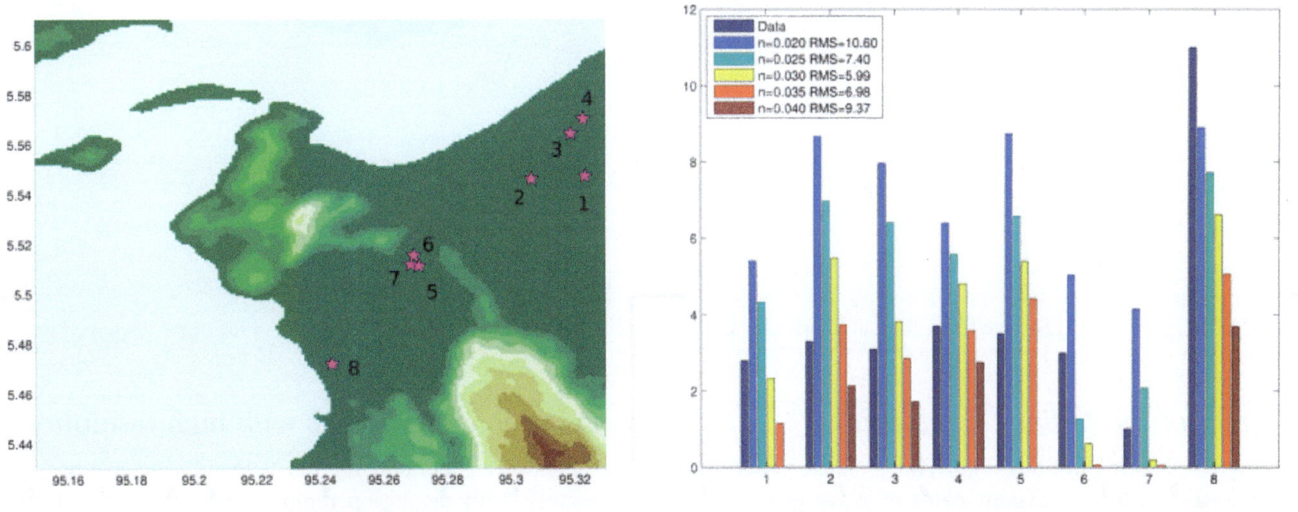

Figure 11. a) Positions of field measurements, b) model results for varying roughness parameters in comparison.

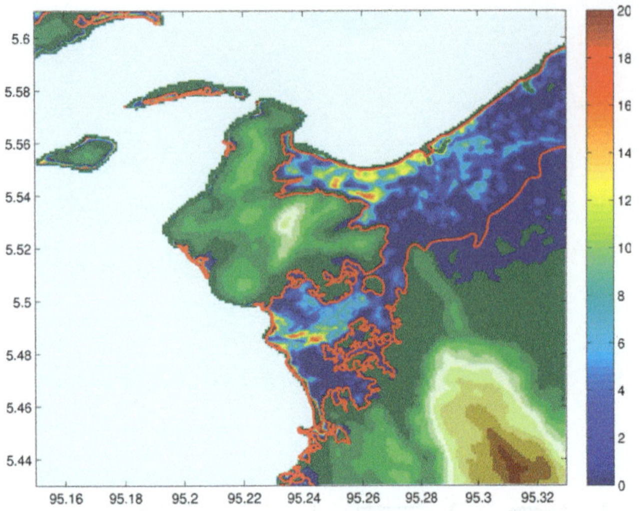

Figure 12. obtained for n=0.025 in Aceh region. The red line indicates the inundated area as it was determined from satellite images and provided by DLR.

Tide gauge records

The Indian Ocean tsunami was observed by tide gauges world wide. Since the arrival time and estimated wave height are among the most important warning products good matching between model results and arrival times and the height of the leading wave crest as they were recorded by tide gauges is desirable. Therefore hindcast experiments with such comparisons are included in the present study as well. Figure 14 summarizes the time series in some of the locations throughout the Indian Ocean where data is available. The comparisons between tide gauge records and model results show generally a good agreement with respect to arrival

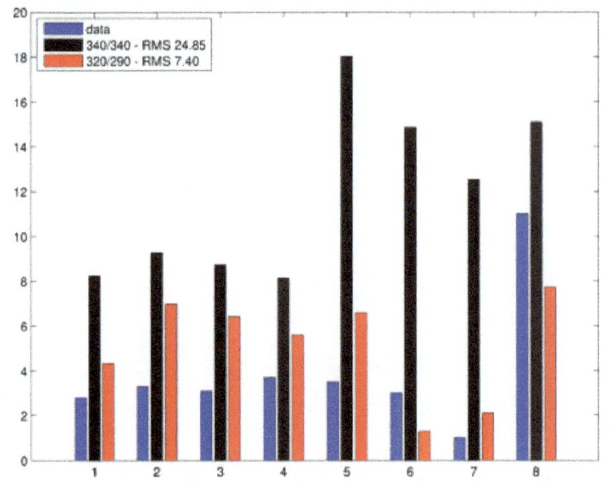

Figure 13. Flowdepth comparison for experiments with strike angles of plates A and C set to 340/340 and 290/320 (compare figure 7). The rms error is considerably improved by the adjustment.

time. The wave height of the leading crest is sometimes underestimated however in most of the locations the agreement is very good. The orientation of the Southern subfaults does not influence the results in far distance as Salalah and Lamu show, the matching in Colombo is slightly better with the uniform values (340/340). This is consistent with the derivation of these strike values as this station was used for optimization in [26].

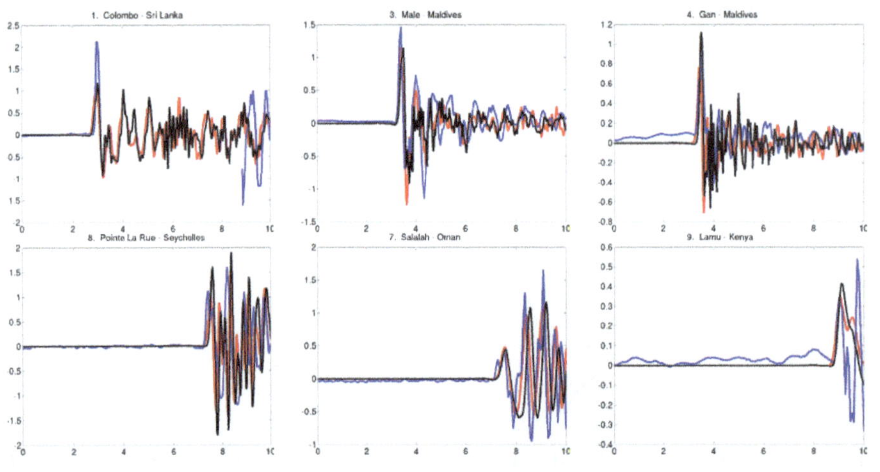

Figure 14. Tide gauge records (blue lines, time in hours, elevation in meter) and corresponding model results in locations displayed in figure 6. The red lines show results for strike angles 290/320 degrees, black lines the corresponding results for 340/340 degrees in subfaults A/C.

Summary of the Indian Ocean study

Summing up, it turns out that TsunAWI is able to reproduce observational results on several scales of wave propagation. Both the large scale propagation and arrival times throughout the Indian Ocean as well as the inundation in a selected area with high resolution is feasible in the approach with unstructured meshes. However the model results depend strongly on the quality of the source model.

Inundation experiments with high resolution

The scenarios generated for the tsunami database are based on a mesh with resolution ranging from 10 km in the deep ocean to 150 m along the global coastline and 50 m in priority regions. This resolution is certainly not high enough for detailed inundation simulations. Hazard maps are based on data sets with much higher resolution and the corresponding model setup must contain a comparable mesh. The shallow water equations are based on very small aspect ratio defined in (9) and as soon as the wave is determined by very fine scale topographic features the validity of the SWE becomes questionable. As long as the focus of such model studies is put on the inundated area alone the limitations of the SWE might still be acceptable and in the following experiments with high resolution are presented. The actual error however has to be investigated and comparisons to more appropriate models have to be performed. Section 2.5 deals with some of the improvements possible with nonhydrostatic corrections in idealized settings. Within the GITEWS project hazard maps were produced by DLR based on model

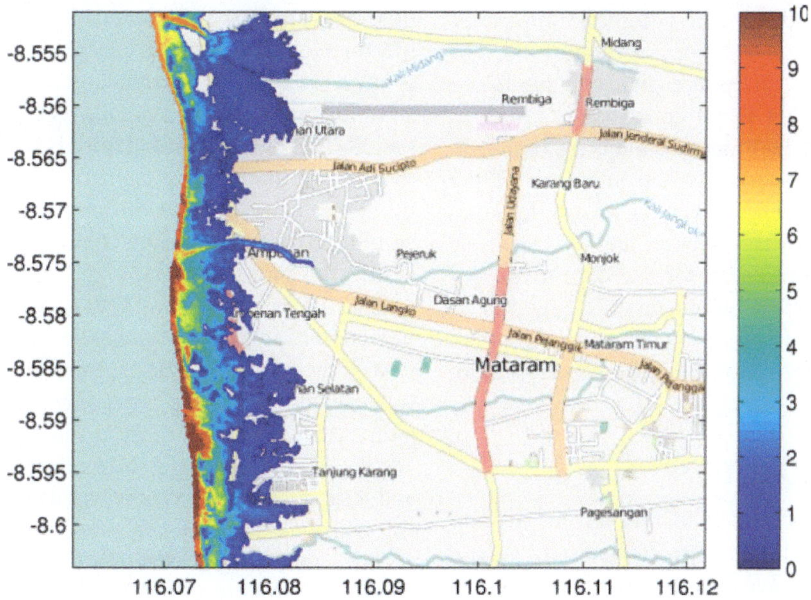

Figure 15. Inundation in Mataram. Flowdepth in the model study based on the Intermap DSM. The bottom roughness is given by a constant Manning value n=0.02.

results with MIKE 21 FM carried out by the German Research Center Geesthacht (GKSS). The Manning roughness parameter in these model runs are varying in space and given by detailed roughness maps. Details on this approach and the model can be found in [27] and citations therein. The focus of this section is to highlight the importance of the bottom roughness parameter. Simulations of tsunamis with identical initial conditions were carried out in a mesh with resolution up to 5 m. The study area is Mataram, on the island of Lombok, Indonesia. The topography data in this case was based on the Intermap data set and provided by DLR. The resolution of the topography data is 5m and two versions, a digital surface map (DSM) and digital terrain map (DTM) were used. The DSM version is a first-reflection data set and contains elevations of vegetation and buildings. The DTM on the other hand is a model where all these features were removed. Figure 15 displays the flow deph based on the DSM topography data, whereas figure 16 shows the same quantity for the DTM data with three different values of Manning's n. Additionally,

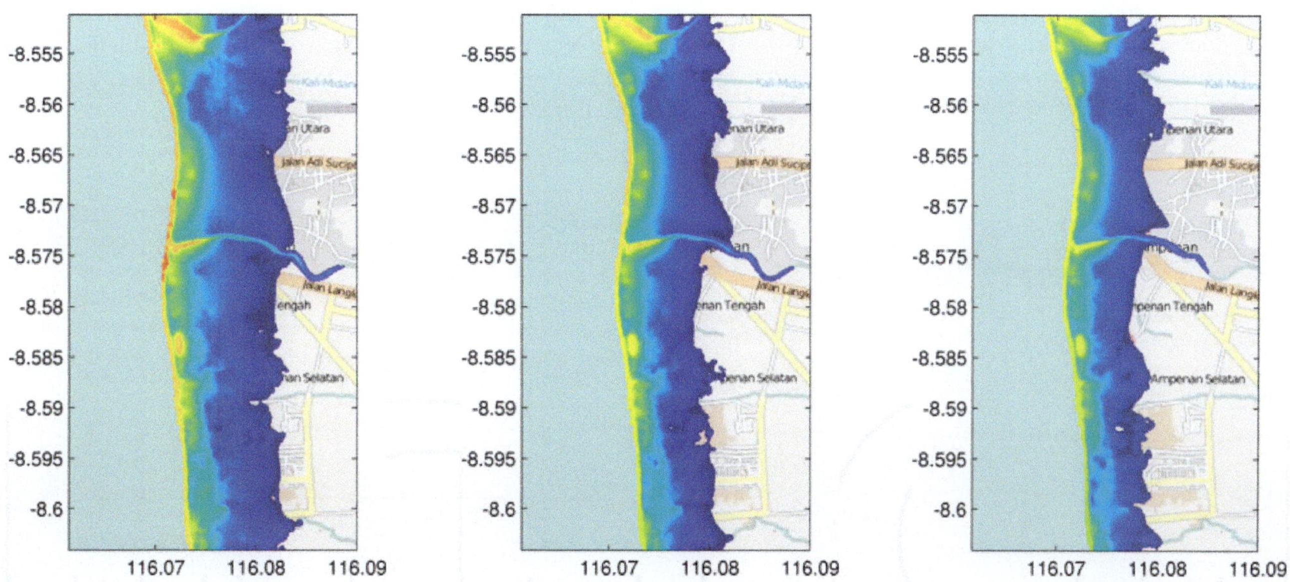

Figure 16. Results as in figure 15, however for Intermap DTM with three different Manning parameters n=0.04, n=0.06, and n=0.08 in the left, middle and right panel.

in case of a tsunami buildings and infrastructure will partly collapse and result in a very heterogenious fluid flow which may be only poorly approximated by a bottom friction parameterization. The result in figure 15 contains small scale features, however since it doesn't take into account the destruction of buildings and vegetation, it may be underestimating the true inundation. [27] suggests the use of DTM data with detailed roughness maps. Results for DTM data and different roughness values are displayed in figure 16. According to [27] the right panel corresponds rather to the situation of an urban area with partly collapsing buildings whereas the left panel corresponds to the situation of the whole area covered by coarse sand. Given the vast differences displayed in figure 15 and 16 this section and its results is mostly meant to raise the awareness for the dependency of inundation results on the friction parameters. The quality of topography data plays an even bigger role and the best suited data set and model parameterization must be chosen before hazard maps are produced.

TIDE-TSUNAMI INTERACTION

For investigating the influence of tidal motion on tsunami wave propagation three model runs were carried out: a) pure tidal motion, b) pure tsunami wave propagation and c) tsunami wave propagation on the background tidal motion. As object of modelling we have chosen the Indonesian coast including south part of Java, Bali, Lombok, Sumbawa and Sumba islands and North Sea. The choice of these two objects is not accidental. The first of these is the object with complex morphometry and sharp bathymetry, which contains a large number of islands and straits. The North Sea is a shelf sea with slowly varying bathymetry.

Interaction between tides and tsunami for Indonesian coast

For calculations, we use the ETOPO 30 sec. morphometry dataset and data on tides in the Indian Ocean derived from the TPXO6.2 dataset of oceanic tides [28]. The calculations were performed on an unstructured mesh (figure 17 a) with 177132 nodes and 347098 elements with a time step of 20 s. The horizontal mesh size varies in the range between 160 m (in the inundation zone) and 29 km (deep ocean). The tsunami wave is initialized by a source model as used in the GITEWS scenario database (figure 17 b). The particular choice of the initial condition is irrelevant for our considerations here. The series of experiments deals with effects of nonlinear interaction between tidal waves and tsunamis in the Indian Ocean. A full solution (propagation of tsunami wave on the tidal background) and composite solution (an arithmetic sum of tsunami wave and tides computed separately) are compared. It turns out that the difference between these solutions is very significant (figure 18) reaching as high as 3.5 m in the coastal region (St.3) where nonlinearity is particularly important. It is noteworthy that the first tsunami wave is only slightly affected by nonlinearity (change in amplitude $\approx 5 - 8\%$, with max. amplitude of tsunami wave 16 m, St.3), while the second wave is affected more essentially, at $\approx 25\%$ (with max. amplitude second tsunami wave 3.5 m). Also, the strong nonlinear interaction leads to phase changes. In search of explanation as to why the impact of tide-tsunami interaction is so significant, we repeated the above cases once more but switching off the momentum advection term in the equation of motion. Figure 18 (Right panel) clearly shows that the evolution of potential and kinetic energy is now significantly different from the full nonlinear case. Hence we conclude that in the near-shore regime, non-linear interaction of the wave-induced velocities contributes greatly to the complex behaviour of tide-tsunami wave phenomena. The exact mechanisms still call for a more careful analysis and suggest a topic for future research.

Tide-tsunami interactions in the North Sea

The last known mega-tsunami to hit rim countries of the North Sea took place over 8,000 years ago. But coastal areas of the North Sea are vulnerable to tsunamis caused by localized underwater landslides. Numerical simulations are performed on the basis of a high-resolution multilayer model to study initial generation of waves by landslides in the Storegga area and with a TsunAWI model to study the wave propagation further in the North Sea. As before, three simulations have been performed to identify the character of the nonlinear interaction of tide and tsunami waves. First simulation dealt with the M2 tidal structure in the North Sea. In the second case only tsunami waves excited by a signal from the landslide model were simulated. The third simulation dealt with a joint solution for the tsunami wave on the

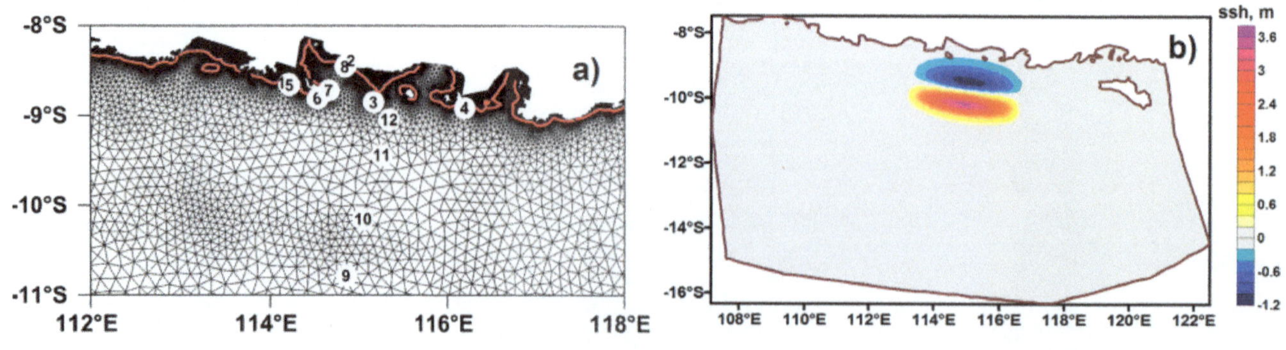

Figure 17. a) Part of the unstructured mesh with control stations. White curve marks coastal line. b) Initial condition for tsunami wave.

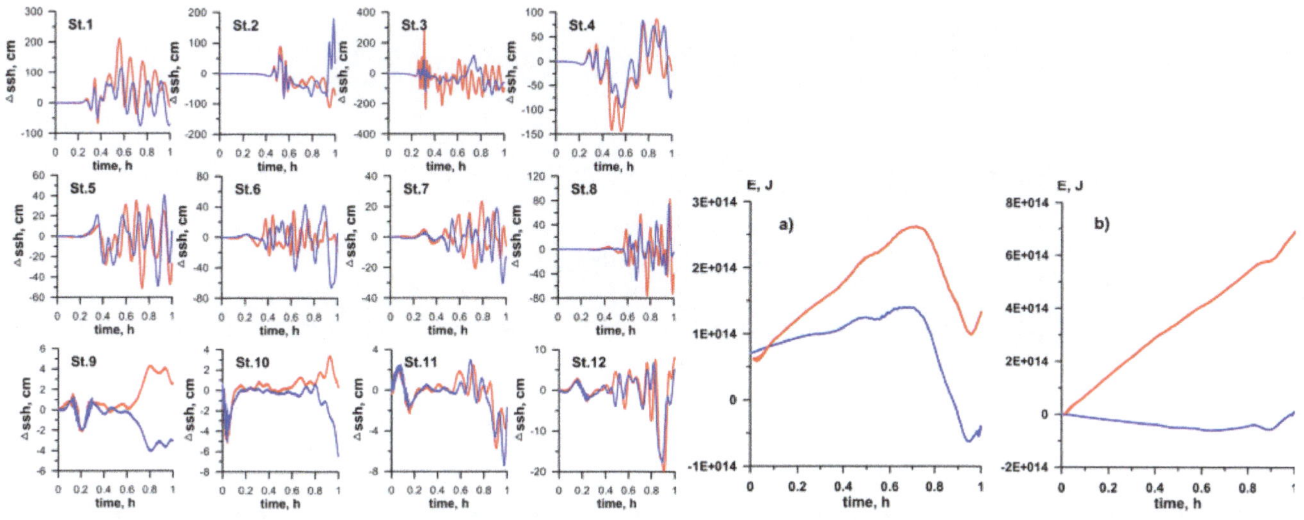

Figure 18. Left panel: Nonlinear interaction for control stations (1-4 - coastal; 5-8 - shelf; 9-12 - deep). Difference in elevation between full solution and composite solution. Red line - high tide; blue line - low tide. Right panel: Difference in potential (a) and kinetic energy (b) between full solution and composite solution. Red line - with advection in momentum equation; blue line - without advection.

background of semidiurnal tide. In this case the boundary information (north boundary) is represented by a superposition of two waves. A combination of two Kelvin waves (one propagating from the northern open boundary along the coast of England and another through the English Channel) produces three nodal lines: on an output from English Channel, near north coast of Germany, and southern coast of Norway [11]. There is reasonable agreement for semidiurnal tidal elevations with respect to observation (figure 19).

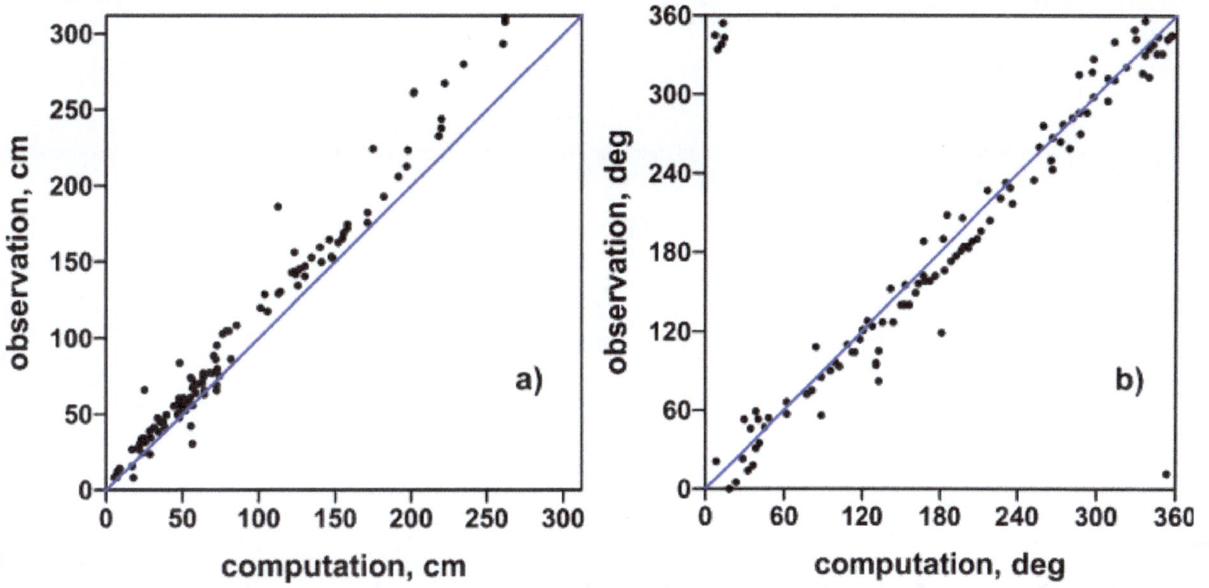

Figure 19. Correlation between the observed and computed values of the amplitudes (a) and phases (b) of the tide wave M2. The RMS error of the computation is 29.8 cm (for 112 stations).

A two-layer model based on shallow water equations written in a curvilinear coordinate system is applied to simulate waves generated by hypothetical submarine slides at Storegga on the Norwegian continental slope (figure 20). The numerical method is based on composite schemes for split operators [29]. The model is initialized by prescribing submarine slides in the Storegga area off western Norway [30, 31]. The signal at the open boundary of the North Sea

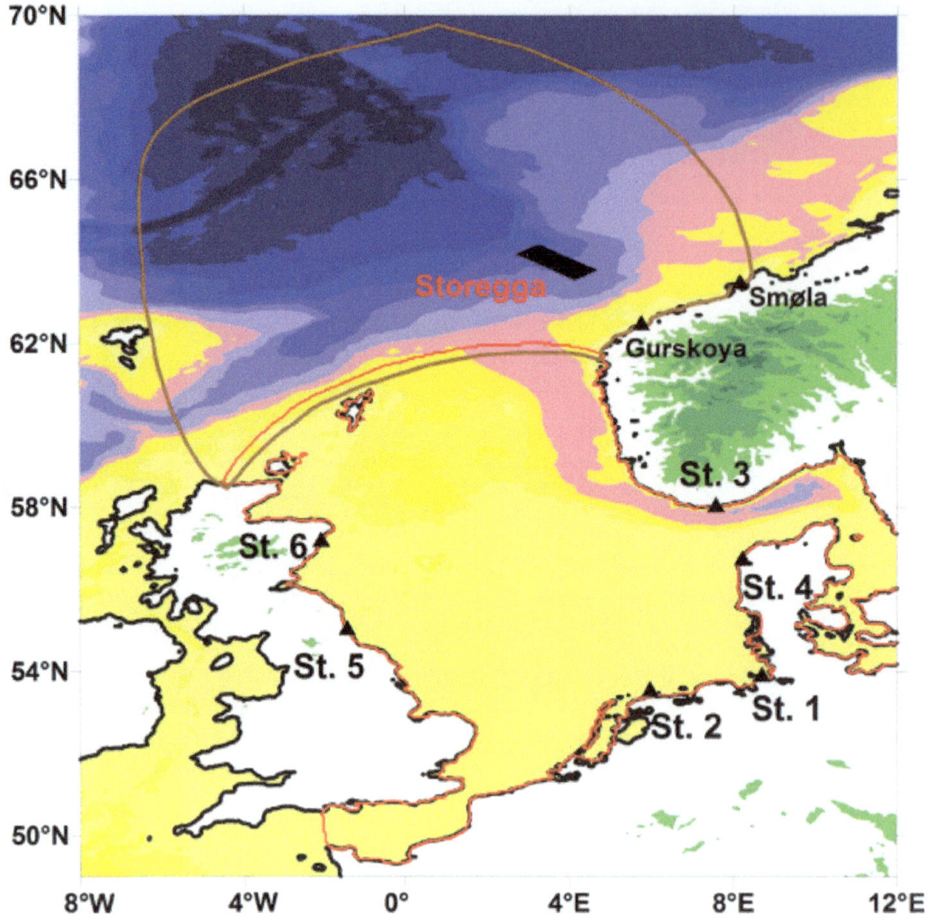

Figure 20. Computational domain with slide positions (Storegga) and control stations. The brown line shows the domain for submarine slides modelling. The red line marks the domain for tsunami wave propagation (TsunAWI model).

simulated by the model serves as the boundary condition for finite-element model (TsunAWI) used to simulate wave propagation further in the North Sea.

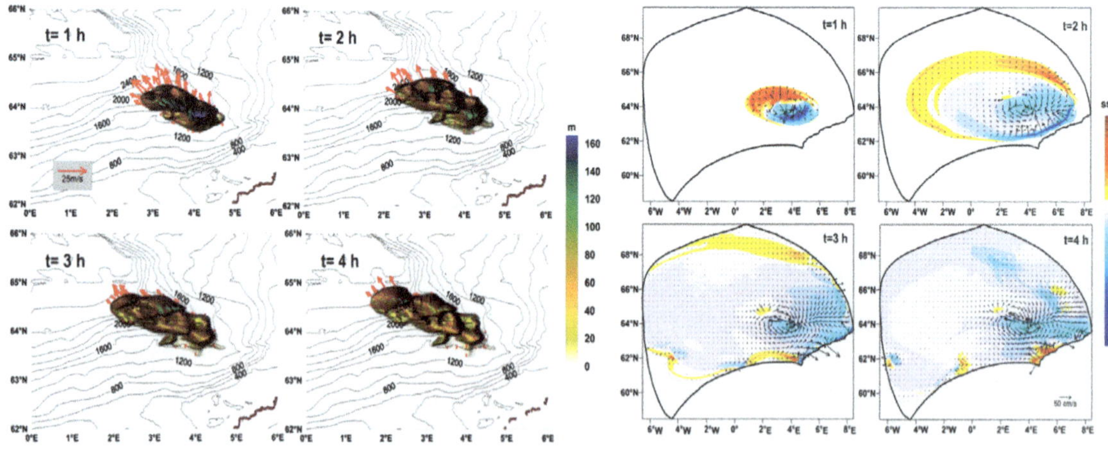

Figure 21. Thickness (m) and velocity (m/s) for Storegga landslide at four subsequent times (Left panel). Sea surface elevation (cm) and velocity (m/s) for Storegga landslide (Right panel) at four time instants.

Figure 21 (Left panel) shows the position, thickness of the submarine slide (Storegga) and velocity field in the bottom layer (landslide) at four different time moments. The slide velocity reaches 25 m/s, in agreement with [30, 31]. The submarine slide leads to long wave on the surface as shown in figure 21 (Right panel) for the same time moments as in figure 21 (Left panel).

A composite solution and full solution are compared. It turns out that the difference between these solutions is fairly large (figure 22, Left panel) reaching as high as 60 cm in the coastal region where nonlinearity is particularly important. Additional experiments showed that the arrival time of tsunami waves in the full solution varies but slightly. In search of explanation

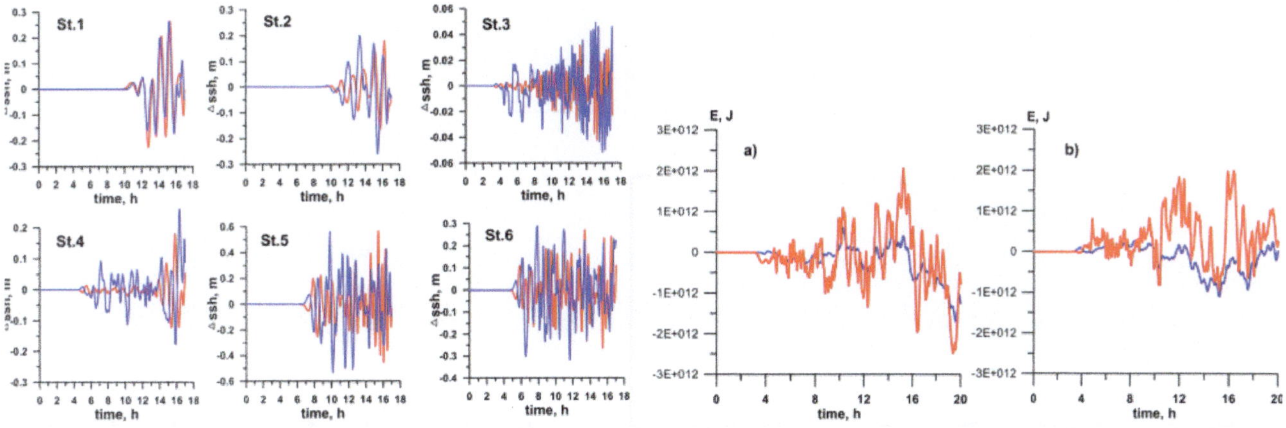

Figure 22. Left panel: Nonlinear interaction. Difference in elevation between a composite solution and full solution. Red line - high tide; blue line - low tide. Right panel: Difference in energy between a solution with advection and solution without (a) momentum advection (blue line) and solution without (b) nonlinearity in the continuity equation (red line).

as to why the impact of tide-tsunami interaction is so significant, we repeated the above cases once more but switching off the momentum advection term in the equation of motion (linear case) or nonlinearity due to change of layer thickness in the equation of continuity (figure 22, Right panel). Here we can see the other mechanism of tide-tsunami nonlinear interaction due to level change.

VALIDATION OF THE NONHYDROSTATIC APPROACH

For comparison of the results delivered by the original shallow water model on the one hand with the nonhydrostatic extension on the other two standard testcases are carried out. The first one is investigated in [18], it describes the behavior of a standing wave within a closed basin in respect of the phase velocity depending on the ratio δ defined in (9). As second application a tank experiment (see [32]) is modeled and the results are compared against observation data at various instants. This example shows dispersive effects that occur when a solitary wave is running up a plane beach.

Standing wave in a basin

The computational domain Ω of this testcase presents a rectangular basin with fixed length and width l = 10 m and w = 4 m, respectively, while its depth h vary in different experiments between 0.25 m and 10.5 m. At the walls of the basin, condition (3) for solid boundaries $\partial\Omega 1$ is applied. With a wave length of λ = 2l a standing wave with an amplitude of a = 0.01 m can be arranged by the initial condition

$$\zeta^0(x,y) = -a\cos(\frac{2\pi x}{\lambda}), \tag{18}$$

as illustrated in Figure 23 a). Both, horizontal and vertical velocity equal zero initially. In the absence of sinks and sources, two waves with same amplitude and frequency emerge. They move in opposite directions and form a standing wave with antinodes at the boundaries, x ∈ {0, l}, and a node in the middle of the basin, x = l/2. An unstructured mesh with a resolution of about 0.125 m covers the domain Ω. Using a time step size of Δt = 0.005 s, several experiments with varying basin depths were carried out with both the original shallow water model and the nonhydrostatically corrected one. Because of the periodicity, the propagation speed can be determined. In this example an inviscid fluid is assumed, so that the pressure gradient is the only force. Modifications of the pressure term due to nonhydrostatic effects entail changes of the motion. In Figure 23 b) the results of both models are compared with the reference propagation speed, estimated by

$$c = \sqrt{\frac{g}{k}\tanh(kh)},$$

in which k = 2π/λ is the wave number. If δ « 1, the argument of the hyperbolic tangent in (19) is very small and the approximation tanh(x) ≈ x can be adopted. So, the propagation speed converges to the phase velocity $c_{sw} = (gh)^{1/2}$ that is independent of wavelength and characterizes motions in the standard shallow water approximation. It is not surprising that the phase velocity curve based on the shallow water model is congruent with csw for all δ. More interesting is the result of the nonhydrostatic approach: although it is simplified by using depth-averaged values, it offers a good approximation to the propagation speed up to ratio δ in excess of about 0.4. Consequently, as δ is increased, the nonhydrostatic correction provides a better approximation to the wave propagation speed.

Solitary wave on a plane beach

In the second testcase the results of the shallow water model and the nonhydrostatic approach are compared to observation data of a tank experiment [32]. The main part of tank is of constant water depth d. Near the shoreline (x = 0), the bathymetry gradually ascents with a constant slope of tan(α) = 1/19.85, as depicted in Figure 24. All spatial quantities in the

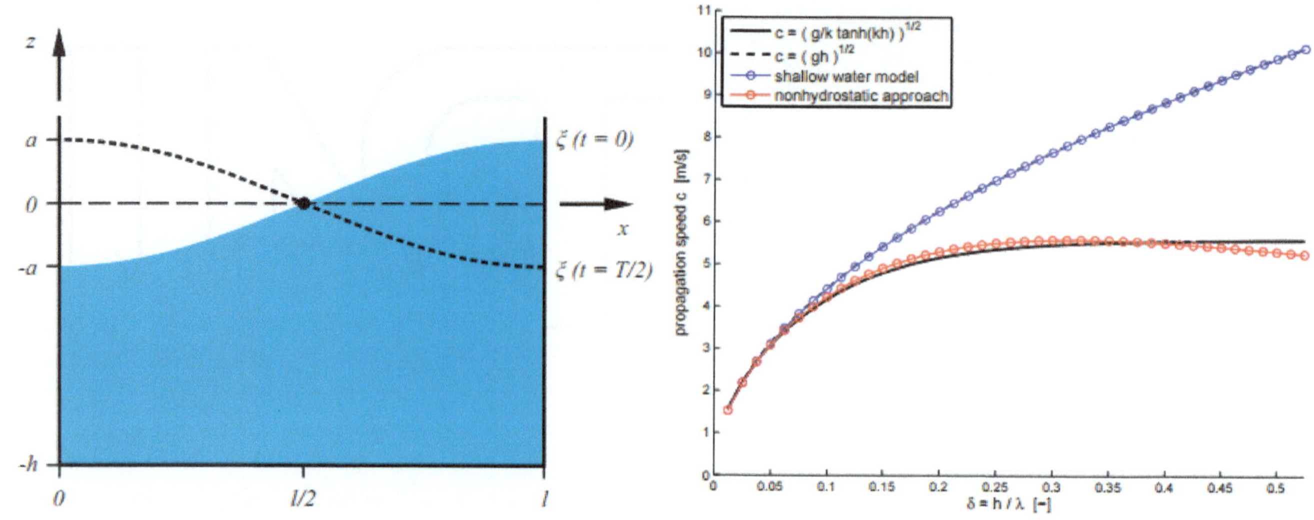

Figure 23. a) Setup of testcase 6.1: Standing wave in a basin. b) Different results in reference to propagation speed c.

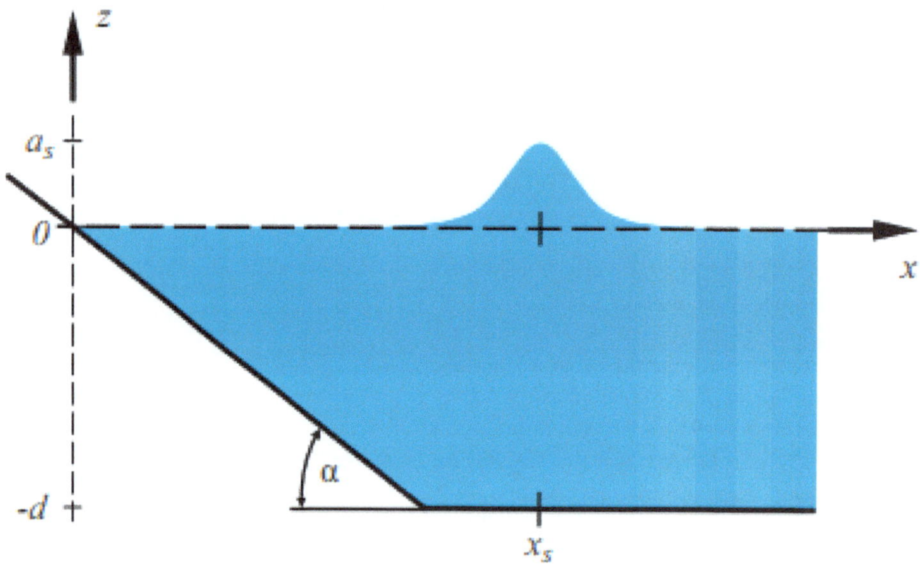

Figure 24. Setup of testcase 6.2: Solitary wave on a plane beach

model setup are made dimensionless with the tank depth d. To simplify matters, d = 1 m is chosen. A solitary wave with maximum height as := a/d is given by

$$\zeta^0(x,y) = \frac{a}{d}\operatorname{sech}^2\left(\sqrt{\frac{3a}{4d}}(x - x_s)\right), \quad x_s = \frac{d}{\tan(\alpha)} + \sqrt{\frac{4d}{3a}}\operatorname{arccosh}(\sqrt{20}). \tag{20}$$

with an initial horizontal velocity of $\mathbf{v}^0 = (-\zeta^0\sqrt{g/d}, 0)^T$. The vertical velocity is prescribed by the boundary conditions (13) and (14). While the ramp can be flooded for $x < 0$, boundary condition (3) on $\partial\Omega 1$ are imposed at the solid walls of the tank. An unstructured mesh with $\Delta x \in [0.1, 0.2]$, in which the finer resolution accords to the ascending part, covers the computational domain $\Omega = [-10, 70] \times [-0.5, 0.5]$. The time step is selected as $\Delta\tau = \Delta t\,(g/d)^{1/2} = 0.004$. With the help of a Manning factor of n = 0.01 the low friction inside of the tank is approximated. By setting the maximum wave height $a_s = 0.0185$, the shallow water model and the nonhydrostatic approach provide similar results which agree very well with the observation data. More interesting is the case of breaking wave with as = 0.3. Figure 25 illustrates different stages of the flow evolution in four snapshots. It is apparent that the nonhydrostatic approach approximates the shape of the solitary wave much better than

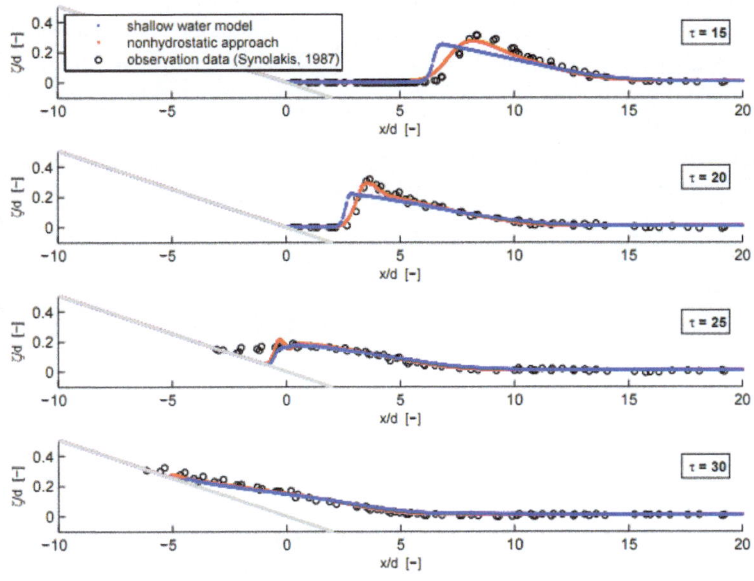

Figure 25. Snapshots at different times $\tau = t\,(g/d)^{1/2}$ for a solitary wave with a maximum wave height of as = a/d = 0.3.

the hydrostatic shallow water model. As a consequence of boundary condition (13) the wave front experiences a vertical displacement. Since this dispersive process cannot be represented by the shallow water model, its wave front steepens in an unnatural way. Furthermore, the snapshot at τ = 30 shows clearly that there are some differences with respect to inundation, which is very important as concerns tsunami warning.

DISCUSSION

The combination of non-conforming velocity with linear elevation suggests a well-rounded choice for shallow-water modelling on unstructured triangular grids, with a particular focus on simulating tsunami wave propagation. Although our approach was initially inspired by the algorithm proposed by Hanert et al. [10] the resulting model is essentially different from it in a number of key directions. First, it is equipped with wetting and drying algorithms and can simulate inundation caused by tsunami. Second, it suggests a choice of stably working discretization of the momentum advection which all improve over the original method of Hanert et al. [10] and differ between themselves in a degree of smoothing applied. Third, it uses the Smagorinsky horizontal viscosity which is crucial for keeping the dissipation on the level that does not affect the quality of the solution. Finally, the explicit time stepping and use of the nodal quadrature (mass matrix lumping) of the time derivative term in the continuity equation ensure numerically efficient performance while providing a straightforward and easy to implement algorithm. Nonhydrostatic effects become important when vertical acceleration is not negligible. For tsunami wave generation and propagation it can happen at the very initial stage and during run up. In this context, we present an algorithm of the nonhydrostatical pressure for vertical averaged equations. We introduced the architecture of the German-Indonesian Tsunami Early Warning System (GITEWS) and methodology for the GITEWS multi-sensor selection. The selection algorithm uses a multi sensor approach combining the different available sensor types (SeisComP3, CGPS and Tide Gauges) to acquire a set of best matching pre calculated scenarios (Tsunami Scenario Repository, TSR) to an earthquake event. The basic principle is to reduce the number of possible tsunami scenarios by using independent measurements of the same event. Only a small number of scenarios can match the independent measurements, even with high uncertainty in each individual set of measurements, since the combination needs to fit. One of the important stages of the successful use of the numerical model is its verification and validation. This work has been performed on test cases and published in [33]. We simulated the tsunami event generated by the Sumatra-Andaman (2004) and Tohoku ¯ (2012) earthquakes. The model results were compared to available data (tide gauge, satellite altimetry, and field measurements in the inundation area). Given still only approximately known parameters of the tsunami source the coincidence between the model and observation is indeed good. Not only the arrival time of the first wave is reliably simulated, but the entire shape of the signal is reproduced reasonably well, and with correct amplitude. The model can be considered as an easy to use and reliable tool which not only serves the purposes of

GITEWS but can be employed for other tasks which can be described in the framework of shallow water equations (with exception of true shock waves for which continuous elevation is a suboptimal choice). Simulations suggest strong nonlinear interaction between the tsunami and tidal waves. The major difference between tsunami simulations with and without tides occurs in the run-up region. Two mechanism of nonlinear interaction were found to be directly related to the morphometry of the object. In areas with high variability in morphometry (sharp bathymetry, complex coastline, etc.), the main role is played by the nonlinear interaction of tidal velocities and tsunamis velocity. Another mechanism of nonlinear interaction operates through the changes in the thickness of the water layer in the presence of tides, which is typical for shallow areas. In this case besides the amplitude of the incoming wave, the arrival time can vary due to the change in wave phase speed. These results lead us to conclude that the account of tidal dynamics may prove to be necessary for the faithful modelling of tsunami waves. For comparison of the results delivered by the shallow water model in the original state and with the nonhydrostatic extension on the other two standard test cases are executed. This example shows dispersive effects that occur when a solution is running up a plain beach. The comparison between the results of the original shallow water model and the model with the nonhydrostatic correction, two test cases were performed. They illustrate the importance of dispersive effects, which may have implication to inundation prediction. The inclusion of nonhydrostatic effects may therefore be necessary for successful modeling of tsunami wave propagation.

BRIEF OVERVIEW

The wave propagation model TsunAWI based on finite elements was presented. We also introduced the architecture of the Tsunami Early Warnin System and methodology for the multi-sensor selection. We simulated the tsunami events generated by the Sumatra-Andaman (2004) and Tohoku (2012) earthquakes. Part of our work addresses the influence of tidal dynamics on tsunami wave propagation in coastal areas. The impact of nonhydrostatic effects is investigated in this study.

ACKNOWLEDGEMENTS

TsunAWI is the result of joint developments of different groups within AWI namely the computing center and the climate dynamics section. We acknowledge Jörn Behrens, Dmitry Sein, and Dmitry Sidorenko for contributions to the model development. We acknowledge Ralf Kiefl from the German Aerospace Agency (DLR) for providung us with inundation areas estimated from satellite images and NOAA/ PMEL/NOAA Center for Tsunami Research for providing satellite altimetry data. We acknowledge Widjo Kongko for providing us with field measurements in Aceh region. The tide gauge data were provided by the Survey of India and National Institute of Oceanography and by the University of Hawaii Sea level Center within the Global Sea Level Observing System (GLOSS-IOC). Matthias Mück from DLR provided us with the Intermap DTM and DSM topography data in Mataram.

REFERENCES

1. Kienle J, Kowalik Z, Murty T. S (1987) Tsunamis generated by eruptions from Mount St.Augustine Volcano, Alaska. Science, 236, 1442–1447.
2. Greenberg D. A, Murty T, Ruffman A (1987) A numerical model for Halifax Harbor tsunami due to the 1917 explosion. Marine Geodesy, 16, 153–167.
3. Baptista A, Priest G, Murty T (1993) Field survey of the 1992 Nicaragua Tsunami. Marine Geodesy, 16, 1692–1703.
4. Mofjeld H, Gonzales F, Titov V, Venturato A, Newman J (2007) Effects of tides on maximum tsunami wave heights: probability distributions. Journal of Atmospheric and Oceanic Technology, 24, 117–123.
5. Weisz R, Winter C (2005) Tsunami, tides and run-up: a numerical study. Papadopoulos G. A Satake K (eds.), Proceedings of the International Tsunami Symposium, Chania, Greece, 27-29 June 2005.
6. Kowalik Z, Proshutinsky T, Proshutinsky A (2006) Tide-tsunami interactions. Science of Tsunami Hazards, 24, 242–256.
7. Titov V. V, Synolakis C. E (1995) Evolution and Runup of Breaking and Nonbreaking waves using VTSC-2. Journal of Waterway, Port, Coastal and Ocean Engineering, ASCE, 121, 308–325.
8. Oliger J, Sundström A (1978) Theoretical and practical aspects of some initial boundary value problems in fluid dynamics. SIAM J. Appl. Math., 35, 419–446.
9. Androsov A, Klevanny K, Salusti E, Voltzinger N (1995) Open boundary conditions for horizontal 2-d curvilinear-grid long-wave dynamics of a strait. Adv. Water Resour., 18, 267–276.
10. Hanert E, Le Roux D, Legat V, Deleersnijder E (2005) An efficient Eulerian finite element method for the shallow water equations. Ocean Modelling, 10, 115–136.
11. Maßmann S, Androsov A, Danilov S (2008) Intercomparison between finite element and finite volume approaches to model North Sea tides. Continental Shelf Research, 30, 680–691.
12. Lynett P, Wu T.-R, Liu P.-F (2002) Modeling wave runup with depth-integrated equations. Coastal Engineering, 46, 89–107.
13. Smagorinsky J (1963) General circulation experiments with the primitive equations. Mon. Wea. Rev., 91, 99–164.
14. Shewchuk J (2006) Triangle: engineering a 2d quality mesh generator and Delaunay triangulator. manocha D. (ed.). In: Lin MC, Applied computational geometry: towards geometric engineering, pp. 203–222, Springer, Heidelberg.
15. Frey W, Field D (1991) Mesh relaxation: a new technique for improving triangulations. Int. J. Numer. Methods Eng., 31, 1121–1133.
16. Casulli V (1999) A semi-implicit finite difference method for non-hydrostatic, free-surface flows. International Journal for Numerical Methods in Fluids, 30, 425–440.
17. Cheung K. F, Yamazaki Y, Kowalik Z (2009) Depth-integrated, non-hydrostatic model for wave breaking and run-up. International Journal for Numerical Methods in Fluids, 61, 473–497.
18. Walters R. A (2005) A semi-implicit finite element model for non-hydrostatic (dispersive) surface waves. International Journal for Numerical Methods in Fluids, 49, 721–737.
19. Fleischer J, Häner R, Herrnkind S, Kloth A, Kriegel U, Schwarting H, Wächter J (2010) An integration platform for heterogeneous sensor systems in GITEWS - Tsunami Service Bus. Natural Hazards and Earth System Science, 10, 1239–1252.
20. Steinmetz T, Raape U, Teßmann S, Strobl C, Friedemann M, Kukofka T, Riedlinger T, Mikusch E, Dech S (2010) Tsunami early warning and decision support. Natural Hazards and Earth System Science, 10, 1839–1850.
21. Behrens J, Androsov A, Babeyko A. Y, Harig S, Klaschka F, Mentrup L (2010) A new multi-sensor approach to simulation assisted tsunami early warning. Natural Hazards and Earth System Science, 10, 1085–1100.
22. Babeyko A, Hoechner A, Sobolev S. V (2008) Modelling tsunami generation for local tsunami early warning

in Indonesia. Geophysical Research Abstracts, 10, eGU2008-A-01523.

23. Harig S, Chaeroni C, Pranowo W. S, Behrens J (2008) Tsunami simulations on several scales: Comparison of approaches with unstructured meshes and nested grids. Ocean Dynamics, 58, 429–440.
24. The general bathymetric chart of the oceans (GEBCO). Webpage (http://www.gebco.net).
25. Shuttle radar topography mission X-SAR / SRTM. Webpage (http://www.dlr.de/srtm/).
26. Tanioka Y, Yudhicara, Kususose T, Kathiroli S, Nishimura Y, Iwasaki S.-I, Satake K (2006) Rupture process of the 2004 great Sumatra-Andaman earthquake estimated from tsunami waveforms. Earth Planets Space, 58, 203–209.
27. Gayer G, Leschka S, Nöhren I, Larsen O, Günther H (2010) Tsunami inundation modelling based on detailed roughness maps of densely populated areas. Natural Hazards and Earth System Sciences, 10, 1679–1687.
28. Egbert G S E (2002) Efficient inverse modeling of barotropic ocean tides. Journal of Atmospheric and Oceanic Technology, 19, 183–204.
29. Androsov A, Voltzinger N (2005) The Straits of World ocean - the general approach to modeling. Nauka, 188 pages, in russian.
30. Harbitz C (1992) Model simulations of tsunamis generated by the Storegga Slide. Marine Geology, 105, 1–21.
31. Løvholt F, Harbitz C, Haugen K. B (2005) A parametric study of tsunamis generated by submarine slides in the Ormen Lange/Storegga area off western Norway. Marine and Petroleum Geology, 22, 219–231.
32. Synolakis C. E, Bernard E. N, Titov V. V, Kânoglu U, González F. I (2007) Standards, ˜ criteria, and procedures for NOAA evaluation of tsunami numerical models. NOAATech. Memo. OAR PMEL-135, NOAA/Pacific Marine Environmental Laboratory, Seattle.
33. Androsov A, Harig S, Behrens J, Schröter J, Danilov S (2008) Tsunami modelling on unstructured grids: Verification and validation. Proceedings of the International Conference on Tsunami Warning (ICTW), Bali, Indonesia.

CHAPTER 3

SOURCE CHARACTERISTICS OF OCEAN INFRAGRAVITY WAVES

Yoko Tono, Kiwamu Nishida, Yoshio Fukao, Akiko To, Narumi Takahashi

Keywords: Infragravity wave Seafloor network Pressure gauge BBOBS DONET Philippine Sea

INTRODUCTION

Ocean infragravity (IG) waves are sea surface gravity waves with periods of several minutes and wavelengths of tens of kilometers. The phase velocity of IG waves observed in deep ocean environments is accurately explained by the gravity wave theory (e.g., Webb et al.1991). These waves are considered to be excited by non-linear interactions between oceanic swells (Longuet-Higgins and Stewart 1962; Herbers et al. 1995) and may be enhanced by tidal modulation in coastal oceans (Guza and Thornton 1982; Okihiro and Guza 1995; Tomson et al. 2006) and deep sea (Sugioka et al. 2010). IG waves also are considered to excite the Earth's hum (Rhie and Romanowicz 2004, 2006; Tanimoto 2005; Nishida et al. 2008; Fukao et al. 2010; Nishida 2013).

Dolenc et al. (2005) compared the power spectra of ocean-bottom seismic records at a station located offshore of the Monterey Bay in California with the wave spectral densities measured by the nearby National Oceanic and Atmospheric Administration (NOAA) buoy. They observed two types of IG wave modulations with short (30- to 40-s) and long (10-day) periods, the latter being correlated with the ocean tides at the station. Sugioka et al. (2010) also observed the tidal modulation of IG waves on the records of broadband ocean-bottom seismometers (BBOBS) at deep seafloors. They further found a remarkable correlation of the IG spectral peak with the seafloor depth. Godin et al. (2013) showed a pronounced dependence of the energy density of IG wave on the frequency and local water depth using pressure gauge records of 28 locations on the seafloor off New Zealand. Harmon et al. (2012) used five differential pressure gauges located off the coast of Sumatra and applied an array analysis. They detected IG waves that propagate along the coast from southeast or south. Crawford et al. (1991) took advantage of a simultaneous measurement of sea-bottom pressure and seafloor displacement at a single station. They measured the sea-bottom pressure changes due to IG waves and the resultant seafloor vertical displacement to take their ratio (compliance) which carries information about the elastic structure of the oceanic crust. When we analyze the sea-bottom pressure and/or seafloor displacement, IG waves, a ubiquitous phenomenon on the sea, can be a strong background noise. Understandings of IG wave-related phenomena are important from the view point of seismology as well as oceanography.

A submarine cable network named the Dense Oceanfloor Network System for Earthquakes and Tsunamis (DONET) was recently deployed offshore of the Kii Peninsula in the Nankai Trough region (Kaneda 2010; Kawaguchi et al. 2010; Nakano et al. 2013). This network is equipped with six instruments at each station, including a three-component BBOBS and an absolute pressure gauge. Thus, DONET provides a good opportunity for array-based compliance analysis to study the crustal structure beneath the network, the generation mechanisms of IG waves, and the IG wave-related seismic phenomena including the Earth's hum. In this study, we investigate the nature of incoming IG waves to DONET on the basis of continuous 3-year observations.

DATA AND METHOD

DONET is an ongoing project and currently consists of five nodes, each of which consists of four stations. Figure 1 shows the station distribution of this network and the seafloor topography in the Izu-northern Bonin region. The station interval is 15 to 20 km, on the order of the wavelengths of

Citation: Yoko Tono, Kiwamu Nishida, Yoshio Fukao, Akiko To, Narumi Takahashi, Source characteristics of ocean infragravity waves, DOI: 10.1186/1880-5981-66-99.

targeted IG waves of our interest. The four stations at C node are set at seafloor depths between 3,500 and 4,400 m, while 16 stations at other nodes are positioned at depths between 1,900 and 2,500 m. The average of theoretical phase velocities of IG wave calculated at a period of 120 s for all stations is 135 m/s with a standard deviation of 14 m/s, while the average of phase velocities at the same period for 16 stations at other nodes is 129 m/s with a standard deviation of 4 m/s. We regarded the 16 stations as an array and analyzed the records of broadband ocean-bottom seismometers (Guralp CMG-3 T, Guralp Systems, Reading, UK) and quartz pressure gauges (Paroscientific, Inc., Redmond, WA, USA) from January 2011 to December 2013. For each station, both records were band-pass-filtered between 0.005 and 0.025 Hz and were corrected for the instrumental response. Both displacement and pressure time series were divided into segments of 3,600 s with an overlap of 1,800 s. Segments including earthquake signals, artificial noise, and boisterous noise disturbances due to extreme weather or nearby typhoon activity were discarded. The record sections at various stations were mutually shifted according to a presumed slowness vector and were then stacked. Each slant-stacked record section was auto-correlated and Fourier-transformed to obtain a power spectrum. The same procedure was repeated with numerous slowness vectors to obtain a two-dimensional frequency-slowness spectrum (e.g., Rost and Thomas 2002; Nishida et al. 2005), and all similarly obtained spectra for 1 day were stacked. The pressure records were similarly processed. The power spectral density (PSD) of either seafloor displacement or pressure was averaged over the analyzed frequency range to obtain the average PSD as a slowness vector function. We searched for the slowness vector that maximizes the average PSD. The direction and amplitude of such a slowness vector were regarded as the incident azimuth and apparent slowness of the incoming IG waves, respectively.

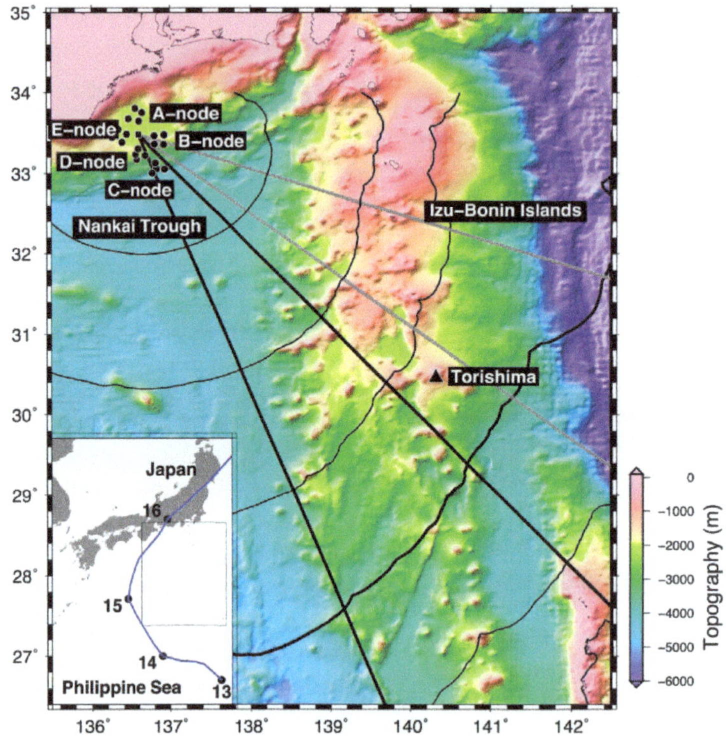

Figure 1 Distribution of DONET stations and submarine topography. The map in the inset shows the location of close-up region, and the blue line shows the track of Typhoon Man-yi from 13 to 17 September, 2013. The color contour is consistent with the marine depth. The black circles show the stations. The A, B, D, and E nodes are located at a depth of approximately 2,000 m, whereas the C node is located approximately 4,000 m deep. The square indicates the midpoint of the array of all the stations of the A, B, D, and E nodes. The wave front of a very long-period IG wave hypothetically emanating from the midpoint of the array is shown for every 15 min of travel time. The two gray and black lines show the azimuthal windows of 110° to 130° (ESE) and 140° to 160° (SSE), respectively.

RESULT

Figure 2a shows an example of the comparison of the seafloor records of vertical displacement (black) and pressure (red). Both were band-pass-filtered between 0.005 and 0.025 Hz and were normalized by the maximum amplitude of each trace, and the polarity of the pressure record was reversed. The agreement between the two traces was remarkable, which was expected if the signal is an IG wave whose phase velocity is much lower than that of a seismic wave (Crawford et al. 1991). Figure 2b shows an example of comparison of the stacked records of vertical displacement (black) and pressure (red). Stacking was accomplished by using a slowness vector that makes the stacked IG wave most energetic. The determined slowness value was consistent with that expected from the seafloor depth and the observed frequency range. This consistency, coupled with the remarkable waveform match shown in Figure 2, offered the most direct support for our interpretation of the observed disturbance as the IG waves. The station configuration and the response function of this array at 0.01 Hz (Rost and Thomas, 2002) are shown in Figure 3. The response function is represented in the slowness vector domain, where its spread around the origin indicates how narrowly in absolute value and how unbiasedly in azimuthal direction the slowness vector can be constrained by the array. The response function of our array is sharp and isotropic enough to warrant further analyses.Figure 4 shows four examples of the average PSD distribution in the slowness vector domain. Highly similar diagrams were also obtained for seafloor displacement fluctuation, as expected from the waveform match between the displacement and pressure records (Figure 2). Figure 4a,b shows typical patterns in winter, Figure 4c shows the typical pattern in spring, and Figure 4d shows a pattern when a typhoon was located in the Philippine Sea approximately 1,000 km to the south of the network. All the figures show a relatively large power between two dashed circles with slowness values of 7 and 9 s/km, or phase velocity values of 140 and 110 m/s, respectively, indicating that the observed waves are IG waves within a frequency range of 0.005 to 0.025 Hz propagating through the ocean at depths around 2,000 m. The IG signal intensity is persistently highest for the incoming waves crossing the Nankai Trough from the southeast (SE). This incoming direction actually consists of two closely adjacent directions: the south-southeast (SSE) and east-southeast (ESE) (see Figure 5 for a histogram of the measured incoming direction). Waves from the SSE direction through the deeper ocean are usually more energetic (Figure 4a) but are at times less energetic (Figure 4b) than those from the ESE direction through the shallower ocean (see Figure 1). These waves, either from SSE or ESE, are particularly strong in winter (see Figure 6).

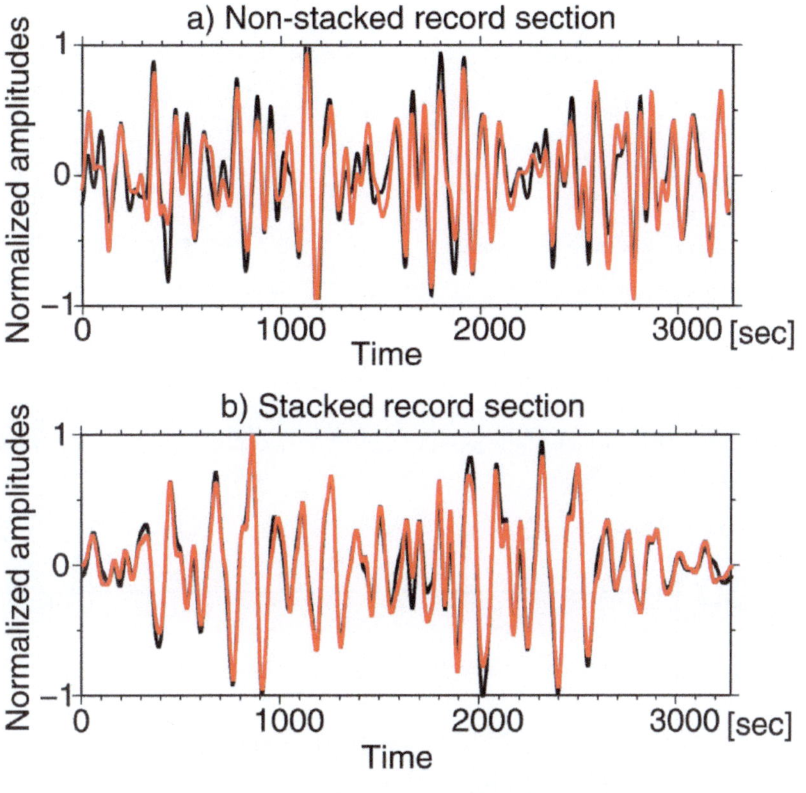

Figure 2 Examples of records sections of vertical displacement (black) and pressure (red). They are band-pass-filtered between 0.005 and 0.025 Hz on 1 January, 2013. **(a)**Single-station records and **(b)** slant-stacked records by an appropriately chosen slowness vector. Each of the records is normalized by its maximum value. The polarity of the pressure record is reversed.

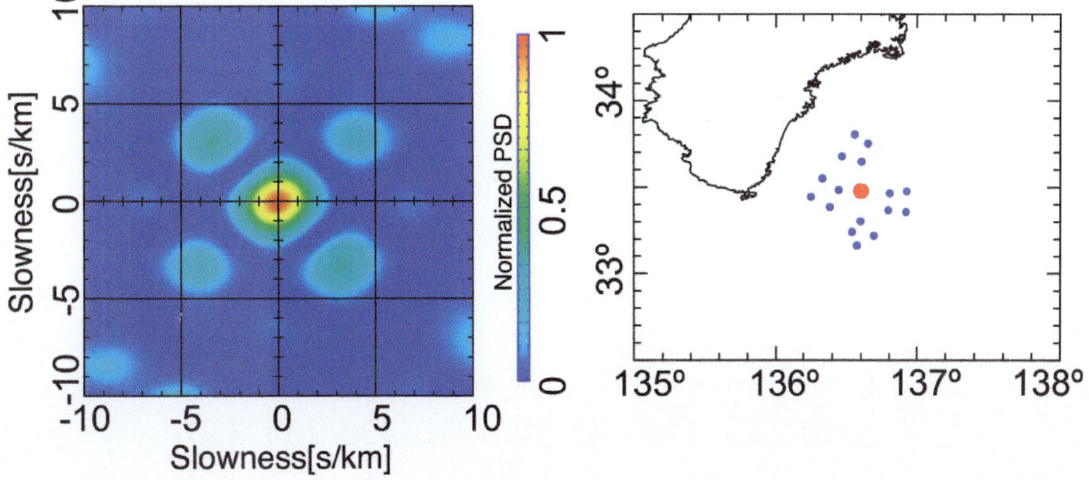

Figure 3 Array configuration (right) and array response function at 0.01 Hz represented in the relative slowness vector domain (left). In the right, the blue and red circles are the 16 stations used in the analysis and the midpoint of this array, respectively. In the left, the amplitude of the PSD is normalized by the maximum value as shown by the color bar.

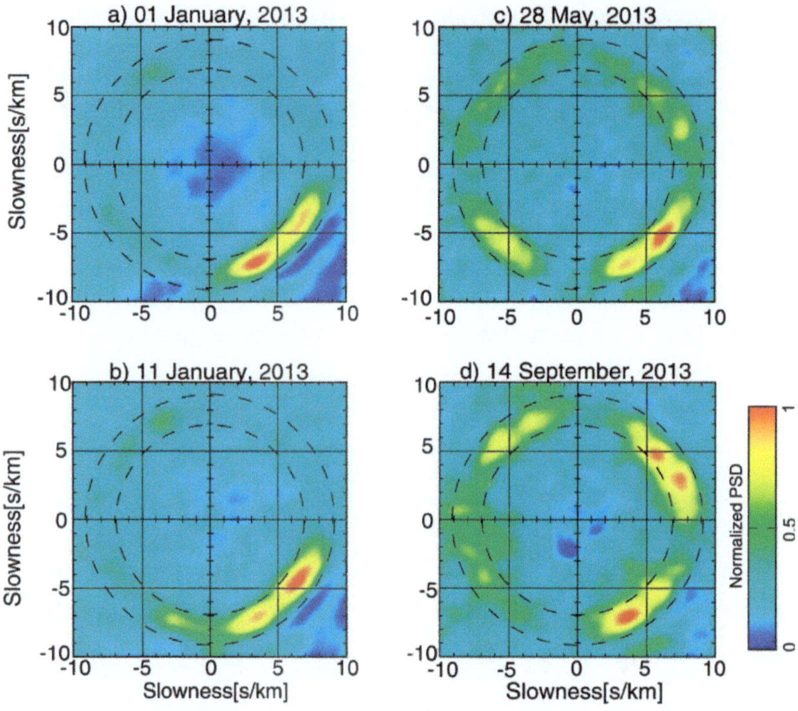

Figure 4 Four examples of the average PSD distribution in the slowness vector domain. The average of PSD is taken over a frequency range between 0.005 and 0.025 Hz. The recorded day is shown at the top of each diagram. The radii of two dashed circles indicate the slowness values of 7 and 9 s/km. The amplitudes of the PSDs in each diagram are normalized by their maximum value. The slowness (incoming direction, velocity) values estimated from the position of the red peak are **(a)** 7.9 s/km (153°, 127 m/s); **(b)** 7.8 s/km (125°, 128 m/s), **(c)** 8.0 s/km (131°, 126 m/s), and **(d)** 7.8 s/km (153°, 128 m/s).

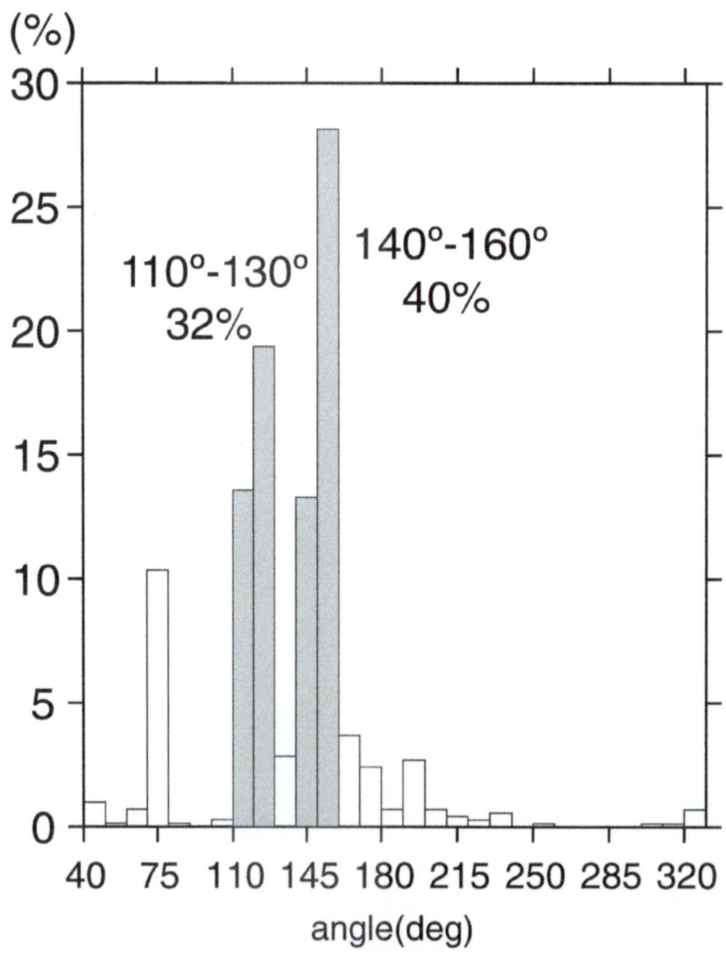

Figure 5 Histogram of the incoming direction of strong IG waves. The incoming directions are estimated from the pressure gauge records for the 3-year period. The shaded bars show two dominant incoming directions of SSE and ESE.

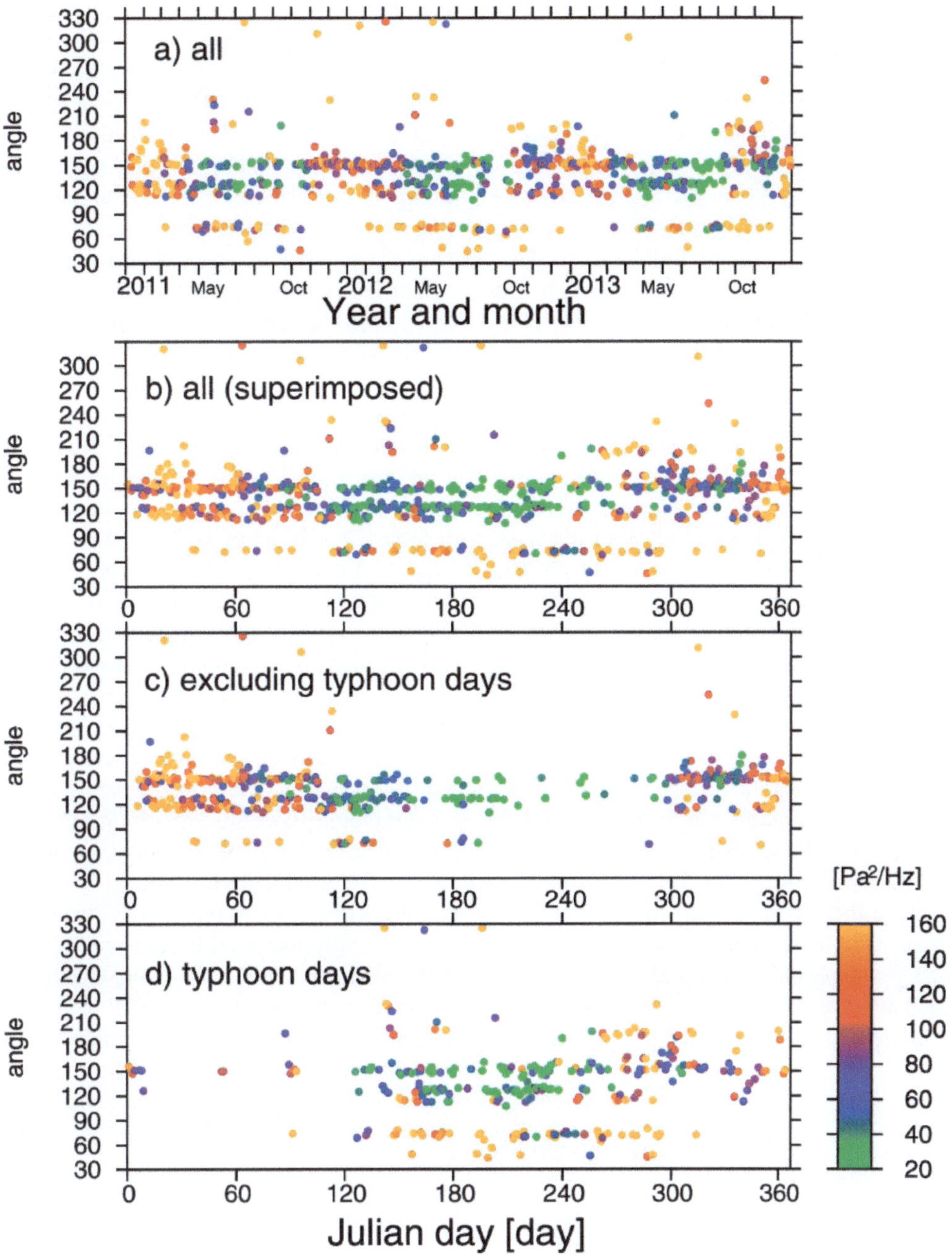

Figure 6 Daily variation of the incoming direction of the maximum-amplitude IG waves. The color indicates the maximum amplitude measured in terms of the average PSD between 0.005 and 0.025 Hz. **(a)** Results through the entire 3-year period. **(b)**Results for each year plotted on the same Julian day. **(c)** Results excluding days in which the Japan Meteorological Agency reported a typhoon or typhoons in the Pacific. **(d)**Results for days in which a typhoon or typhoons existed in the Pacific.

An example of the typical pattern in spring (and summer) is shown in Figure 4c, where the waves from the SE reduce their intensity so that wave signals from the other directions, particularly from the southwest (SW) and from the northeast (NE) to east-northeast (ENE), become more visible. These waves may be interpreted as edge waves trapped by reflections from the coast and refractions backward from the Nankai Trough to propagate through the corridor in between. Figure 4d shows the impact of Typhoon Man-yi on 14 September, 2013, where it was located on the Philippine Sea far to the south of the network as sketched in the inset of Figure 1(Japan Meteorological Agency (JMA); http://www.jma.go.jp/jma/jma-eng/jma-center/rsmc-hp-pub-eg/trackarchives.html). This typhoon moved from a back azimuth of 175° to 193° during the day and amplified the IG waves incoming from the SSE direction, and even more greatly from the NE to ENE direction, whereas those incoming from the ESE remained weak. Strengthening of IG waves from the NE to ENE direction is commonly observed when a typhoon is reported to be on the Philippine Sea, regardless of the exact incoming direction of IG waves originated by the typhoon, which varies widely in a range between the SSE and SW. These observations suggest that the amplified IG waves from the NE to ENE direction were generated secondarily at far distances from the primary (typhoon) source.We measured the incoming direction of the most energetic IG waves during the 3-year period. In the subsequent discussion, we show only the results from the pressure records, which are in agreement with those from the displacement records within an error of 5°. Figure 5 shows a histogram of the measured incoming direction. The SE direction dominates, although the detail is composed of two closely adjacent directions with a clear gap in between: one is SSE in a range of 140° to 160° (measured clockwise from the north) and the other is ESE in a range of 110° to 130°, which account for 40% and 32% of the total measurements, respectively. Together, these two directions account for 72% of the measurements. Figure 6a shows the daily variations of the incoming direction of the most energetic IG waves throughout the 3 years, where the color indicates the PSD value. The seasonal variations of the incoming directions are clarified by Figure 6b, in which the results for each of 3 years are superimposed. Several features can be observed from this figure: (1) The SSE and ESE persist as the dominant incident directions; (2) the energy of the IG wave incoming from these two directions shows the same seasonal variations of strong in winter and weak in summer; and (3) in summer, strong waves occasionally are incoming from the other directions. To explore observation (3) in detail, we divided the plots in Figure 6b into two as Figure 6c,d. The plots in Figure 6d are limited to the days on which the JMA reported a typhoon or typhoons on the Philippine Sea, whereas the plots in Figure 6c exclude such days. Figure 6c reinforces observations (1) and (2), which imply two distinct excitation sources of IG waves which are spatially stable and seasonally varying synchronously. Figure 6d indicates that the typhoon-associated IG waves are incoming from separate directions, one more or less directly from the typhoon in the SSE (as shown in Figure 4d) to SW and the other apparently from the submarine topographic high in the NE to ENE. IG waves incoming from this direction are often more energetic than those directly from the source on the typhoon's track. It is noted that the wave amplitudes from the two stationary sources in the SSE and ESE directions remained largely unchanged, even in the days where a typhoon was reported to be on the Philippine Sea.

DISCUSSION

We have identified two distinct excitation sources of IG waves in the SSE and ESE directions away from the network. Figure 7 shows that the IG waves from these two sources have different spectral characteristics. If we search for the source using the lower-frequency components (0.005 to 0.01 Hz), it is dominantly located in the SSE window (140° to 160°) as shown in Figure 7a, which accounts for 60% of the total measurements. If, on the other hand, search is made for the source using the higher-frequency components (0.01 to 0.025 Hz), it is dominantly located in the ESE window (110° to 130°) as shown in Figure 7b, which accounts for 52% of the total measurements. Clearly, lower-frequency IG waves are incoming more dominantly from the SSE than from the ESE. Higher-frequency IG waves are incoming more dominantly from the ESE than SSE but also include other incoming directions of 40° to 80° (16%), 190° to 230° (13%), and 300° to 330° (2%). The signal intensities in the higher frequency range are in general weaker than those in the lower frequency range as shown by the difference of detected IG wave PSD amplitude. This is in part because the higher-frequency IG wave decreases its amplitude more rapidly with depth than the lower-frequency IG wave. Although Figure 4c,d shows relatively large power in the NW direction (300° to 330°), the waves from this direction are usually very weak as compared to those from the SE direction (Figure 4a,b). The most energetic waves incoming from the SE direction are poorly accompanied by reflections from the nearest coast, where wave scattering, wave-wave conversion, and dissipation may be significant. As illustrated in Figure 1, the ocean in the ESE window is shallower with larger and more densely populated islands, whereas the ocean in the SSE window is deeper with smaller and more sparsely populated islands. We suggest that the spectral difference between the two azimuthal windows (Figure 7) reflects these differences in seafloor topography, which should affect the processes of generation and propagation of IG waves. In Figure 1, we added the inverse refraction diagram for a very long IG wave incoming to the array, which was obtained by combining the Tsunami Travel Times (TTT) software (Wessel 2009) with the bathymetry data of ETOPO1 (http://www.ngdc.noaa.gov/mgg/global/global.html). This diagram demonstrates that the wave front of the long IG wave in the ESE widow is more strongly distorted by the seafloor complexity than that in the SSE window. Although the DONET alone cannot locate the sources in the ESE and SSE windows, installation of a temporal seafloor array in the southeast window, e.g., near the Torishima Island (Figure 1), may suffice to determine these two source locations simultaneously.Besides the stationary sources discussed above, IG waves can be generated sporadically by extreme weather, typically by typhoons. The incoming direction of the typhoon-originated IG waves depends in a seemingly complex way on the past track of the moving typhoon, as shown in the inset of Figure 1. Most unexpectedly, a typhoon to the south far from the network always enhances IG waves incoming from the NE to ENE direction (Figure 6d). The enhanced IG waves are often more energetic than those directly from the source region (Figure 4d), suggesting some amplification mechanism at the secondary source at distances far away from the primary source.

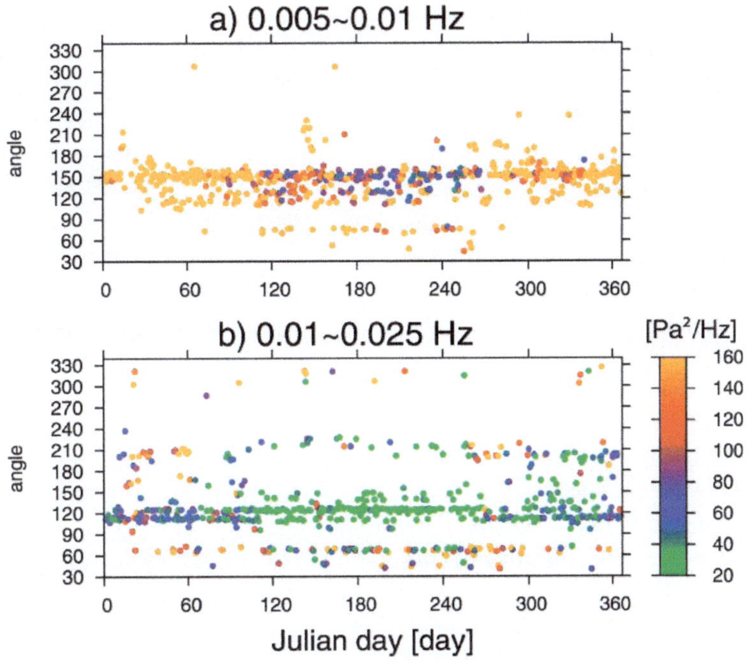

Figure 7 Daily variation of the incoming direction of the maximum-amplitude IG waves. (a) Waves band-pass-filtered between 0.005 and 0.01 Hz and **(b)** waves band-pass-filtered between 0.01 and 0.025 Hz. See Figure 6 for other explanations.

CONCLUSION

Although IG waves are known to be a ubiquitously observable phenomenon (Webb et al. 1991), this does not necessarily mean that IG waves are generated everywhere in the ocean. We have identified persistent, energetic IG sources in two azimuthal windows SSE and ESE of the network. The identified sources are rather localized and remain geographically stationary but show seasonally varying intensities of strong in winter and weak in summer. Higher-frequency waves are more dominantly incoming from the shallower ocean with more complex seafloor topography in the ESE direction. Lower-frequency waves are more dominantly incoming from the deeper ocean with less complex seafloor topography in the SSE direction. As shown in Figures 4c and 7b, an additional remarkable observation was the persistence of feeble IG waves incoming from the SW direction and from the NE-ENE directions, which may be interpreted as edge waves generated by reflections from the coast and refractions backward from the Nankai Trough to propagate as trapped waves in between. In particular, IG waves incoming from the NE-ENE directions are greatly amplified when IG waves originated by a typhoon on the Philippine Sea are incident. The amplified amplitudes often well exceed the amplitudes of incident waves from the primary origin. A seafloor network equipped with broadband ocean-bottom seismometers and pressure gauges, such as the DONET, is highly useful for detecting IG waves and observing the related phenomena.

ACKNOWLEDGEMENTS

The Generic Mapping Tools (Wessel and Smith 1991) and Seismic Analysis Code (Goldstein et al. 1998) were used in this study. A part of the DONET data is available fromhttp://www.jamstec.go.jp/donetevent/NINJA/top.do. We thank Dai Suetsugu for all his help. We also thank two anonymous reviewers and the associate editor (Azusa Nishizawa) for many constructive comments.

AUTHORS' CONTRIBUTIONS

YT, KN, and YF conceived and designed the study; AT and NT acquired the data; YT, KN, and YF performed the analysis and interpretation of data; YT drafted the manuscript; and YF made a critical revision of the manuscript. All authors read and approved the final manuscript.

REFERENCES

1. Crawford WC, Webb SC, Hildebrand JA: Seafloor compliance observed by long-period pressure and displacement measurements. J Geophys Res 1991, 96: 16,151–16,160. doi:10.1029/91JB01577
2. Dolenc D, Romanowicz B, Stakes D, McGill P, Neuhauser D: Observations of infragravity waves at the Monterey ocean bottom broadband station (MOBB). Geochem Geophys Geosyst 2005, 6: Q09002. doi:10.1029/2005GC000988
3. Fukao Y, Nishida K, Kobayashi N: Seafloor topography, ocean infragravity waves, and background Love and Rayleigh waves. J Geophys Res 2010, 115: B04302. doi:10.1029/2009JB006678
4. Godin OA, Zabotin NA, Sheehan AF, Yang Z, Collins JA: Power spectra of infragravity waves in a deep ocean. Geophys Res Lett 2013, 40: 2159–2165. doi:10.1002/grl.50418
5. Goldstein P, Dodge D, Firpo M, Ruppert S: What›s new in SAC2000? Enhanced processing and database access. Seismological Research Lett 1998, 69: 202–205.
6. Guza RT, Thornton EB: Swash oscillations on a natural beach. J Geophys Res 1982, 87: 483–491. doi:10.1029/JC087iC01p00483
7. Harmon N, Henstock T, Srokosz M, Tilmann F, Rietbrock A, Barton P: Infragravity wave source regions determined from ambient noise correlation. Geophys

Res Lett 2012, 39: L04604. doi:10.1029/2011GL050414

8. Herbers THC, Elager S, Guza RT: Generation and propagation of infragravity waves. J Geophys Res 1995, 100: 24,863–24,872. doi:10.1029/95JC02680

9. Kaneda Y: The advanced ocean floor real time monitoring system for mega thrust earthquakes and tsunamis. Application of DONET and DONET2 data to seismological research and disaster mitigation 2010. MTS/IEEE OCEANS 2010 MTS/IEEE OCEANS 2010

10. Kawaguchi K, Araki E, Kaneko S, Nishida T, Komine T: Subsea engineering ROV and seafloor observatory construction. Proceedings of international symposium on underwater technology 2010/international workshop on scientific use of submarine cables and related technologies 2010, SSC11–1019, CD-ROM 2010.

11. Longuet-Higgins MS, Stewart RW: Radiation stress and mass transport in gravity waves, with application to 'surf beat'. J Fluid Mech 1962, 13: 481–504.

12. Nakano M, Nakamura T, Kamiya S, Ohori M, Kaneda Y: Intensive seismic activity around the Nankai trough revealed by DONET ocean-floor seismic observations. Earth Planets Space 2013, 65: 5–15. doi:10.5047/eps.2012.05.013

13. Nishida K: Earth›s background free oscillations. Annu Rev Earth Planet Sci 2013, 41: 719–740. doi:10.1146/annurev-earth-050212–124020

14. Nishida K, Kawakatsu H, Fukao Y, Obara K: Background Love and Rayleigh waves simultaneously generated at the Pacific Ocean floors. Source distribution of Earth›s background oscillations. Geophys Res Lett 2008, 35: L16307.

15. Nishida K, Fukao Y, Watada S, Kobayashi N, Tahira M, Suda N, Nawa K, Oi T, Kitajima T: Array observation of background atmospheric waves in the seismic band from 1 mHz to 0.5 Hz. Geophys J Int 2005, 162: 824–840. doi:10.1111/j.1365–246X.2005.02677.x

16. Okihiro M, Guza RT: Infragravity energy modulation by tides. J Geophys Res 1995, 100: 16,143–16,148. doi:10.1029/95JC01545

17. Rhie J, Romanowicz B: A study of the relation between ocean storms and the Earth›s hum.Geochem Geophys Geosyst 2006, 7: Q10004. doi:10.1029/2006GC001274

18. Rhie J, Romanowicz B: Excitation of Earth›s continuous oscillations by atmosphere–ocean-seafloor coupling. Nature 2004, 431: 552–556. doi:10.1038/nature02942

19. Rost S, Thomas C: Array seismology: methods and applications. Rev Geophys 2002, 40(3):1008. doi:10.1029/2000RG000100

20. Sugioka H, Fukao Y, Kanazawa T: Evidence for infragravity wave-tide resonance in deep oceans. Nat Commun 2010, 1: 84. doi:10.1038/ncomms1083

21. Tanimoto T: The oceanic excitation hypothesis for continuous oscillations of the Earth.Geophys J Int 2005, 160: 276–288. doi:10.1111/j.1365–246X.2004.02484.x

22. Tomson J, Elgar S, Raubenheimer B, Herber THC, Guza RT: Tidal modulation of infragravity waves via nonlinear energy losses in the surfzone. Geophys Res Lett 2006, 33: L05601. doi:10.1029/2005GL025514

23. Webb SC, Zhang X, Crawford W: Infragravity wave in the deep ocean. J Geophys Res 1991, 96: 16,151–16,160. doi:10.1029/90JC02212

24. Wessel P: Analysis of observed and predicted tsunami travel times for the Pacific and Indian oceans. Pure Appl Geophys 2009, 166(1–2):301–324.

25. Wessel P, Smith WHF: Free software helps map and display data. Eos Trans AGU 1991, 72: 441.

CHAPTER 4 | ACOUSTIC-GRAVITY WAVES

Oleg A Godin, Nikolay A Zabotin, and Terence W Bullett

Keywdords: Acoustic-gravity waves Infragravity waves Air-sea interaction Ionosphere Radio sounding

INTRODUCTION

Ionospheric manifestations of tsunamis (Artru et al. 2005; Makela et al. 2011; Galvan et al.2012; Komjathy et al. 2012; Occhipinti et al. 2013) provided dramatic evidence of coupling between wave processes in the ocean and in the atmosphere. But tsunamis are only a transient feature of a more general phenomenon: infragravity waves (IGWs) (Webb et al. 1991; Herbers et al. 1995; Aucan and Ardhuin 2013; Godin et al. 2014). IGWs are permanently present surface gravity waves in the ocean with periods longer than the longest periods (approximately 30 s) of wind-generated waves. IGWs propagate transoceanic distances and, because of their long wavelengths (from approximately 1 km to hundreds of km), offer a mechanism for coupling wave processes in the ocean, atmosphere, and the solid Earth (Hines 1972; Crawford et al.1998; Rhie and Romanowicz 2004; Bromirski and Stephen 2012). The notion that background ocean waves may generate waves in the atmosphere has existed for a long time (Hines 1972; Peltier and Hines 1976; Livneh et al. 2007), but no quantitative theory of this effect has been suggested. Here, we provide a strict physical justification for this hypothesis. Using a coupled atmosphere-ocean model taking into account both compressibility of the air and the gravity, we show that there exist two distinct regimes of IGW penetration into the atmosphere. At frequencies above a transition frequency, there are evanescent waves in the atmosphere propagating horizontally along the ocean surface. At frequencies below the transition frequency, IGWs continuously radiate their energy into the upper atmosphere in the form of acoustic-gravity waves (AGWs). The transition frequency is a function of ocean depth; it varies slowly near 3 mHz for typical depth values and drops sharply to zero only for extreme depth values. Using a semi-empirical model of the IGW power spectrum (Godin et al. 2013), we derive an estimate of the flux of the mechanical energy and mechanical momentum from the deep ocean into the atmosphere due to IGWs and predict specific forcing on the atmosphere in coastal regions. We conclude that AGWs of oceanic origin may have an observable impact on the upper atmosphere.

Citation: Oleg A Godin, Nikolay A Zabotin, Terence W Bullett, Acoustic-gravity waves, DOI: 10.1186/s40623-015-0212-4.

THEORY

Infragravity waves and atmospheric waves

IGWs are usually modeled as surface gravity waves in an incompressible ocean with a free surface and a rigid seafloor. In this approximation, IGWs have the dispersion equation

$$\omega^2 = gk \tan hkH, \tag{1}$$

where ω and k are the wave frequency and wavenumber, $g \approx 9.8$ m/s^2 is the acceleration due to gravity, and H is the ocean depth (Gill 1982). IGW phase speed $c = \omega/k$ does not exceed $(gH)^{1/2}$and is much smaller than sound speed in water and compressional and shear wave speeds in the ocean bottom. Therefore, corrections to the dispersion equation due to water compressibility and seafloor compliance are rather small (Tsai et al. 2013; Watada et al. 2014) and do not change the nature of IGWs as horizontally propagating surface waves, where both k and ω are real-valued (Brekhovskikh and Godin 1998).

The free-surface boundary condition implies that the wave-induced motion of the air has no effect on oceanic waves. For a simple description of coupling between waves in the ocean and atmosphere, we model the latter as an isothermal gas half-space. Continuous waves with horizontal wave vector **k** on the sea surface generate plane waves in the half-space. Horizontal and vertical components of the wave vector of the atmospheric waves are **k** and m, where

$$m^2 = \frac{\omega^2}{c_s^2} - k^2 - \frac{\gamma^2 g^2}{4c_s^4} + \frac{(\gamma-1)g^2k^2}{\omega^2 c_s^2}, \tag{2}$$

$k^2 = \mathbf{k}\cdot\mathbf{k}$, c_s is the sound speed in air, and $\gamma \approx 1.4$ is the ratio of specific heats at constant pressure and constant volume (Gill 1982). The atmospheric waves have a finite impedance, i.e., the ratio of pressure perturbation in a fluid parcel to its vertical velocity. All these quantities must be continuous on the air-sea interface (Gill 1982). For oceanic waves, the impedance is small as long as the ratio $R \sim 1.3 \times 10^{-3}$ of the mass densities of the air and water at the water-air interface is small. We use a perturbation theory to quantify the coupling between oceanic and atmospheric waves.

Perturbation theory

Our theoretical analysis of the effects of the atmospheric load on infragravity waves is based on modeling the ocean surface as a boundary with finite impedance and application of the recently developed perturbation theory (Godin 2012) for normal modes of acoustic-gravity waves. The perturbation theory predicts the change in the dispersion relation of normal modes of acoustic-gravity waves in generic waveguides in horizontally stratified fluids, which results from small perturbations of the waveguide's parameters.

Introduce a Cartesian coordinate system with horizontal coordinates x and y and a vertical coordinate z increasing upwards. Consider a fluid layer with a rigid boundary at $z=-H$ and a compliant boundary that coincides with the horizontal plane $z=0$ in the absence of waves. The fluid is stationary in the absence of waves. Let $W_0(z, \omega)\exp(ik(\omega)x - i\omega t)$ and $P_0(z,\omega)\exp(ik(\omega)x - i\omega t)$ describe the wave-induced vertical displacement of fluid parcels and pressure variation, which satisfy linearized (with respect to the wave amplitude) governing equations and boundary conditions $W(z, \omega)=0$, $z=-H$ on the rigid boundary and $P(z, \omega)=\rho g W(z, \omega)$, $z=0$ on the free surface (Gill 1982). Here, t is time, ω is wave frequency, k is wavenumber, and ρ is fluid density. The functions $W_0(z, \omega)$ and $P_0(z, \omega)$ describe a normal mode with the dispersion relation $k(\omega)$ in an unperturbed waveguide and are assumed to be known.

Now, consider a waveguide that differs from an unperturbed waveguide by having a different upper boundary. For waves with harmonic dependence $\exp(ikx - i\omega t)$ on horizontal coordinates and time, the boundary condition can be written as

$$P(z,\omega) - \rho g W(z,\omega) = -i\omega Z(k,\omega) W(z,\omega), \quad z=0. \tag{3}$$

Here, $P - \rho g W$ and Z have the physical meaning, respectively, of the wave-induced pressure variation in a moving fluid parcel (i.e., Lagrangian pressure perturbation) and the boundary impedance (Godin 2012). Impedance of the free boundary equals zero. When Z is small, there exists a normal mode with the dispersion relation $k_1(\omega) = k(\omega) + \delta k(\omega)$ (Godin 2012), where

$$\delta k = \frac{i\omega^3}{2k} W_0^2(0,\omega) \left[\int_{-H}^{0} \frac{dz}{\rho} P_0^2(z,\omega) \right]^{-1} Z(k(\omega),\omega) + O(Z^2). \tag{4}$$

Note that the impedance in the right-hand side of Equation (4) is calculated for the unperturbed value of the normal mode wavenumber.

IGWs are usually modeled as surface gravity waves in an incompressible ocean with a constant water density, a free surface, and a rigid seafloor. In this approximation, one has (Gill 1982)

$$P(z,\omega) = P(0,\omega)\frac{\cosh k_0(z+H)}{\cosh k_0 H}, \quad W(z,\omega) = \frac{P(0,\omega)}{\rho g}\frac{\sinh k_0(z+H)}{\sinh k_0 H}, \tag{5}$$

and the unperturbed dispersion relation (1).

Lagrangian pressure perturbations and normal displacement are continuous at fluid-fluid interfaces (Gill 1982). Therefore, the impedance $Z(k, \omega)$ is also continuous and can be calculated from the ratio of the Lagrangian pressure perturbation and the vertical displacement in an atmospheric wave with the frequency ω and horizontal wavenumber k. The ocean surface impedance depends on atmospheric conditions. The simplest atmospheric model is that of a stationary, isothermal, perfect gas of constant composition occupying the half-space $z>0$. In the isothermal atmosphere, acoustic-gravity waves (AGWs) have the dispersion relation (2). The lower boundary of the isothermal gas half-space has the impedance (Godin 2012; Godin and Fuks 2012)

$$Z(k,\omega) = \frac{R\rho(0)}{\omega(\omega^2 c_s^{-2} - k^2)}\left[\omega^2 m(k,\omega) + ig(k^2 - 0.5\omega^2 \gamma c_s^{-2})\right], \tag{6}$$

where $\rho(0)$ and $R \sim 1.3 \times 10^{-3}$ are the water density and the ratio of the air and water densities at the ocean surface. In Equation (6), the sign of the vertical component $m(k, \omega)$ of the wave vector should be chosen as follows: $m = i|m|$, when $m^2 < 0$; m=|m|sgn(ω2−(γ−1)g2c−2s),m=|m|sgn(ω2−(γ−1)g2cs−2),when $m^2 \geq 0$ (Godin and Fuks 2012).

From Equations (1) and (4) to (6), we find the perturbation to the IGW dispersion relation (1) caused by the atmospheric loading:

$$\delta k = \frac{Rk\left(2k^2c_s^2-\gamma\omega^2-2i\omega^2c_s^2m/g\right)\sinh 2kH}{2\left(k^2c_s^2-\omega^2\right)\left(\sinh 2kH + 2kH\right)}. \tag{7}$$

Here, only terms of the first order in R are retained, and ω and k satisfy the unperturbed dispersion relation (1). Equation (7) reduces to known results in the previously considered case of an infinitely deep ocean (Godin and Fuks 2012).

An equivalent of Equation (7) can be obtained, although with much more effort, without using the perturbation theory and by deriving instead an exact dispersion equation for acoustic-gravity waves in a coupled system consisting of a liquid layer and an isothermal gas half-space.

Equation (7) indicates that there are two distinct wave-propagation regimes, which are separated by a transition frequency such that $m = 0$. When $m^2 < 0$ in Equation (2), waves in the air are evanescent, their energy density decreasing with height z above the ocean surface as $\exp(-2z|m|)$. Then, correction δk (7) is real, and the IGW remains a true surface wave. When$m^2 > 0$, waves in air propagate away from the ocean surface and carry energy and momentum into the upper atmosphere. For propagating waves, the amplitude of the velocity oscillations increases with height as $[\rho_a(0)/\rho_a(z)]^{1/2}$, where $\rho_a(z)$ is the air density and $\rho_a(0)$ refers to the mean sea level. Since the IGW phase speed is always smaller than sound speed in air, m^2 (2) can be positive only at frequencies ω smaller than the buoyancy frequency $N = (\gamma - 1)^{1/2}g/c_s$. At$\omega < N$, it follows from the radiation condition at $z \to +\infty$ that $m < 0$ when $m^2 > 0$ (Godin and Fuks 2012). Then, δk (7) is complex, and the imaginary part Im $\delta k > 0$. This means that, due to the atmospheric loading, the IGW slowly attenuates in the direction of its horizontal propagation and thus becomes a leaky wave (Brekhovskikh and Godin 1998). This is a natural corollary of the energy conservation and radiation of IGW energy into the atmosphere. Radiation into the atmosphere is only one of the mechanisms of IGW decay. As discussed in the section 'IGW attenuation' below, comparison with observed tsunami decay rates (Rabinovich et al. 2013) shows that the decay rate predicted by Equation (7) is smaller than measured decay rates.

Radiation of atmospheric waves

An analysis of Equations (1) and (2) shows that, within our simple environmental model, radiation of IGW energy into the atmosphere occurs at frequencies $0 < \omega < \omega_b(H)$, where the transition frequency ω_b steadily decreases with depth from $\omega_b(0) = N$ in shallow water (about 3.0 mHz for $c_s = 330$ m/s) to zero at H=Hb≡4(γ−1)γ−2c2s/g≈0.82c2s/gH=Hb≡4(γ−1)γ−2cs2/g≈0.82cs2/g (about 9,100 m for $c_s = 330$ m/s; see Figure 1a). When a more realistic environmental model is used, $\omega_b(H)$ andH_b also depend on the temperature and wind-velocity profiles in the atmosphere. For most of the ocean, the transition frequency is between 2.5 and 3 mHz (Figure 1a). Only small areas of the ocean with the greatest depths, such as the Mariana, Kermadec, and Puerto Rico trenches, have transition frequencies below 2.25 mHz.

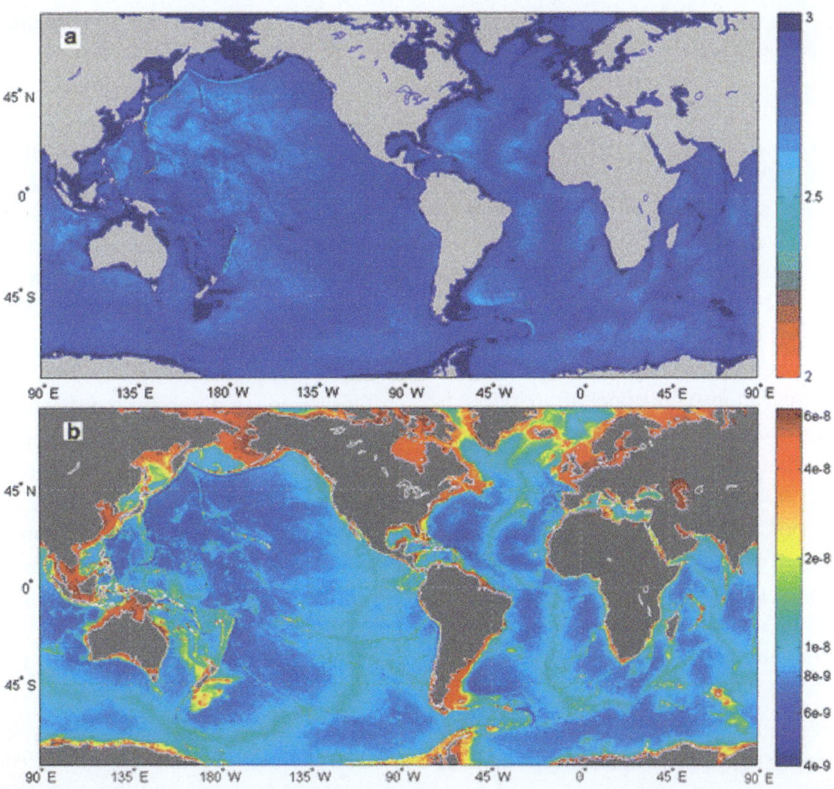

Figure 1 Geography of atmospheric wave radiation by infragravity waves in the ocean. (a) The transition frequency ω_b from surface to leaky infragravity waves (in mHz) is shown by color for various geographical locations. At frequencies below the transition frequency, infragravity waves continuously radiate acoustic-gravity waves into the upper atmosphere. **(b)** The surface density I of the vertical flux of mechanical power radiated into the atmosphere (in W/m²) is shown by color on the logarithmic scale. The transition frequency and the power flux density are calculated for an isothermal atmosphere with sound speed $c_s = 330$ m/s.

Prediction of the upper cutoff frequency ω_b is consistent with an earlier analysis (Peltier and Hines 1976) of the limiting case of non-dispersive tsunami waves, where $kH \ll 1$. Indirect experimental support for our predictions is provided by the observations at the Infrasound Laboratory on the Island of Hawaii that the noise level, i.e., the spectral level of ambient pressure fluctuations in air, increases sharply at frequencies below approximately 4 mHz (Garces et al. 2011).

When investigating the conditions for IGW radiation into the atmosphere, it is imperative to consider the atmospheric waves as acoustic-gravity waves, i.e., to take into account effects due to buoyancy and compressibility simultaneously. When surface waves are considered as a potential motion of an incompressible fluid, which is an excellent approximation for wind seas (Gill 1982), orbital velocities in the wave exponentially decrease with distance from the sea surface at the same rate in the water and the air, and no radiation of wave energy into the atmosphere is predicted. In the opposite case, when buoyancy is neglected, we have acoustic waves in air, and radiation into the atmosphere occurs when the phase speed c of surface waves exceeds the sound speed (Brekhovskikh and Godin 1998). That would require the ocean depth H>c2s/gH>cs2/g (Eckart 1953). Perhaps surprisingly, an account of the buoyancy effects results in converse dependence on the ocean depth. As discussed above, there is no radiation into the atmosphere from a horizontally homogeneous ocean with depth H>Hb≈0.82c2s/g.H>Hb≈0.82cs2/g. The reason for the qualitatively different behavior of acoustic and acoustic-gravity waves lies in the fact that m^2 decreases with increasing k^2, when $g=0$, but increases when $g\neq 0$ and $\omega<N$, see Equation (2).

Observations of ionospheric manifestations of recent catastrophic tsunamis (Artru et al. 2005; Makela et al. 2011; Galvan et al. 2012) serve as a spectacular manifestation of radiation of acoustic-gravity atmospheric waves by IGWs in the ocean. Background IGWs, which are believed to be generated by the nonlinear interaction of sea swell with bathymetry close to shore, have much smaller amplitude than destructive tsunamis but are continuously present throughout the ocean (Webb et al. 1991; Herbers et al. 1995; Aucan and Ardhuin 2013; Godin et al. 2014). Over extended time periods, the contribution of the background IGWs into the flux of mechanical energy from the ocean into the upper atmosphere is expected to dominate over the contributions of tsunamis. For a rough estimate, we assume that powerful tsunamis with ocean surface displacement of 0.2 m account for half of the tsunami energy, occur on average every 5 years, last 2 h at open ocean locations, and affect 10% of the ocean surface. For RMS surface elevations in background IGWs, we conservatively take 3 mm (Webb et al. 1991; Aucan and Ardhuin 2013; Godin et al. 2013). Then, the time-averaged tsunami energy is about 2% of the background IGW energy.

The mechanical power radiated into the atmosphere can be calculated as an integral over the vertical component of the power flux density on the ocean surface (Godin and Fuks 2012). Random IGWs with power spectrum S of the ocean surface displacement radiate into the upper atmosphere the power

$$I = \rho_a(0) \int_{\omega_{\min}}^{\omega_b(H)} \frac{\omega^3 S(\omega, H)|m(\omega)|d\omega}{m^2(\omega) - \omega^2 c_s^{-2}} \quad (8)$$

per unit area of the ocean surface. Here ω_{min} is the lowest frequency of IGWs present and $\rho_a(0)$ is the air density at sea level; $I=0$, when $\omega_{min} \geq \omega_b(H)$. When AGWs dissipate in the upper atmosphere, they exert a force on the mean flow (wind), which can be calculated as the vertical flux of the AGW pseudomomentum (Fritts and Alexander 2003). AGW forcing plays an important role in large-scale circulation of the middle and upper atmosphere (Fritts and Alexander 2003; Geller et al. 2013). The horizontal force per unit area of the ocean surface is given by the same integral as in Equation (8) but with $\omega^2 \langle k(\omega) \rangle$ substituted for ω^3 in the numerator in the integrand. Here, $\langle k(\omega) \rangle$ is the horizontal wave vector of IGWs averaged over all propagation directions. The horizontal force is non-zero only for anisotropic IGW fields. The magnitude of the horizontal force reaches its upper bound, when all IGWs propagate in the same direction; then, $|\langle k(\omega) \rangle|$ equals the wavenumber $k(\omega, H)$ of IGWs from the dispersion equation (1). This limiting case can be thought of as corresponding to IGWs generated near and propagating away from a straight shoreline. A force of the same order of magnitude in the direction perpendicular to the shoreline is expected to be applied to the upper atmosphere above the shore due to AGW radiation by IGWs with generic directivity. In the latter case, the requisite anisotropy arises from the absence of IGWs on the shore side of the coastline.

We apply the recently developed semi-empirical model $S(\omega, H) = B \tanh^{-3/2} kH$, where $B = 2.1 \times 10^{-5}$ m²/Hz, of the IGW power spectrum (Godin et al. 2013) to estimate the mechanical energy and momentum radiated by the ocean into the upper atmosphere. The spectrum was obtained for IGWs with frequencies 0.3 mHz and higher. The efficiency of AGW radiation decreases rapidly at frequencies below approximately 0.5 mHz (Figure 2a,b), which makes the surface densities of radiated power [Equation (8)] and pseudomomentum insensitive to unknown IGW spectral levels below 0.3 mHz. The IGW spectrum and the IGW and AGW dispersion relations (1) to (7) combine to make AGW radiation more efficient at shallower ocean depths and at frequencies close to but smaller than the transition frequency ω_b (Figure 2a,b). The effect of the ocean depth is particularly pronounced for the momentum flux. Atmospheric waves have the unusual property that the horizontal and vertical components of their group velocity depend on frequency in rather different ways. The horizontal component is insensitive to frequency. In contrast, the vertical component (Figure 2c) and the direction of wave propagation (Figure 2d) vary strongly with frequency. The maximum vertical velocity exceeds 80 m/s and is attained at frequencies close to $0.8\ \omega_b$ (Figure 2c). Geographical distribution of the mechanical energy radiation is illustrated by Figure 1b. The total power radiated into the middle and upper atmosphere by IGWs globally is estimated to be approximately 0.3 MW. The sound speed value$c_s = 330$ m/s, which is used in our model calculations, corresponds to air temperature of about 271 K. The modeling results presented in Figures 1, 2, 3, and 4 are insensitive to the particular choice of c_s within the range of possible sound speed values above the ocean surface.

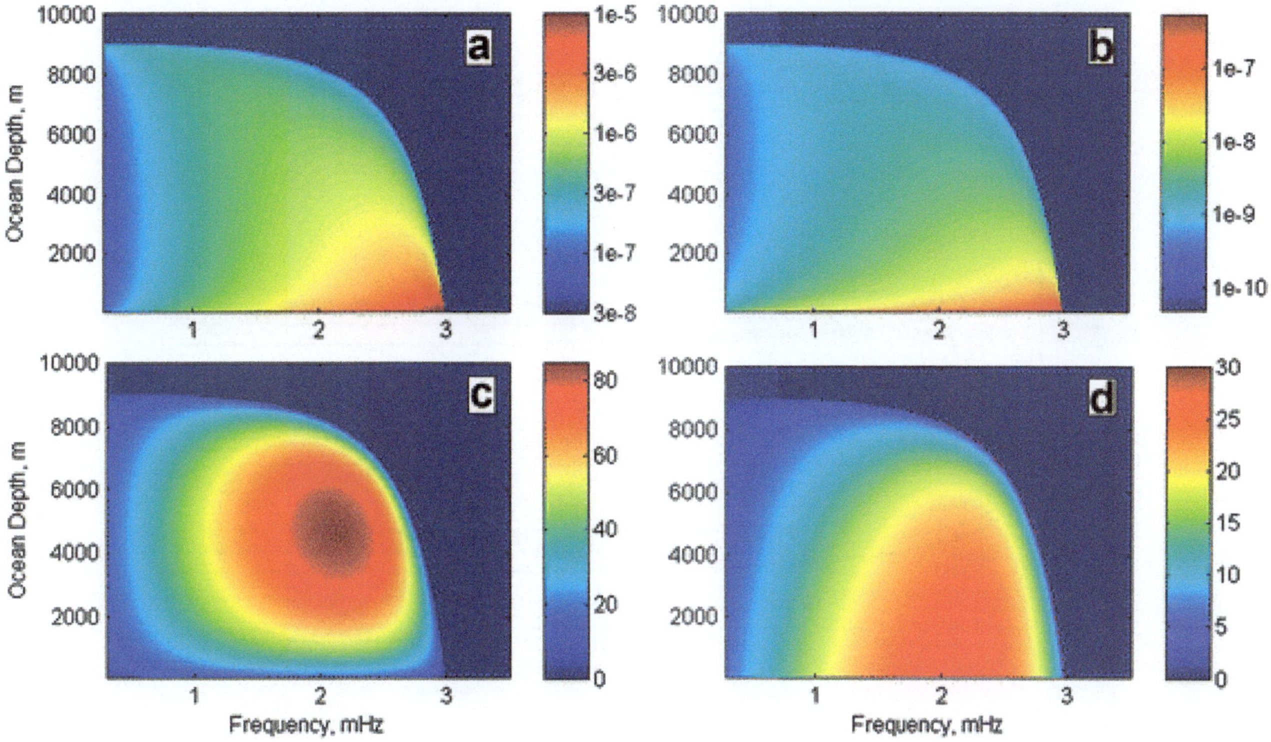

Figure 2 Spectra of atmospheric waves radiated by infragravity waves in the ocean. **(a)** Power spectrum of the surface density of the mechanical energy flux is shown on the logarithmic scale relative 1 W/m^2 Hz. **(b)** Power spectrum of the vertical flux of horizontal pseudo-momentum is shown for the straight coast model on the logarithmic scale relative 1 Pa/Hz. AGW power spectra are calculated for various ocean depths using a semi-empirical model (Godin et al. 2013) of the IGW power spectrum. **(c)** Vertical component of the group velocity of AGWs in m/s. **(d)** Angle (in degrees) between the ocean surface and the direction of AGW propagation.

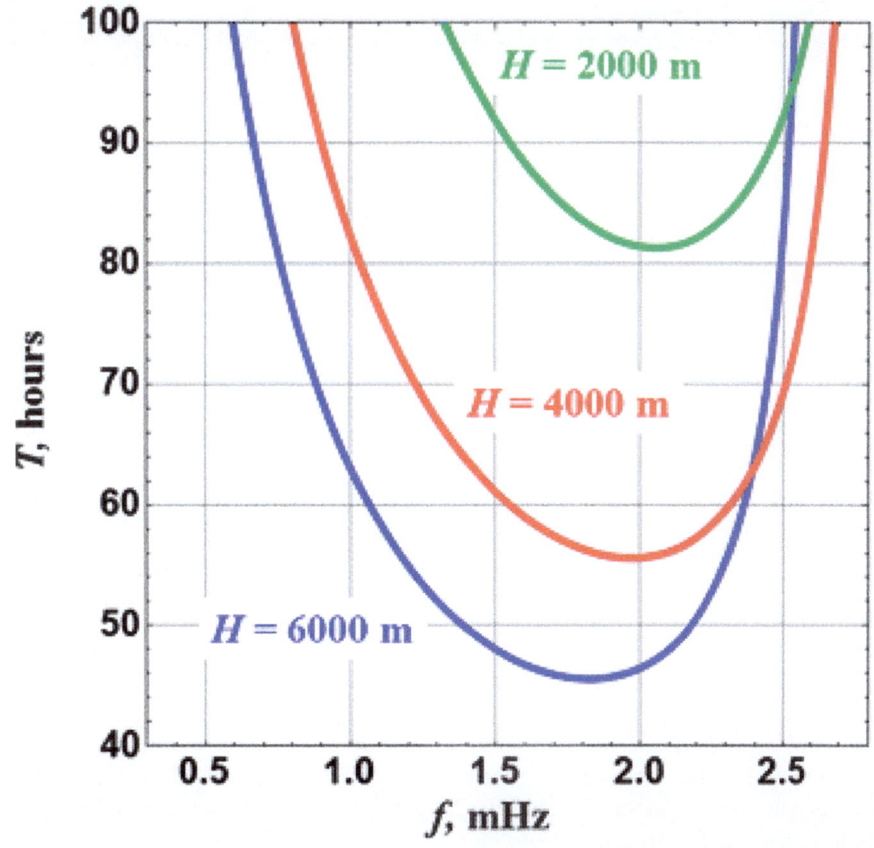

Figure 3 Attenuation of infragravity waves in the ocean due to radiation of atmospheric waves. Decay time T of infragravity waves is shown as a function of wave frequency for ocean depths H = 2,000, 4,000, and 6,000 m. IGW attenuation is calculated assuming an isothermal atmosphere with sound speed c_s = 330 m/s.

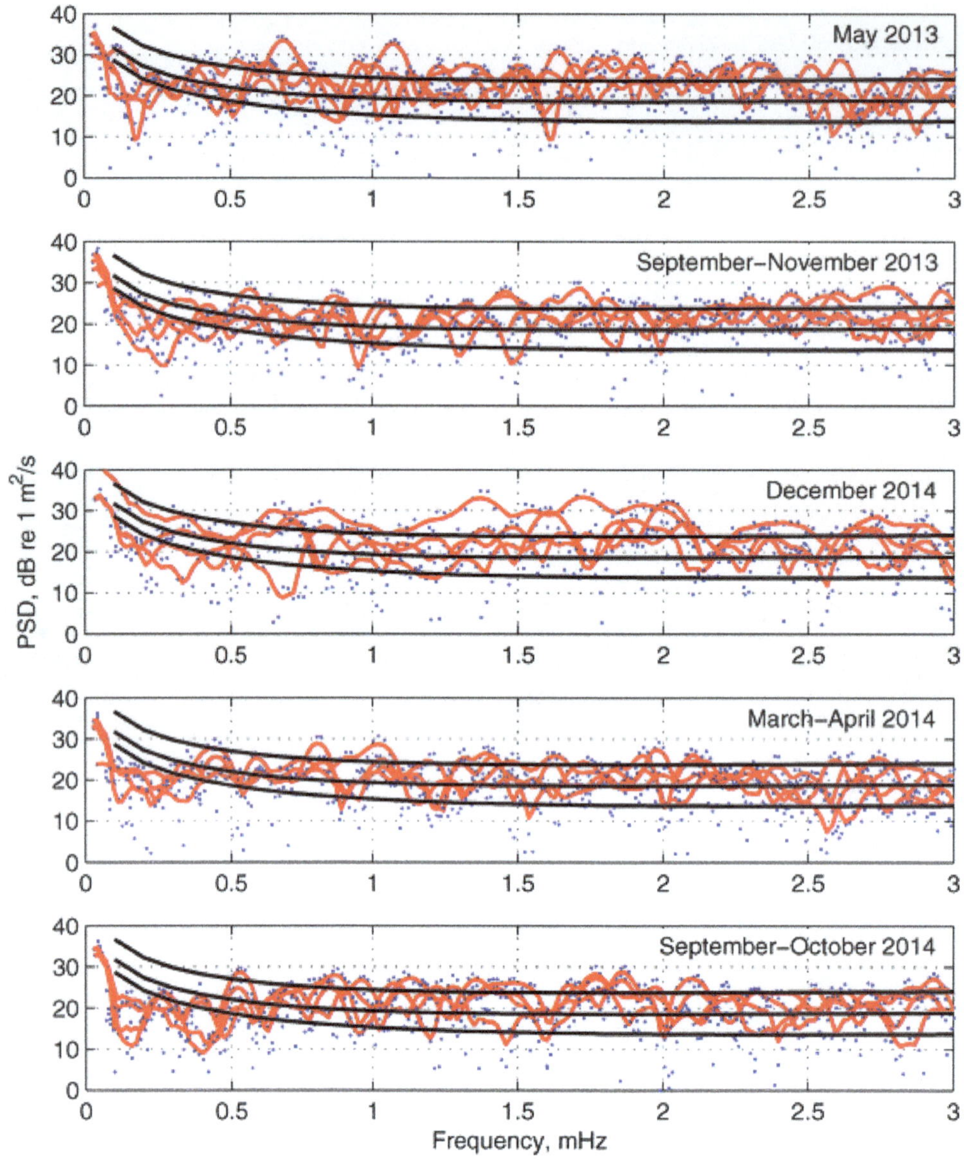

Figure 4 Measured power spectra. Measured power spectra of velocities' fluctuations along the magnetic field line at an altitude of 150 km are compared with predicted contributions due to atmospheric waves radiated by infragravity waves in the ocean. The spectra are measured by radio sounding of the ionosphere using an onshore Dynasonde (Wright and Pitteway 1999; Zabotin et al. 2006) HF radar system located at NASA's Wallops Island Flight Facility, Virginia, USA. The five panels refer to five distinct seasons in 2013 to 2014, and each panel contains four experimental spectral estimates obtained from the data series, specific dates and times of which are listed in the text. Blue dots show the results of spectral measurements. Red curves represent the results of smoothing the experimental spectra using a 0.5-mHz sliding window. Black curves show contributions of the atmospheric waves of oceanic origin, which are estimated for various ocean depths (200, 1,000, and 5,000 m, top to bottom) using a semi-empirical model (Godin et al. 2013) of the IGW spectra.

IGW attenuation

Atmospheric loading results in the perturbation δk, which is given by Equation (7), to the dispersion relation (1) of IGWs in an ocean with a free surface. From Equation (7) it follows that, at frequencies $0<\omega<\omega_b(H)$, IGW amplitude attenuates with propagation distance L as $\exp[-L\mathrm{Im}(\delta k)]$ due to continuous radiation of mechanical energy into the atmosphere. Since $R \ll 1$, an IGW needs to propagate a long distance of the order of R^{-1} of its wavelengths to radiate half of its energy into the atmosphere. As a function of time t, the energy E of free IGWs decreases exponentially: $E(t)=E(0)\exp(-t/T)$, where the e-folding time $T=[2c_g\mathrm{Im}(\delta k)]^{-1}$ is usually referred to as the decay time. From Equations (1) and (7), for the IGW group speed c_g and decay time, we find

$$c_g = \left(\frac{\partial \omega}{\partial k}\right)_H = \frac{\omega}{k}\cdot\frac{2kH + \sinh 2kH}{2\sinh 2kH} \tag{9}$$

and

$$T = \frac{g\left(k^2 - \omega^2 c_s^{-2}\right)}{R\omega^3|\mathrm{Re}(m)|}. \tag{10}$$

Here, k is the real-valued IGW wavenumber; small relative corrections of the order of R are neglected in Equations (7) and (10). IGW attenuation due to AGW radiation is small over the wave period, and the decay time is usually large since the air-to-water density ratio $R \ll 1$ (Figure 3). As expected, the decay time tends to infinity at the transition frequency ω_b, where $m = 0$.

In addition to radiation of atmospheric waves, which takes place only at frequencies $\omega < \omega_b(H)$, many other processes (viscous attenuation, bottom friction and percolation, scattering at small-scale bathymetric features, nonlinear interaction with other ocean wave types) can potentially contribute to IGW attenuation. Energy loss at reflection from coastlines is usually considered to be the dominant decay mechanism, at least for long-period IGWs such as tsunamis (Munk et al.1962; Munk 1963; Van Dorn 1984, 1987). Recent measurements by Rabinovich et al. (2013) give decay rates of 18.5 to 20.0 h and 20.5 to 21.5 h for the 2010 Chile tsunami at wave periods 6 to 20 min (wave frequency 0.8 to 2.8 mHz) and 20 to 60 min (wave frequency 0.3 to 0.8 mHz), respectively. For the 2011 Tohoku tsunami, the measured decay rates (Rabinovich et al.2013) are 17.0 to 19.0 h and 21.3 to 25.0 h at wave periods 6 to 20 min and 20 to 60 min, respectively. Similar decay rates were derived previously by Munk (1963) and Van Dorn (1984) from much scarcer data sets than those employed in the recent study by Rabinovich et al. (2013).

Comparison of the predicted contribution of the atmospheric wave radiation (Figure 3) and measured (Munk 1963; Van Dorn 1984; Rabinovich et al. 2013) IGW decay rates shows that (i) IGW energy loss due to AGW radiation is consistent with available experimental data; (ii) radiation of atmospheric waves is a non-negligible mechanism of attenuation of IGWs away from shore; and (iii) depending on the average ocean depth, AGW radiation can account for up to one third of the observed decay of IGWs with periods from 6 to 20 min.

EXPERIMENT

Methods

We use Dynasonde techniques to measure spectra of vertical velocity disturbances in the upper atmosphere. Dynasonde is a method of precision ionospheric radio sounding, which is based on detection of radio echoes, i.e., radio wave reflections from the inhomogeneous ionospheric plasma, in a broad frequency range of 1 to 20 MHz. Various analysis routines specific for Dynasonde applications perform inversions of measured parameters (range, angles of arrival, polarization, line-of-sight Doppler speed, amplitude) of the echoes to retrieve a number of ionospheric parameters (vertical profile of the plasma density, vector velocities of the plasma structures, spectra of small-scale irregularities, etc.) attributed to real altitudes (Wright and Pitteway 1999; Zabotin et al. 2006). The transition from the echo ranges measured in the so-called virtual space (where the difference between the group speed of the high-frequency (HF) electromagnetic waves in plasma and the vacuum speed of light is neglected) to actual spatial coordinates is essential for accuracy of our analysis. A component of Dynasonde data analysis software, the 3-D plasma density inversion procedure NeXtYZ (Zabotin et al. 2006), finds 3-D spatial location of the reflection point of each echo. The inversion procedure possesses the accuracy and sensitivity, which are necessary to visualize and measure wave activity in the upper atmosphere. Averaging of the vertical component of the Doppler velocities for all echoes reflected within a narrow altitude range during a sounding session produces a single data point for our analysis.

We have used data obtained with the Wallops Island VIPIR/Dynasonde HF radar system in 2013 to 2014 in the regime where the sounding sessions were repeated every 2 min. The data screening has been performed using web-based visualization tools both provided by the standard Dynasonde software suite (Zabotin et al. 2005) and developed specifically for this project. Figure 5 presents an example of the standard output of the visualization software: a day-long scan of the vertical Doppler speed values shown by color as a function of the time of the day and of the real altitude. One can see a typical transition from night to day conditions near 13 UT and a transition from day to night conditions near 21 UT. The slightly inclined color strips are indications of the traveling ionospheric disturbances with their phase fronts propagating downward and therefore most likely caused by the acoustic-gravity waves propagating upward. Several criteria were used for the data selection. First of all, errors in ionogram trace identification and screening made by autonomous analysis must be infrequent in the data sets to not influence the statistics. Second, data series used for the spectral calculations must be long enough (at least several hours) because the frequencies of the wave disturbances of interest are as low as 0.1 mHz. We have adopted the coupling model between ionized and neutral components of the atmosphere according to which the velocity of the ion motion is estimated as a projection of the AGW-induced velocity perturbations onto the magnetic field line direction (Nicolls et al. 2014). This means that our analysis requires data from the upper ionosphere (above the E region). On the other hand, we cannot go too high because we would like to minimize the uncertainties and complications associated with the AGW absorption due to air viscosity and thermal conductivity. Only daytime ionospheric sounding data can provide information about the desired 130- to 180-km altitude range. Ionospheric reflections from the altitudes below 150 km are often blanketed by the underlying regular and sporadic E layers. The gray lacuna in Figure 5 between approximately 14 and 20 UT and between roughly 110 and 140 km is an example of manifestation of such blanketing. One can tolerate infrequently missing data points due to, for example, occasional high radio interference at some narrow frequency bands or due to short-lasting sporadic layers. But long interruptions in data series should be avoided. That is why summer seasons, when distortions to the data caused by sporadic E layers are common, are not suitable for this analysis.

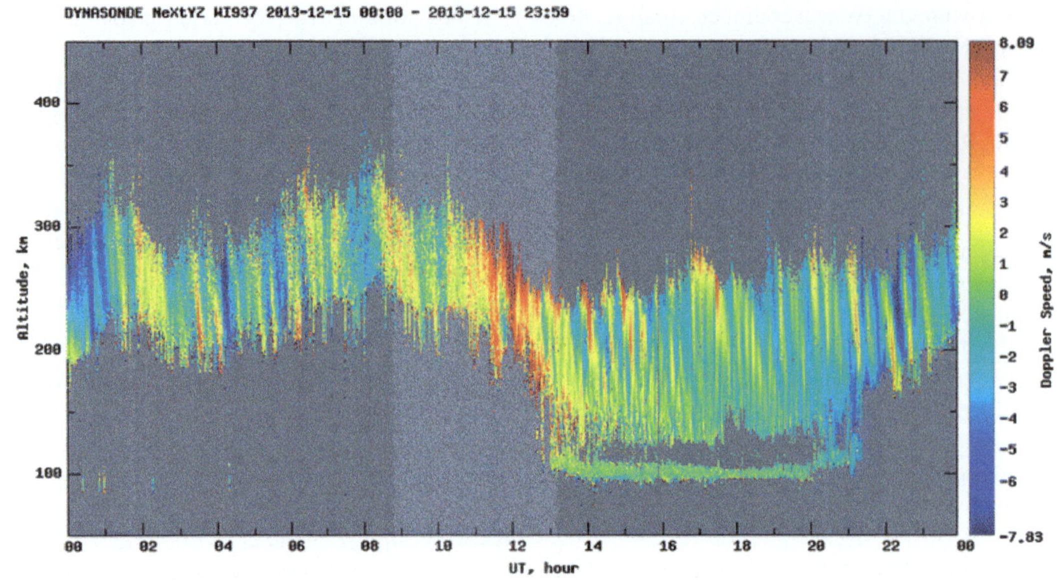

Figure 5 An example of the standard output of the Dynasonde analysis visualization software. A day-long scan of the vertical Doppler speed values is shown by color as a function of the time of the day (in UT hours) and of the real altitude (in km) for Wallops Island, Virginia, USA on 15 December 2013.

Combination of all the factors stated above makes 150 km the optimal altitude for data collection. Specifically, the following 20 time intervals, each one between 6 and 11 h long, were selected: 12 to 23 UT on 8 May 2013, 11 to 21 UT on 9 May 2013, 12 to 22 UT on 14 May 2013, 11 to 23 UT on 18 May 2013, 12 to 22 UT on 29 September 2013, 12 to 21 UT on 24 October 2013, 14 to 20 UT on 17 November 2013, 13 to 22 UT on 30 November 2013, 13 to 21 UT on 15 December 2013, 13 to 21 UT on 22 December 2013, 13 to 21 UT on 27 December 2013, 13 to 21 UT on 29 December 2013, 12 to 22 UT on 10 January 2014, 12 to 22 UT on 15 March 2014, 12 to 22 UT on 18 March 2014, 12 to 22 UT on 22 April 2014, 13 to 20 UT on 26 April 2014,12 to 21 UT on 26 September 2014, 13 to 22 UT on 29 September 2014, and 12 to 22 UT on 30 September 2014.

Every data point was obtained by averaging the vertical projections of the line-of-sight Doppler speed data for the 150- to 151-km altitude range. All the time intervals are affected by the diurnal variations of the Doppler speed; there is a background motion of the ionization layers downward after sunrise and upward before sunset. To suppress the spectral harmonics related to the diurnal variations, the Hann window was used for every data segment, with subsequent application of restoration coefficient 2 to the resulting power spectra. Some of the data points are missing, because of the reasons stated above, so the data series are inherently non-uniform. To account for this, the spectrum evaluation was performed with Hocke's (1998) modification of the Lomb-Scargle algorithm.

The experimental Doppler speed spectra were converted into the spectra of the ion movements by applying the factor $\sin^{-2}\varphi$, where φ is the magnetic dip angle. These are compared to the power spectra of the IGW-induced velocities parallel to the magnetic field line at an altitude of 150 km, which are calculated using the semi-empirical model spectra $S(\omega, H)$ (Godin et al. 2013) of the ocean surface displacements caused by the infragravity waves. The altitude chosen for the comparison is low enough (Hines 1960) to neglect effects of absorption of IGW-radiated acoustic-gravity waves, which have frequencies below approximately 3 mHz and horizontal spatial scales of tens and hundreds of kilometers. From AGW polarization relations (e.g., Godin2012), we find the relation

$$\langle v_m^2 \rangle = \langle v_3^2 \rangle \left[\sin^2\varphi + \frac{(\gamma-1)g^2\omega^{-2}-c_s^2}{2\left(c_s^2-\omega^2k^{-2}\right)} \cos^2\varphi \right] \tag{11}$$

between the wave-induced fluctuations v_m of the velocity along the magnetic line and the vertical velocity v_3 in a stationary atmosphere. Here, angular brackets denote statistical averaging. The first and second terms in the brackets in Equation (11) come from the wave-induced vertical and horizontal motions, respectively. We assume here that the IGW field is isotropic. The power spectrum $S_v(\omega, z)$ of the vertical velocity fluctuations v_3 at height z in the atmosphere is related to the spectrum of the ocean surface elevations as follows: $S_v(\omega, z) = \omega^2[\rho_a(0)/\rho_a(z)]S(\omega, H)$. The power spectra of the IGW-induced fluctuations v_m of the velocity along the magnetic line, which are calculated from $S_v(\omega, z)$ using Equation (11), are utilized below for comparisons with experimental v_m power spectra that are derived from Doppler speed measurements (Figure 4). Note that in a vertically stratified atmosphere with the horizontal winds, the total vertical power flux carried by an acoustic-gravity wave remains unchanged until the wave reaches a turning point, and existence and altitude of the turning points depend on the direction of wave propagation in the horizontal plane (Fritts and Alexander 2003). Therefore, the details of the vertical stratification are much less significant for the integral quantities considered in this paper than, for example, for calculating the spatial distribution of the wave field.

Observations

Atmospheric waves are known to be generated by a number of physical processes, associated most notably with convection, wind shear, and air-flow interaction with orography (Fritts and Alexander 2003; Geller et al. 2013). To assess the significance of the IGW contribution to the atmospheric wave field in the middle and upper atmosphere above oceans, we compare statistics of measured fluctuations of velocity with the velocity fluctuations due to random AGWs radiated by background IGWs in the ocean (Figure 4). To our knowledge, detailed measurements of velocity fluctuations in the middle and/or upper atmosphere above the ocean are not available at this time. As a proxy for the atmosphere above the ocean, we use experimental results obtained with HF radar systems at a coastal location in the North Atlantic.

Specifically, we use the data obtained with the VIPIR/Dynasonde HF radar system located at NASA's Wallops Flight Facility in Virginia, USA. Projections of the velocity fluctuations onto the magnetic field line direction at ionospheric altitudes and their power spectra, which are shown in Figure 4, are retrieved from Doppler shifts of radio echoes reflected by the ionosphere in the real altitude range of 150 to 151 km (Wright and Pitteway 1999). The choice of this altitude range is determined by specifics of the ionospheric radio sounding as explained in the previous section. Figure 2d shows that AGWs of oceanic origin propagate obliquely away from the ocean surface. AGWs observed by the radar would be radiated by free IGWs at different distances from shore depending on the wave frequency and bathymetry. For comparison with the measured spectra, in Figure 4, we present predictions for oceans with three different depths H, which encompass the range of the relevant H values in the vicinity of Wallops. The IGW contribution to velocity fluctuations in the atmosphere is expected to be a weighted average of spectra corresponding to different H.

Although measured velocity fluctuations include contributions of both atmospheric waves of various origins and turbulence, Figure 4 shows that the expected velocity fluctuations due to the waves of oceanic origin are within the range covered by the natural variability of the measured spectra. Comparison of the smoothed measured spectra (red lines in Figure 4) and theoretical predictions (black lines in Figure 4) indicates that there are extended periods of time when the bulk of the velocity variance at the thermospheric altitudes observed at Wallops can be attributed to AGWs of oceanic origin. These observations suggest the feasibility of direct measurements that would confirm correlation (coherence) between the wave phenomena in both media. A project of this kind is underway with a support from the Basic Research Challenge program of the US Office of Naval Research (Zabotin et al. 2013).

From tropospheric to mesospheric heights, wave processes are believed to be dominated by AGWs that are generated by moist convection, shear flow instabilities, jets, atmospheric fronts, and interaction of winds with the earth's topography (Gossard and Hooke 1975; Fritts and Alexander 2003; Geller et al. 2013). The power flux in these AGWs rapidly decreases with height due to viscous dissipation, wave breaking, and blocking at the critical levels (Gossard and Hooke 1975; Fritts and Alexander 2003; Geller et al. 2013). Only a very small (and unknown) fraction of the energy of these AGWs reaches ionospheric heights. AGWs of oceanic origin, and especially the AGWs radiated by IGWs in deep water, are characterized by considerably higher horizontal phase speeds, which can exceed the maximum wind speeds. This leads to larger horizontal spatial scales and, correspondingly, to weaker viscous absorption (Godin 2014). Fast AGWs do not have critical levels and escape the critical level filtering (Hines 1960; Gossard and Hooke 1975). Therefore, it should not be surprising that the AGWs of oceanic origin, which have small amplitudes and are hardly observable in the troposphere and stratosphere, play a much larger role at ionospheric heights.

CONCLUSION

Atmospheric loading leads to a qualitative change in properties of infragravity waves in the ocean. These waves cease to be surface ones and become leaky waves at frequencies below a certain depth-dependent transition frequency of the order of 3 mHz. The leaky waves continuously radiate a small fraction of their energy into the atmosphere as acoustic-gravity waves. The acoustic-gravity waves of oceanic origin propagate into the middle and upper atmosphere, where they appear to account for a significant part of the observed velocity fluctuations. As the atmospheric waves dissipate at high altitudes, they impact the atmospheric dynamics through deposition of mechanical energy and momentum transported from the ocean. Further research is needed to refine the simple model presented in this paper by account of geographic and seasonal variability of IGWs in the ocean and three-dimensional inhomogeneity of the atmosphere, in order to quantify the energy and momentum transport due to radiation of long atmospheric waves by background infragravity waves.

ACKNOWLEDGEMENTS

The authors are grateful to L. Zabotina for development of web-based visualization software for results of the Dynasonde data analysis. Insightful comments by Nozomu Nishitani and two anonymous referees helped to improve presentation. The work reported in this paper was supported by the Office of Naval Research through award N00014-13-1-0348. This project utilized the Janus supercomputer, which is supported by the National Science Foundation (award number CNS-0821794) and operated by the University of Colorado Boulder. The VIPIR/Dynasonde system is hosted by NASA's Wallops Flight Facility.

AUTHORS' CONTRIBUTIONS

OG and NZ conceived this project and wrote the paper. OG developed the analytical theory. NZ performed numerical calculations and Dynasonde data analysis. TB performed radar experiments and collected data. All authors read and approved the final manuscript.

REFERENCES

1. Artru J, Ducic V, Kanamori H, Lognonné P, Murakami M (2005) Ionospheric detection of gravity waves induced by tsunamis. Geophys J Int 160:840–848
2. Aucan J, Ardhuin F (2013) Infragravity waves in the deep ocean: an upward revision. Geophys Res Lett 40:3435–3439

3. Brekhovskikh LM, Godin OA (1998) Acoustics of layered media. 1: plane and quasi-plane waves. Springer, Heidelberg
4. Bromirski PD, Stephen RA (2012) Response of the Ross Ice Shelf, Antarctica, to ocean gravity-wave forcing. Ann Glaciol 53:163–172
5. Crawford WC, Webb SC, Hildebrand JA (1998) Estimating shear velocities in the oceanic crust from compliance measurements by two-dimensional finite difference modeling. J Geophys Res 103:9895–9916
6. Eckart C (1953) The theory of noise in continuous media. J Acoust Soc Am 25:195–199
7. Fritts DC, Alexander MJ (2003) Gravity wave dynamics and effects in the middle atmosphere. Rev Geophys 41:1003
8. Galvan DA, Komjathy A, Hickey MP, Stephens P, Snively J, Song YT, Butala MD, Mannucci AJ (2012) Ionospheric signatures of Tohoku-Oki tsunami of March 11, 2011: model comparisons near the epicenter. Radio Sci 47:RS4003
9. Garces MA, Perttu A, Williams B (2011) Infrasonic troposphere-ionosphere coupling in Hawaii, Paper SA23A-1897 presented at the 2011 AGU Fall Meeting, San Francisco, CA, USA, 5–9 December 2011
10. Geller MA, Alexander MJ, Love PT, Bacmeister J, Ern M, Hertzog A, Manzini E, Preusse P, Sato K, Scaife AA, Zhou T (2013) A comparison between gravity wave momentum fluxes in observations and climate models. J Clim 26:6383–6405
11. Gill AE (1982) Atmosphere–ocean dynamics. Academic, New York
12. Godin OA (2012) Acoustic-gravity waves in atmospheric and oceanic waveguides. J Acoust Soc Am 132:657–669
13. Godin OA (2014) Dissipation of acoustic-gravity waves: an asymptotic approach. J Acoust Soc Am 136:EL411–EL417
14. Godin OA, Fuks IM (2012) Transmission of acoustic-gravity waves through gas–liquid interfaces. J Fluid Mech 709:313–340
15. Godin OA, Zabotin NA, Sheehan AF, Yang Z, Collins JA (2013) Power spectra of infragravity waves in a deep ocean. Geophys Res Lett 40:2159–2165
16. Godin OA, Zabotin NA, Sheehan AF, Collins JA (2014) Interferometry of infragravity waves off New Zealand. J Geophys Res Oceans 118:1103–1122
17. Gossard EE, Hooke WH (1975) Waves in the atmosphere. Elsevier, Amsterdam
18. Herbers THC, Elgar S, Guza RT (1995) Generation and propagation of infragravity waves. J Geophys Res 100:24863–24872
19. Hines CO (1960) Internal atmospheric gravity waves at ionospheric heights. Can J Phys 38:1441–1481
20. Hines CO (1972) Gravity waves in the atmosphere. Nature 239:73–78
21. Hocke K (1998) Phase estimation with the Lomb-Scargle periodogram method. Ann Geophys 16:356–358
22. Komjathy A, Galvan DA, Stephens P, Butala MD, Akopian V, Wilson VO, Mannucci AJ, Hickey M (2012) Detecting ionospheric TEC perturbations caused by natural hazards using a global network of GPS receivers: the Tohoku case study. Earth Planets Space 64:1287–1294
23. Livneh DJ, Seker I, Djuth FT, Mathews JD (2007) Continuous quasiperiodic thermospheric waves over Arecibo. J Geophys Res 112, A07313
24. Makela JJ, Lognonné P, Hébert H, Gehrels T, Rolland L, Allgeyer S, Kherani A, Occhipinti G, Astafyeva E, Coïsson P, Loevenbruck A, Clévédé E, Kelley MC, Lamouroux J (2011) Imaging and modeling the ionospheric airglow response over Hawaii to the tsunami generated by the Tohoku earthquake of 11 March 2011. Geophys Res Lett 38:L13305
25. Munk WH (1963) Some comments regarding diffusion and absorption of tsunamis. In: Proceedings of the Tsunami Meeting, X Pacific Science Congress, IUGG Monographs, 24th edn. IUGG, Paris, pp 53–72
26. Munk WH, Miller GR, Snodgrass FE (1962) Long-period waves over California's continental borderland, III. The decay of tsunamis and the dissipation of tides. J Mar Res 20:119–120
27. Nicolls MJ, Vadas SL, Aponte N, Sulzer MP (2014) Horizontal parameters of daytime thermospheric gravity waves and E region neutral winds over Puerto Rico. J Geophys Res 119:575–600
28. Occhipinti G, Rolland L, Lognonné P, Watada S (2013) From Sumatra 2004 to Tohoku-Oki 2011: the systematic GPS detection of the ionospheric signature induced by tsunamigenic earthquakes. J Geophys Res 119:3626–3636
29. Peltier WR, Hines CO (1976) On the possible detection of tsunamis by monitoring of the ionosphere. J Geophys Res 81:1995–2000
30. Rabinovich AB, Candella RN, Thomson RE (2013) The open ocean energy decay of three recent trans-Pacific tsunamis. Geophys Res Lett 40:3157–3162
31. Rhie J, Romanowicz B (2004) Excitation of Earth's continuous free oscillations by atmosphere–ocean-seafloor coupling. Nature 431:552–556
32. Tsai VC, Ampuero J-P, Kanamori H, Stevenson DJ (2013) Estimating the effect of Earth elasticity and variable water density on tsunami speeds. Geophys Res Lett 40:492–496
33. Van Dorn WG (1984) Some tsunami characteristics deducible from tide records. J Phys Oceanogr 14:353–363
34. Van Dorn WG (1987) Tide gage response to tsunamis. Part II: other oceans and smaller seas. J Phys Oceanogr 17:1507–1516
35. Watada S, Kusumoto S, Satake K (2014) Traveltime delay and initial phase reversal of distant tsunamis coupled with the self-gravitating elastic Earth. J Geophys Res 119:4287–4310
36. Webb SC, Zhang X, Crawford W (1991) Infragravity waves in the deep ocean. J Geophys Res 96:2723–2736

37. Wright JW, Pitteway MLV (1999) A new data acquisition concept for digital ionosondes: phase-based echo recognition and real-time parameter estimation. Radio Sci 34:871–882

38. Zabotin NA, Wright JW, Bullett TW, Ye Zabotina L (2005) Dynasonde 21 principles of data processing, transmission, storage and web service. In: Proceedings of the ionospheric effects symposium 2005. Alexandria, VA, USA., pp 7B3-1–7B3-3

39. Zabotin NA, Wright JW, Zhbankov GA (2006) NeXtYZ: three-dimensional electron density inversion for dynasonde ionograms. Radio Sci 41:RS6S32

40. Zabotin NA, Godin OA, Bullet TW (2013) Tracing acoustic-gravity waves from the ocean into the ionosphere: ONR BRC Project., University of Colorado Boulder. http://surf.colorado.edu. Accessed 29 Dec 2014

CHAPTER 5

DEEP WATER OBSERVATIONS OF EXTREME WAVES WITH MOORED AND FREE GPS BUOYS

Takuji Waseda, Masato Sinchi, Keiji Kiyomatsu, Tomoya Nishida, Shunsuke Takahashi, Sho Asaumi, Yoshimi Kawai, Hitoshi Tamura and Yasumasa Miyazawa

Keywords: GPS wave sensor Deep ocean slack-moored buoy Wave hindcast Extreme waves Nonlinear wave Wave group

INTRODUCTION

Observing waves in the open ocean is still a challenge. The only instrument that can map the significant wave height globally is the satellite altimeter. However, altimeters cannot detect wave direction and the observation interval is rather long. Satellite synthetic aperture radar (SAR) can provide an estimate of the directional spectrum, but it is not practical to use SAR to monitor waves regularly. For these reasons, waves are mapped globally based on numerical wave model. Therefore, altimeter and wave forecast/hindcast data ought to be validated by moored wave riders or bottom mounted wave sensors located mostly in relatively shallow waters (e.g., NDBC buoys, Swail et al. 2010, Nowphas system in Japan, Nagai et al. 2005). There are, however, a number of meteorological and tsunami monitoring buoys in deep waters such as TAO array and DART buoys. Conceivably, these buoys can be used to measure waves by monitoring the motion of the platform assuming that the buoy follows the surface. A point-positioning GPS sensor was attached to the meteorological buoy (K-TRITON) of Japan Agency for Marine-Earth Science and Technology in the North-West Pacific in 2009 and in 2012 (Waseda et al. 2011a). The advantage of the GPS wave sensor over conventional accelerometer is the ease of the analysis. The accelerometer data can be contaminated by low-frequency noise and the high-pass filter applied to remove the noise can artificially enhance the extreme wave height (Collins et al. 2014).

Extreme waves or the freak waves have been studied extensively in the past few decades. Freak waves are statistically rare waves defined as waves exceeding twice or 2.2 times the significant wave height. Extreme waves, on the other hand, may refer to freak waves, freak and giant waves, giant but not freak waves, and possibly unexpected waves. In this paper, we use the term extreme waves referring to large waves but not necessarily exceeding twice the significant wave height. The key to understanding extreme wave generation mechanism is reliable observational evidence. One of the most well studied freak wave is the Draupner Wave observed in January 1 1995 (Haver 2004). However, the wave was measured remotely by down-looking laser which might be subject to uncertainty (Magnusson et al. 2013). Gigantic waves observed in Taiwan during passage of typhoon Krosa (Liu et al. 2008) was measured by an accelerometer which might produce anomalously large wave height if not properly processed (Collins et al. 2014). Therefore, a direct measurement of extreme wave by GPS sensor might become an attractive alternative for observing extreme waves offshore. The accuracy of the measurement depends on how well the platform follows the wave motion. Tulin and Landrini (2001) documented the kinematic properties of breaking wave and showed that when the particle velocity exceeds the group velocity the waves will inevitably undergo breaking. The implication is that even the orbital velocity of non-breaking wave can reach the group velocity without undergoing severe breaking event. Extreme waves are not necessarily a breaking wave, but the horizontal motion of the particle can accelerate to reach the group velocity or it might even exceed the group velocity. To understand the ability of a tethered wave to detect extreme waves, its horizontal motion will be studied.

The principle of the GPS wave measurement and dynamic analysis of buoy motion will be outlined in Section 2. The mean wave statistics from the 2009 and 2012–2013 observations will be compared with wave model estimates in Section 3. During the total of about 12 months of observation, two freak waves

Citation: Takuji Waseda, Masato Sinchi, Keiji Kiyomatsu, Tomoya Nishida, Shunsuke Takahashi, Sho Asaumi, Yoshimi Kawai, Hitoshi Tamura, Yasumasa Miyazawa, Deep water observations of extreme waves with moored and free GPS buoys, DOI: 10.1007/s10236-014-0751-4.

around 12 m in height and three giant waves of around 20 m in height were observed. The horizontal motion of the buoy will be analyzed including these giant waves in Section 4. The kinematic properties of the nonlinear waves inferred from the observations will be discussed in the context of wave tank experiment in Section 5. Conclusions follow.

Principle of GPS-based wave observations in the deep ocean

The first open-ocean buoy measurement was conducted from August 30 to December 6, 2009 utilizing the JAMSTEC K-TRITON buoy at the JKEO site (JAMSTEC Kuroshio Extension Observatory, 38°05′N, 146°25′E, and 5,400 m deep, phase 4, Fig. 1). The K-TRITON is a slack-moored buoy system whose mooring cable length is approximately 7,700 m long, the lower 2,500 m is buoyant, and the 700 m wire cable at the top and 4,500-m Nylon rope in the middle are sinkable (Fig. 2). In stagnant water, the cable is designed to form an S shape, but when there is a current field, the mooring cable can straighten completely and elongate. The second successful open-ocean observation was conducted at the New Kuroshio Extension Observatory (NKEO, 144.8E 33.8 N, Fig. 1) from June 20, 2012 to March 23, 2013. The particulars of the K-TRITON buoy and the mooring cables were nearly identical for the JKEO and NKEO systems[1].

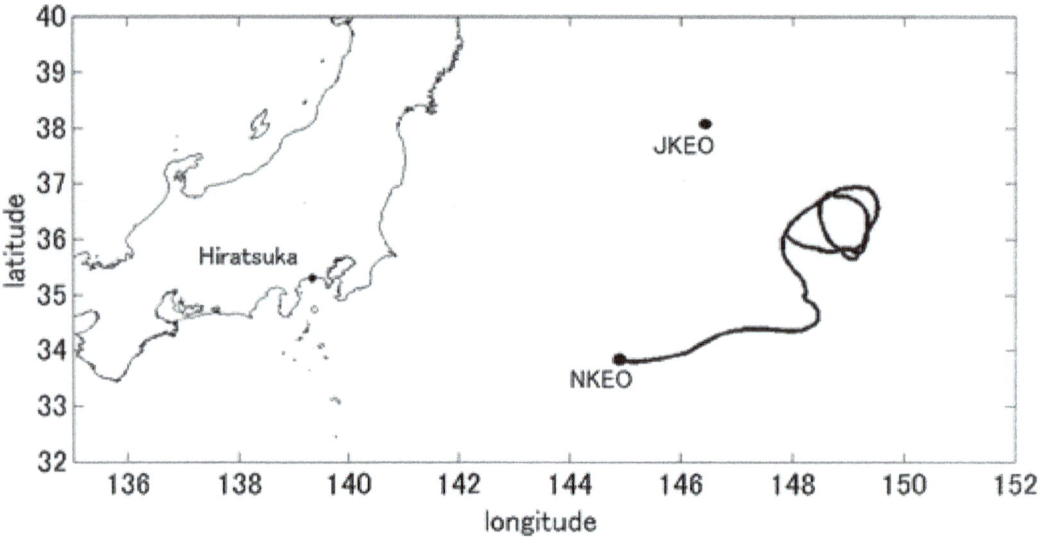

Figure 1 Locations of the JKEO (Jamstec Kuroshio Extension Observatory) and NKEO (New Kuroshio Extension Observatory) observation sites, and the Hiratsuka observation tower. A line originating from the NKEO buoy indicates the trajectory of the buoy which drifted from March 8 to 23, 2013

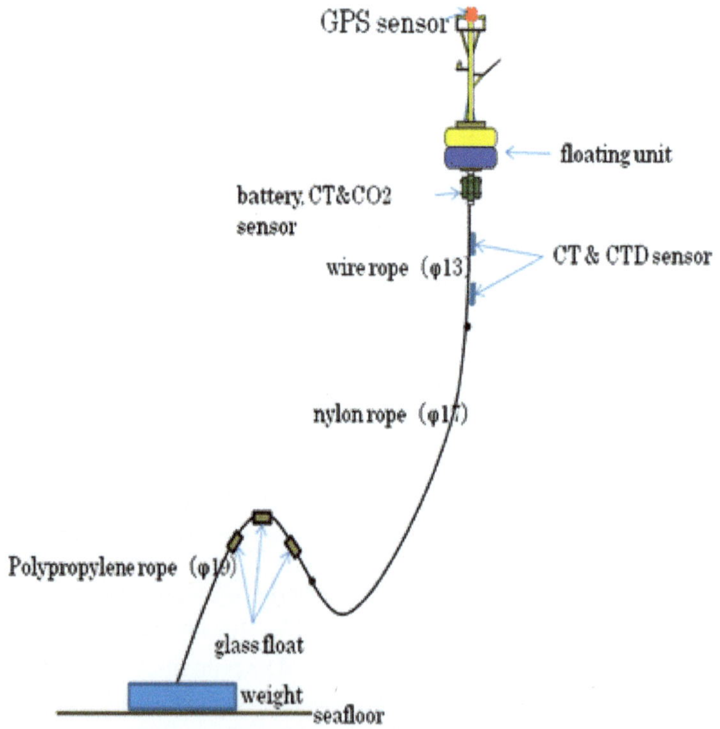

Figure 2 K-TRITON Buoy slack-mooring configuration. The top part (wire and nylon ropes) is sinkable and the bottom part up to the glass float (Polypropylene rope) is buoyant

The GPS longitude, latitude and altitude were recorded at 2.5 Hz for 20 min on the hour. The raw time series are recorded on storage and when an event occurs (e.g., significant wave height over 3 m), are transmitted by Iridium communication. The derived data that are processed on board are transmitted using Iridium satellite telemetry every hour. Thus the buoy motion during the observation period can be monitored by the GPS locations. When the NKEO mooring cable got cut loose in March 2013, the buoy started to drift away, and within 2 weeks, it was about 400 km apart from the original location (Fig. 1). During this time, the buoy was freely floating without a constraint of the mooring cable.

The open-ocean observation at JKEO and a reference measurement at stationary point on land (Table 1) with the same GPS sensor were analyzed to determine coefficients of filters to remove GPS noise and effect of unwanted pitch/roll resonance. Unlike the commonly used Real Time Kinematic (RTK) GPS wave sensor, the point-positioning GPS wave sensor does not require a reference station. Although infamous for large errors, it is known that the noise of the point-positioning GPS sensor is mostly confined to low frequencies. Taking advantage of the spectral characteristics of the GPS noise, Yamaguchi et al. (2005) proposed to use a high-pass filter to decipher the wave signal from GPS noise. A number of point-positioning GPS wave buoys were developed recently for the use in the open ocean (see Waseda et al. 2011a). A typical GPS noise spectrum from a stationary platform is shown in Fig. 3. The GPS noise level (red line) drops off at roughly f^{-2}, and at the frequency range of typical ocean waves, the noise level is 30 dB lower than the wave energy. Thus, a simple high-pass filter with a 30-s cutoff period would work to decipher the wave signal from point-positioning GPS record.

Table 1 Summary of observed wave parameters

Observation	**JKEO**		**NKEO**		**NKEO/drift**	
Period	August 30, 2009–December 6, 2009		June 20, 2012–March 8, 2013		March 8–23, 2013	
Number of records	366[a]		5,913		347	
Remarks (sections where analyzed)	Section 2, 3, 4		Section 3, 4		Section 3, 4	
	Zero-up	Zero-down	Zero-up	Zero-down	Zero-up	Zero-down
Number of freak waves	31	22	199	203	6	6
H1/3 H1/3‾	3.8 m	3.8 m	2.7 m	2.7 m	3.3 m	3.3 m
T1/3 T1/3‾	10.2 s	10.2 s	9.4 s	9.4 s	9.7 s	9.7 s
ak1/3 ak1/3‾	0.067	0.067	0.056	0.056	0.061	0.061
max $H_{1/3}$	8.4 m	8.3 m	13.3 m	13.4 m	8.1 m	8.3 m
max H	14.3 m	13.2 m	19.0 m	22.8 m	15.0 m	14.0 m
Number of records w. extreme waves of KC > 40	64	70	445	472	28	26
H1/3 H1/3‾	5.1 m	4.9 m	4.9 m	4.9 m	5.3 m	5.6 m
T1/3 T1/3‾	10.9 s	10.9 s	10.6 s	10.7 s	10.9 s	11.1 s
ak1/3 ak1/3‾	0.077	0.075	0.079	0.078	0.080	0.082
Hext. Hext.‾	8.4 m	8.2 m	8.2 m	8.1 m	8.9 m	9.3 m
Text. Text.‾	11.2 s	11.2 s	10.8 s	10.9 s	10.2 s	10.6 s
akext. akext.‾	0.145	0.141	0.153	0.148	0.175	0.168
(U/cp/ak)ext. (U/cp/ak)ext.‾	1.76	1.74	1.81	1.78	1.61	1.52

[a]The wave sensor failed in December 2009 and stopped transmitting the data but spontaneously recovered from March to September 2010. The statistical analysis presented in Section 2 includes 32 additional data obtained in 2010.

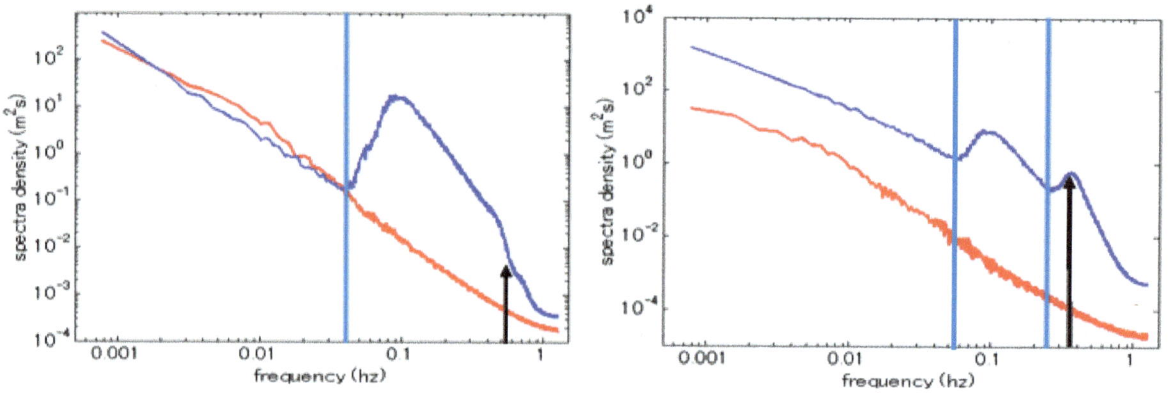

Figure 3 The noise spectrum of the point-positioning GPS heave signal (red line) and the averaged spectrum derived from the JKEO (JAMSTEC Kuroshio Extension Observatory) observation (blue line). Left figure is heave motion and right figure is East–west motion. The vertical blue lines are the cutoff frequency for the removal of GPS noise and roll resonance. The cutoff frequency of the high-pass filter for the heave motion is 0.03 Hz and those of the band-pass filter for the east–west motion are 0.045 and 0.28 Hz. Thearrows indicate the numerically estimated resonance frequency of heave and pitch/roll

The quality of the GPS wave observation depends on the Response Amplitude Operators (RAOs) or the transfer functions of the buoy. Because of lack of data and difficulty in including the effect of cable constraint, the RAOs were numerically estimated by empirically determining the unknown buoy parameter (see Appendix 1 for detail). The natural frequency of the estimated heave motion was around 0.55 Hz but the effect was hardly detectable in the observed heave spectrum (indicated by black arrow in Fig. 3). On the other hand, a noticeable peak around 0.36 Hz appears in the spectrum of the horizontal motion (east–west or north–south), indicated by black arrow in Fig. 3, right. It turns out that the anomalous peak appears as a result of pitch or roll resonance but not because of the surge or sway resonance. The natural frequency of the roll/pitch motion of the buoy explains well the observed anomalous peak in the spectrum (Appendix 1).

Based on these analyses, the filter coefficients were determined: the cutoff frequency of the high-pass filter for the heave motion is set to 0.03 Hz; the cutoff frequencies of the band-pass filter for the pitch/roll motion are set to 0.045 and 0.28 Hz. With these filters, the bias of the stationary record reduced to about 0.01 m and the root-mean-square error to be 0.02 m, thus, successfully removing the low frequency GPS noise error. The analysis conducted in this study did not apply the RAO corrections (Sinchi 2011). The impact of RAOs was mostly negligible for our observation (Appendix 1). In the range of 0.05 to 0.2 Hz, the heave amplitude response was nearly 1 and phase shift was negligible (5° at 0.47 Hz). On the other hand, the surge response amplitude gradually decreases to around 0.93 at 0.2 Hz.

For the analysis of horizontal velocity (Section 4), further quality control was made based on the estimated Keulegan-Carpenter number (KC number hereafter). Based on the estimated horizontal velocity of the buoy U for the given wave, the KC number is estimated as KC=UT/Dwhere T is the corresponding zero-up/down-crossing wave period and D is the diameter of the K-TRITON buoy (2.1 m). The threshold was set to KC=40 which limits the analysis of the horizontal motion of the buoy to waves of length 200 m in average (Table 1). Therefore, the horizontal excursion of the buoy is assured to be sufficiently large compared to the buoy itself.

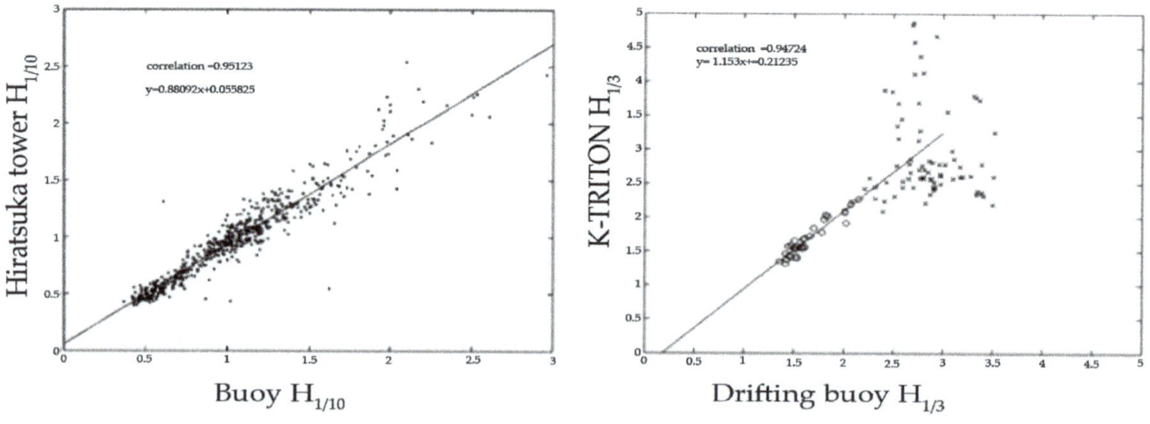

Figure 4 (Left) Scatter diagram of the 1/10 significant wave heights of the drifting GPS wave buoy compared with tower measurements (correlation 0.95). (Right) Scatter diagram of the 1/3 significant wave heights of the drifting GPS wave buoy and the K-TRITON buoy (correlation 0.95 when buoys are less than 100 km apart). The circles represent observations of waves when the distance between the buoys was less than 100 km

The observed wave heights were calibrated in two steps. First, a small drifting type GPS buoy was moored near an observation tower at Hiratsuka (see Fig. 1 for location) and the observed wave heights $H_{1/10}$ were compared to the ultra-sonic wave sensor data attached to the tower (Fig. 4, left). Then, the observed significant wave height, $H_{1/3}$, from the K-TRITON buoy was compared against $H_{1/3}$ from the GPS drifting buoy deployed simultaneously at the JKEO site. They compared well until the drifting buoy was over 100 km away (Fig. 4, right, Waseda et al.2011a). Therefore, the altitude observed by K-TRITON buoy seems to represent well the surface elevation. In the next section, we further validate the observed wave statistics.

OBSERVED WAVE STATISTICS AND COMPARISON WITH WAVE MODEL

To assure that the buoy had indeed followed the path of a water particle at the surface, basic wave parameters were estimated from the buoy position record of the JKEO. First, the statistics of the elevation were estimated (Fig. 5, upper left); the probability density function (pdf) is nearly Gaussian, and the pdf of the extremum is well approximated by the analytical formula of Cartwright and Longuet-Higgins (1956), Fig. 5, upper right. Thus, the observed surface elevation displays the characteristics of a random signal with narrow spectral bandwidth. The Fourier spectrum S(f) from the elevation (or the buoy altitude) records were ensemble averaged (796 degrees of freedom). The saturation spectrum $B(f) = S(f)f^4$ in the range of 0.1 to 0.3 Hz (or 3 to 10-s wave period) is nearly constant corresponding to the f^{-4} equilibrium spectral tail (e.g., Toba 1973), Fig. 5, lower left.[2] Consequently, the relationship between the non-dimensional significant wave height and the wave age agrees quite well with the Toba's 3/2 law with a slight difference in the value of the constant B (Fig. 5, lower right). From these comparisons, we can conclude that the free surface elevation is accurately traced by the moored K-TRITON buoy.

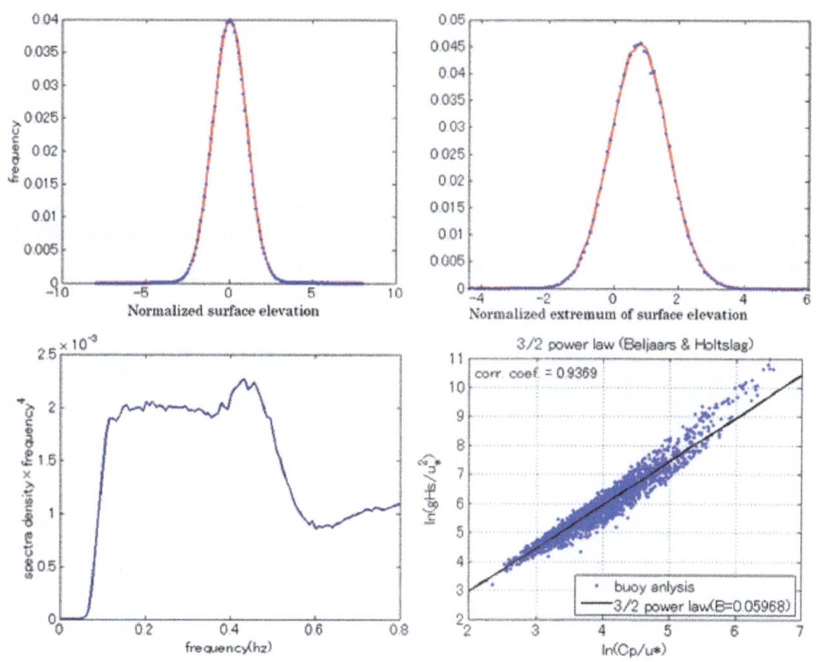

Figure 5 The probability density function (pdf) of the surface elevation (upper left) and of the extreme value of the surface elevation (upper right). The Gaussian pdf and the Cartwright and Longuet-Higgins theoretical estimate for the extreme value pdf are shown together in red line. Saturated spectrum $B(f) = S(f)f^4$ estimated from all the records is shown in lower left diagram (396 degrees of freedom). Frequency (in hertz) is shown on the horizontal axis and saturation on the vertical axis. The non-dimensional significant wave height is plotted against wave age in lower right diagram. The solid line is a fit corresponding to Toba's 3/2 law

Next, the observed significant wave height will be validated against a wave hindcast. The time records of the significant wave height ($H_{1/3}$) from the 2009 JKEO and 2012–2013 NKEO observations are plotted together with hindcasted H_{m0} in Fig. 6. JKEO $H_{1/3}$ is compared against the hindcast significant wave height H_{m0} of an original model and H_{m0} of the JMA operational Coastal Wave Model (Japan Meteorological Agency/CWM, e.g., Tauchi et al. 2007). The developed hindcast model (2007–2013) based on NOAA WaveWatchIII embeds the 0.1° × 0.1° (124.9–148.1° E, 27.9–44.1° N) Japan model within the 1° × 1° Pacific model (100–290° E, −65 to 65° N); the wave direction is discretized at 10°interval and the frequency is discretized for 35 frequencies at variable intervals between 0.0412 and 1.0521 Hz. First, the reanalysis and analysis wind products (ERA-interim, NCEP-CFSR, NCEP-GFS and JMA-GSM[3]) were validated against observed wind records (NDBC buoys, TAO-TRITON buoys, Nowphas buoys and JKEO buoys[4]), and then the Pacific wave model outputs forced by these winds were validated against the observed wave records (NDBC buoys, Nowphas buoys and JKEO buoy); the bias tended to be smallest with ERA-interim wind but the correlation was highest with JMA-GSM wind near Japan. Eventually, JMA-GSM was chosen to force the Pacific model and JMA-MSM was chosen to force the Japan model (Waseda et al. 2014). The hindcasted wave height compares well with the observations (Fig. 6), except at extreme events when the model tends to underestimate theH_s. The JKEO observation was compared against CWM estimates as well. The resolution of CWM is

0.05° around Japan (120–150° E, 20–50° N) and 72-h wave forecast was available. The time series of the significant wave heights among wave products compares reasonably well. The standard deviation (around 1 m), the centered root-mean-square differences and the correlation (around 0.95) are depicted in the Taylor diagram (Fig. 7). The centered root-mean-square difference is larger and the correlation is lower between CWM and the observation (point C in Fig. 7, left) than our simulation and the observation (point B in Fig. 7, left). The model was validated against NKEO observation as well (Fig. 7, right). The correlation is slightly smaller (0.90) than JKEO, and both the standard deviation (around 1.2 m) and centered rms difference are slightly larger than JKEO case, because of seasonal variation. Overall, we conclude that both the JKEO and NKEO wave observations compare reasonably well with the Hindcast simulations.

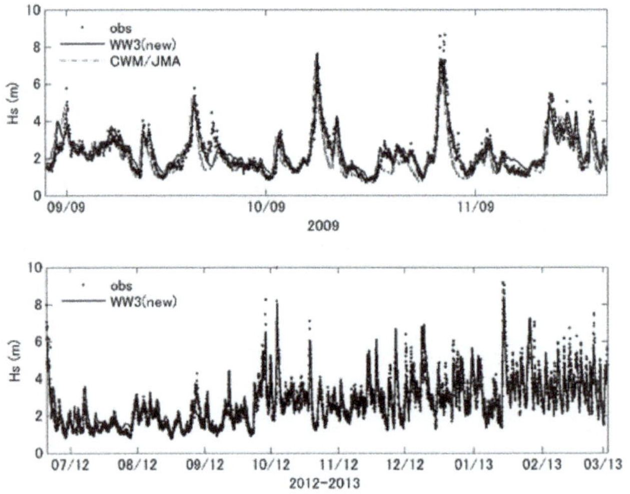

Figure 6 Time series of significant wave height compared against wave hindcast estimates. (top) The JKEO observation $H_{1/3}$ (dots) with original WaveWatchIII hindcast H_{m0} (solid line) and JMA/CWM hindcast H_{m0} (dash-dotted line). (Bottom) The NKEO observation $H_{1/3}$(dots) with original WaveWatchIII hindcast H_{m0} (solid line)

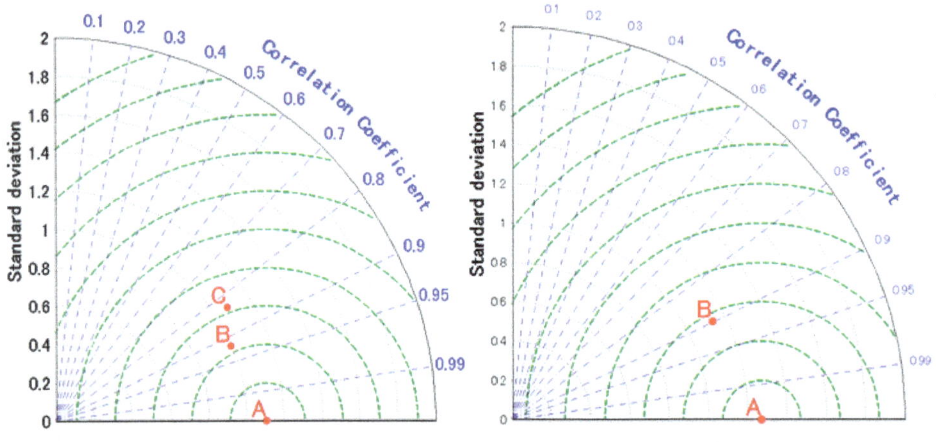

Figure 7 Taylor diagram comparing the significant wave height (H_s) from the observation (A) and wave hindcasts. In the left diagram, JKEO observed $H_{1/3}$ is compared against WaveWatchIII hindcast H_{m0} (B), and JMA/CWM hindcast (C). In the right diagram, NKEO observed $H_{1/3}$ is compared against the WaveWatchIII hindcast H_{m0} (B)

BUOY TRAJECTORIES AND INFERRED ORBITAL MOTIONS OF EXTREME WAVES IN A WAVE GROUP

Moored at a depth of about 5,700 m, the K-TRITON buoy at NKEO has moved as much as 7,000 m or equivalently 0.07°in latitude from the sinker location (Fig. 8, left). The colored dots in Fig. 8 are 20-min-long trajectories of the buoy position; the color represents maximum wave height of each record. The buoy position is not always on the circumference of the circle and is moving around within the circle. Similar buoy motion was observed at the JKEO K-TRITON buoy as well. During the observations in 2009 and in 2012–2013, a few notable extreme waves were observed. In October 26, 2009, 19:00 (UTC) and in October 27, 2009, 16:00 (UTC) during passage of typhoon 20, freak waves exceeding 10 m were observed; 12.3 m wave height ($H_{1/3}$ = 5.8 m) and 13.2 m wave height ($H_{1/3}$ = 6.6), respectively. These two freak waves have distinct directional characteristics, former being narrow and latter being broad (e.g., Waseda et al.2011b). On October 4, 2012, extreme waves of 22.8-m wave height ($H_{1/3}$ = 13.4), and 17.3-m wave height ($H_{1/3}$ = 10.3) were observed during passage of typhoon 19. And on January 14, 2013, an extreme wave height 17.7 m ($H_{1/3}$ = 10.0) was observed during passage of a bomb cyclone. These waves were not freak waves. The buoy trajectory of the largest waves observed by the K-TRITON buoy (22.8 m on October 4, 2012) will be studied in more detail.

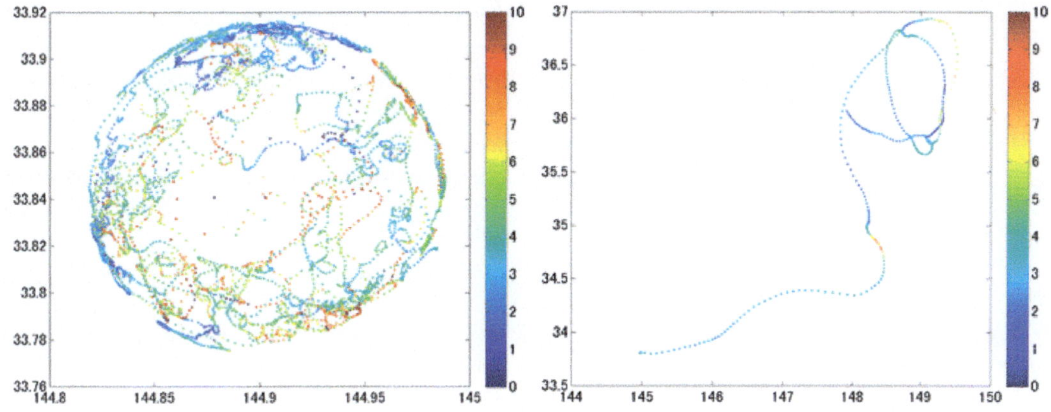

Figure 8 NKEO buoy locations during June 20, 2012–March 8, 2013 (left) and during March 8–23, 2013 (right). The color represents maximum wave height for each 20-min record. The buoy locations are distributed in a circle of approximately 6,000 m radius, in the left figure

The buoy trajectory from the 20-min record at 01:00 October 4, 2012 (UTC), shows an almost linear translation (~18 cm/s) and a random motion due to wave (Fig. 9, left). In the middle of the record, corresponding to the largest wave, the buoy makes a leap in its horizontal position. Comparing the time series of elevation, zonal position and meridional position, it is apparent that the buoy position shifted largely at the front-face of the wave (Fig. 9, right column). Between the zero-up-crossing point (circle) and the zero-down-crossing point (down-pointing triangle) in the surface elevation time series, the horizontal position changes about 36 m (10 m zonally and 34 m meridionally), which corresponds to 11.5 m/s horizontal speed of the buoy. Since the corresponding phase speed and wave period are 23.7 m/s and 15 s, respectively, the ratio of the maximum horizontal motion and the phase speed is 0.48. This means that the horizontal speed of the water particle reached group velocity. On the other hand, for the 17.3 m maximum wave height ($H_{1/3} = 10.3$) case which was observed just an hour later (01:00 October 4, 2012 [UTC]), the maximum horizontal speed was much smaller (4.8 m/s), Fig. 10. Nevertheless, because the steepness, ak, of the corresponding wave is around 0.17, the estimated horizontal speed exceeds that of the weakly non-linear wave $U = (ak)\,C_p$ where $ak = 0.2$.

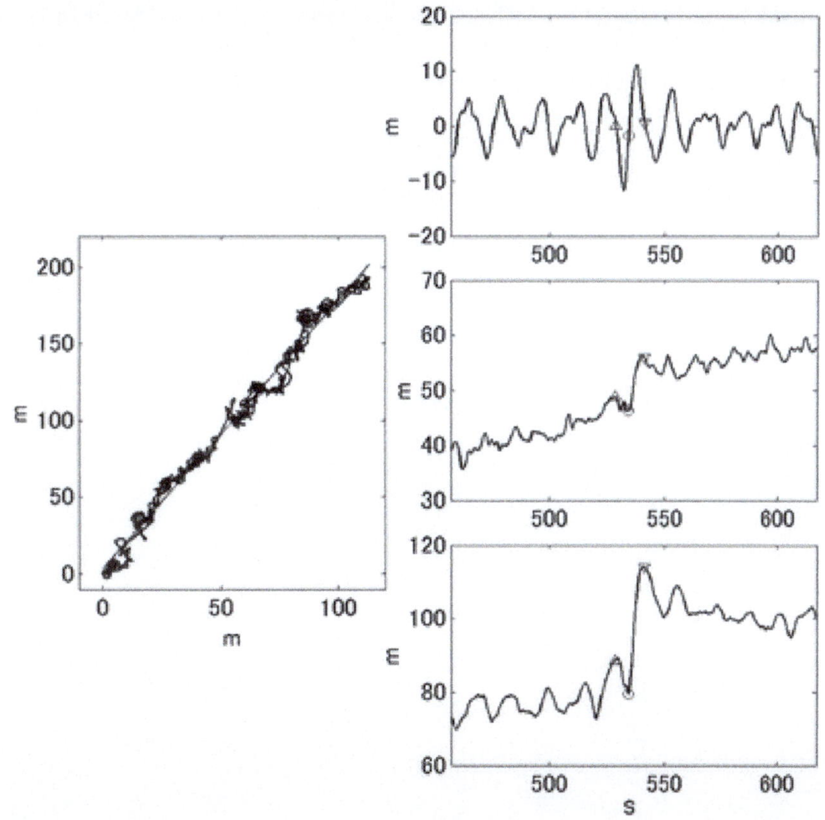

Figure 9 Plane view of the 20-min trajectory of the K-TRITON buoy when the maximum wave height of 22.8 m was observed (October 4, 2012, 1 AM UTC; left). Time series of the filtered surface elevation (top right) and un-filtered zonal (middle right) and meridional (lower right) positions. Upper and lower triangles correspond to the zero-up crossing points indicating the individual wave with largest down-crossing wave height in record

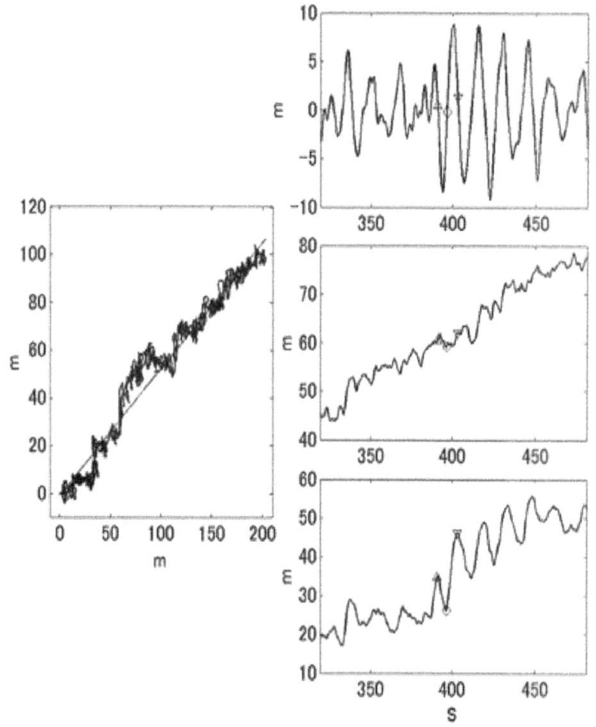

Figure 10 Plane view of the 20-min trajectory of the K-TRITON buoy when the maximum wave height of 17.3 m was observed (October 4, 2012, 2 AM UTC; left). Time series of the filtered surface elevation (top right), and un-filtered zonal (middle right), and meridional (lower right) positions

For a narrow banded wave system, a wave group is formed and the waves break when the individual wave passes through the peak of the envelope (Donelan et al. 1972). As the wave undergoes breaking, the horizontal velocity of the fluid particle accelerates and exceeds the phase speed. It is known that even for a non-breaking wave, a wave group is formed, and the velocity of the water particle exceeds the weakly non-linear orbital velocity (Tulin and Landrini2001). The maximum horizontal velocities of the highest zero-up crossing and down-crossing waves in the 20-min records were analyzed (Fig. 11). Note that the data presented here are quality controlled by the KC number as described in Section 2, hence cases where the buoy did not follow the orbital motion of the wave are removed[5]. Because the horizontal excursion of the buoy tends to exceed the wave height (upper left), the horizontal speed must be higher than the weakly non-linear orbital speed $U = (ak)\,C_p$, where ak is derived from individual wave height and period. The maximum horizontal speed was estimated from the buoy motion removing the slow translation speed due to current and wind (Fig. 11, upper right). The data does not show any indication of the wave height limiting the horizontal speed, hence assuring that the buoy motion is not constrained by the mooring cable. For each individual wave, the zero-crossing period is used to estimate the phase velocity. The maximum horizontal speed normalized by the phase speed U/C_p is mostly below 0.4, but was quite scattered and reached as high as 1 (Fig. 11, lower left). The U/C_p equals the steepness ak, according to weakly non-linear theory. Hence the exceedance of U/C_p to ak represents the degree of nonlinearity of the wave. The ratio U/C_p/akranges mostly below 3 but at times can reach almost 5 (Fig. 11, lower right). The peak of the distribution is slightly over 1 and therefore considerable amount of waves are nonlinear but not necessarily breaking since U/C_p is less than 0.5. The average value of U/C_p/ak was around 1.78 for the mean wave period of around 11 s. The steepness of these extreme waves were in average around 0.147, see Table 1. Because the significant steepness was around 0.078, occurrence probability of freak wave is considered to be normal. Thus, the current analysis indicates that the particle speed of extreme waves in a group, but not necessarily statistically rare, exceeds the weakly non-linear estimate of the orbital speed.

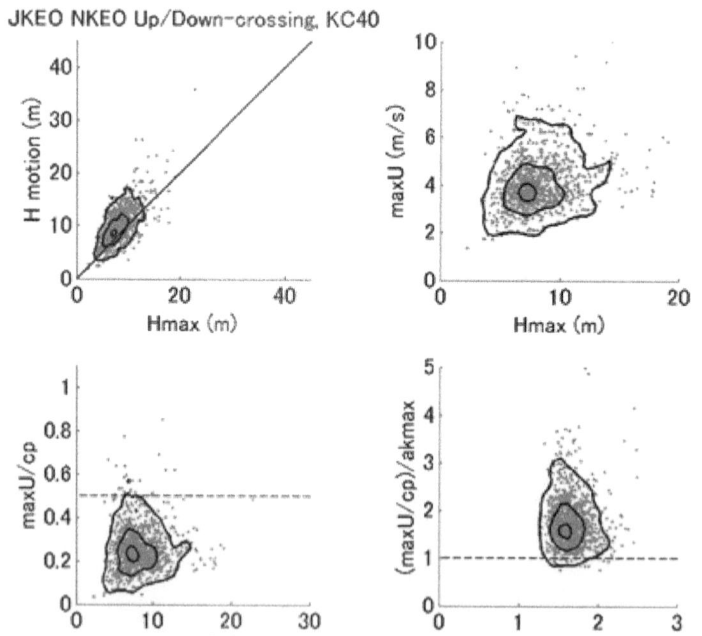

Figure 11 The magnitude of the horizontal excursion of the buoy corresponding to the largest wave in each 20-min record is plotted against maximum wave height (H_{max}) (upper left). TheU_{max} (upper right) and the U_{max}/cp (lower left) are plotted against H_{max}; note that the drift speed of the buoy is subtracted from the horizontal speed. The ratio of the normalized maximum wave height and the wave steepness of the corresponding wave is plotted against AI (abnormality index H_{max}/H_s; lower right). The contour lines indicate exceedance probability of 10, 50, and 90 %

In order to validate the analysis of the horizontal speed observed by the K-TRITON buoy, the same analysis was conducted for the period when the buoy was untethered and freely drifting (March 8–23, 2013). Apparently, the horizontal excursion of the buoy was much larger than the wave height (Fig. 12, upper left). As can be seen from the buoy track after the mooring cable got cut loose (Fig. 8, right), the translation speed of the untethered buoy is larger. To eliminate the ambiguity of the speed estimate caused by this large horizontal displacement, the translation speed due to buoy drift was subtracted. As a result, the distribution of the observed and normalized maximum horizontal velocity against maximum wave height (Fig. 12, upper right, lower left), and the relationship between the degree of nonlinearity $U/C_p/ak$ and the abnormality index $AI = H_{max}/H_{1/3}$ (Fig. 12, lower right), are consistent with the earlier analysis of data obtained when the buoy was moored. Since the number of data is significantly lower than the tethered case, the distribution at high values of U/C_p is sparse.

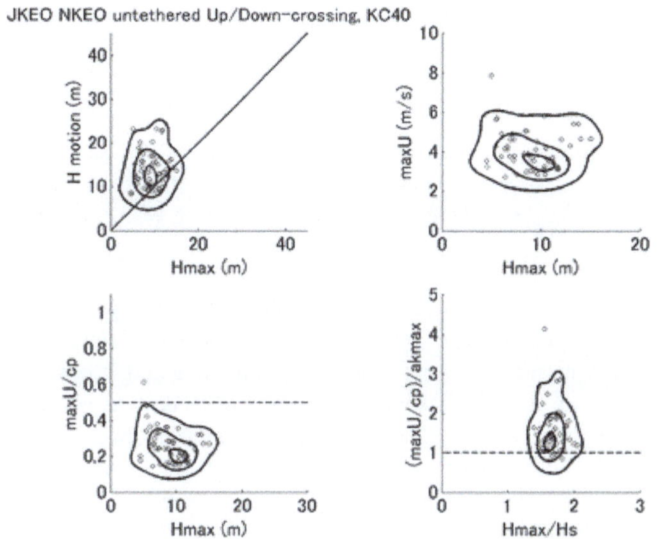

Figure 12 The same as Fig. 8 but for an untethered buoy during March 8–23, 2013

The tethered and untethered buoy records gave us a unique opportunity to evaluate the influence of mooring cable to the motion of the buoy in Deep Ocean. The comparison showed that the buoy motion is not constrained by the mooring cable. Dynamic analysis of the mooring system is necessary to assure this conclusion. We have conducted a preliminary computation with a finite element model but it was difficult because of the elasticity of the cable and possible loss of cable tension which made the system numerically unstable. An alternative method, if cable curvature is small, is to use a lamped-mass mooring cable model which is more stable.

DISCUSSION

When waves break, the speed of the particle at the crest accelerates to $U = O(C_p)$. As a result, the fluid particle orbit completely opens. Tulin and Landrini (2001) showed numerically that even if the waves are not breaking, the particle at the peak of a wave group accelerates. The fluid particle motion was visualized in a small wind-wave flume (10.0 m long, 60 cm wide, and 80 cm deep) at the University of Tokyo, Kashiwa Campus by tracking a marker floating on the free surface (Takahashi 2012). The float is a 5.8-mm-diameter polystyrene sphere of specific gravity 0.98. The images were taken 24 frames per second and the maker float was tracked digitally. The Stokes drift of a 1-m-long regular wave was estimated with an error of less than 1 %.

By controlling the steepness, uni-directional non-breaking wave groups and breaking wave-groups of carrier wave wavelength 1 m were produced in the flume following the method outlined in Tulin and Waseda (1999). Even when the waves are not breaking, the particle repeatedly accelerates and the orbit completely opens (Fig. 13, left). Magnitude of the horizontal excursion of the particle is much larger when the waves are breaking (Fig. 13, right). The estimated horizontal speed of the particle of the non-breaking case was 46.4 cm/s (37 % of phase speed) and of the breaking case was 82.8 cm/s (66 % of phase speed).

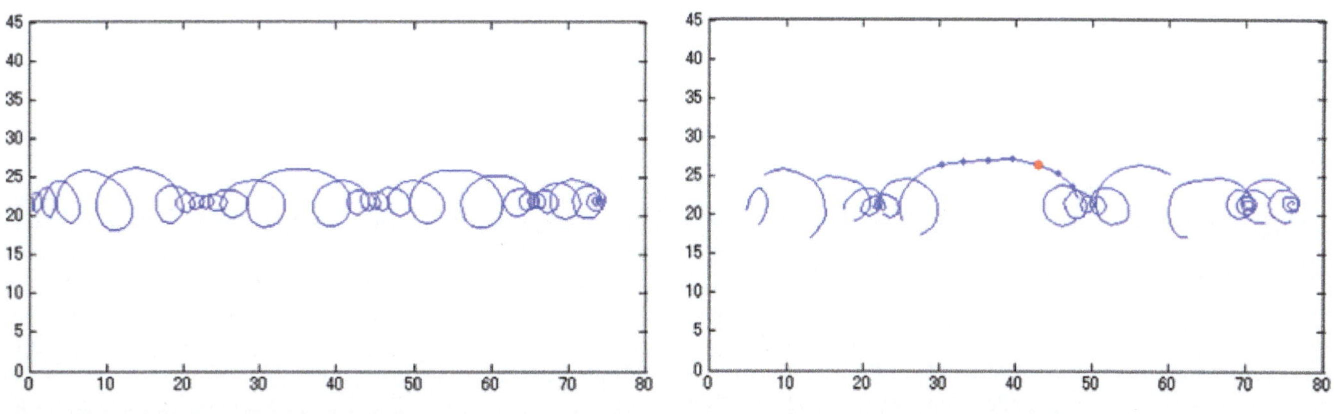

Figure 13 A particle trajectory of modulated wave train for non-breaking (left) and breaking (right) cases. The vertical and horizontal axes are length scale in centimeter

According to Tulin and Landrini (2001), the waves will inevitably undergo breaking when the particle speed exceeds the group velocity. To test this hypothesis, the measured particle speeds for all the cases ($ak = 0.05 \sim 0.2$) are plotted against the initial wave steepness of the modulated wave trains (Fig. 14). As the initial steepness increases, the particle speed accelerates at the peak of the modulation, and when the speed exceeds the group velocity, the waves broke (red circles). The measured particle speed did not reach the phase speed but the particle speed is much larger than the speed estimated based on the weakly nonlinear theory (solid line); the local wave amplitude was obtained numerically by the modified Nonlinear Schroedinger equation (mNLS, Trulsen and Dysthe 1990). In other words, the particle speed is larger than the quasi-linear estimate using the maximum amplitude of the envelope ($U_{observed} > (A(x,t)k)_{local} C_p$). The observed maximum wave amplitudes a_{max} reached about twice the initial wave amplitudes $a_0$but were less than (1+2√)a0(1+2)a0 corresponding to the analytical solution of NLS (Akhmediev et al. 1987) (figure not shown). Therefore, the tank experiment indicates that the particle velocity can reach a much higher value than that estimated from the local wave amplitude.

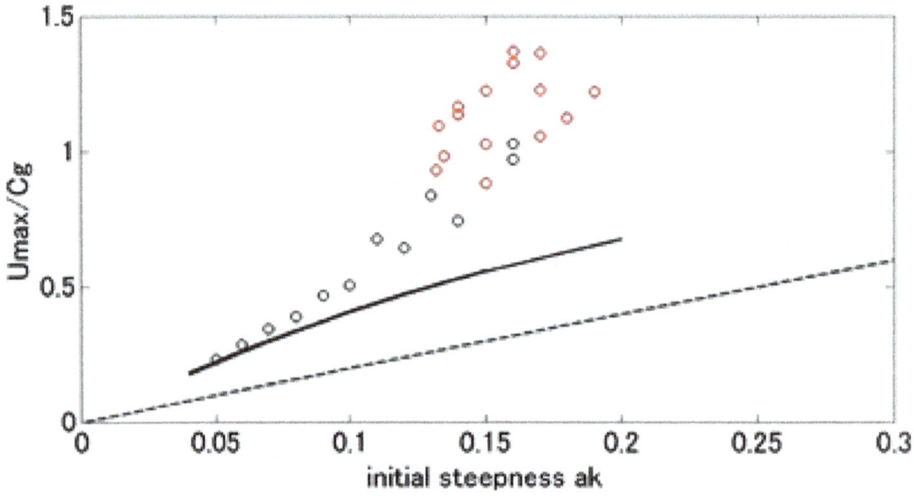

Figure 14 The normalized maximum horizontal particle speed of the uni-directional modulated wave train is plotted against the initial steepness. The circles are the measured values from the tank experiment; red denotes breaking events. The broken line corresponds to$U = (ak) C_p$ using the initial steepness; the solid line corresponds to $U = (ak)_{NLS} C_p$ where the steepness is based on weakly nonlinear estimate of modified Nonlinear Shroedinger equation

Because the experiment was conducted following a single float, it was unlikely to have recorded the largest particle speeds which can approach the phase speed. Likewise, the K-TRITON buoy would have likely missed observing the largest speed of the propagating waves. Nevertheless, the observed normalized horizontal speed of the buoy U/C_p reached as high as 4 to 5 times the wave steepness $(ak)_{local}$ (Fig. 11, lower right), which corresponds to maximum horizontal speed around 60 to 80 % of the phase speed. According to Tulin and Landrini (2001), waves whose particle velocity exceeded the group speed should undergo breaking. It is likely that the K-TRITON buoy had encountered a number of breaking waves. Whether the freak waves necessarily break or not is an open question. From our observations, a large population of data for the freak waves (e.g., $H_{max}/H_s > 2$) are distributed around $U_{max}/C_p/(ak)_{max} = 1.0$-2.0. Therefore, the observed freak waves were most likely not breaking.

CONCLUSION

A GPS sensor was attached to a slack-moored oceanographic/meteorological buoy in the North West Pacific near Japan at a depth 5,000 m. The buoy was not originally designed to measure ocean waves, but through data analysis and comparison with wave hindcast model, the buoy motion was proven to represent orbital motion of the ocean waves. In addition to the surface elevation, special attention was paid to the horizontal motion of the buoy. Conventional Eulerian observation by fixed sensors cannot measure the Lagrangian motion of the water particle. This study demonstrated the usefulness of Lagrangian wave observations based on GPS positioning of a moored, surface following buoy. By analyzing the highest waves in each 20-min records, we have shown that the fluid particle speed can accelerate at the peak of the wave group and far exceed the phase speed estimated by weakly nonlinear theory. The maximum wave height observed was 22 m and the associated horizontal speed was about 12 m/s. In the last few decades, freak waves have been studied extensively from a statistical point of view and the community seemed to have reached to a consensus on the significance of weak nonlinearity. However, what the sea farers care about is whether freak waves are dangerous “monster wave” or not. Whether such waves can be a threat to ships navigating in seas or offshore platforms wait for further research. This study demonstrated the usefulness of Lagrangian wave observation based on GPS positioning of tethered buoy.

Footnotes

1 http://www.jamstec.go.jp/iorgc/ocorp/ktsfg/data/jkeo/JKEO4mooring.htm

2 The slight reduction of B(f) at 0.35 Hz and increase at 0.45 Hz do not correspond to heave or pitch/roll resonance frequencies and are overly emphasized compared to the spectral shape shown in Fig. 3. The drop off beyond 0.5 Hz cannot be explained by the heave RAO. Thus, these characteristics most likely represent the true wave signal.

3 ERA-interim: European ReAnalysis-interim (Dee et al. 2011)

NCEP-CFSR: NCEP-Climate Forecast System Reanalysis (Saha et al. 2010)

NCEP-GFS: National Centers for Environmental Prediction—Global Forecast System (NCEP Office Note 442 2003)

JMA-GSM: Japan Meteorological Agency—Global (Mizuta et al. 2006)

[4] NDBC: National Data Buoy Center (Meindl and Hamilton 1992)

TAO: Tropical Atmosphere Ocean (McPhaden et al. 2010)

TRITON: Triangle Trans-Ocean Buoy Network (Kuroda and Amitani 2001)

Nowphas: Nationwide Ocean Wave information network for Ports and HArbourS (Nagai et al. 2005)

JKEO: Japan Kuroshio Extention Observatory (Tomita et al. 2010)

[5] For a typical laboratory particle tracking velocimetry, the KC number is over 100. The threshold value of KC number was set to 40 so as not to exclude too many data points for the analysis.

ACKNOWLEDGEMENTS

The first author acknowledges H. Tomita, A. Nagano, and K. Taniguchi of JAMSTEC for their assistance in conducting the K-TRITON buoy observation. He also acknowledges S. Komori and H. Yoshida of Zeni Lite Buoy Co. for their assistance in setting up the GPS wave sensor. The research was supported by Grant-in-Aid for Scientific Research. WAFO (2000) was used in the analysis.

APPENDIX 1: K-TRITON BUOY AND ITS RESPONSE AMPLITUDE OPERATOR

The K-TRITON buoy was developed based on the m-Triton system which is a low-cost lightweight buoy system developed for use in the Indian Ocean (Ishihara et al. 2010). The slack-mooring system (Fig. 2) was composed of three segments: a polypropylene rope with positive buoyancy in the lower part, a nylon rope with negative buoyancy in the upper part, and a wire rope segment directly beneath the buoy. The cable forms an S shape without external horizontal force. The particulars of the K-TRITON buoy and the overview of the JKEO site observations are summarized in:http://www.jamstec.go.jp/iorgc/ocorp/ktsfg/data/jkeo/index.html.

Ideally, the RAOs of each platform should be used to translate buoy motion to the wave signal. However, quite often the RAOs are not known, especially when the observational buoy is moored. Unlike discus buoys that measure the roll-pitch motion of buoy, GPS wave buoy measures the displacements directly. Therefore, the possible constraint of the horizontal motion by mooring cable is a concern.

The heave, pitch, and surge RAOs of the buoy were estimated using a conventional three-dimensional boundary element method and are shown in Fig. 15. In the range of frequencies of our interest (0.03–0.3 Hz), the gain of the heave is nearly equal to 1 and the phase difference is null. The pitch and surge motions are 90° out of phase, and their gain increase gradually with frequency. Pitch resonance frequency is around 0.36 Hz. Because the distribution of weight is not known exactly, to compute the RAOs, the radius of gyration K_{XX} and longitudinal metacentric height GML were adjusted; Fig. 15, lower right. From the possible combinations of GML and K_{XX} that produce the observed pitch resonance frequency of 0.36 Hz, GML = 0.5 K_{XX} = 0.65 were chosen.

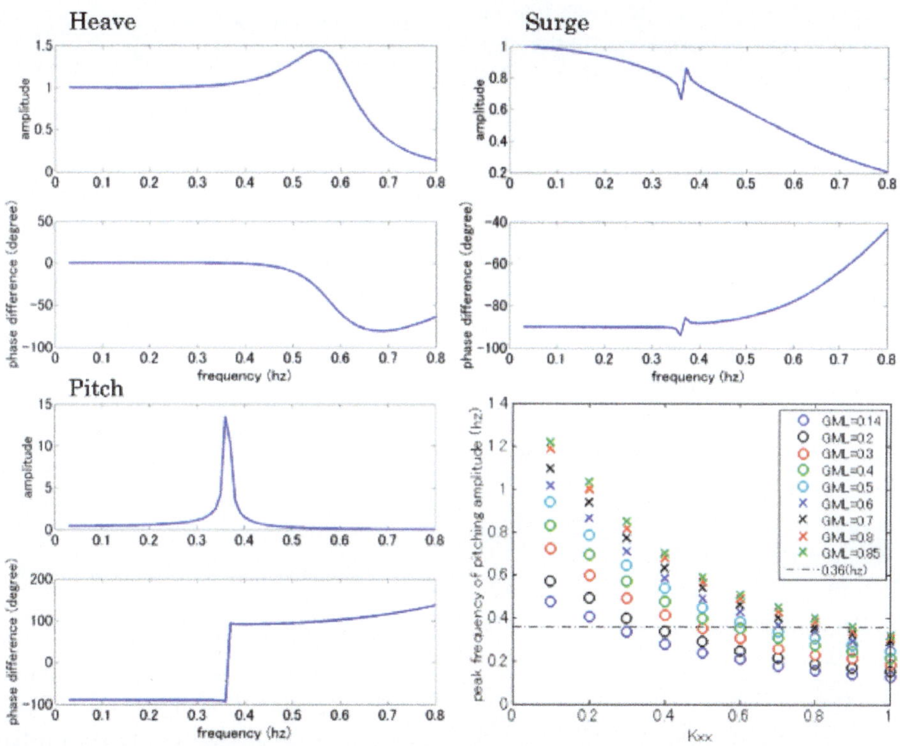

Figure 15 The Response Amplitude Operators of the heave, surge, and pitch motion of the K-TRITON buoy. The pitching resonance frequency of various combinations of radius of gyration and longitudinal metacentric height is plotted in lower right figure

When the buoy at the New KEO site (NKEO) was untethered because of cut mooring cable by accident from March 8 to 24, 2013, the resonant peak appeared at a much lower frequency and larger amplitude (Fig. 16). Our comparison of tethered and untethered buoy motion proves that the empirical adjustment of the radius of gyration effectively took care of the additional tension and damping effect of the mooring cable.

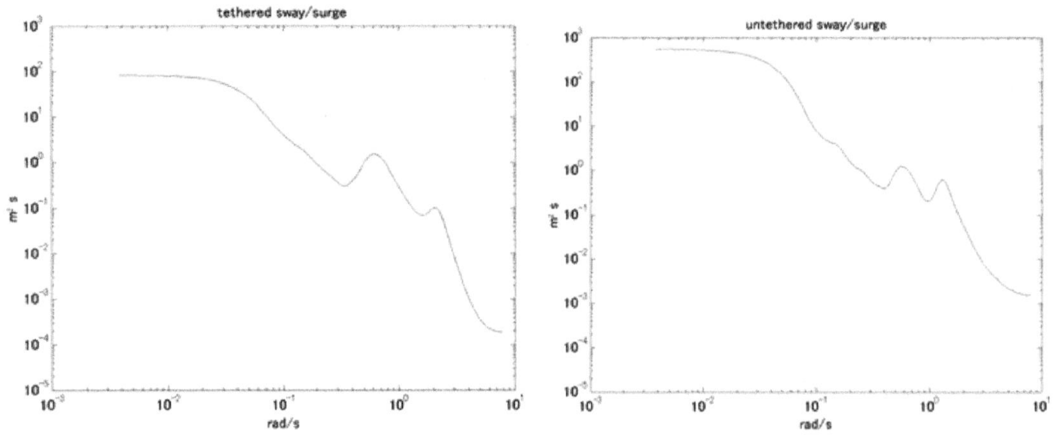

Figure 16 The spectrum of the east–west motion; tethered (left) and untethered (right) cases from the NKEO observation

REFERENCES

1. Akhmediev NN, Eleonskii VM, Kulagin NE (1987) Exact first-order solutions of the nonlinear Schrödinger equation. Theor Math Phys 72(2):809–818
2. Cartwright DE, Longuet-Higgins MS (1956) The statistical distribution of the maxima of a random function. Proc Roy Soc London Series A 237:212–232
3. Collins CO III, Lund B, Waseda T, Graber HC (2014) On recording sea surface elevation with accelerometer buoys: lessons from ITOP (2010). Ocean Dyn 64(6):895–904
4. Dee DP, Uppala SM, Simmons AJ, Berrisford P, Poli P, Kobayashi S, ..., Vitart F (2011) The ERA Interim reanalysis: configuration and performance of the data assimilation system. Q J R Meteorol Soc 137(656), 553–597
5. Donelan M, Longuet-Higgins MS, Turner JS (1972) Periodicity in whitecaps. Nature 239:449–451
6. Haver S (2004) A possible freak wave event measured at the Draupner jacket January 1 1995. Proc. Rogue Waves 2004
7. Ishihara Y, Yamaguchi M, Fukuda T, Matsunaga H, Murashima T (2010) m-TRITON and TRITON buoy system, Tropical Moored Buoy Implementation Panel—10, Scotland, 26 September, http://www.pmel.noaa.gov/tao/proj_over/tip/tip10_presentations.html
8. Kuroda Y, Amitani Y (2001) TRITON: new ocean and atmosphere observing buoy network for monitoring ENSO. Umi no Kenkyu 10:157–172 (in Japanese with English abstract)
9. Liu PC, Chen HS, Doong DJ, Kao CC, Hsu YJG (2008) Monstrous ocean waves during typhoon Krosa. Ann Geophys 26:1327–1329
10. Magnusson AK, MET-Norway B, Meteorology M (2013) Variability of sea state measurements and sensor dependence, Workshop: Statistical models of the Metocean environment for engineering uses, IFREMER 30.09-01.10.2013
11. McPhaden M, Busalacchi AJ, Anderson DLT (2010) A TOGA Retrospective. Oceanography 23(3):86–103. doi:10.5670/oceanog.2010.26
12. Meindl EA, Hamilton GD (1992) Programs of the National Data Buoy Center. Bull Am Meteorol Soc 73(7):985–993
13. Mizuta R, Oouchi K, Yoshimura H, Noda A, Katayama K, Yukimoto S, ..., Nakagawa M (2006) 20-km-mesh global climate simulations using JMA-GSM model—mean climate states. Journal of the Meteorological Society of Japan, vol 84(1), 165–185
14. Nagai T, Satomi S, Terada Y, Kato T, Nukada K, Kudaka M (2005, June) GPS buoy and seabed installed wave gauge application to offshore tsunami observation. In Proceedings of the Fifteenth International Offshore and Polar Engineering Conference (pp 292–299)
15. NCEP Office Note 442 (2003) The GFS Atmospheric Model, 14.,www.weather.gov/ost/climate/STIP/AGFS_DOC_1103.pdf
16. Saha S, Moorthi S, Pan HL, Wu X, Wang J, Nadiga S, ..., Reynolds RW (2010) The NCEP climate forecast system reanalysis. Bulletin of the American Meteorological Society, 91(8), 1015–1057
17. Sinchi M (2011) Data analysis of the ocean wave observation for solving the generation mechanism of freak wave, Master's thesis, the University of Tokyo, Graduate School of Frontier Sciences (in Japanese)
18. Swail V, Jensen R, Lee B, Turton J, Thomas J, Gulev S, Yelland M, Etala P, Meldrum D, Birkemeier W, Burnett W, Warren G (2010) Wave measurements, needs and developments for the next decade. Proceedings of the OceanObs, '09
19. Takahashi S, (2012) Particle motions of nonlinear water waves, Master's thesis, the University of Tokyo, Graduate School of Frontier Sciences (in Japanese)
20. Tauchi T, Kohno N, Kimura M (2007) The improvement of JMA operational wave models,ftp://ftp.wmo.int/Documents/PublicWeb/amp/mmop/documents/JCOMM-TR/J-TR-44/WWW/Papers/Full_WaveW2007Tauchi.pdf
21. Toba Y (1973) Local balance in the air-sea boundary processes. J Oceanographical Soc Japan 29(5):209–220
22. Tomita H, Kubota M, Cronin MF, Iwasaki S, Konda M, Ichikawa H (2010) An assessment of surface heat fluxes from J-OFURO2 at the KEO and JKEO sites, J Geophys Res -Oceans 115:C03018, 10.1029/2009jc005545.

23. Trulsen K, Dysthe KB (1990) Frequency down-shift through self modulation and breaking. In Water Wave Kinematics (pp 561–572). Springer Netherlands
24. Tulin MP, Landrini M (2001) Breaking waves in the ocean and around ships. In Twenty-Third Symposium on Naval Hydrodynamics
25. Tulin MP, Waseda T (1999) Laboratory observations of wave group evolution, including breaking effects. J Fluid Mech 378:197–232
26. WAFO-group (2000) "WAFO—A Matlab Toolbox for Analysis of Random Waves and Loads—A Tutorial" Math. Stat., Center for Math. Sci., Lund Univ., Lund, Sweden. ISBN XXXX, URLhttp://www.maths.lth.se/matstat/wafo
27. Waseda T, Sinchi M, Nishida T, Tamura H, Miyazawa Y, Kawai Y, Ichikawa H, Tomita H, Nagano A, Taniguchi K (2011a) GPS-based wave observation using a moored oceanographic buoy in the deep ocean, Proceedings, June, ISOPE-Maui
28. Waseda T, Hallerstig M, Ozaki K, Tomita H (2011b) Enhanced freak wave occurrence with narrow directional spectrum in the North Sea. Geophys Res Lett 38, L13605. doi:10.1029/2011GL047779
29. Waseda T, Asaumi S, Kiyomatsu K (2014) Improving resource assessment of wave power based on spectral wave model, OMAE 2014, June 9–13, San Francisco, USA
30. Yamaguchi I, Kasai T, Igawa H, Harigae M, Komori S, Shigenaga T, Hosaka Y (2005) Ocean wave sensing system using point-positioning GPS receiver, Space Engineering Conference, 14, 29–34, 2005-12-15

CHAPTER 6

OPTICAL DARK ROGUE WAVE

Benoit Frisquet, Bertrand Kibler, Philippe Morin, Fabio Baronio, Matteo Conforti, Guy Millot &Stefan Wabnitz

INTRODUCTION

During the last couple of decades, the study or rogue waves (also known as freak waves), which are high amplitude surface waves in open water, has attracted a great deal of attention since the first scientific measurement, by means of a laser beam, of an exceptionally large "Draupner wave" at an oil platform in the north sea on new year's day of 1995[1]. Although rogue waves are generally referred to as giant waves of exceptional elevation, it is known to sailors that rogue waves are often accompanied by deep troughs or, as the nautical lore goes, "holes in the sea". In spite of their extensive studies, the mechanisms at the origin of rogue waves remain to date largely unexplained. Yet, it is known that big waves form with greater regularity in ocean regions swept by powerful currents, e.g., the Agulhas around South Africa, the Kuroshio off Japan, the Gulf Stream off the eastern United States and in the famous Bermuda triangle. In these waters, surface waves meet currents running in the opposite direction. As a matter of fact, scientists who examined satellite images of the Agulhas current in the Indian Ocean recently revealed the formation of "eddies", or transient coherent whirlpools that remain intact for months[2]. These rogue waves have been renamed as oceanic black holes, since they represent the black holes of turbulence: they are fully analog to photonic gravitational black holes[2].

Quite remarkably, significant progress in our understanding of rogue wave formation has been enabled in recent years by simple photonic test bench emulators, thanks to analog models based on laser light propagation in optical fibers[3,4,5]. So far, optical rogue wave experiments have demonstrated different mechanisms for the generation of large amplitude transient optical waves, such as for example the Peregrine soliton[6]. In this work, we theoretically and experimentally consider the optical analog of crossing sea states, which is described by means of the coupled nonlinear Schrödinger equations (NLSEs)[7,8]. We provide what is, to our knowledge, the first experimental demonstration, by using standard telecom components and fiber, of the generation of an optical dark rogue wave (ODRW), which is a transient hole of light where the optical intensity is almost zero at its nadir.

The interest of our study extends beyond the field of rogue wave phenomena, since it permits us to speculate an intriguing connection between optical rogue waves in fibers and analogue gravity models. In recent years, technological advances in the field of photonics have stimulated researchers to develop a few analogues of general relativistic gravitational phenomena within compact lab benches, thus evidencing observations of event horizons, gravitational lensing and Hawking radiation[9,10,11,12,13,14]. With the aim of gaining new insights into inaccessible astronomical objects, the implementation of such powerful tools remains a major challenge when studying the coupling between different gravitational features[15]. Previous experiments were based on the observation of laser light propagation through an effective moving medium or space flow, which is obtained by exploiting the intensity-dependent refractive index or Kerr effect, for example induced by a co-propagating optical pulse.

In fact, a high-power femtosecond optical pulse was previously launched in the anomalous dispersion regime of a photonic crystal fiber to form an optical soliton[10]. Such pulse generated, via cross-phase modulation (XPM), a time-varying refractive index shift, hence providing an effective moving medium for the propagation of a frequency tunable quasi-continuous wave (cw) probe that was injected at a

Citation: Benoit Frisquet, Bertrand Kibler, Philippe Morin, Fabio Baronio, Matteo Conforti, Guy Millot & Stefan Wabnitz, Optical Dark Rogue Wave, doi:10.1038/srep20785.

widely different wavelength in the normal dispersion regime of the fiber. Dispersion engineering permits to match the group velocities of the soliton pulse and the probe, so that the probe is slightly slower (or faster) than the pulse. When a collision with the leading edge (trailing edge) of the pulse occurs, via XPM the cw probe experiences a frequency shift towards the red (blue). This leads to probe acceleration (slowing down), until a full matching of the pulse and probe group velocities is achieved. This corresponds to reaching "points of no return" – probe light cannot cross the pulse and an event horizon is thus realized. Although purely classical, such effect can be interpreted as mimicking the kinematics of light at a black hole or a white hole optical event horizon, respectively[10,12] (see Fig. 1b,c). Moreover, the collision between optical solitons and dispersive waves at event horizons has also been recently proposed as a possible mechanism for the generation of rogue wave events[16,17].

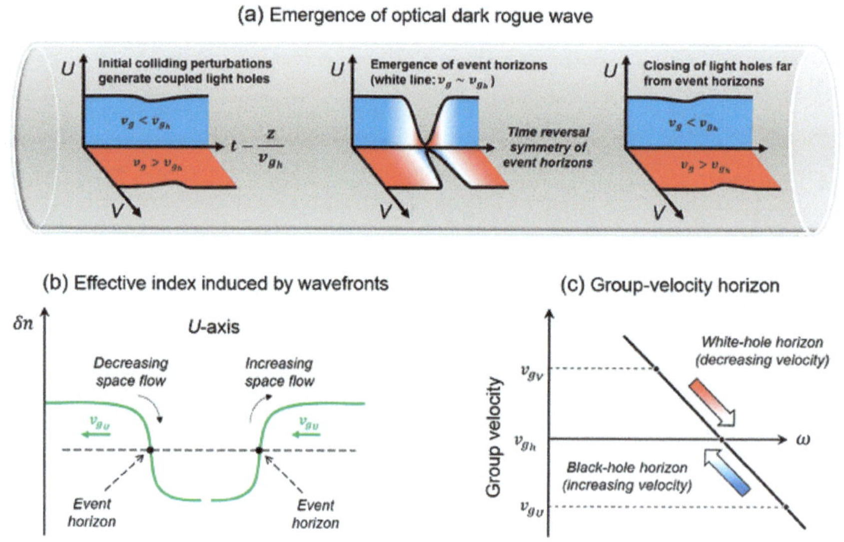

Figure 1: Illustration of the optical ODRW solution of the vector nonlinear fiber-analog system. (**a**) Evolution of the temporal wave profiles in the two propagation channels (orthogonal polarizations U and V) of the optical fiber, plotted in a reference frame moving with the group velocity at horizon. (**b**) Event horizons generated in the propagation channel U in analogy to the flowing river of space in astrophysical black and white hole horizons[12]. The gradient of the gravitational field across the horizons is analogous to the gradient of the refractive index induced by orthogonally polarized modulated wavefronts. (**c**) Corresponding laboratory-frame group velocity plotted as a function of frequency, channels U and V exhibit linear group velocities with opposite signs relative to the effective group velocity at horizons.

A complementary and fully equivalent description of the temporal collision between the soliton pulse and the weak probe may be provided in the frequency domain. The permanent red and blue frequency shifts of the probe that occur at the black and white hole horizons correspond to the frequency conversion of the probe signal into an idler[14]. Recent experiments have further demonstrated that it is not necessary to use high power femtosecond pulse sources to induce the time localized refractive index perturbations that provide the effective flowing medium for the probe[14]. In fact, a mode-locked short pulse laser source emits a frequency comb in the Fourier domain. An equivalent frequency comb may be dynamically generated along the fiber via cascaded four-wave mixing, by simply launching two intense cw pumps at the fiber input[14]. In this way analog event horizon phenomena can be reproduced by means of standard fiber optics telecom equipment.

So far, analogue gravity kinematics was limited to scalar optical wave propagation test models. On the other hand, there is currently much interest in the study of multimode nonlinear optical effects[18]. Complex nonlinear mode coupling interactions exhibit a much richer spectrum of excitations with respect to the scalar case, thus potentially expanding the range of space-time analogies, for example to the case of multi-particle interactions and quantum gravity[19].

In this work, we introduce the concept of multi-component analog gravity, by exploiting the nonlinear polarization interaction of light beams in a randomly birefringent telecommunication optical fiber. We point out that the generation of spatiotemporal localized event horizons is associated with the sudden emergence (and disappearance) of the ODRW solution of the coupled NLSEs. Our results may provide important physical insights in the phenomenology of other implementations of analog gravity models, e.g., in condensed matter physics and hydrodynamics[19,20,21,22,23,24,25].

To gain a simple intuition of the physical mechanism leading to multicomponent horizons in optical fibers, let us consider the following reasoning. As it is well known, if a light pulse propagates through a normally dispersive medium, its higher frequency components travel slower than its lower frequency components. Therefore the instantaneous frequency grows larger with time across the pulse profile. Moreover, if two pulses are injected in the fiber with different carrier frequencies, the high frequency pulse will travel with a slower group velocity with respect to the low frequency pulse. Suppose that the two frequency-shifted pulses are injected along orthogonal state of polarizations (SOPs). Optical fibers have the remarkable property that, if the relative frequency shift is sufficiently low, the two pulses maintain well their orthogonality along the entire propagation length, even though the local fiber birefringence varies randomly along the propagation distance and in time. In a spatially rotating frame of reference such that the fast linear evolution of the SOP along the fiber is removed, the two pulses represent linearly uncoupled propagation channels, where light moves at different speeds.

When using a reference frame travelling with the average of the velocities of the two polarization channels, the situation is completely analogous to the flow of water in two oppositely directed currents7·8. Now as the optical intensity of the pulses grows larger, the optical Kerr effect leads to a power-dependent contribution to the fiber refractive index. This induces an instantaneous frequency shift or chirp across the intensity profile of the pulses, hence, in combination with dispersion, to a temporal advance or retardation of the pulse fronts with respect to the average group velocity (which is zero in the travelling frame). Ultimately, the interplay of the Kerr effect and dispersion forces the light to stop at well-defined event horizons that surround a space-time localized ODRW as summarized in Fig. 1. In hydrodynamics, this situation corresponds to the growth of the crest or the deepening of the trough of a mass of water, as a result of a non-uniform water wave velocity profile and nonlinearity7·8. Quite remarkably, the nonlinear fiber optics situation is also completely analogous to the propagation of two-component Bose-Einstein condensates, which exhibit black or white hole horizon phenomena19·20·21·22·23.

Note that, in contrast with earlier fiber optics experiments on analog event horizons10·14, in our case there is no soliton-induced black hole or white hole horizon experienced by a probe pulse. On the contrary, a dark rogue wave is formed from the polarization modulation instability (PMI) of the two input orthogonally polarized and frequency shifted pumps. Yet, quite strikingly, we find that the ODRW may be associated with the presence of ephemeral (or spatiotemporal localized) event horizons.

RESULTS

Two-component Manakov system

The growth of temporal patterns such as bright solitons26 and rogue waves3·4·5·6 from an initial quasi-cw background is activated in optical fibers by the MI process. In the normal dispersion regime (which corresponds to the shallow water regime in hydrodynamics), there is no MI for scalar waves. However the nonlinear wave coupling between two orthogonal polarization modes may still lead to the MI of an intense cw field. In telecommunication fibers with a random rotation of the local birefringence axes, it is possible to average out the short-scale linear polarization oscillations in order to describe nonlinear pulse propagation effects27. This leads to the integrable, incoherently coupled NLSEs, also known as Manakov equations28. The propagation of two orthogonally polarized optical pumps at a relative frequency offset, say, $\Delta=\Delta\omega/2\pi$, is described by the two-component Manakov system

$$i\frac{\partial U}{\partial z}+i\frac{\delta}{2}\frac{\partial U}{\partial t}-\frac{\beta_2}{2}\frac{\partial^2 U}{\partial t^2}+\gamma(|U|^2+|V|^2)U=0$$
$$i\frac{\partial V}{\partial z}-i\frac{\delta}{2}\frac{\partial V}{\partial t}-\frac{\beta_2}{2}\frac{\partial^2 V}{\partial t^2}+\gamma(|U|^2+|V|^2)V=0 \quad (1)$$

Here z and t denote the propagation distance and retarded time (in a frame travelling at the group-velocity evaluated at the central carrier frequency ω_o) coordinates; U and V are the complex slowly varying amplitudes of the two pump waves at frequencies $\omega_{u,v}=\omega_o\pm\pi\Delta$, respectively, and $d=\beta_2$ D is associated with their group-velocity mismatch (GVM) owing to normal group-velocity dispersion. In fact, the U (V) pump is a slow (fast) wave with respect to waves at the carrier frequency ω_o. β_2 and γ are the group-velocity dispersion and the effective Kerr nonlinear coefficient at frequency ω_o, respectively. As discussed in the Supplementary Section, whenever two cws with different carrier frequencies and orthogonal SOPs are injected in a fiber, they are subject to polarization MI (PMI)29. As we shall demonstrate next, the evolution of PMI may lead to the creation of an ODRW (see Fig. 1a).

Polarization rogue wave

The variation of optical power and phase with time and space coordinates for the two orthogonally polarized, frequency-shifted waves composing an ODRW solution of the Manakov system is illustrated in Fig. 2. As can be seen, the power of both polarization components exhibits a sudden and isolated dip all the way down to almost zero (see panels (a-d)). Dark rogue wave solutions have a simple analytic representation as the ratio between two polynomials, which can be obtained by the Darboux dressing method based on the inverse scattering transform formalism29. Panels (e-g) in Fig. 2 show the time dependence of power, instantaneous frequency (or chirp) and frequency spectrum in both polarizations at the point where quasi-zero intensity is reached. Panel (f) shows that short-lived, very large frequency shifts (with ~3 THz amplitude, and with opposite signs for the two polarization modes) occur at the center of the ODRW. Note that the power spectrum of the high and low frequency polarization components down-shifts or up-shifts its center of mass, respectively, as shown in Fig. 2(g).

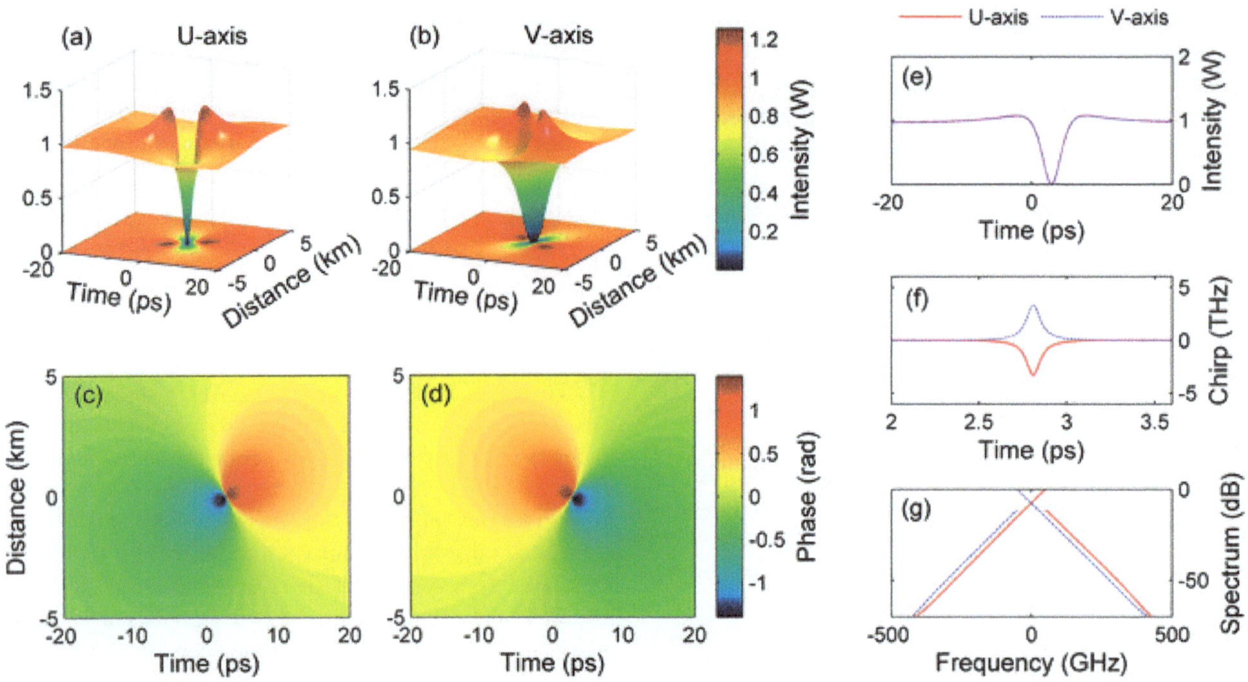

Figure 2: ODRW solution of the vector NLSE in the normal dispersion regime. (**a,b**) 3D and contour plots showing the dependence of optical power on time and propagation distance in the U and V polarization modes (note that the distance of the light hole center is arbitrarily shifted at the origin). (**c,d**) Contour plots of the phase in both polarization modes. (**e**) Time dependence of power in each polarization at z = 0, showing that zero intensity is reached at t near 2.8 ps. (**f**) Associated instantaneous chirp or frequency shift with opposite sign in each polarization. (**g**) Corresponding asymmetric power spectrum of the field in each of the two polarization modes.

Figure 3 provides the plot of the instantaneous (or local) light velocity $\Delta v_g(z,t)$ in each axis of polarization of the ODRW, as a function of propagation distance z. The velocity Δv_g is obtained as the sum of the constant linear group velocity at the frequency of each pump wave, and the local phase velocity induced by the intensity-dependent phase shift. As it can be seen, in each polarization around the notch where the light intensity is close to zero, there is a localized event horizon in space-time (black curves), where indeed light slows down (or speeds up) until it reaches the zero speed (or $\Delta v_g = v_{gh} = 0$). Note that the zero velocity horizons encircle the minimum intensity point at the middle of the notch.

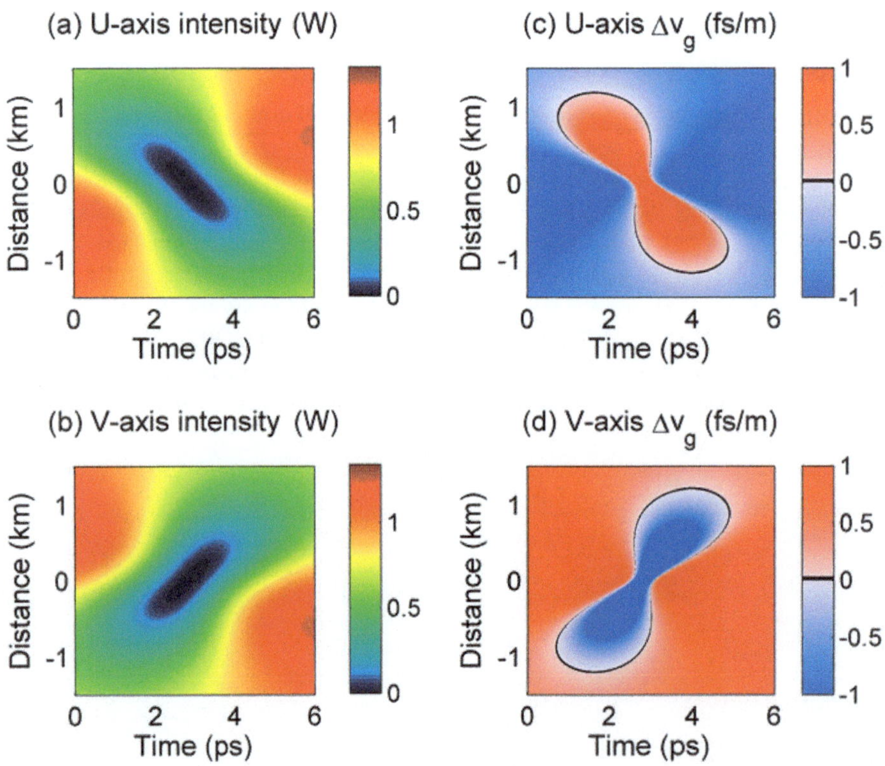

Figure 3. Instantaneous (or local) light velocity along the ODRW. Left column: intensity of orthogonal U and V components of ODRW solution of the vector NLSE in the normal dispersion regime for the pump power P = 1.9 W; Right column: corresponding plot of the instantaneous or local light velocity Δv_g versus propagation distance, showing that there is a localized event horizon in space-time (black curve) defined by

the condition $\Delta v_g = 0$, where indeed light slows down (or speeds up) until it reaches a zero speed. Note that the value range of color-bars is arbitrarily restricted to the range [−1,1] to better highlight the event horizon curve.

For the U polarization, away from the ODRW the light speed is negative, and equal to its linear value $\Delta v_g = -2/\delta$. When approaching the event horizon from the outside of the notch, the high-frequency (slow) wave with amplitude U is frequency down (or red) shifted, thus experiencing a speed boost or acceleration until it stands still as the horizon is reached (in the average group velocity reference frame). A transient black hole event horizon (a region in space time that light cannot cross) is thus formed. Similarly, for the low-frequency (fast) wave V, away from the ODRW the light speed is positive, and equal to its linear value $\Delta v_g = 2/\delta$. As the event horizon is approached from the outside, the intensity dependent frequency up (or blue) shift leads to a light slowing down, until a white hole event horizon results, where again light stands still. Remarkably, Fig. 3 also shows that inside the intensity notch in regions encircled by the horizons, the local light speed is positive for the U polarization and negative for the V polarization. Still, as the event horizons are approached from the inside of the notch, light stops and cannot exit from the hole. See Supplementary Section for more details about the analysis of emergence of event horizons as a function of the ODRW features.

Differently with the case of bright-pulse or soliton induced event horizons10, here the horizons are spatiotemporal localized curves, which surround the intensity notch of the ODRW (see Fig. 3). Therefore the red and blue frequency shifts that are observed at the ODRW event horizons are of transient character, and they fully disappear after the ODRW has vanished (for better overview of the full wave kinematics, see also the spectrogram evolution for both polarization modes along propagation in the supplementary video).

Experimental setup

We used the experimental configuration of Fig. 4 for creating a ODRW of light in the optical fiber. The two colliding cw pumps (red and green lines) were provided by wavelength-division-multiplexed, orthogonally polarized diode laser pumps. To induce the initial intensity perturbations, we used an electro-optic modulator (i.e., seeding stage). The orthogonal, frequency-shifted and modulated pumps were amplified and injected in a 3 km long standard telecommunication fiber with normal chromatic dispersion, and very low polarization mode dispersion. The last condition is important to guarantee the strong linear polarization mixing that is necessary to arrive to the Manakov system. For further details about the experimental setup, see the Methods section. To generate an optical ODRW, it is necessary that the input modulation breaks the stability of the orthogonal polarization pumps. The stability analysis of the Manakov system leads to the small-signal sideband gain shown in Fig. S1a: when the total pump power P grows larger than a certain critical value $P = P_0$, the band of unstable sideband frequency shifts extends from the pump spacing down to zero, the baseband MI condition. Correspondingly, perturbations with an arbitrarily low frequency still experience exponential gain, albeit at progressively lower rates. It turns out that it is precisely such condition that enables the growth of an isolated ODRW30. In our experiments, the initial modulation frequency (35 GHz) was chosen to be half of the value that leads to peak PMI gain. This choice stems from the trade-off of imposing the slowest modulation on the one side, while still having sufficient sideband gain to observe the emergence of the ODRW within the 3 km fiber length. Such value of fiber length was also chosen to satisfy the requirement that fiber losses have virtually no impact on the propagation dynamics.

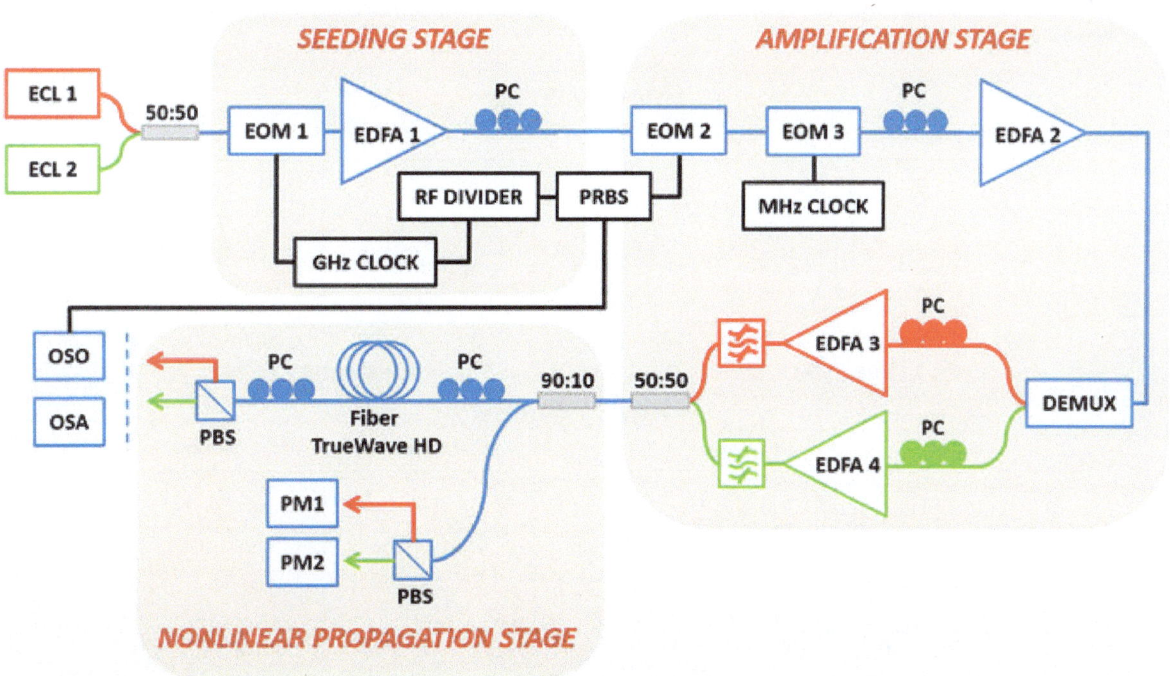

Figure 4: Experimental setup, red and green lines depict the two wavelength-division-multiplexed and orthogonally polarized pumps. ECL: external-cavity diode laser; 50/50 & 90:10: fiber couplers; EOM: (intensity or phase) electro-optic modulator; EDFA: Erbium doped fiber amplifier; PRBS: pseudo-random binary sequence generator; PC: polarization controller; PM: power-meter; OSO: optical sampling oscilloscope; OSA: optical spectrum analyzer.

Experimental results

To validate the use of the Manakov system, first we carried out an extensive analysis of spontaneous (i.e., quantum-noise induced) PMI in the fiber. Experimental spectra showing the emergence of noise sidebands, which agree quantitatively well with the predictions of the linear stability analysis, are shown in the Supplementary Section. When the modulating signal is present, the nonlinear reshaping of the counter-propagating (in a reference frame which travels with the average group velocity of the two pumps) polarizations leads to the generation of a time periodic train of black intensity notches, as shown in the numerical intensity plots of Fig. 5a,b. Quite remarkably, Fig. 5c,dshow that each of the generated notches closely matches the shape of a single, isolated ODRW which corresponds to the exact solution (see alsoFig. 5e,f). This is clearly confirmed by the experimental results of Fig. 6: here we compare input and output (after 3 km of optical fiber) intensities and power spectra from the experiments with an input periodic intensity modulation, and their corresponding analytical ODRW solutions. As can be seen, an overall excellent quantitative agreement (with no adjustable parameters) is obtained between theory and experiments both in the time and in the frequency domains. Only slight discrepancies appear due to the non-ideal initial conditions used for the generation of the ODRW. The observed spectral asymmetry provides a clear signature of the strong nonlinear chirping which prevents light from exiting the dark notch.

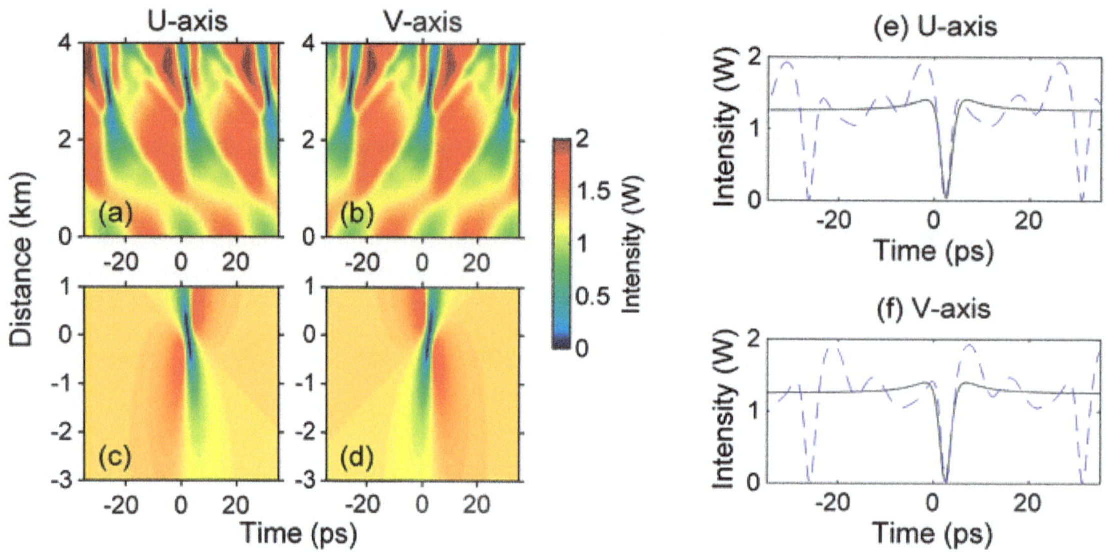

Figure 5: Contour plots of optical intensity in orthogonal polarization modes. **(a,b)** numerical solutions showing the generation a periodic train of optical dark rogue waves from an initially temporally modulated background (note that distance of light hole center is arbitrarily shifted at origin). **(c,d)** Exact solution of an isolated dark rogue wave. **(e,f)** Comparison of temporal intensity profiles obtained in the previous panels at the dark rogue wave centers.

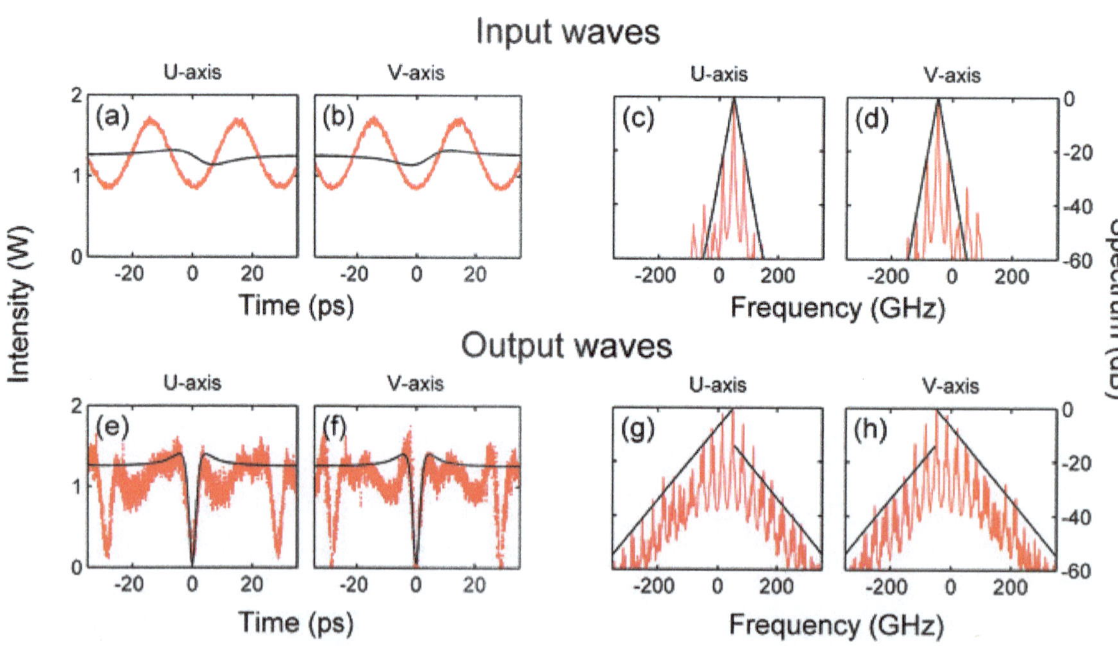

Figure 6: Experimental observation of the ORRW. **(a,b)** Temporal profiles of power in the U and V polarization modes at the fiber input (red solid traces). **(c,d)** Power spectra at the fiber input. **(e,f)** Output intensities after 3 km of optical fiber. **(g,h)** Power spectra at the fiber output. Solid black lines represent the analytical dark rogue wave solution.

In addition to Fig. 6 showing the comparison between experimental results and the corresponding analytical dark rogue wave solution for each of the polarization modes, Fig. 7 reports the comparison for the sum of optical intensities of both polarization modes. Panel (a) highlights the temporal symmetry during the evolution of the total intensity of the ODRW along the propagation distance, which reveals a constant average group velocity. As can be seen in panels (b) and (c), both temporal and spectral profiles of the total optical intensity at the dark rogue wave center are symmetric. The observed spectral symmetry provides a clear signature of an effective event horizon, formed by coupled black- and white-hole horizons. The sum of optical intensities of both polarization modes can be considered as the global physical picture of the dark rogue wave.

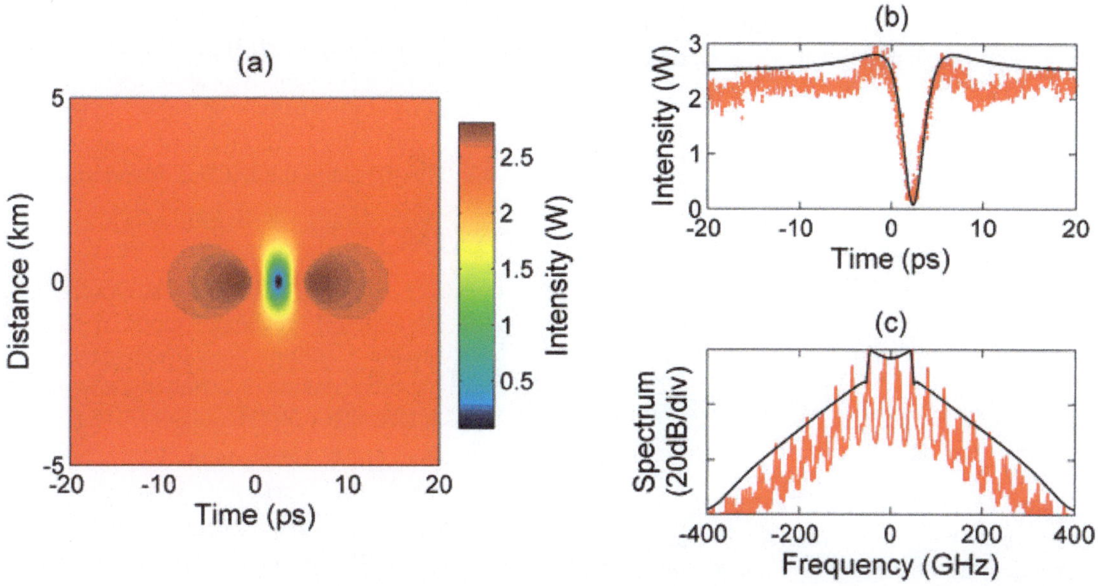

Figure 7 (**a**) Contour plot of the total intensity of the ODRW solution, note that distance of the dark notch center is arbitrarily shifted at the origin. (**b,c**) Comparison between analytical (black lines) and experimental results (red lines) at the optical ODRW center in both time and spectral domains, respectively.

DISCUSSION

In summary, by using standard telecom components, we have been able to provide the first experimental observation of an optical dark rogue wave, that is, a spatio-temporal localized hole of light that becomes nearly black at its middle. This result opens the way to the challenging frontier of rogue wave research on multi-component wave systems. For example, consider two interacting Bose-Einstein condensates, which are described by a pair of Gross-Pitaevskii equations formally identical to the Manakov system (1). The two-component NLSEs (1) are also obtained in oceanography as a result of the crossing of two opposite currents, which may lead to the controlled water-tank based experimental demonstration of water dark rogue waves.

The presence of event horizons connected with the dark rogue wave suggests us to the possibility to introduce a multi-component analog model for describing the light kinematics around gravitation-induced black or white holes. The two polarization components of the ODRW could provide the optical analogs of gravitational black or white holes, originating from fiber nonlinearity-induced horizons experienced by colliding and orthogonally polarized modulated wavefronts. The light wave kinematics of the ODRW exhibits a fundamental difference with respect to the soliton induced horizons which have been investigated before. Indeed, the present black and white hole horizons are localized both in time and in space, as they are inherently linked with the transient appearance and disappearance of the ODRW, and not with a space stationary object such as an optical (bright or dark) soliton. Hence, in stark contrast with soliton-induced horizons that lead to permanent frequency conversions of probe beams, the red and blue frequency shifts that are associated with the ODRW event horizons are of purely ephemeral character: they appear from nowhere and disappear without leaving a trace, just as the rogue wave does.

Finally, note that the spontaneous creation of photon pairs induced by PMI amplification of quantum noise (as discussed in details in part 1 of the Supplementary section), which unavoidably accompanies the process of ODRW generation induced by the input signal seed, could lead to the study of the optical analog of Hawking radiation. Although the ODRW is an ephemeral wave structure, the amplification of quantum noise is not, hence Hawking radiation will persist even after the disappearance of the ODRW. The rigorous investigation of the analogy between ODRWs and gravitational black or white holes, and the generation of optical Hawking radiation, are beyond the scope of the present paper and left for further work.

METHODS

Experiments

The two pump waves were superposed with a polarization maintaining fiber optical coupler with a 50:50 coupling ratio. An intensity modulator (EOM1) driven by a 35-GHz RF clock was used to generate sidebands on either side of both pumps. An erbium-doped fiber amplifier (EDFA1) was used to compensate the insertion loss introduced by the electro-optic modulator. The sinusoidal perturbation on each pump with adjustable frequency and amplitude then seeds the PMI process. Here the perturbation frequency was chosen to be half the peak PMI gain for the following two-pump system: 100 GHz-frequency spacing and 2.5 W total input power. The perturbation amplitude of the continuous-wave field was set to $\varepsilon = 0.16$ in order to observe the ODRW at precisely 3-km in our fiber with the available pump power. In order to reach the required peak power in our experiments, we temporally carved both pump waves thanks to an electro-optic modulator (EOM2) driven by a pseudo-random binary sequence generator. More precisely, 100-ns square pulse trains were generated with a duty cycle of 1:10 and then amplified (EDFA2). Such pulses provide a quasi-cw background condition, since the pulse duration is about four orders of magnitude larger than both the spontaneous and the induced modulation period. Moreover, to suppress stimulated Brillouin scattering (SBS) that may occur in the optical fiber, a phase modulator (EOM3) was inserted into the setup in order to increase the spectral linewidth of the two pump waves. The phase modulator was driven by a 67-MHz RF signal, thus enabling us to work at relatively high pump powers, still being far below the SBS threshold. The two pump waves were spectrally separated by means of a programmable optical filter (DEMUX). A pair of polarization controllers (PCs) allowed us to obtain two pumps with orthogonal linear states of polarization. The pumps were finally recombined after their independent amplification by high power erbium-doped fiber amplifiers (EDFA3&4) and before injection into the optical fiber. The optical fiber used in our experiment was a reverse-TrueWave fiber with the chromatic dispersion of −14 ps/nm/km, the nonlinear coefficient $\gamma = 2.4\ W^{-1}.km^{-1}$ and the attenuation of 0.25 dB/km at $\lambda_0 = 1554.7$ nm. This fiber has a very low PMD equal to $0.017\ ps.km^{-1/2}$. Two fiber spans were used in our experiments. A 3 km-long span was used to generate the ODRW (main manuscript). A longer span of 5 km was also used to validate the use of the Manakov model by analyzing the spontaneous PMI (supplementary information). At the fiber output, a polarization beam splitter (PBS) selects the output light propagating in the two orthogonal linear polarizations. The output light was simultaneously analyzed both in the spectral and temporal domain by means of an optical spectrum analyzer (OSA) and an optical sampling oscilloscope (OSO). Spectral measurements (Yokogawa - AQ6370 OSA) were carried out with 0.02 nm resolution bandwidth. The optical sampling oscilloscope (Picosolve PSO-101) had 0.8 ps resolution. The orthogonal polarization states were continuously tracked before injection into the fiber thanks to power measurements after their separation into a polarization beam splitter. We performed cut-back measurements (not shown here) to verify the kinematics of the ODRW generation and disappearance at distances before and after the ODRW centre point (3 km).

Theoretical solution and numerical simulation

The ODRW solution of the Manakov system that is presented in theSupplementary Section is depicted in Figs 2 and 3 by using a total pump power P = 1.9 W, a pump spacing equal to 100 GHz, and the fiber parameters given above. Figures 5, 6, 7 and S2 report a similar theoretical solution but using the pump power used in the experiments (P = 2.5 W). The Manakov equations were numerically solved by using the split-step Fourier method26. The simulation shown in Fig. 5considered the experimental input conditions, namely, a weakly modulated continuous-wave field with modulation frequency and depth corresponding to our experiment. The numerical results in Fig. S1b,calso included fiber loss and a −70 dB noise background.

ACKNOWLEDGEMENTS

The authors thank J. Fatome for comments and advice on light polarization issues during the experimental work. The authors gratefully acknowledge support from the Italian Ministry of University and Research (MIUR, Project Nb. 2012BFNWZ2), from the French National Research Agency (Grants No. ANR-12-BS04-0011 OPTIROC and No. ANR-14-ACHN-0014 NoAWE), the Labex ACTION program (contract ANR-11-LABX-0001-01), and the iXCore Foundation.

REFERENCES

1. Haver, S. Freak wave event at Draupner jacket January 1 1995.Statoil, Technical report. PTT-KU-MA (2003). Available at:http://www.ifremer.fr/web-com/stw2004/rogue/pres/Session_3.2/Haver_Draupner.pdf(Accessed: 14th December 2015)
2. Haller, G. & Beron-Vera, F. J. Coherent Lagrangian vortices: the black holes of turbulence. J. of Fluid Mechanics 731, R4 (2013).
3. Solli, D. R., Ropers, C., Koonath, P. & Jalali, B. Optical rogue waves.Nature 450, 1054–1057 (2007).
4. Onorato, M., Residori, S., Bortolozzo, U., Montina, A. & Arecchi, F. T. Rogue waves and their generating mechanisms in different physical contexts. Phys. Reports 528, 47–89 (2013).
5. Dudley, J. M., Dias, F., Erkintalo, M. & Genty, G. Instabilities, breathers and rogue waves in optics. Nat. Photon. 8, 755–764 (2014).
6. Kibler, B. et al. The Peregrine soliton in nonlinear fibre optics.Nat. Phys. 6, 790–795 (2010).
7. Onorato, M., Osborne, A. R. & Serio, M. Modulational Instability in Crossing Sea States: A Possible Mechanism for the Formation of Freak Waves. Phys. Rev. Lett. 96, 014503 (2006).
8. Onorato, M., Proment, D. & Toffoli, A. Freak waves in crossing seas. Eur. Phys. J. Special Topics 185, pp 45–55 (2010).
9. The power of analogies. Nat. Photon. 8, 1 (2014).
10. Philbin, T. G. et al. Fiber optic analog of the event horizon. Science319, 1367 (2008).
11. Belgiorno, F. et al. Hawking radiation from ultrashort laser pulse filaments. Phys. Rev. Lett. 105, 203901 (2010).
12. Faccio, D. Laser pulse analogues for gravity and analogue Hawking radiation. Contemporary Physics 53, 97–112 (2012).
13. Sheng, C., Liu, H., Wang, Y., Zhu, S. N. & Genov, D. A. Trapping light by mimicking gravitational lensing. Nat. Photon. 7, 902–906 (2013).
14. Webb, K. E. et al. Nonlinear optics of fibre event horizons. Nat. Comm. 5, 4969 (2014).

15. Faccio, D., Arane, T., Lamperti, M. & Leonhardt, U. Optical black hole lasers. Class. Quantum Grav. 29, 224009 (2012).
16. Demircan, A. et al. Rogue events in the group velocity horizon.Sci. Reports 2, 00850 (2012).
17. Demircan, A. et al. Rogue wave formation by accelerated solitons at an optical event horizon. Appl. Phys. B 115, 343–354 (2014).
18. Wright, L. G., Christodoulides, D. N. & Wise, F. W. Controllable spatiotemporal nonlinear effects in multimode fibres. Nat. Photon. 9, 306 (2015)
19. Liberati, S., Visser, M. & Weinfurtner, S. Analogue quantum gravity phenomenology from a two-component Bose–Einstein condensate. Class. Quantum Grav. 23, 3129–3154 (2006).
20. Garay, L. J., Anglin, J. R., Cirac, J. I. & Zoller, P. Sonic Analog of Gravitational Black Holes in Bose-Einstein Condensates. Phys. Rev. Lett. 85, 4643–4647 (2000)
21. Kevrekidis, P. G., Frantzeskakis, D. J. & Carretero-Gonzalez R.Emergent Nonlinear Phenomena in Bose-Einstein Condensates(Springer-Verlag, Berlin, 2008).
22. Larré, P.-E. & Pavloff, N. Hawking radiation in a two-component Bose-Einstein condensate. Europhysics Letters 103, 60001 (2013).
23. Steinhauer, J. Observation of self-amplifying Hawking radiation in an analogue black-hole laser. Nat. Phys. 10, 864–869 (2014).
24. Rousseaux, G., Mathis, C., Maïssa, P., Philbin, T. G. & Leonhardt, U.Observation of negative-frequency waves in a water tank: a classical analogue to the Hawking effect? New J. Phys. 10 053015 (2008).
25. Unruh, W. G. Has Hawking Radiation Been Measured? Found. Phys. 44, 532–545 (2014).
26. Agrawal, G. P. Nonlinear Fiber Optics (5th Edition, Academic Press, Oxford 2013).
27. Wai, P. K. A. & Menyuk, C. R. Polarization mode dispersion, decorrelation and diffusion in optical fibers with randomly varying birefringence. J. Lightwave Technol. 14, 148–157 (1996).
28. Manakov, S. V. On the theory of two dimensional stationary self-focusing of electromagnetic waves. Sov. Phys. JETP 38, 248–253 (1974).
29. Baronio, F. et al. Vector rogue waves and baseband modulation instability in the defocusing regime. Phys. Rev. Lett. 113, 034101 (2014).
30. Baronio, F., Chen, S., Grelu, P., Wabnitz, S. & Conforti, M.Baseband modulation instability as the origin of rogue waves.Phys. Rev. A 91, 033804 (2015).

CHAPTER 7

LONG-PERIOD GROUND MOTIONS: OBSERVATIONS AND MODEL SIMULATIONS OF LONG-PERIOD SURFACE WAVES

Shunsuke Takemura, Mai Akatsu, Kei Masuda, Kengo Kajikawa, and Kazuo Yoshimoto

Keywords: Long-period ground motion Surface wave propagation Sedimentary S-wave velocity Kanto Basin

BACKGROUND

Long-period ground motions with dominant periods of several to 10 s are observed frequently during shallow local and/or regional earthquakes, especially in large sedimentary basins (e.g., Beck and Hall 1986; Shin and Teng 2001; Koketsu et al. 2005). Long-period ground motion has the potential to cause significant resonance in basins, and this can lead to severe damage of large-scale, man-made structures such as high-rise buildings, oil storage tanks, and suspension bridges (e.g., Koketsu and Miyake 2008). Therefore, to assist disaster mitigation efforts for future large earthquakes, it is necessary to evaluate the characteristics of long-period ground motions in relation to the three-dimensional (3D) heterogeneous structure of large sedimentary basins.

In the Kanto Basin (see Figure 1), where the biggest metropolitan area of Japan is located, many studies on long-period ground motions have been conducted from both the scientific and engineering point of views (e.g., Yamanaka et al. 1989; Kinoshita et al. 1992; Yamazaki et al.1992; Miyake and Koketsu 2005; Furumura and Hayakawa 2007). The dominant period of the long-period ground motions observed in the Kanto Basin tends to be approximately 7 s for shallow moderate-to-large earthquakes (e.g., Miyake and Koketsu 2005; Furumura and Hayakawa 2007; Furumura et al. 2011), which is strongly controlled by the velocity structure at depths of less than 2 km (e.g., Yoshimoto and Takemura 2014b). Using seismic data from dense strong motion networks in the Kanto Basin (Kinoshita 1998; Okada et al. 2004), previous studies have revealed that the long-period ground motions are generated by surface waves propagating over the thick sediments in the Kanto Basin (e.g., Koketsu and Kikuchi 2000; Miura and Midorikawa 2001; Furumura and Hayakawa 2007; Hayakawa 2009). The nature of the surface wave propagation reported by these studies is very complex, and current data imply that there is strong lateral inhomogeneity in the sedimentary structure of the Kanto Basin.

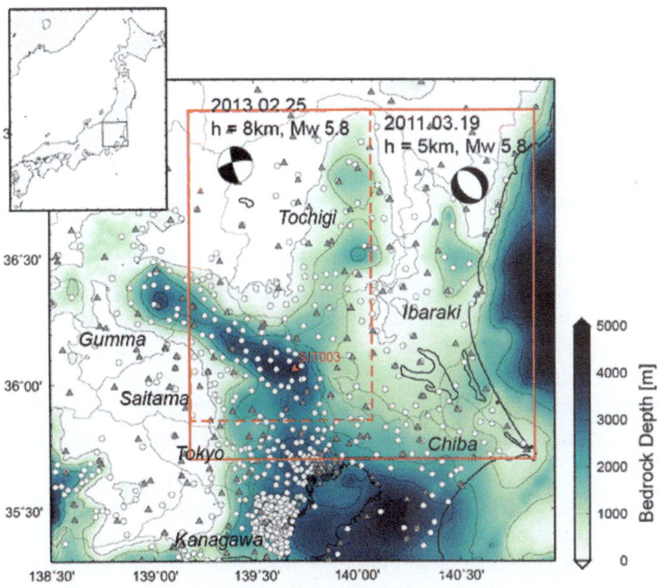

Figure 1 Spatial distributions of JIVSM bedrock depth and strong motion stations in the Kanto region, Japan. Gray triangles and white circles are K-NET/KiK-net (Kinoshita 1998; Okada et al. 2004) and SK-net stations, respectively. Dashed and solid red rectangles show the simulation areas of 3D waveform modeling and validation of the constructed velocity structure model, respectively. Plotted focal mechanisms were used in 3D FDM simulations.

Citation: Shunsuke Takemura, Mai Akatsu, Kei Masuda, Kengo Kajikawa, Kazuo Yoshimoto, Long-period ground motions: observations and model simulations of long-period surface waves, DOI: 10.1186/s40623-015-0201-7.

Extensive efforts to construct a 3D structure model of the Kanto Basin have been put forth by many researchers (e.g., Yamanaka and Yamada 2006; Koketsu et al. 2008, 2009; Fujiwara et al.2009). Using available geophysical and geological information in Japan, Koketsu et al. (2008,2009) proposed the Japan Integrated Velocity Structure Model (JIVSM), which is one of the latest and most widely used structure models for the evaluation of strong and long-period ground motions of large earthquakes (e.g., Iwaki et al. 2013; Maeda et al. 2013b), the estimation of rupture complexities for large earthquakes (e.g., Koketsu et al. 2011; Asano and Iwata 2012), and the simulation of seismic wave scattering (e.g., Maeda et al. 2014; Takemura and Yoshimoto 2014). The JIVSM represents the 3D heterogeneous subsurface structure using 23 constant-velocity layers including those of the crust, the mantle, the Philippine Sea (PHS) Plate, and the Pacific (PAC) Plate, and also the complex basement shapes of sedimentary basins. Despite such a detailed model of the subsurface structure, precise evaluations of the long-period ground motions for the Kanto Basin still have room for improvement, compared to the similar trials for other large sedimentary basins in Japan, e.g., the Osaka Basin (e.g., Iwaki and Iwata2010) and the Nobi Basin (e.g., Horikawa et al. 2008).

Recently, Yoshimoto and Takemura (2014a) suggested that realistic modeling of the sedimentary velocity structure is necessary for the precise evaluation of long-period ground motions in the northern Kanto Basin. Using a simple velocity gradient function (Ravve and Koren 2006), they proposed the construction of a smooth-varying 3D velocity structure model of the sediments from the vertical seismic profile (VSP) measurements of 14 deep boreholes in the Kanto Basin. Although their 3D finite-difference method (FDM) simulations successfully reproduced the excitation and propagation of long-period surface waves in a limited area of the northern edge of the Kanto Basin, the Kanto Basin is too large to achieve a detailed modeling of the 3D velocity structure using data from only 14 boreholes.

In this study, we construct a new 3D velocity structure model for the northern Kanto Basin (area outlined by the red square in Figure 1), which can be used for the precise evaluation of long-period ground motions with periods of 4 to 8 s. The model construction employs waveform analysis and available geophysical data in this region. The model constructed in this study exhibits strong lateral variations of S-wave velocities in the northern Kanto Basin, which may have large effects on the long-period surface wave excitation and propagation. In order to validate this model, we conducted large-scale 3D FDM simulations of the long-period ground motion for two shallow moderate earthquakes in northern Tochigi and northern Ibaraki and confirmed that there was good agreement between our simulation results and the observed seismograms. Based on the simulation results, we discuss the characteristics of surface wave propagation in the laterally inhomogeneous sedimentary basin.

METHODS

Depth-varying properties of sedimentary S-wave velocity

Yoshimoto and Takemura (2014a) pointed out that the depth-varying properties of S-wave velocity in the Kanto Basin can be practically characterized by a simple velocity gradient function (Ravve and Koren 2006). The depth-dependent S-wave velocity function is represented by the following equation:

$$V_S(z) = V_0 + \Delta V\left[1\text{-}\exp\left(-\frac{\alpha z}{\Delta V}\right)\right] \quad (1)$$

where V_0 is the S-wave velocity at the surface ($z=0$), ΔV is the increment of the S-wave velocity at infinite depth, and α is the positive constant that determines the velocity-depth gradient. After applying the assumption of $\Delta V=3.2$ km/s, which is the S-wave velocity of the bedrock according to the JIVSM, the local S-wave velocity structure in the Kanto Basin can be represented using only the bedrock depth and two parameters (V_0 and α) of this function. Figure 2 shows the V_0 and α dependency of the S-wave structure for sediment with a fixed $\Delta V=3.2$ km/s. As shown in Figure 2, a variation of the two parameters can result in a wide variety ofS-wave velocities at different depths.

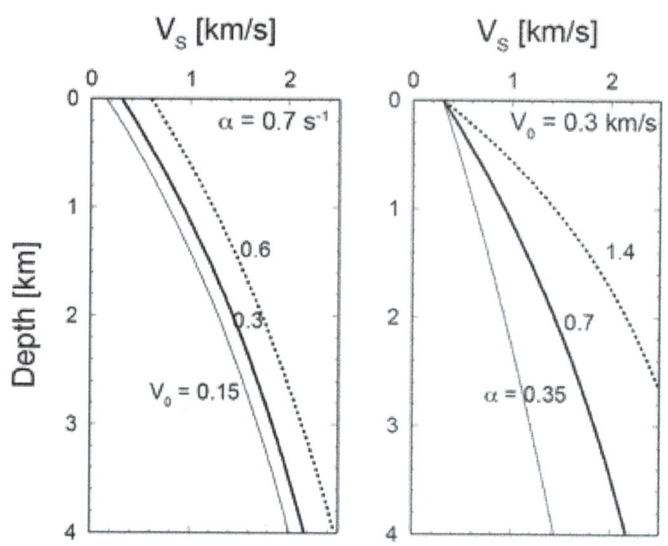

Figure 2 V_0andαdependency on the sedimentaryS-wave velocity at depth. Each curve is calculated based on a simple velocity gradient function, Equation 1. ΔV is assumed to be 3.2 km/s.

Conversion of the layered structure to a smooth depth-varying structure

It may be useful to consider the conversion of the layered structure model to a smooth depth-varying structure model when constructing a 3D velocity model of sedimentary structure for long-period ground motion simulations. This is because many conventional structure models were constructed by assuming a layered structure and it is difficult to compile these models, which are often represented by different medium parameters and numbers of layers.

From Equation 1, the one-way travel time of S wave $t(z)$ from $z=0$ to a certain depth z is analytically calculated as follows:

$$t(z) = \frac{\Delta V}{\alpha(V_0+\Delta V)}\ln\frac{(V_0+\Delta V)\exp(\alpha z/\Delta V)\text{-}\Delta V}{V_0} \quad (2)$$

To determine the appropriate values of V_0 and α for the conversion of a layered structure model, we defined a residual for the one-way S-wave travel times by the following equation:

$$\text{Residual} = \frac{1}{N}\sum_{k=1}^{N}\left(t^{\text{LSM}}(z_k) - t(z_k, V_0, \alpha)\right)^2 \quad (3)$$

where z_k is the depth, $t^{LSM}(z_k)$ is the one-way S-wave travel time of the layered structure model at the depth of z_k, $t(z_k, V_0, \alpha)$ is the one-way S-wave travel time with assumed V_0 and α at the depth of z_k that is calculated using Equation 2, and N is the total number of data points down to the depth of the bedrock. A simple grid search for V_0 and α can be performed to find the minimum residual and best-fit parameters, which represent the one-dimensional (1D) smooth depth-varying S-wave velocity structure of the sediments.

P-wave velocity, density, and attenuation structure

To construct velocity structure models for 3D FDM simulations, we adopted the modeling procedure used in Yoshimoto and Takemura (2014a). The P-wave velocity (V_P) and density (ρ) of sediment were assumed from the S-wave velocity using empirical relations reported by the Ministry of Education, Sports, Culture, Science and Technology (MEXT) (2007) and Shiomi et al. (1997), respectively.

Here, we employ the frequency-dependent Q^{-1} model for P and S waves based on the formulations for FDM simulations of Robertsson et al. (1994). We adopt a single relaxation mechanism with a central frequency of $f_0 = 1$ Hz, which gives a peak Q^{-1} value at this frequency. Anelastic attenuation (Q) values of the sediments for P and S waves were assumed using the empirical relation of Brocher (2008):

$$Q_S = -16 + 104.13V_S - 25.225V_S^2 + 8.2184V_S^3$$
$$Q_P = 2Q_S \quad (4)$$

Other structures beneath the sediment, including the geometry of bedrock, the Moho, the PHS Plate, and the PAC Plate, were obtained from the JIVSM. The physical parameters of the JIVSM are shown in Table 1.

Table 1 P- andS-wave velocities, densities, and anelastic attenuation of the JIVSM

Layer	V_P[km/s]	V_S[km/s]	ρ[g/cm^3]	Q_P	Q_S
Sedimentary layer 1	1.8	0.5	1.95	170	100
Sedimentary layer 2	2.3	0.9	2.10	306	180
Sedimentary layer 3	3.0	1.5	2.25	510	300
Basement	5.5	3.2	2.65	680	400
Upper crust	5.8	3.4	2.70	680	400
Lower crust	6.4	3.8	2.80	680	400
Upper mantle	7.5	4.5	3.20	850	500
Oceanic crust layer 2 of PHS	5.0	2.9	2.40	340	200
Oceanic crust layer 3 of PHS	6.8	4.0	2.90	510	300
Oceanic mantle of PHS	8.0	4.7	3.20	850	500
Oceanic crust layer 2 of PAC	5.4	2.8	2.6	340	200
Oceanic crust layer 3 of PAC	6.5	3.5	2.8	510	300
Oceanic mantle of PAC	8.1	4.6	3.4	850	500

Each parameter was obtained from Koketsu et al. (2008, 2009).

Estimation of V_0 and α near the edge of the northern Kanto Basin

For modeling the velocity structure in the northern edge of the Kanto Basin where available geophysical data are sparse, we conducted 3D FDM simulations of the long-period ground motions during the northern Tochigi earthquake (Mw = 5.8) on 25 February 2013 and estimated the local values of V_0 and α. We supposed that the local values of V_0 and α at a certain station near the basin edge could be estimated from an existing velocity structure model if the simulated waveforms from the model successfully agree with the observed long-period ground motions.

The model used for the 3D FDM simulation covered an area of 135.6 × 82.8 km^2 in the horizontal directions (red dashed rectangle in Figure 1) and 52.5 km in depth, which was discretized by grid intervals of 0.15 km in the horizontal directions and 0.075 km in the vertical direction. The wave propagation in each grid point was calculated by solving equations of motions using a staggered-grid FDM with fourth-order and second-order in space and time, respectively. With a minimum S-wave velocity (0.3 km/s) and grid size (0.15 km), the FDM simulations can examine seismic wave propagation for periods longer than 4 s with sampling of eight grid points per minimum S wavelength.

To perform an effective simulation, we employed a parallel 3D FDM simulation code based on a domain-partitioning procedure that utilizes a large number of processors with a message-passing interface (MPI) (e.g., Furumura and Chen 2004). We used a split-type perfectly matched layer (PML)

absorbing boundary condition at the boundaries around the 3D model (e.g., Moczo et al. 2007; Maeda and Furumura 2013; Maeda et al. 2013a) to avoid artificial refractions from the 3D model boundaries, and a smoothing technique for layer interfaces near the PML boundaries was also applied for numerical stability (Maeda et al. 2013a).

The source depth of 8 km, focal mechanism of the strike/dip/rake = 168/86/−10°, and seismic moment of $M_0 = 5.54 \times 10^{17}$ Nm were assumed by referring to the F-net centroid moment tensor (CMT) solution (Fukuyama et al. 1998; Okada et al. 2004). A source time function represented by the asymmetric cosine function (Ji et al. 2003) with $t_s = 0.3$ and $t_e = 2.7$ was adopted for a point seismic source. After the calculation, the M_0 value was adjusted to 77% of the CMT estimation to explain the observed S-wave amplitude at rock site TCG011 (K-NET Kuzuu) (Yoshimoto and Takemura 2014a).

RESULTS

Model construction

Local structure from VSP measurements

Yoshimoto and Takemura (2014a) proposed a simple sedimentary velocity structure model of the Kanto Basin using VSP measurements at 14 deep boreholes (Yamamizu 1996, 2004). They reported that the 3D FDM simulations using this model were able to successfully reproduce the observed Love waves at seismic station SIT003 (K-NET Kuki; red triangle in Figure 1) in the northern Kanto Basin. Thus, we adopted the local S-wave velocity structures at 14 deep boreholes and SIT003 for our model construction. The values of V_0 and α at each site are listed in Table 2.

Table 2 Estimated values ofV_0andαat 14 deep boreholes and one K-NET station

Site	Longitude	Latitude	V_0[km/s]	α[s^{-1}]
Mohka	139.955	36.446	0.40	1.44
Isesaki	139.188	36.311	0.67	0.32
Iwatsuki	139.735	35.929	0.29	0.79
Hidaka	139.387	35.896	0.49	0.84
Tokorozawa	139.538	35.780	0.40	0.69
Edosaki	140.333	35.949	0.11	1.68
Narita	140.301	35.828	0.30	0.73
Shimohsa	140.024	35.793	0.30	0.69
Chiba	140.245	35.546	0.35	0.55
Yohro	140.156	35.283	0.56	0.59
Futtsu	139.859	35.341	0.38	0.73
Fuchu	139.470	35.654	0.46	0.53
Yokohama	139.523	35.496	0.43	0.74
Atsugi	139.357	35.401	0.17	2.67
Kuki (SIT003)	139.700	36.068	0.37	0.80

Each value was estimated by Yoshimoto and Takemura (2014a).

Using these parameters, we constructed sedimentary seismic velocities, density, and the anelastic attenuation structure based on the procedure discussed in the previous section. For spatial interpolation of the values of V_0 and α, we used a gridding algorithm 'surface' in the Generic Mapping Tools (Wessel and Smith 1998) software with a tension factor of 0.3. To construct a 3D velocity structure model for whole depths, the structures beneath the sediments, including the topography of the bedrock, Moho, subducted oceanic plates, and mantle, were obtained from the JIVSM. Hereafter, we call this velocity structure model the 'initial model,' which was mainly constructed from the data of VSP measurements.

The spatial variations of V_0 and α of the initial model shown in Figure 3 seemed to be too smooth to represent the lateral variations of the sedimentary structure of the Kanto Basin. In spite of this, as shown in Figure 4, it was confirmed by our 3D FDM simulation that this model can be used practically to explain the observed long-period ground motions (period of 4 to 8 s) at eight stations near the basin edge (bold circles in the map of Figure 4). Figure 4 shows seismograms at only four stations, because simulated waveforms at four stations located near TCH.009 are very similar to those at TCH.009. Thus, we adopted the values of V_0 and α at these eight stations for the remaining process of our model construction.

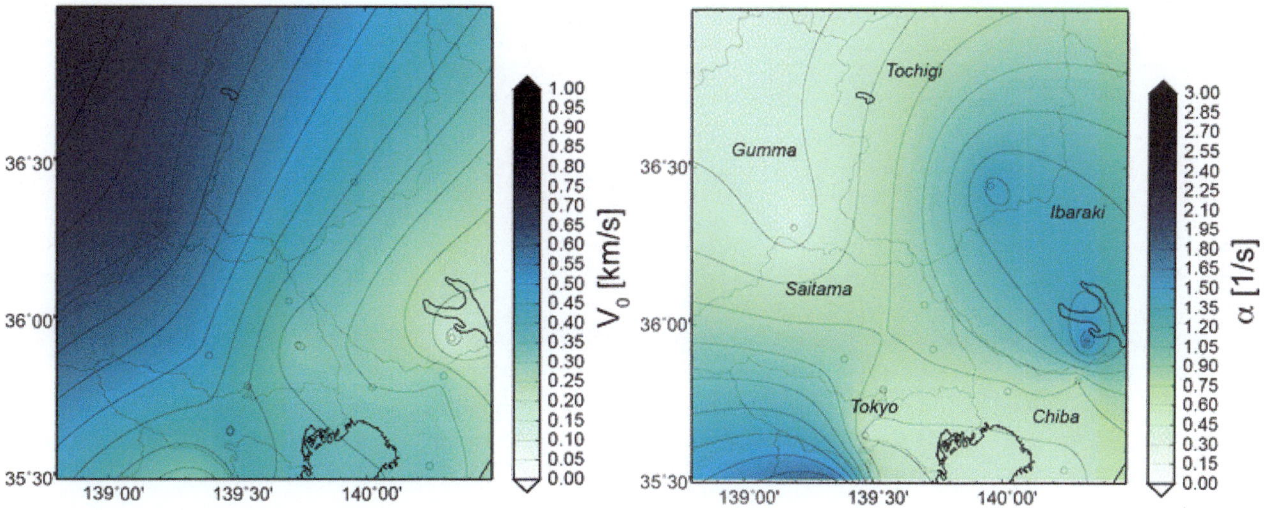

Figure 3 **Spatial variations of V_0 and α in the initial model.** The values for V_0 and α were constructed by the interpolation of 15 local S-wave velocity structures (Yoshimoto and Takemura 2014a).

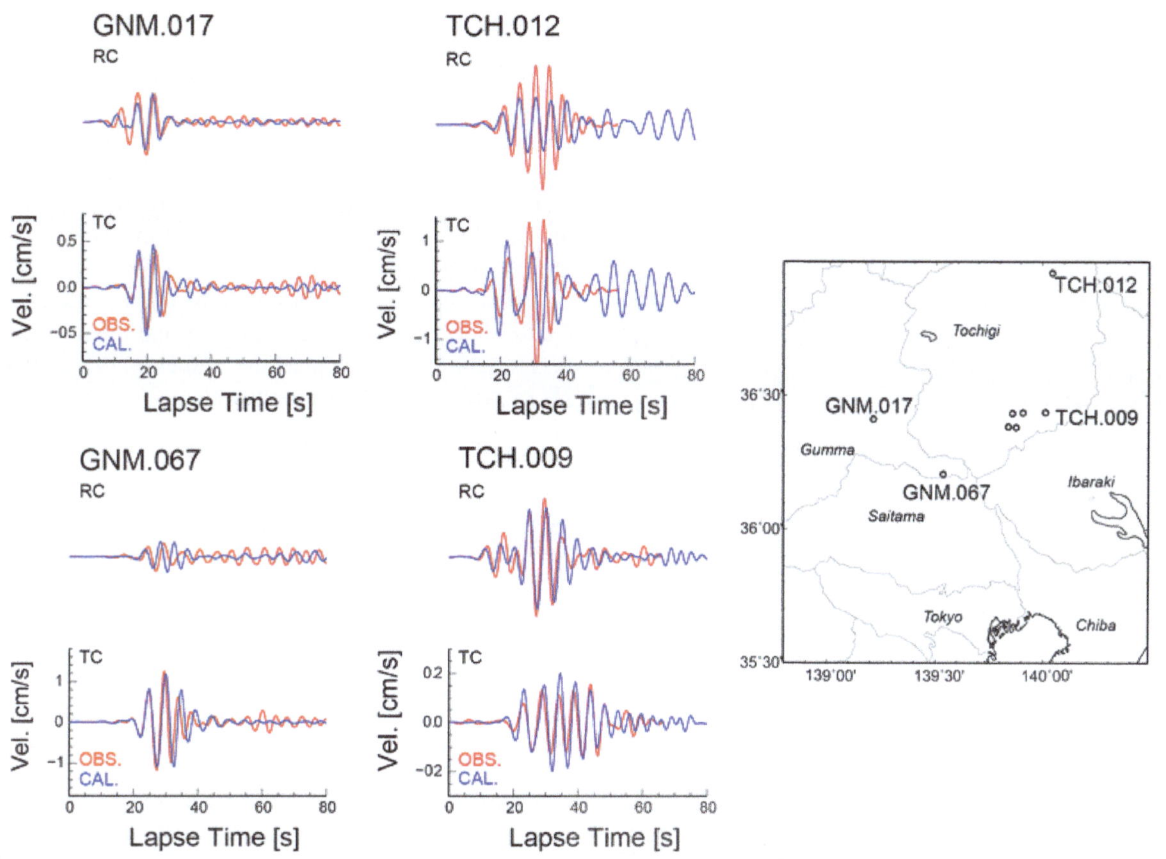

Figure 4 **Comparison of observations and simulations for horizontal long-period ground motions during the northern Tochigi earthquake.** The simulated seismograms of radial and transverse components (RC and TC, respectively) were derived from simulation results of the initial model shown in Figure 3. A band-pass filter for the period of 4 to 8 s was applied into both observed and simulated seismograms. Bold circles in the map are stations, where the simulation results practically reproduced the observed long-period ground motions.

Local structure from JIVSM

We converted the JIVSM layered sedimentary structure using the conversion procedure mentioned above. As an example, we briefly explain here how we derived the estimates of V_0 and α at site GNM.070. The JIVSM was constructed by three sedimentary layers (black line in Figure 5a), and its one-way S-wave travel time was a piecewise linear function (black line in Figure 5b). We employed a grid search procedure to find the minimum residual value using Equation 3. Figure 5c shows the distribution of normalized residuals in the model parameter space. Although there was trade-off involved in optimizing the two different parameters, the point of global minimum could be clearly detected in this plot. The one-way S-wave travel time curve and the S-wave velocity structure from the estimated parameters are shown in Figures 5b,a, respectively.

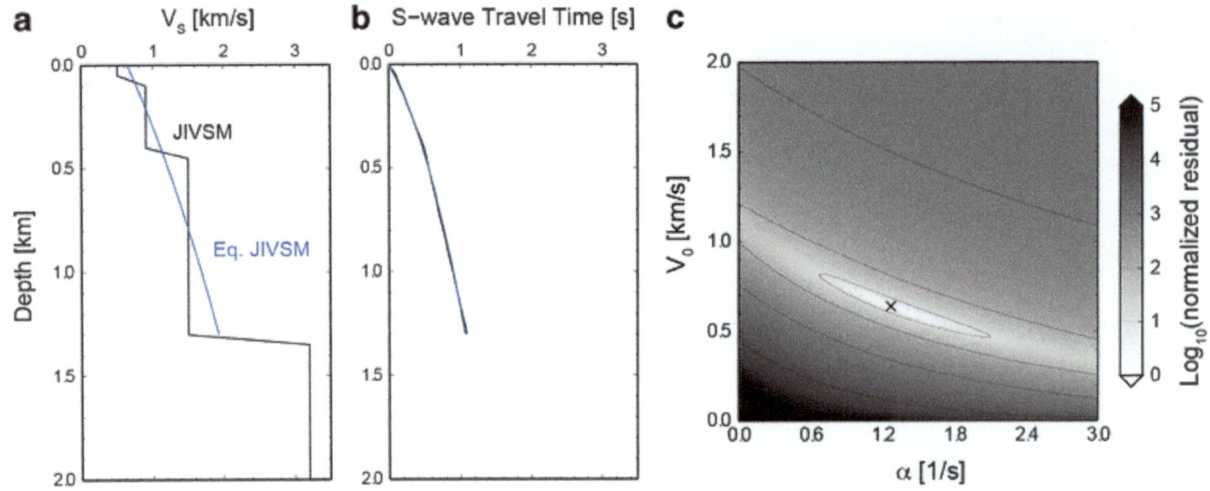

Figure 5 Estimation of V_0 and α from a layered S-wave velocity structure at site GNM.070. **(a)** S-wave velocity structure of the layered structure (black line) and best-fit velocity gradient function (blue line). **(b)** One-way travel time of the layered structure (black line) and best-fit velocity gradient function (blue line). **(c)** Distribution of the normalized residual values calculated by Equation 3 for the model parameter space.

Based on this conversion procedure, we estimated V_0 and α of the JIVSM in northern Kanto, where bedrock depths are larger than 1 km. The estimated parameters were used to construct a smoothed sedimentary velocity structure model using the same interpolation technique mentioned in the construction of the initial model. Hereafter, we call the constructed smoothed sedimentary velocity structure model from the JIVSM the 'equivalent JIVSM (E. JIVSM)'.

We conducted a 3D FDM simulation of the long-period ground motion for the northern Tochigi earthquake using the E. JIVSM. Figure 6 shows the comparison of the horizontal long-period ground motions (4 to 8 s) between the simulation of this model and the observational data. The simulation results of the E. JIVSM (Figure 6a) reproduced the observed long-period ground motions at seven stations near the edge of the northern Kanto Basin (bold circles in the lower right map of Figure 6b). Simulated waveforms of the E. JIVSM also showed high similarity to those of the JIVSM (Figure 6b). These results indicate that our conversion procedure is useful for the simulation of long-period ground motions. From these investigations, we adopted the values of V_0 and α at these seven stations for further analyses.

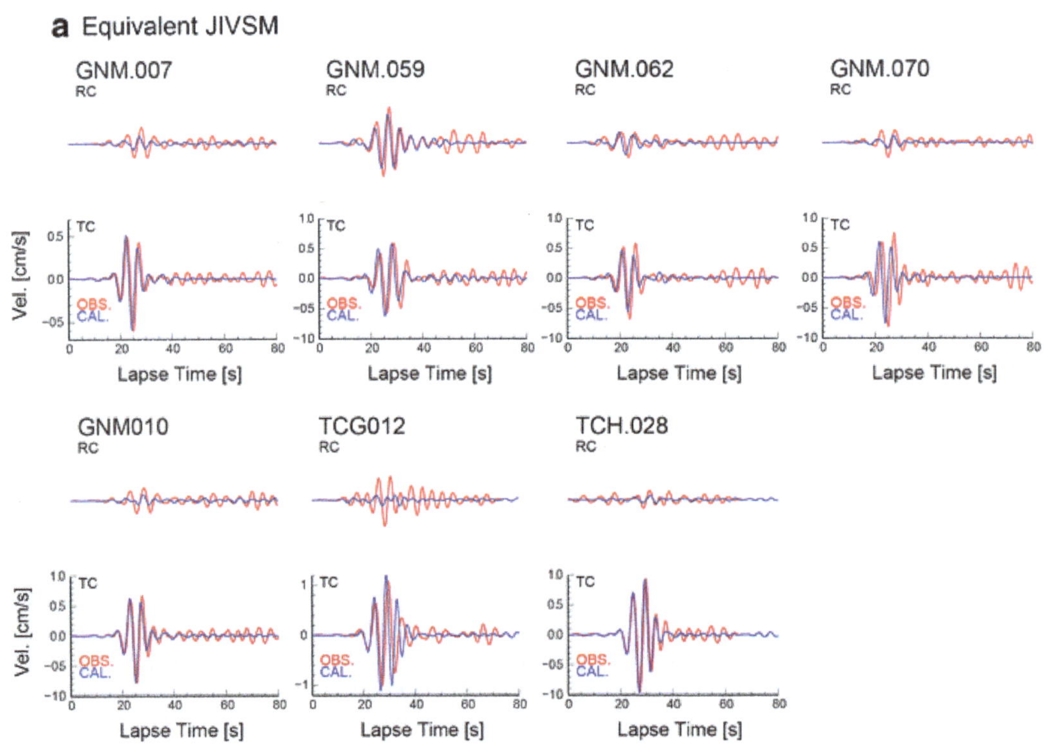

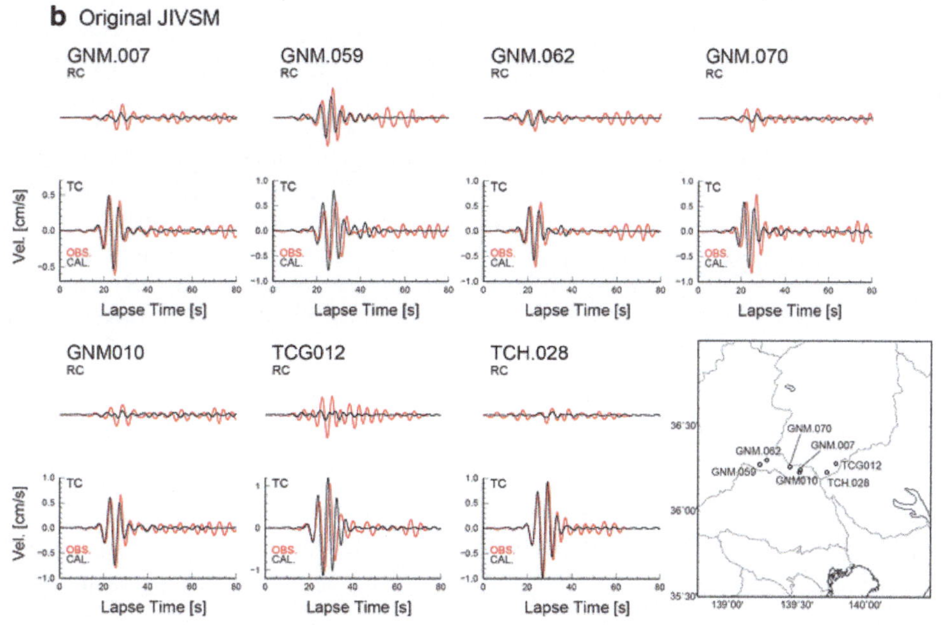

Figure 6 Comparisons of observations and simulations for horizontal long-period ground motions during the northern Tochigi earthquake. Simulated seismograms of radial and transverse components (RC and TC, respectively) for a period of 4 to 8 s were derived from the **(a)** E. JIVSM and **(b)** JIVSM.

Local structure from microtremor surveys

We converted the layered sedimentary structure models from dense microtremor surveys in Saitama, Ibaraki, and Chiba (Matsuoka and Shiraishi 2002; Yamanaka et al. 2005; Senna et al.2013) using the same conversion procedure discussed above. Since these models were derived from the dispersion analysis of surface waves, the incorporation of these models may improve the accuracy of long-period ground motion simulations. In this study, we converted 160 models (diamonds in Figure 7a), where the bedrock depths are larger than 0.5 km, to allow for stable estimations of α.

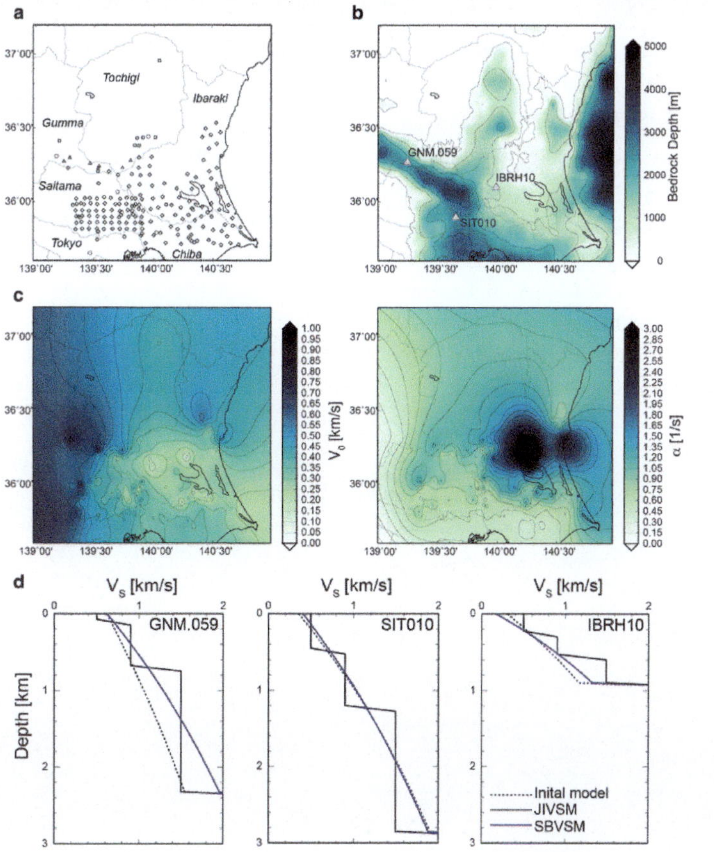

Figure 7 Sedimentary structure in the area of the red solid square in Figure1. **(a)**Distribution of the 190 stations that were used for the estimation of the spatial variations of V_0 and α, **(b)** bedrock depth, **(c)** spatial variations of V_0 and α, and **(d)** examples

of local S-wave structures at three sites. Open circles, squares, bold triangles, and bold diamonds are data points from VSP measurements (Yamamizu 1996, 2004; Yoshimoto and Takemura 2014a), the waveform analysis using the initial model (Figure 3), the JIVSM (Koketsu et al. 2008, 2009), and microtremor surveys (Matsuoka and Shiraishi2002; Yamanaka et al. 2005; Senna et al. 2013), respectively.

3D velocity structure model by integration of local structures

We integrated the estimated parameters at all of the sites mentioned above (Figure 7a) to construct a 3D velocity structure model of the sediment in the northern Kanto Basin (Figure 7b). Figure 7c shows the spatial variations of V_0 and α estimated by the interpolation, which used a total of 190 local site models. In Figure 7c, we found very complex lateral variations of the two parameters compared to the initial model (Figure 3). For example, high surface S-wave velocities (V_0) appeared at the western side of the Kanto Basin and strong velocity increments against depth (α) were found in central Ibaraki. By adopting the same procedure that we used for constructing the initial model, we constructed a 3D velocity structure model for the 3D FDM simulation of long-period ground motions in this region. Hereafter, we call this model the 'smoothed basin velocity structure model (SBVSM)'. Figure 7d shows a comparison of the local S-wave velocity structures of the JIVSM, initial model, and SBVSM at three sites.

LARGE-SCALE 3D FDM SIMULATION OF LONG-PERIOD GROUND MOTION IN THE NORTHERN KANTO BASIN

Performance of the constructed model for the evaluation of long-period ground motion

To demonstrate the performance of the SBVSM for the evaluation of long-period ground motions, we conducted large-scale 3D FDM simulations of long-period ground motions for two shallow moderate earthquakes. The model used for the 3D simulation covered an area of 153.6×153.6 km^2 in the horizontal directions (red solid square in Figure 1) and 60.0 km in depth. The scheme for the simulation was the same as that used for the previous simulations in this study.

One earthquake used in the simulations was the northern Tochigi earthquake, which was used for the parameter estimation near the basin edge. Here, we examine the propagation of the long-period surface wave in a wide area of the Kanto Basin, including the areas of Ibaraki, Saitama, Tokyo, and northern Chiba. Another earthquake was the northern Ibaraki earthquake (Mw = 5.8) that occurred on 19 March 2011 at a depth of 5 km. We employed these earthquakes to validate the SBVSM in Ibaraki and northern Chiba. For the latter earthquake, a focal mechanism of strike/dip/rake = 141/48/−94° and seismic moment $M_0 = 6.35 \times 10^{17}$ Nm were assumed by referring to the F-net CMT solution. A source time function represented by the asymmetric cosine function (Ji et al. 2003) with $t_s = 0.3$ and $t_e = 2.7$ was adopted. After the calculation, the M_0 value of this event was adjusted to 77% of the CMT estimation to explain the observed S-wave amplitude at rock site IBRH19 (KiK-net Tsukuba).

Figure 8 shows a comparison of the simulated and observed horizontal long-period ground motions for the northern Tochigi earthquake. We also conducted a FDM simulation using the JIVSM as a reference. The simulation result of the JIVSM (black lines) showed large discrepancies with the observed seismograms in terms of arrival times, amplitudes, and durations. Large-amplitude surface waves arrived earlier compared to the observations, especially at TCG001, SIT003, and IBRH10. In addition, the amplitudes of the surface waves were overestimated at SIT.015, SIT010, and SIT003. Furthermore, we confirmed that the JIVSM did not reproduce the observed excitation of the long-period surface waves at the northern edge of the Kanto Basin (TCG001, TCGH06, and SIT003). In contrast, simulation results of the SBVSM accurately reproduced the observed amplitudes and arrival times of surface waves at not only stations near the basin edge (GNM.059, TCGH06, TCG001, and SIT003) but also at stations far from the edge (SIT.015, SIT010, and TKY.1220). A high reproducibility of the observed long-period ground motion at the middle and eastern parts of the Kanto Basin may be achieved by incorporating the S-wave velocity structures derived from dense microtremor surveys in Saitama, Ibaraki, and northern Chiba.

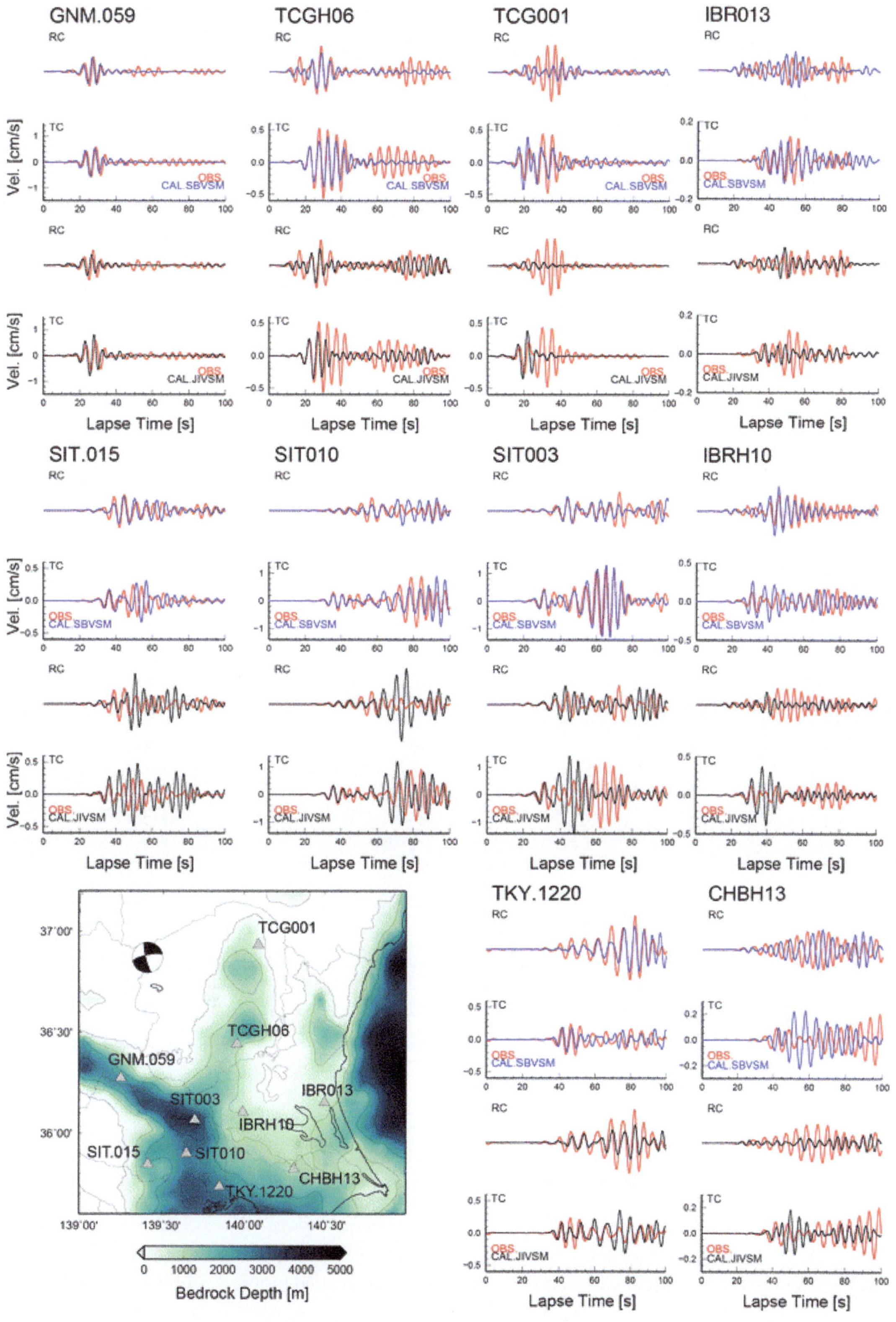

Figure 8 Observations and SBVSM simulations of horizontal long-period ground motions for the northern Tochigi earthquake. A band-pass filter for the period of 4 to 8 s was applied into both observed and simulated seismograms. Blue, black, and red lines show simulations of the SBVSM, simulations of the JIVSM, and observations, respectively.

The SBVSM also performed well during the evaluation of the long-period ground motion for the northern Ibaraki earthquake, which was not used in the parameter estimations; additionally, this earthquake occurred in a different back azimuth compared to the northern Tochigi earthquake. Figure 9 shows a comparison of the simulated and observed horizontal long-period ground motions for this earthquake. Significant differences were not found between the results from the two structure models in the area of shallow bedrock depths (IBR012 and IBRH19). However, simulated waveforms of the JIVSM in deep sediment areas (SIT.081, SIT003, SITH01, and CHB002) were overestimated. Simulated waveforms of the SBVSM reproduced the observed amplitudes and arrival times of the long-period surface waves in the northern Kanto Basin very accurately.

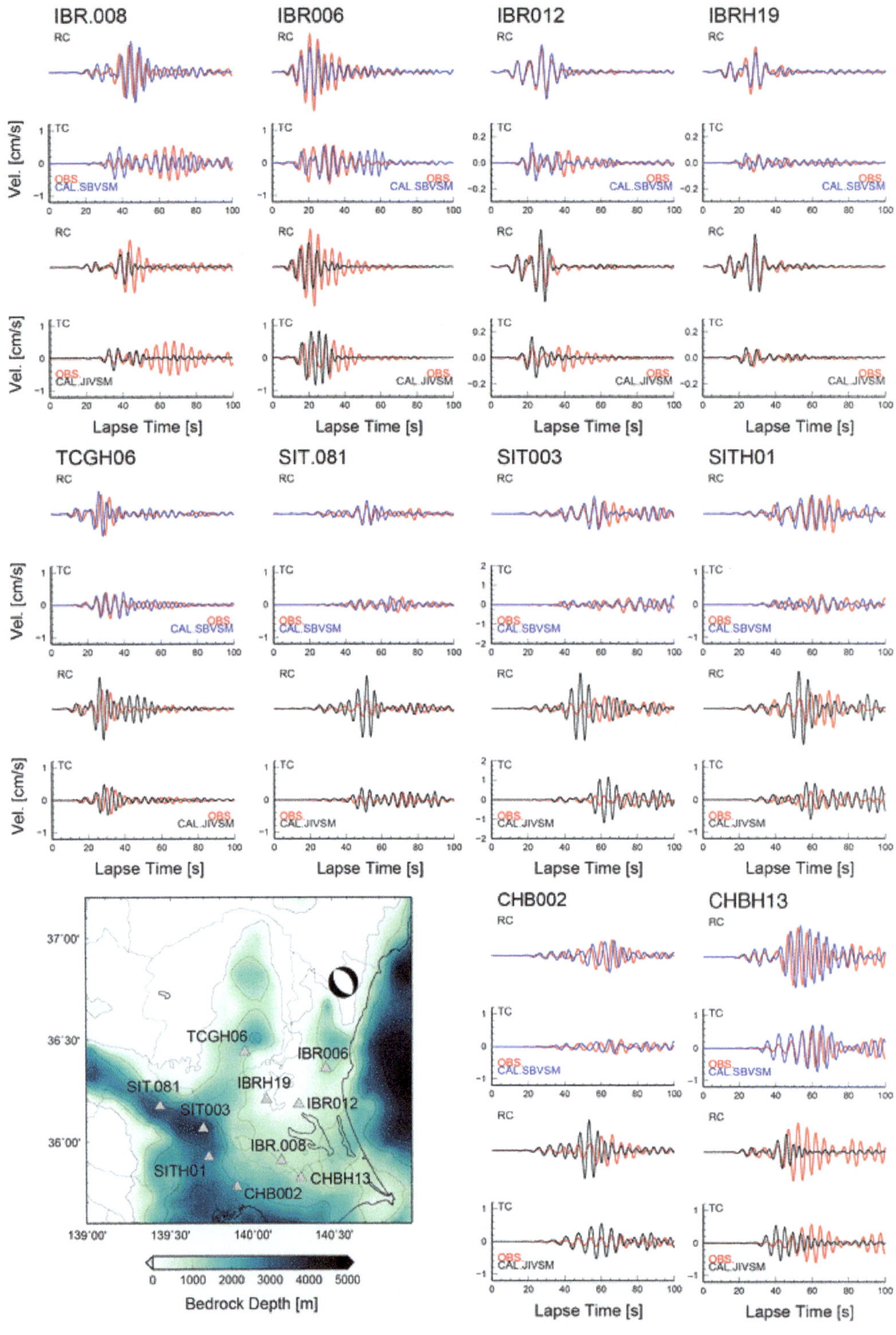

Figure 9 Observations and SBVSM simulations of horizontal long-period ground motions for the northern Ibaraki earthquake. A band-pass filter for the period of 4 to 8 s was applied into both observed and simulated seismograms. Blue, black, and red lines are simulations of the SBVSM, simulations of the JIVSM, and observations, respectively.

Surface wave propagation in the northern Kanto Basin

Since simulated waveforms of the SBVSM accurately reproduced the observed waveforms in the northern Kanto Basin (Figures 8 and 9), it was concluded that our simulations succeeded in reproducing the excitation and propagation of surface waves in this area. Thus, using these simulated and observed waveforms, we investigated the characteristics of the excitation and propagation of the long-period surface waves during two shallow moderate earthquakes in detail.

Figure 10 shows snapshots of the seismic wave propagation for a period of 4 to 8 s derived from simulations of the SBVSM for the two earthquakes. The spatial variation of the amplitudes at each time step was calculated by the interpolation of the root mean square amplitude of the simulated horizontal seismograms observed at virtual seismic stations, which were uniformly distributed at an interval of 3 km.

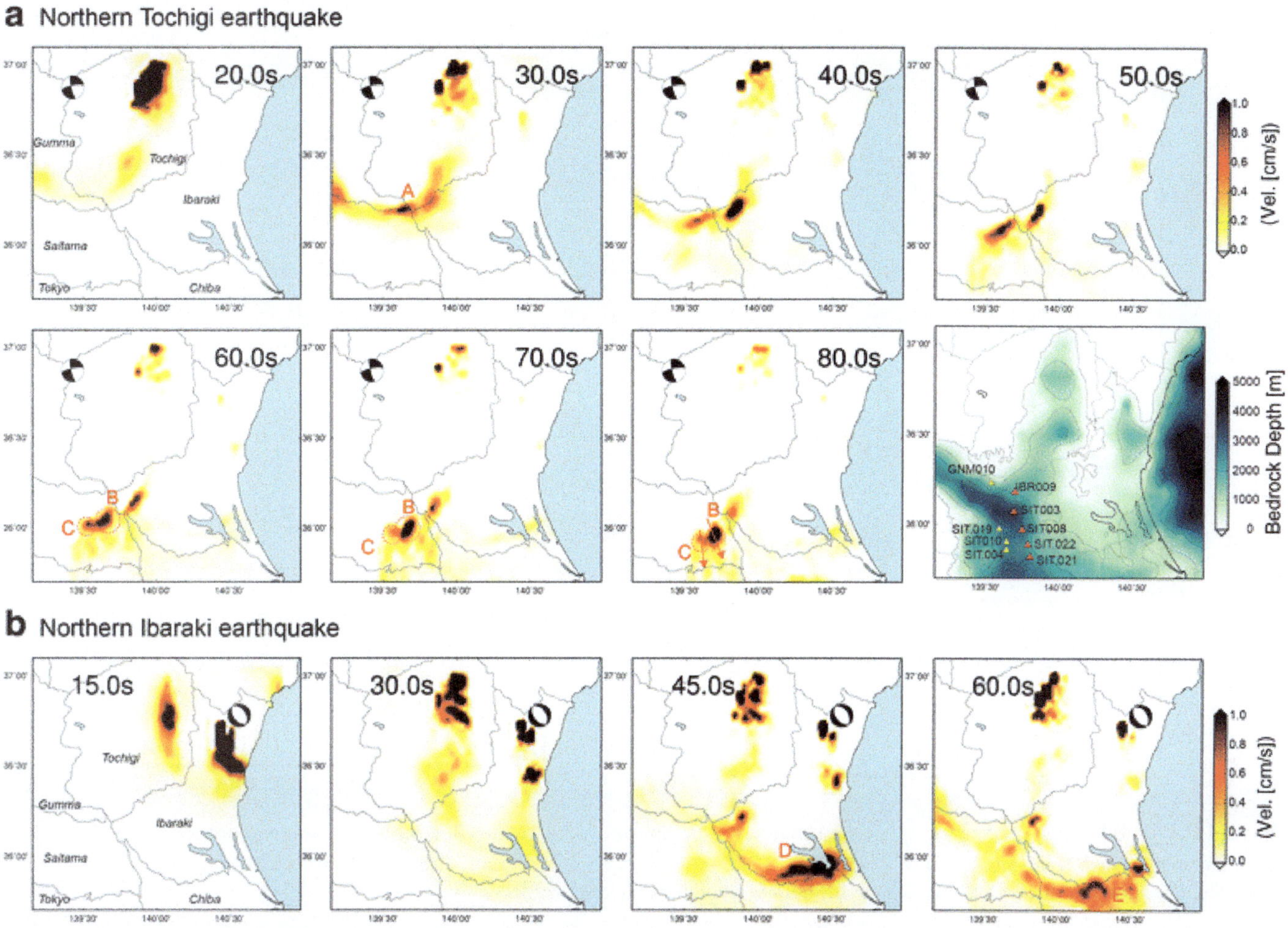

Figure 10 Snapshot of the propagation of simulated long-period ground motions (period of 4 to 8 s). Simulations of the **(a)** northern Tochigi earthquake and **(b)**northern Ibaraki earthquake. The spatial variation of bedrock depth is also shown in lower right part of **(a)**. The red arrows denote the propagation directions of phases B and C.

In an early snapshot of the northern Tochigi earthquake (t = 30 s; Figure 10a), strong surface wave excitation around the Gumma-Saitama border (mark A) was observed. The excited surface waves propagated towards the middle part of the Kanto Basin, and they had a complex propagation pattern. At a lapse time of t = 50 to 80 s, the surface waves near the Saitama-Chiba border (mark B) were propagating much slower than the surface waves propagating in the midst of Saitama (mark C). In the later frames (t = 70 to 80 s), very slow surface waves (mark B) propagated towards Tokyo and Chiba while preserving their large amplitudes and shapes of wave packets.

Figure 10b shows the seismic wave propagation during the northern Ibaraki earthquake. At a lapse time of t = 45 s, large-amplitude surface waves were excited at the Ibaraki-Chiba border (mark D), and consequently, these waves propagated towards central Chiba. There was a significant dispersion of surface waves at northern Chiba (mark E), and these showed a delay of the large-amplitude arrival, which was confirmed with the observed seismograms at CHBH13 (lower right part of Figure 9).

From the snapshots of both earthquakes, we determined that the excitation of the surface waves occurred effectively at the area with a bedrock depth of approximately 1 km (e.g., at mark A and mark D in Figure 10). The excited surface waves propagated in the laterally inhomogeneous sedimentary basin and displayed complicated behaviors of propagation, especially for the northern Tochigi earthquake (mark B and mark C in Figure 10a). A conventional Fourier analysis revealed that the dominant period of the surface waves was approximately 6 s.

Figure 11 shows the velocity seismograms of the transverse component at the stations aligned along the propagation path of phases C and B in the middle and eastern parts of Saitama. In the aligned seismograms, we confirmed both the faster phase C and slower phase B as seen in Figure 10a, and these phases were reproduced by our 3D FDM simulation.

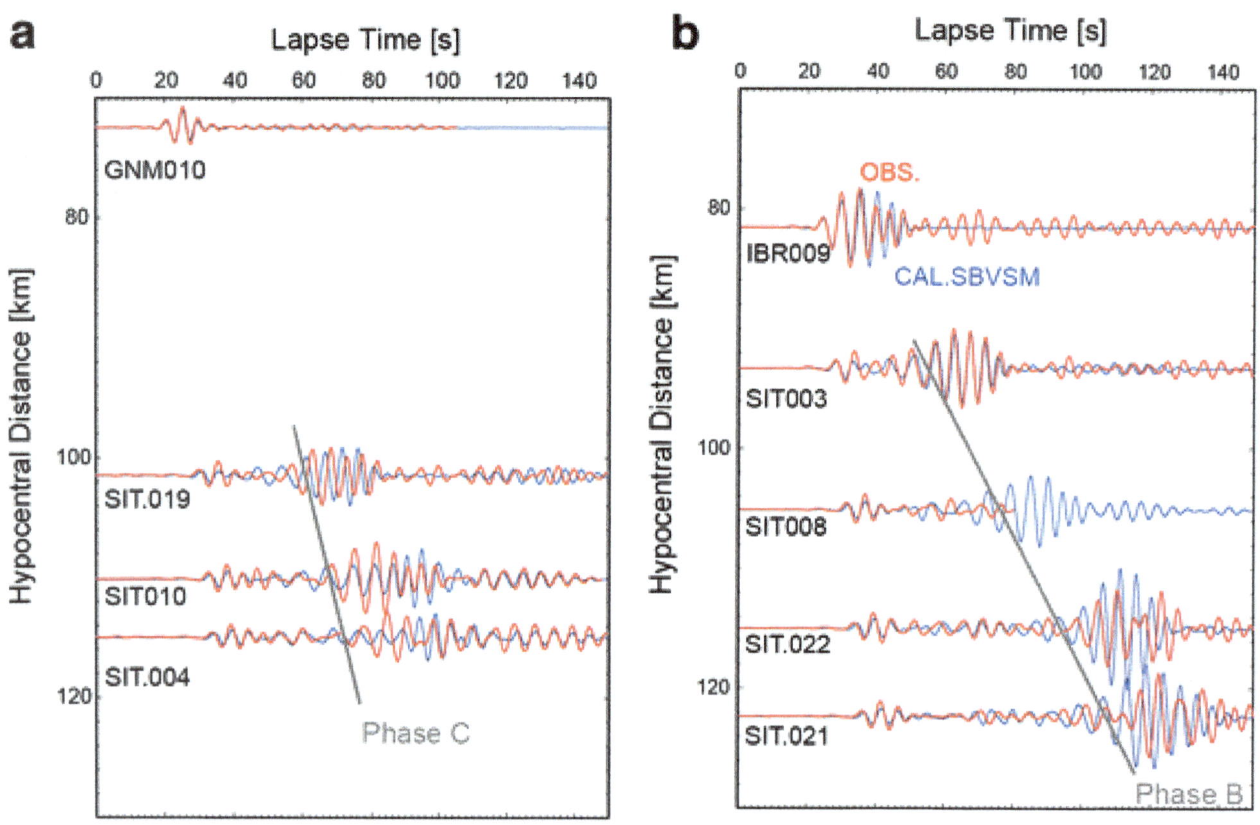

Figure 11 Aligned transverse velocity seismograms for the period of 4 to 8 s.Seismograms were recorded at **(a)** stations in middle Saitama (yellow triangles in Figure 10a) and **(b)** stations in eastern Saitama (red triangles in Figure 10a) during the northern Tochigi earthquake. Blue and red lines are simulations of the SBVSM and observations, respectively.

To interpret this observation, we evaluated the two-dimensional (2D) map of the group velocity of fundamental Love waves for the periods of 4, 5, 6, 7, and 8 s using the codes developed by Herrmann (2013). The 1D velocity structure of the SBVSM at each site was used for constructing this map. Figure 12 illustrates the 2D map of the group velocities for each period, and these maps show strong lateral variation in the Kanto Basin. The map of group velocity for the period of 6 s successfully explains the propagation speed of both phases in the observations and simulations.

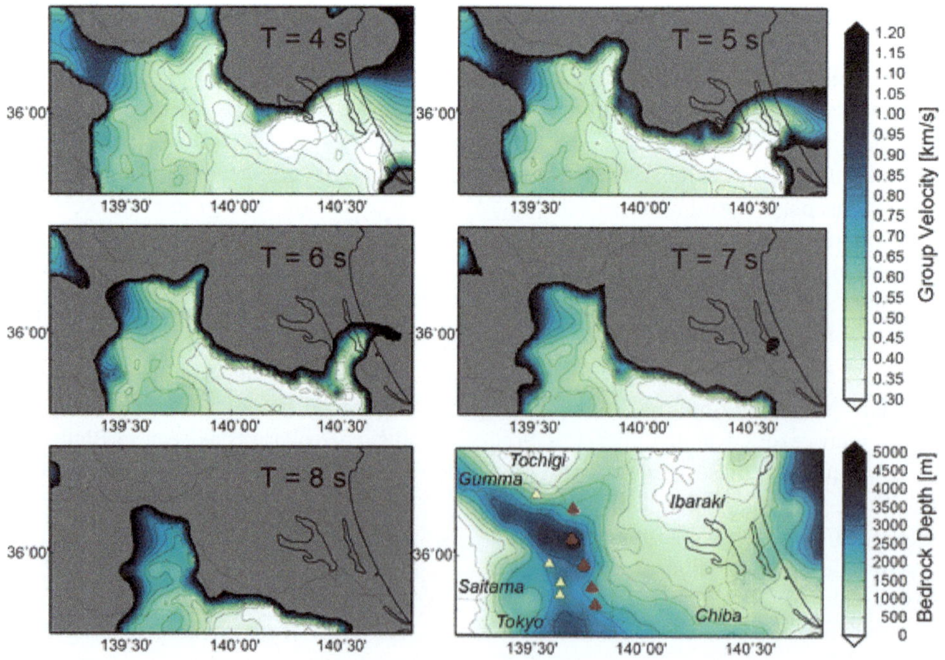

Figure 12 Spatial variation of Love wave group velocity for periods (T) of 4 to 8 s. The local group velocities of Love waves were derived from the SBVSM. The spatial variation of the bedrock depth is also shown in the lower right panel. The areas, where group velocities are larger than 1.2 km/s, were masked. The group velocities for each period were calculated using codes developed by Herrmann (2013). Plotted stations in the map of bedrock depth are the same as those used in Figure 10.

In Figures 10 and 11, large-amplitude Love waves stably propagated towards the southern part of the Kanto Basin, preserving large amplitudes and shapes of distinct wave packets. Similar observations of distinct surface waves have been reported in previous studies (e.g., Furumura and Hayakawa 2007; Hayakawa 2009). To clarify the cause of the observed stable propagation of Love waves in the inhomogeneous Kanto Basin, we examined the detailed period dependency of the Love wave group velocity beneath the linear array. Figure 13 shows the Love wave group velocities as a function of the period derived from the local S-wave velocity structures beneath the linear array of eastern Saitama. The evaluated group velocities displayed very weak period dependency at and around the dominant period of the observed Love waves (6 s), especially at SIT.021 and SIT.022. This may be the reason why the observed and simulated Love waves stably propagated without strong dispersion and, consequently, preserved their shapes of wave packets.

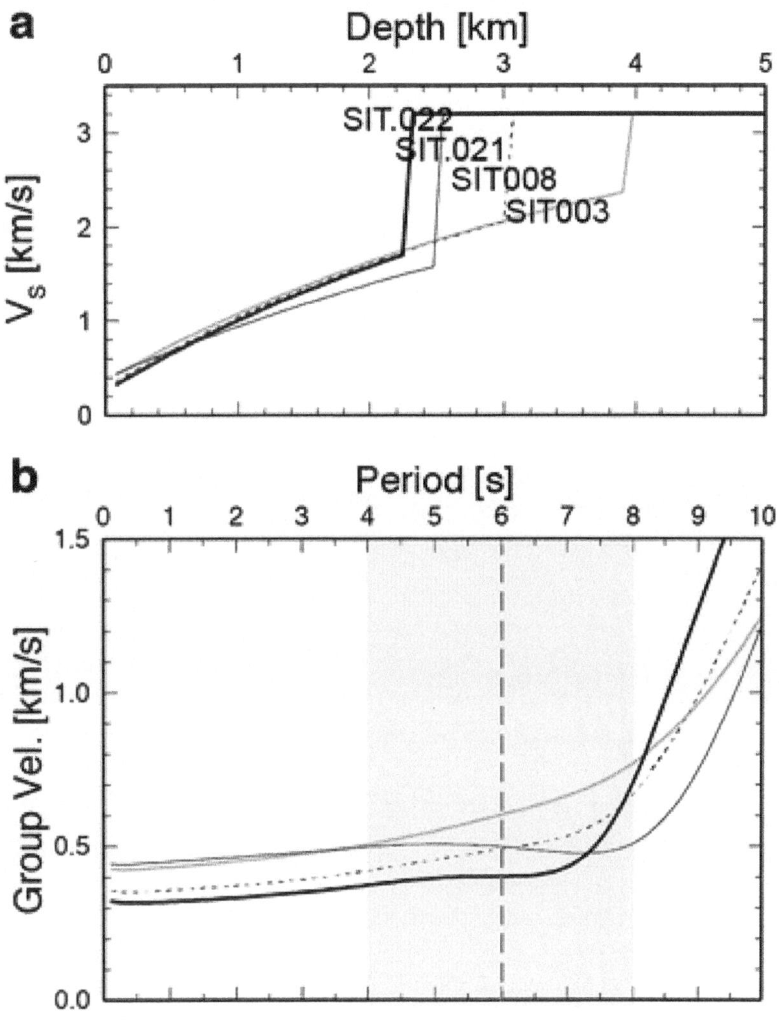

Figure 13 Group velocity of Love waves as a function of the period. (a) Local S-wave velocity structures from the SBVSM and **(b)** period-dependent group velocity of the fundamental Love wave at each station. Stations are shown in the lower right part of Figure 10a. Red dashed line represents the dominant period of the distinct Love wave packet during the northern Tochigi earthquake. Group velocities were calculated using codes developed by Herrmann (2013).

DISCUSSION

Anelastic attenuation of sediments is an important factor for the amplification of long-period ground motion in sedimentary basins. In this study, we employed Q values estimated by a simple empirical relation (Brocher 2008) and a frequency-dependent Q^{-1} model for P and Swaves with a central frequency of $f_0 = 1$ Hz based on the formulations for FDM simulations of Robertsson et al. (1994). Here, we discuss the effect of the anelastic attenuation of the sediments on the modeling of long-period ground motions.

Figure 14a shows the frequency-dependent Q_S^{-1} (black line) assumed in the simulations, which was obtained from the JIVSM sedimentary layer 2 (see Table 1) and Brocher›s empirical relation with various central frequencies (navy blue lines). Brocher Q_S^{-1} values were calculated using $V_S = 0.9$ km/s, which corresponds to the S-wave velocity of JIVSM layer 2. The attenuation of Brocher Q_S^{-1} is approximately three times stronger than that of the JIVSM Q_S^{-1} if the same central frequency is assumed. If we employ a lower central frequency ($f_0 = 0.5$ Hz), very strong attenuation can be expected for the frequency range of long-period ground motion (pink area in Figure 14a).

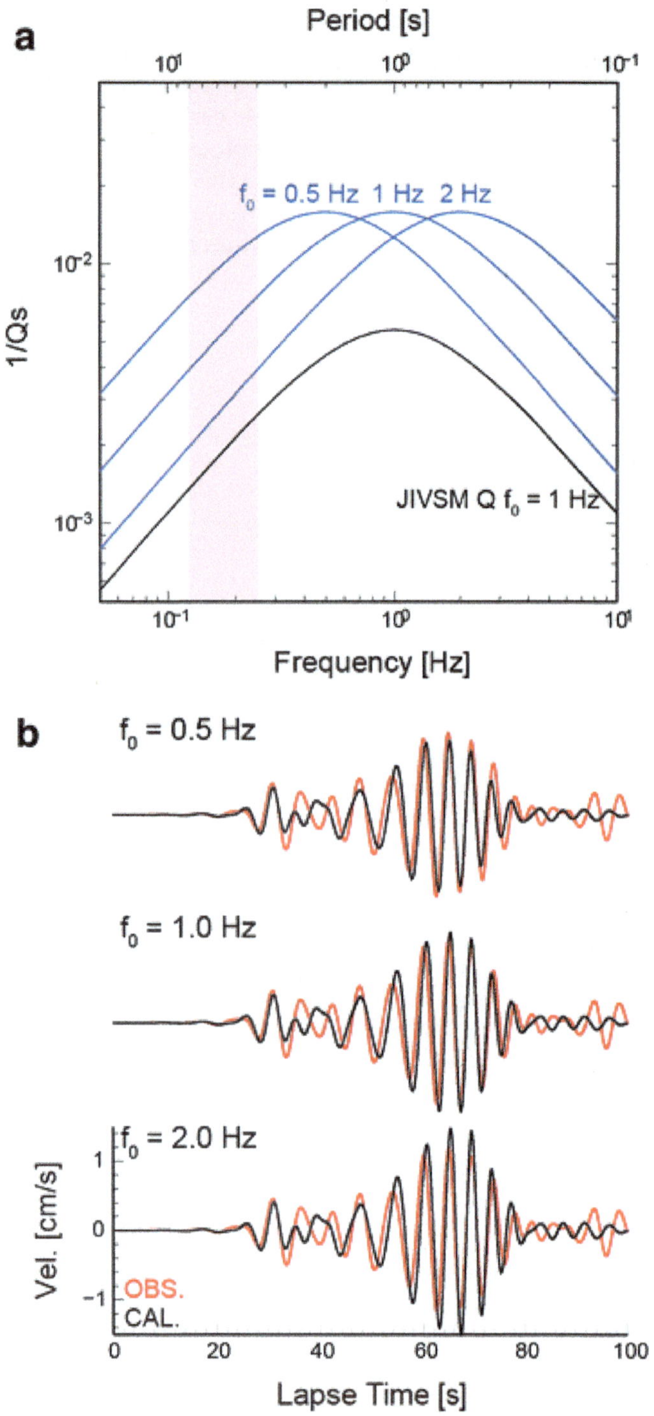

Figure 14 Effect of frequency-dependent Q_S^{-1} on simulated seismograms. **(a)** Frequency dependence of Q_S^{-1} derived from the JIVSM Q_S of sedimentary layer 2 (black line) and the Brocher Q_S relation assuming $V_S = 0.9$ km/s (navy blue lines). **(b)** Velocity seismograms of the transverse component at SIT003 for frequencies of 0.125–0.25 Hz (period of 4 to 8 s) derived from simulation results of the northern Tochigi earthquake with various central frequencies of Q_S^{-1}. The pink area in **(a)** represents the target frequency range in our study.

To investigate the optimal Q_S values for long-period ground motions, we conducted additional FDM simulations of long-period ground motions using the central frequencies of $f_0 = 0.5$, 1, and 2 Hz. Figure 14b shows a comparison of the simulated and observed waveforms of the transverse component (period of 4 to 8 s) at SIT003 (red triangle in Figure 1) during the northern Tochigi earthquake. The simulation results demonstrated that the central frequency of Q_S affected the amplitude of the later part of the surface wave packet (period of approximately 4 s), whereas no apparent effects appeared for the earlier part and dispersion feature of the Love wave. By comparing these results with the observed seismogram, the most preferable value of the central frequency was 1 Hz, which was used in the 3D FDM simulations in this study.

Although we proposed a new 3D velocity structure model for the evaluation of the long-period ground motions in the northern Kanto Basin, it may not be acceptable for short-period (<1 s) seismic wave propagation because of the lack of small-scale velocity heterogeneities. The original S-wave velocity structure obtained by VSP measurements has rich shorter wavelength components of the velocity heterogeneities (Yamamizu 1996, 2004; Shiomi et al. 1997). To achieve precise simulations that are capable of reproducing shorter-period seismic waves, such as for the evaluation of the peak ground acceleration and scattering in the Kanto Basin, we should revisit the original results from the VSP measurements and the small-scale velocity heterogeneities should be embedded in the model.

In this study, we used only one earthquake for the estimation of the local structural parameters (V_0 and α) near the basin edge. It would have been better to use more earthquakes distributed in a wide azimuthal range for precise estimations. However, there were very few relevant moderate earthquakes (Mw approximately 6) that occurred near the basin edge. Hence, it would be useful to validate and reconstruct the SBVSM using various azimuthal events in future studies.

In the Kanto Basin, the sedimentary structure of the JIVSM was mainly constructed using geophysical information from refraction surveys (Koketsu et al. 2008, 2009). However, the refraction surveys were not densely carried out in the northern part of the Kanto Basin compared with the middle and southern parts. Thus, the JIVSM did not reproduce the observed long-period ground motion very well, especially for excitation near the basin edge. In contrast, by introducing dense geophysical data around the northern Kanto Basin (see Figure 7a), the SBVSM was able to reproduce the observed excitation and propagation of long-period surface waves. We believe that the incorporation of data from denser microtremor surveys has the potential to improve the reproducibility of the observed long-period ground motions in the Kanto Basin. In addition, precise information on the bedrock depths, which was not discussed in this study, should be required for more confident evaluations in future studies.

CONCLUSIONS

For practical evaluations of the long-period ground motions (period of 4 to 8 s), we constructed a 3D velocity structure model of the northern Kanto Basin using a simple velocity gradient function. Our structure model SBVSM, which was constructed from a waveform analysis and dense geophysical data from VSP measurements, the JIVSM, and the microtremor surveys showed strong lateral inhomogeneities in this basin. The large-scale 3D FDM simulations using the SBVSM model accurately reproduced the observed long-period ground motions during two moderate earthquakes that occurred in the northern Kanto region. Hence, these results show the adequacy of our modeling procedure.

By comparing the simulations and observations, it was found that the observed long-period ground motions were generated by the long-period surface waves excited at the basin edge, where the bedrock depths are approximately 1 km. The excited surface waves propagated towards the central or southern parts of the Kanto Basin, and they showed complicated propagation patterns with large-amplitude, distinct, long-period wave packets. Our results strongly indicate that it is necessary to consider the strong lateral variation of the sedimentary velocity structure to better understand the complex propagation characteristics of the surface waves in this basin.

ACKNOWLEDGEMENTS

We would like to thank two anonymous reviewers and the editor, H. Takenaka, for constructive comments that improved an earlier draft of this manuscript. We acknowledge the National Research Institute for Earth Science and Disaster Prevention, Japan (NIED) for providing the KiK-net and K-NET waveform data and the CMT solutions from the F-net. We also used SK-net waveform data that were provided by the Earthquake Research Institute, the University of Tokyo. The FDM simulations of seismic wave propagation were conducted on the computer system of the Earthquake and Volcano Information Center at the Earthquake Research Institute, the University of Tokyo. One of the authors, ST, is grateful for the financial support provided by a Grant-in-Aid from JSPS (the Japan Society for Promotion of Science; No. 24.5704). All figures in the present study were drawn using the Generic Mapping Tools software package developed by Wessel and Smith (1998).

AUTHORS' CONTRIBUTIONS

ST conducted the large-scale 3D FDM simulations, analyzed both simulated and observed seismograms, and drafted this manuscript. MA conducted 3D FDM simulations for modeling the velocity structures near the northern edge of the Kanto Basin. KM carried out the conversions of layered structures derived from microtremor surveys to smoothed depth-gradient structures. KK carried out the theoretical analysis of long-period surface wave propagation in the Kanto Basin. KY surveyed the available geophysical data and helped draft this manuscript. All authors approved the final manuscript.

REFERENCES

1. Asano K, Iwata T (2012) Source model for strong ground motion generation in the frequency range 0.1-10 Hz during 2011 Tohoku earthquake. Earth Planets Space 64:1111–1123, doi:10.5047/eps.2012.05.003
2. Beck JL, Hall JF (1986) Factors contributing to the catastrophe in Mexico City during the earthquake of September 19, 1985. Geophys Res Lett 13:593–596, doi:10.1029/GL013i006p00593
3. Brocher TM (2008) Key elements of regional seismic velocity models for long period ground motion simulations. J Seismol 12:217–221, doi:10.1007/s10950-007-9061-3
4. Fujiwara H, Kawai S, Aoi S, Morikawa N, Senna S, Kudo N, Ooi M, Hao KX-S, Hayakawa Y, Toyama N, Matsuyama N, Iwamoto K, Suzuki H, Ei R (2009) A study on subsurface structure model for deep sedimentary layers of Japan for strong-motion evaluation. Technical note of the National Research Institute for Earth Science and Disaster Prevention, No 337 (in Japanese)
5. Fukuyama E, Ishida M, Dreger DS, Kawai H (1998) Automated seismic moment tensor determination by using on-line broadband seismic waveforms. Zisin 51:149–156, in Japanese with English abstract
6. Furumura T, Chen L (2004) Large scale parallel simulation and visualization of 3-D seismic wavefield using Earth simulator. Comput Model Eng Sci 6:153–168, doi:10.3970/cmes/2004.006.153
7. Furumura T, Hayakawa T (2007) Anomalous propagation of long-period ground motions recorded in Tokyo during the 23 October 2004 M_w 6.6 Niigata-ken Chuetsu, Japan, Earthquake. Bull Seismol Soc Am 97:863–880, doi:10.1785/0120060166
8. Furumura T, Takemura S, Noguchi S, Takemoto T, Maeda T, Iwai K, Padhy S (2011) Strong ground motions from the 2011 off-the Pacific-Coast-of-Tohoku, Japan (Mw = 9.0) earthquake obtained from a dense nationwide seismic network. Landslides 8:333–338, doi:10.1007/s10346-011-0279-3
9. Hayakawa T (2009) Generation and propagation of the long-period ground motion in the Kanto basin: dense seismic-network analysis and large-scale numerical simulations. PhD Dissertation, The University of Tokyo, Japan (in Japanese with English abstract)
10. Herrmann RB (2013) Computer programs in seismology: an evolving tool for instruction and research. Seismol Res Lett 84:1081–1088, doi:10.1785/0220110096
11. Horikawa H, Yoshimi M, Sekiguchi H, Yoshida K, Sugiyama Y, Satake K, Fukuwa N, Suzuki H, Matsuyama H, Ying L, Takizawa F (2008) A three-dimensional subsurface structure model of the Chukyo area, central Japan. Annual report on active fault and paleoearthquake researches. Geol Surv Jpn 8:203–254, in Japanese with English abstract
12. Iwaki A, Iwata T (2010) Simulation of long-period ground motion in the Osaka sedimentary basin: performance estimation and the basin structure effects. Geophys J Int 181:1062–1076, doi:10.1111/j.1365-246X.2010.04556.x
13. Iwaki A, Morikawa N, Maeda T, Aoi S, Fujiwara H (2013) Finite-difference simulation of long-period ground motion for the Sagami Trough megathrust earthquakes. J Disast Res 8(5):926–940
14. Ji C, Helmberger DV, Wald DJ, Ma KF (2003) Slip distribution and dynamic implication of the 1999 Chi-Chi, Taiwan earthquake. J Geophys Res 108:2412, doi:10.1029/2002JB001764
15. Kinoshita S (1998) Kyoshin net (K-NET). Seismol Res Lett 69:302–332
16. Kinoshita S, Fujiwara H, Mikoshiba T, Hoshino T (1992) Secondary Love waves observed by a strong-motion array in the Tokyo lowlands, Japan. J Phys Earth 40:99–116

17. Koketsu K, Kikuchi M (2000) Propagation of seismic ground motion in the Kanto basin, Japan. Science 288(5469):1237–1239, doi:10.1126/science.288.5469.1237
18. Koketsu K, Miyake H (2008) A seismological overview of long-period ground motion. J Seismol 12:133–143, doi:10.1007/s10950-007-9080-0
19. Koketsu K, Hatayama K, Furumura T, Ikegami Y, Akiyama S (2005) Damaging long-period ground motions from the 2003 Mw 8.3 Tokachi-oki, Japan earthquake. Seismol Res Lett 76:67–73, doi:10.1785/gssrl.76.1.67
20. Koketsu K, Miyake H, Fujiwara H, Hashimoto T (2008) Progress towards a Japan integrated velocity structure model and long-period ground motion hazard map. In: Proceedings of the 14th World Conference on Earthquake Engineering, S10-038, Beijing, October 12–17, 2008
21. Koketsu K, Miyake H, Afnimar TY (2009) A proposal for a standard procedure of modeling 3-D velocity structures and its application to the Tokyo metropolitan area, Japan. Tectonophysics 472:290–300, doi:10.1016/j.tecto.2008.05.037
22. Koketsu K, Yokota Y, Nishimura N, Yagi Y, Miyazaki S, Satake K, Fujii Y, Miyake H, Sakai S, Yamanaka Y, Okada T (2011) A unified source model for the 2011 Tohoku earthquake. Earth Planet Sci Lett 310:480–487, doi:10.1016/j.epsl.2011.09.009
23. Maeda T, Furumura T (2013) FDM simulation of seismic waves, ocean acoustic waves and tsunamis based on tsunami-coupled equations of motion. Pure Appl Geophys 170:109–127, doi:10.1007/s00024-011-0430-z
24. Maeda T, Furumura T, Noguschi S, Takemura S, Sakai S, Shinohara M, Iwai K, Lee SJ (2013a) Seismic- and tsunami-wave propagation of the 2011 off the Pacific coast of Tohoku earthquake as inferred from the tsunami-coupled finite-difference simulation. Bull Seismol Soc Am 103:1456–1472, doi:10.1785/0120120118
25. Maeda T, Morikawa N, Iwaki A, Aoi S, Fujiwara H (2013b) Finite-difference simulation of long-period ground motion for the Nankai Trough megathrust earthquakes. J Disast Res 8(5):912–925
26. Maeda T, Furumura T, Obara K (2014) Scattering of teleseismic P-wave by the Japan Trench: a significant effect of reverberation in the seawater column. Earth Planet Sci Lett 397:101–110, doi:10.1016/j.epsl.2014.04.037
27. Matsuoka T, Shiraishi H (2002) Application of an exploration method using microtremor array observations for high resolution surveys of deep geological structures in the Kanto plains: estimation of 3-D S-wave velocity structure in the southern part of Saitama prefecture. Butsuri Tansa 55(2):127–143, in Japanese with English abstract
28. Ministry of Education, Sports, Culture, Science and Technology, Japan (MEXT) (2007) Deep seismic exploration. In: MEXT (ed) Annual report of special project for earthquake disaster mitigation in urban areas, I. Regional characterization of the crust in metropolitan areas for prediction of strong ground motion. MEXT, Tokyo, pp. 215–230
29. Miura H, Midorikawa S (2001) Effects of 3-D deep underground structure on characteristics of rather long-period ground motion: examination in and around Yokohama City. Zisin 54:381–395, in Japanese with English abstract
30. Miyake H, Koketsu K (2005) Long-period ground motions from a large offshore earthquake: the case of the 2004 off the Kii peninsula earthquake, Japan. Earth Planets Space 57:203–207
31. Moczo P, Robertsson OA, Einster L (2007) The finite-difference time-domain method for modeling of seismic wave propagation. In: Wu R-S, Maupin V (eds), Advances in wave propagation in heterogeneous earth. Adv Geophys 48:421–516, doi:10.1016/S0065-2687(06)48008-0
32. Okada Y, Kasahara K, Hori S, Obara K, Sekiguchi S, Fujiwara H, Yamamoto A (2004) Recent progress of seismic observation networks in Japan—Hi-net, F-net, K-NET and KiK-net—. Earth Planets Space 56:xv–xxviii
33. Ravve I, Koren Z (2006) Exponential asymptotically bounded velocity model: part I—effective models and velocity transformations. Geophysics 71:T53–T65, doi:10.1190/1.2196033
34. Robertsson J, Blanch JO, Symes WW (1994) Viscoelastic finite-difference modeling. Geophysics 59:1444–1456, doi:10.1190/1.1443701
35. Senna S, Maeda T, Inagaki Y, Suzuki H, Matuyama H, Fujiwara H (2013) Modeling of the subsurface structure from the seismic bedrock to the ground surface for a broadband strong motion evaluation. J Disast Res 8(5):889–903
36. Shin T-C, Teng T-L (2001) An overview of the 1999 Chi-Chi, Taiwan, Earthquake. Bull Seismol Soc Am 91:895–913, doi:10.1785/0120000738
37. Shiomi K, Sato H, Ohtake M (1997) Broad-band power-law spectra of well-log data in Japan. Geophys J Int 130:57–64, doi:10.1111/j.1365-246X.1997.tb00987.x
38. Takemura S, Yoshimoto K (2014) Strong seismic wave scattering in the low-velocity anomaly associated with subduction of oceanic plate. Geophys J Int 197:1016–1032, doi:10.1093/gji/ggu031
39. Wessel P, Smith WHF (1998) New improved version of generic mapping tools released. EOS, Trans Am Geophys Union 79(47):579, doi:10.1029/98EO00426
40. Yamamizu F (1996) Down-hole measurements of seismic wave velocities in deep soil deposits beneath the Tokyo metropolitan area. Rep Natl Res Inst Earth Sci Disaster Prev 56:1–32
41. Yamamizu F (2004) Seismic wave velocity structures in Kanto area as revealed by the crustal activity observation well VSP. Tech Note Natl Res Inst Earth Sci Disaster Prev 251:1–75, in Japanese with English abstract
42. Yamanaka H, Yamada N (2006) Estimation of 3D S-wave velocity model of the Kanto basin, Japan, using

Rayleigh wave phase velocity. Bull Earthq Res Inst Univ Tokyo 81:295–301

43. Yamanaka H, Seo K, Samano T (1989) Effects of sedimentary layers on surface-wave propagation. Bull Seismol Soc Am 79:631–644

44. Yamanaka H, Komada N, Yamada N (2005) Estimation of 3D S-wave velocity model of the Kanto basin, Japan, for prediction of long-period ground motion. In: Proceedings of 2nd international conference on urban earthquake engineering, Tokyo, 7-8 March 2005

45. Yamazaki K, Minamishima M, Kudo K (1992) Propagation characteristics of intermediate-period (1–10 seconds) surface waves in the Kanto Plain, Japan. J Phys Earth 40:117–136

46. Yoshimoto K, Takemura S (2014a) Surface wave excitation at northern edge of Kanto basin, Japan. Earth Planets Space 66:16, doi:10.1186/1880-5981-66-16

47. Yoshimoto K, Takemura S (2014b) A study on the predominant period of long-period ground motions in the Kanto Basin, Japan. Earth Planets Space 66:100, doi:10.1186/1880-5981-66-100

CHAPTER 8

LONG-PERIOD GROUND MOTION FROM EARTHQUAKES

Arben Pitarka, Abdullah Al-Amri, Michael E. Pasyanos, Arthur J. Rodgers And Robert J. Mellors

INTRODUCTION

The Persian/Arabian Gulf (hereafter referred to as the Arabian Gulf, or simply Gulf) lies near the boundary of the Arabian and Eurasian Plates. While the Gulf is mostly aseismic, the Arabian Peninsula is surrounded by regional seismic sources in the tectonically active areas of Iran and Turkey to the northeast, the Red Sea Rift bordering the Shield to the southwest, and the Dead Sea Transform fault zone to the north (Johnson1998; Al-Amri2013). These seismic sources control the majority of the seismic hazard in and around the Arabian Peninsula. Throughout recorded history, many damaging earthquakes have occurred along the Arabian Plate boundaries. These events have damaged buildings, and resulted in injuries and fatalities (Ambraseys1988; Ambraseys*et al.* 1994; Alsinawi1986).

The Zagros Mountains, which are located on the northeast side of the Arabian Gulf, is a very seismically active region. Large earthquakes in this region are capable of producing significant ground motion on the southwest side of the Gulf. Recent examples are the 2005 M5.7 and the 2008 M6.1 Qeshm Island earthquakes, which were particularly strongly felt in high-rise buildings in the urban centers along the Gulf. The Makran subduction zone, located at the southern end of the plate boundary, where the Arabian plate subducts underneath the Eurasian plate, is capable of magnitude 7 and 8 events. These will also produce strong shaking along the Gulf. A repeat of the November 1945 M8.1 Makran earthquake will cause long-duration and high-amplitude ground motion along the Gulf.

In this paper, we analyze ground motions recorded on the Arabian side of the Gulf using a mix of observations and full waveform 3D modeling. In particular, we examine the amplitude of long-period surface waves, which may present a hazard to large structures with resonant periods of 2–10 s. A number of large structures, such as high-rise buildings and bridges, exist in populated areas along the Gulf, and more are planned to be build. As the population increases and new areas are developed, the seismic risk to human life and infrastructure increases. In order to fully capture the seismic hazard of the Arabian Peninsula, we must understand the effects of the regional 3D underground structure on wave propagation. These effects cannot not be fully predicted by simply adopting GMPEs that are developed for other regions.

In this article we briefly describe the tectonic setting and crustal structure of the region. Then we discuss characteristics of ground motion from recent events. Finally, we investigate structural effects on ground motion by modeling 3D anelastic wave propagation using a finite difference method.

TECTONICS AND SEISMICITY OF THE ARABIAN PLATFORM REGION

Figure 1 shows the location of the study area which includes the Arabian Gulf. The Arabian Plate is under-thrusting the southern Eurasian margin along the Zagros Mountains Thrust on the northwest side of the Arabian Gulf. Further southeast the oceanic crust continues to subduct under the southern region of the Eurasian plate. The transition zone between the continental underthrusting and oceanic subduction consists of complex strike-slip faulting associated with some thrust faulting.

Citation: Pitarka, A., Al-Amri, A., Pasyanos, M. E., Rodgers, A. J., & Mellors, R. J. (2015). Long-Period Ground Motion from Earthquakes. DOI: http://dx.doi.org/10.1007/s00024-014-0858-z.

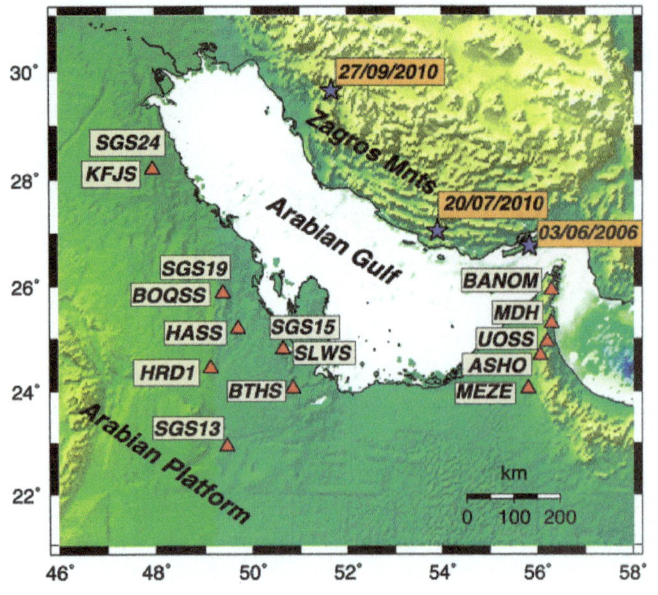

Figure 1: Map of the study region. *Triangles* show the location of seismic stations, and *stars* show the location of three earthquakes used in analysis of recorded ground motion.

The Arabian Peninsula consists of exposed Paleozoic shield in the west, but is covered by sediments in the east (Arabian Platform). These sediments dip E–NE gently into the thick basin sediments of the Arabian Gulf. In general, the crust is 40–45 km thick and may be slightly thicker in the east. The velocity of the upper mantle is about 8.2 km/s (Al-Amri1998; Al-Amri*et al.*1999; Rodgers*et al.* 1999; Tkalčić*et al.* 2006). Gök (2008) showed that sedimentary basin structure varies across the Arabian Platform, but crystalline crust is largely uniform in thickness and wavespeed.

Regional seismic phases display considerable azimuthal variation. High-frequency regional S-wave phases are quite different for paths crossing the Arabian Shield than those crossing the Arabian Platform (Mellors*et al.* 1999; Al-Damegh*et al.* 2004). In particular, the mantle Sn phase is nearly absent for paths crossing parts of the Arabian Shield, while the crustal Lg phase has abnormally large amplitude. The situation is reversed for paths crossing the Platform and Gulf, where Lg tends to show lower amplitudes, but Sn propagates efficiently. These variations appear to be due to a combination of variations in mantle attenuation and crustal structure (Rodgers*et al.* 1997). Al-Damegh*et al.* (2004). observed low Q_{Pn} (<100 at 1.5 Hz) along the western coast of the Arabian Plate and along the Dead Sea fault system. Higher Q_{Pn} values of the order of 400 were observed within the Arabian Shield and Platform at the same frequency. Similar patterns of crust and upper mantle Q_p and Q_s in the Arabian Shield and Arabian Platform are found by Pasyanos*et al.* (2009) in their model of *P*-wave and *S*-wave attenuation in the Middle East region using multiphase (Pn, Pg, Sn, Lg) attenuation tomography.

Because of the limited amount of recorded ground motion data and poor seismic station coverage, the seismicity of the Gulf region is not well characterized. The seismicity is very low in the Arabian Gulf. In contrast, the seismicity is very high in the Zagros Mountains thrust belt that extends for a distance of about 1,500 km in a northwest–southeast direction. The earthquakes in the Zagros define a zone of about 200 km wide that runs parallel to the fold belt. The majority of earthquakes occur in the crustal part of the Arabian plate that is underthrust along the folded belt. Magnitude 5 earthquakes are frequent and magnitude 6 may occur every year. The nature of deformation across this zone is complex, involving both thrust and strike-slip as indicated by earthquake focal mechanisms (Talebian and Jackson2004).

ANALYSIS OF RECORDED LONG-PERIOD GROUND MOTION IN THE GULF REGION

The Arabian Gulf lies on a deep sedimentary basin. The sedimentary layers dip eastward, reaching a depth of up to 8 km adjacent to the Zagros. Figure 2 shows a map of depth to basement in the region proposed by Laske and Masters (1997). The deep structure of the Gulf is composed of geologically old and consolidated sediments with moderately high shear velocities (e.g., Pasyanos*et al.* 2014). However, younger sediments near the surface have much lower velocities and probably high attenuation. Rodgers*et al.* (2006) and Al-Amri*et al.* (2008) demonstrated that the sedimentary layers of the Arabian Gulf cause higher amplitude and longer duration ground shaking then normally observed in a continental crustal structure. The complex basin structure is of particular concern for the earthquake hazard in the Gulf.

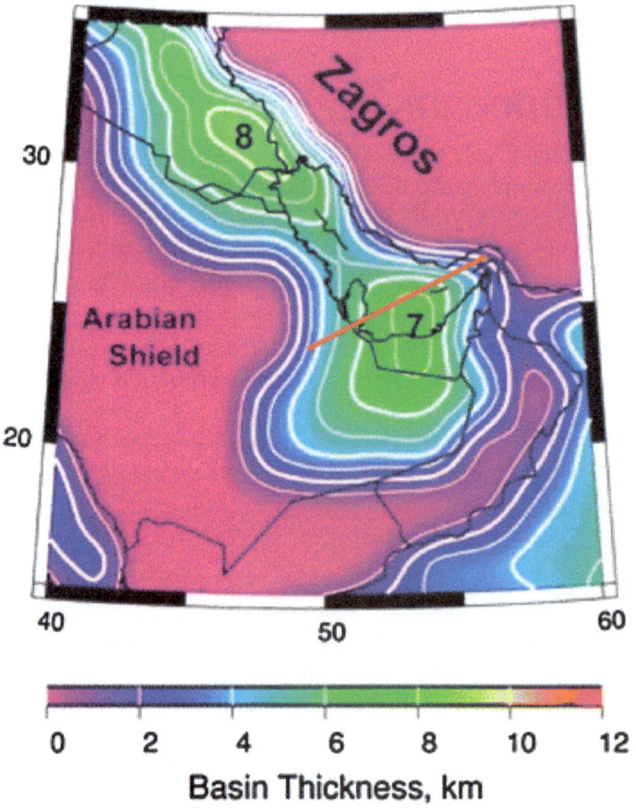

Figure 2: Sediment thickness of the Arabian Gulf and surrounding region (Laske and Masters1997).*Red line* indicates the location of the velocity model cross section shown in Fig. 7. *Bold numbers* indicate maximal basin depth in km

This study is an attempt to analyze quantitatively ground motion characteristics observed in the western side of the Arabian Gulf from earthquakes in the Zagros Mountains and Iranian Plateau. In particular, we focused on long-period (1–10 s) ground motion waveforms. Figure 3shows typical time histories of ground motion velocity recorded at seismic stations in eastern Saudi Arabia from the M5.1 June 3, 2006 earthquake (see Fig. 1 for its location). The north–south component, corresponding most naturally with the transverse component of ground motion is shown for two frequency ranges: (a) 0.02–0.05 Hz (50–20 s) and 0.1–1.0 Hz (10–1 s). The very long period band (20–50 s) shows the surface waves arriving between 3.5 and 3.0 km/s, as expected for normal continental paths at regional distance. The surface wave is relatively simple with a short duration, and is normally dispersed, as expected. However, at shorter periods of 1–10 s the ground motion shows an unexpectedly long duration of up to hundreds of seconds and at lower group velocities. This long duration shaking will cause increased cyclical loading on large engineered structures, such as high-rise buildings, bridges, and pipelines, that are sensitive to seismic motion in the period range of 1–10 s. Similar observations have been made for other earthquakes recorded at station HASS for which the wave path crosses the thick sedimentary cover in the Persian/Arabian Gulf. The observed, extraordinarily long duration is likely due to a waveguide effect from the deep sedimentary structure of the basin which enhances long-period ground motion (e.g., MELLORS*ET AL.* 1999). Seismic waveforms are affected by both wave path and source effects (e.g., PITARKA*ET AL.* 1998). Path effects are a consequence of heterogeneous underground structure and surface topography. Lateral and vertical variations in velocity cause dispersion, diffraction, and reflections, all of which affect wave amplitude and duration. The effect of wave scattering due to small-scale heterogeneities (e.g., FRANKEL and CLAYTON1986) is significant at intermediate and high frequencies (0.05–1 Hz). In general, wave attenuation is stronger in sedimentary basins than in basement and crystalline crust. However, the increased duration and relatively large amplitude of ground motion recorded along the west coast of the Gulf region suggests that seismic attenuation in the Gulf basin is very low for these periods.

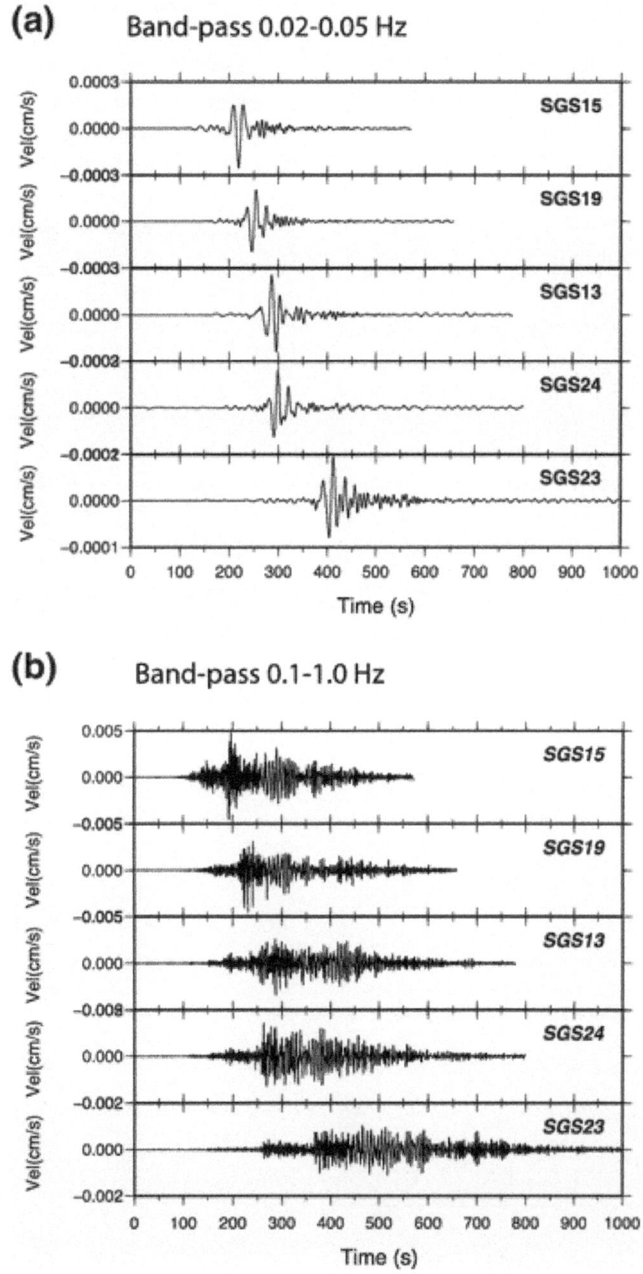

Figure: 3: Seismograms for June 3, 2006 earthquake recorded at stations in eastern Saudi Arabia filtered **a** 0.02–0.05 Hz (50–20 s), and **b** 0.1–1.0 Hz (10–1 s). Note that the 1–10 s period band shows an unusual long duration signal of more than 300 s

ACCELERATION RESPONSE SPECTRA

We used ground motions from several earthquakes in the Zagros thrust belt recorded at stations on the west coast of the Gulf to investigate the effects of the Gulf basin structure on long period wave propagation. The location of the seismic stations is shown in Fig. 1. Although the number of stations is limited, their location allows for a comparison of wave propagation effects along different E–W directions across the basin.

In the absence of calibrated GMPEs for the Gulf region we analyzed the spectral characteristics of ground motion in the region by comparing recorded spectral acceleration with the GMPEs developed by Campbell and Bozorgnia (2008) and Boore and Atkinson (2006), hereafter referred to as CB08 and BA06, respectively. These equations are based on ground motion data from earthquakes around the world. CB08, a Next-Generation Attenuation equation, was developed for earthquakes in active, shallow crustal structures. It is controlled by various parameters, including magnitude, distance, type of rupture, and basin depth (depth to layer with V_s = 2.5 km/s). Because of the limited amount of good quality data at long distances the standard deviation of this equation is large at distances longer than 200 km and periods shorter than 3 s. On the other hand, BA06 is constrained better at large distances. BA06 was developed for eastern North America, which in some respects resembles the sediments/shield and tectonically stable structure of eastern Arabian Platform.

We compared the spectral ratio between the recorded and predicted spectral acceleration for 5 % damping at different periods. Figure 4 shows the comparison with CB08 as a function of distance. Not all the earthquakes considered in this comparison were well recorded at all stations. We only show results for stations that produced good quality data from at least two earthquakes. The predicted spectral acceleration was computed for a depth to basement of 3.5 km. The spectral ratios clearly show different behaviors at different periods. The recorded spectral acceleration is much higher than the one predicted by the GMPEs in the 10 s period. It is comparable in the 5 s period and much lower in the 1 s period. The comparison with BA06 has a similar trend (Figs. 5, 6). In relative terms these results suggest that, in general, waves crossing the Gulf basin from the east are attenuated much less than expected at long periods. They are attenuated more than expected at shorter periods. Some of the amplification could be due to the basin structural effects, which for a deep basin such as the Gulf basin, are expected to be stronger at long periods. Long-period surface waves, which dominate the 10-s spectral response, are affected by the entire basin structure since they penetrate through the upper 10 km of the crust. On the other hand, shorter period waves (periods shorter than 1 s) have energy concentrated in the shallow basin sedimentary layers, and their shorter wavelengths are more susceptible to wave scattering effects compared to longer period waves. Consequently, short-period waves generated by earthquakes on the east coast of the Gulf are more prone to attenuation. Their amplitude is relatively small when they reach the west coast of the Gulf. The departure of the recorded spectral amplitudes from the GMPEs does not seem to correlate with source distance.

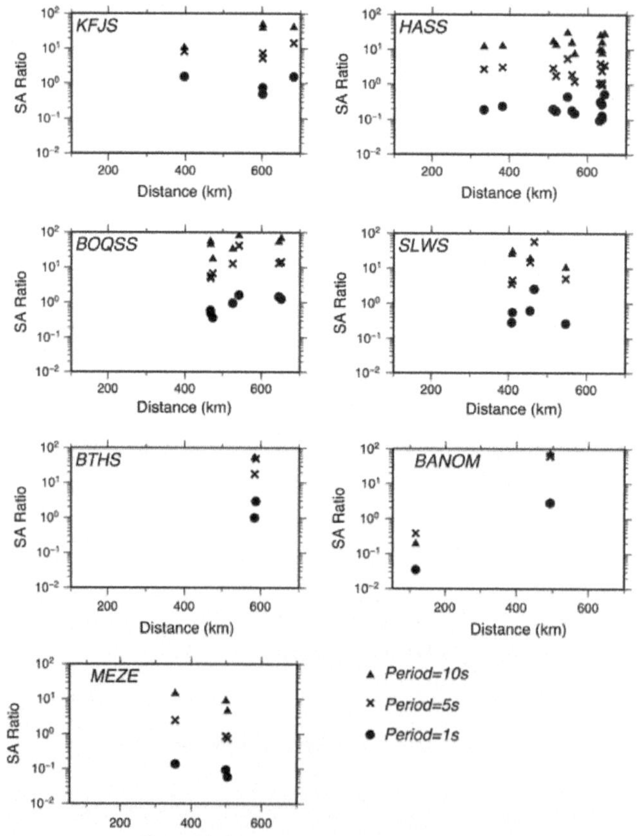

Figure 4: Spectral acceleration ratios of recorded and predicted ground motion for the geometric mean horizontal component at periods 1 s (*solid circle*), 5 s (*x*'s), and 10 s (*triangles*), and 5 % damping. The predicted ground motion was computed using the GMPE of (Campell and Bozorgnia2008). The station names are shown in each panel

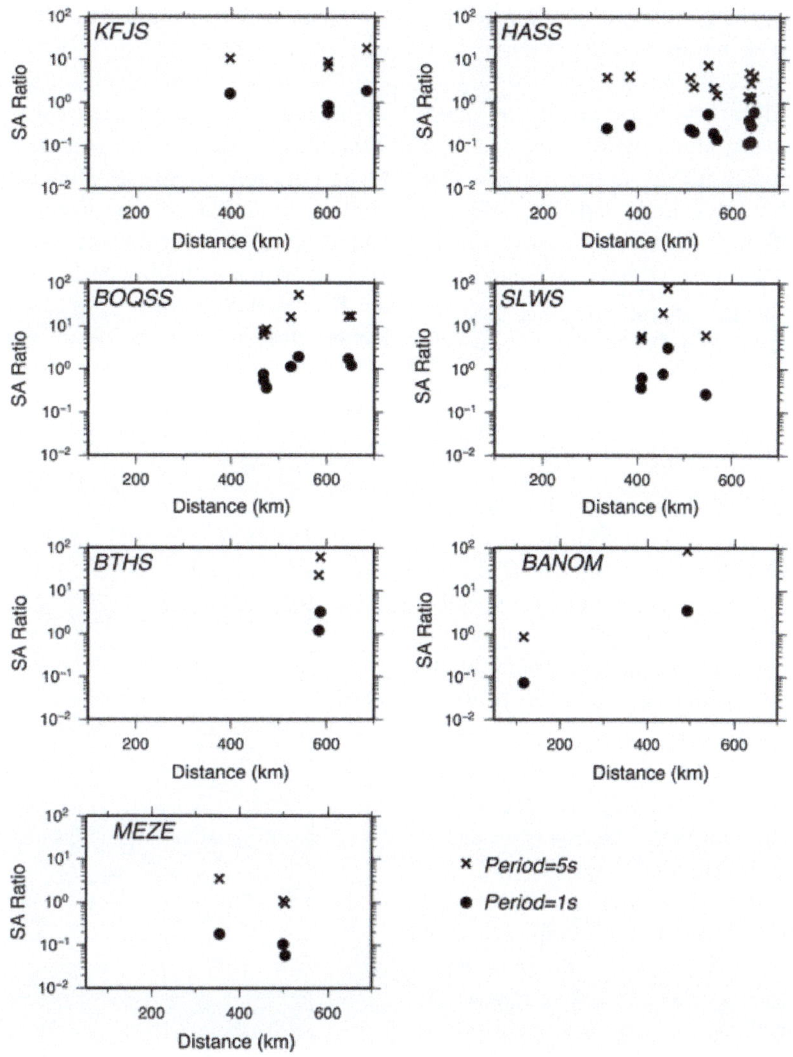

Figure 5: Spectral acceleration ratios of recorded and predicted ground motion for the geometric mean horizontal component at periods of 1 s (*solid circles*), 5 s (*x*'s), and 5 % damping. The predicted ground motion was computed using the GMPE of Boore and Atkinson (2006), developed for eastern North America. The station names are indicated in each panel

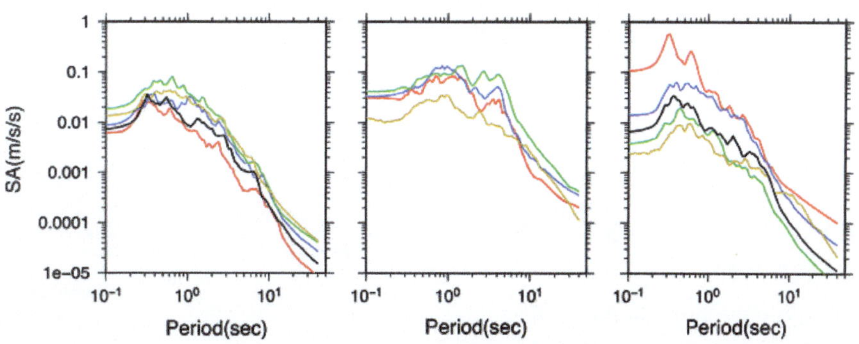

Figure 6: Geometrical mean of horizontal ground motion acceleration response spectra from M5.3 2006/06/03 earthquake (*left panel*), M5.7 2010/07/20 earthquake (*middle panel*), and M5.9 2010/09/27 earthquake (*right panel*), recorded at selected stations in the Gulf region. The spectra are corrected for the radiation pattern and geometrical spreading. The names of the stations are shown in each panel

We also observed a significant dissimilarity among acceleration response spectra from three selected earthquakes recorded at different stations. This is illustrated using ground motion time histories from the M5.3 2006/06/03, M5.7 2010/07/20, and M5.9 2010/09/27 earthquakes, corrected for geometrical spreading and radiation pattern. The earthquake locations are indicated in Fig. 1, and the comparison is shown in Fig. 8. In the case of the 2006/06/03 earthquake, the response spectra are similar for periods longer than 3 s, except for station SGS24, which recorded much lower ground motion. The similarity breaks down at shorter periods. The largest difference is obtained in the period range of 0.5–3 s between stations SGS15 and SGS13 which have a similar direction from the event, but very different source distances. Besides having a longer distance, SGS13 is located outside the basin. Our interpretation of the observed similarities between the stations for periods longer than 3 s and dissimilarities for periods shorter than 3 s is that, while the energy carried by longer period waves leaks outside the basin as they propagate toward the west, short-period energy is trapped in the basin sedimentary layers. Consequently, long-period waves on the west coast of the Gulf will maintain their large amplitude and duration regardless of the station's location relative to the basin. In contrast, the amplitude of short-period waves will mostly depend on whether the site is within or outside the basin.

The most significant feature of the response spectra from the 2010/07/20 earthquake is the difference in spectral amplitudes between stations SLWS, KFJS, and BOQS. The three stations are on wave paths that cross the deepest parts of the basin. However, at SLWS the spectral acceleration is much larger in the period range 2–20 s. In contrast, the spectral amplitudes at BOQSS are very similar to spectral amplitudes at KFJS, which has a much longer epicentral distance, and is located on different wave path. The spectral differences between SLWS and BOQSS are an indication of path-specific wave propagation effects on ground motion along different paths across the basin. The basin depth, and probably its geometry, along the two wave paths changes considerably.

A similar trend in ground motion variability is seen for the 2010/09/27 earthquake. Although the epicentral distances of stations BOQSS, SLWS, and HRD1 are twice as large as that of UOSS, the observed spectrum at these stations is at least a factor of 10 larger, especially in the period range of 0.1–3 s. At periods longer than 5 s the spectral amplitudes at UOSS exceed those at other stations. UOSS is located in the shallow part of the basin has a relatively flat spectrum on a broad period range. The similarities and dissimilarities among stations suggest that path-specific effects can selectively amplify certain periods at sites in the western Gulf coast.

Based on these observations, we conclude that the deep sedimentary structure in the Gulf acts as a waveguide that enhances long-period seismic wave amplification. In contrast, the shallow sedimentary layers attenuate high frequency waves, either due to wave scattering or due to low quality factor in the shallow sedimentary layers. Therefore, the knowledge of the basin structure on a broad scale range and validated modeling of its effects on wave propagation is crucial in predicting ground motion in the west coast of the Gulf.

SIMULATION OF BASIN EFFECTS ON LONG-PERIOD GROUND MOTION

The current knowledge of crustal structure in the Gulf region is limited to large-scale features that are expected to affect the amplitude and duration of long-period ground motion. New crustal models of the region are under development. PASYANOS*ET AL*. (2014) have recently developed a crustal model that has a higher resolution compared to existing models such as CRUST5.1 (MOONEY*ET AL*. 1998) and CRUST2.0 (BASSIN*ET AL*. 2000). In order to test our hypothesis of a basin wave guide to explain observed features of ground motion we investigated 3D wave propagation in the Gulf region using different basin models. We developed three experimental basin models representing different basin edge geometries and spatial extensions of surface sedimentary layers in the eastern and western parts of the Gulf. These features are not resolved in the current crustal models. The thickness of the basin was constrained by the basin structure reported by (LASKE and MASTERS1997). Figure 7 shows E–W vertical cross sections of the velocity models. In Model 1 the surface sedimentary layer continues across the entire model. In Model 2 the lateral extent of the surface sedimentary layer is limited toward the west. In Model 3 the geometry of the eastern basin edge is modified so that it can better channel the seismic energy coming from the seismic source below. The seismic velocity is the same for each model. The model dimensions are 800 km × 300 km × 90 km, and the top sedimentary layer has the properties: V_p = 2.2 km/s, V_s = 1.2 km/s, density 2 g/cm^3, Q_p = 200 and Q_s = 100.

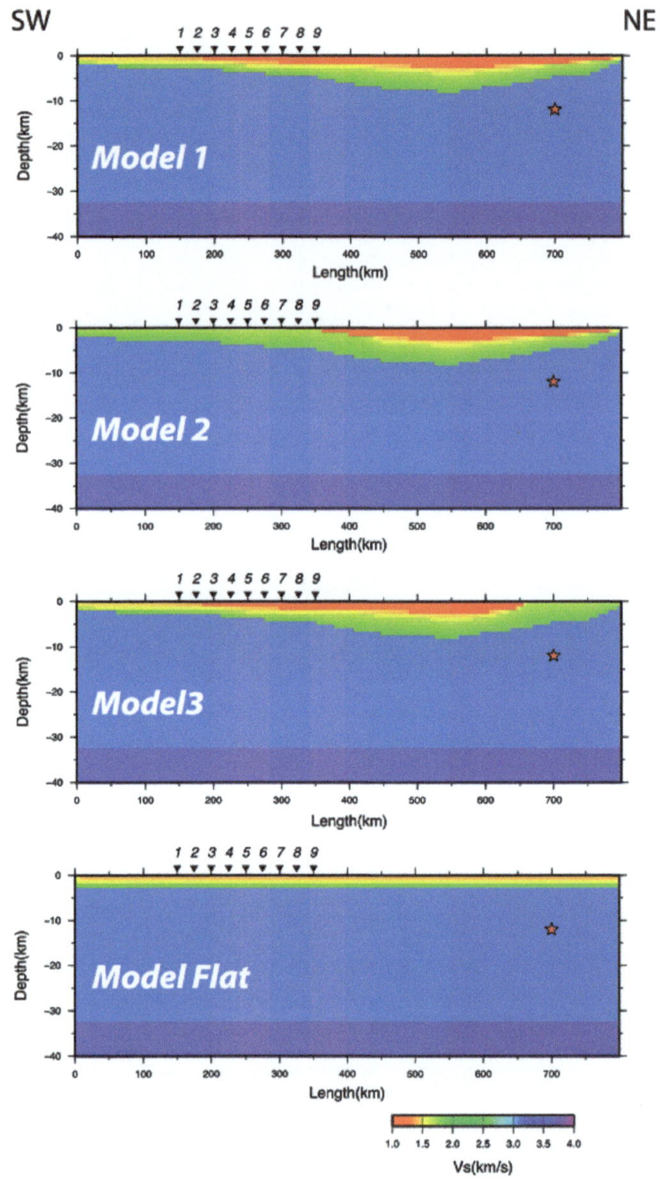

Figure 7: Vertical cross-sections of four velocity models used in the simulation of basin generated waves. The *star* indicates the location of the double-couple point source, and *triangles*indicate the location of receivers used in computing velocity ground motion. The location of the cross-section is indicated in Fig. 2

We used WPP, a computer program developed at Lawrence Livermore National Laboratory (PETERSSON and SJOGREEN2010A), to model three-dimensional anelastic wave propagation. WPP uses a second-order anelastic finite difference method (NILSSON*ET AL*. 2007) with mesh refinement (PETERSSON and SJOGREEN2010B) and a curvilinear grid for treating non-planar, free-surface boundary conditions (Appelo and Petersson2008). The code is parallelized, and has been used to model ground motion from large earthquakes in California, such as the 1906 San Francisco earthquake (e.g., Aagaard*et al.* 2008).

We used a grid spacing of 200 m in the top 10 km of the model, and 400 m in the region below. The numerical scheme used in the finite difference code is proven to be accurate for a minimum of 15 grid nodes per wavelength, which corresponds to a maximum frequency of 0.4 Hz. Nevertheless, in our analysis of simulated ground motion we used a maximum frequency of 0.3 Hz. We computed velocity seismograms at a linear array of stations, indicated in Fig. 7, for an M5.4 earthquake using a double-couple point source with a thrust mechanism, which is typical for earthquakes in the Zagros region. The source has a depth of 12 km, and a Gaussian-type slip velocity function with a corner frequency of 1 Hz.

Figure 8a compares ground motion velocity seismograms computed with Model 1, Model 2, and Model 3, band-pass filtered at 0.01–0.3 Hz. Figure 8b and c show the same comparison, but in the frequency ranges of 0.1–0.3 Hz (2.5–10 s period range) and 0.02–0.05 Hz (20–50 s period range), respectively. We show the radial and vertical components of ground motion, which is dominated by basin-induced waves. Because of the predominant thrust mechanism of the source, the transverse component, not shown here, has much smaller amplitude. Note that the epicentral distances of receivers 1 and 2 are longer than 500 km. They fall within the source distance range of the seismic stations located on the west side of the Gulf. The other receivers were used to analyze the propagation of basin surface waves.

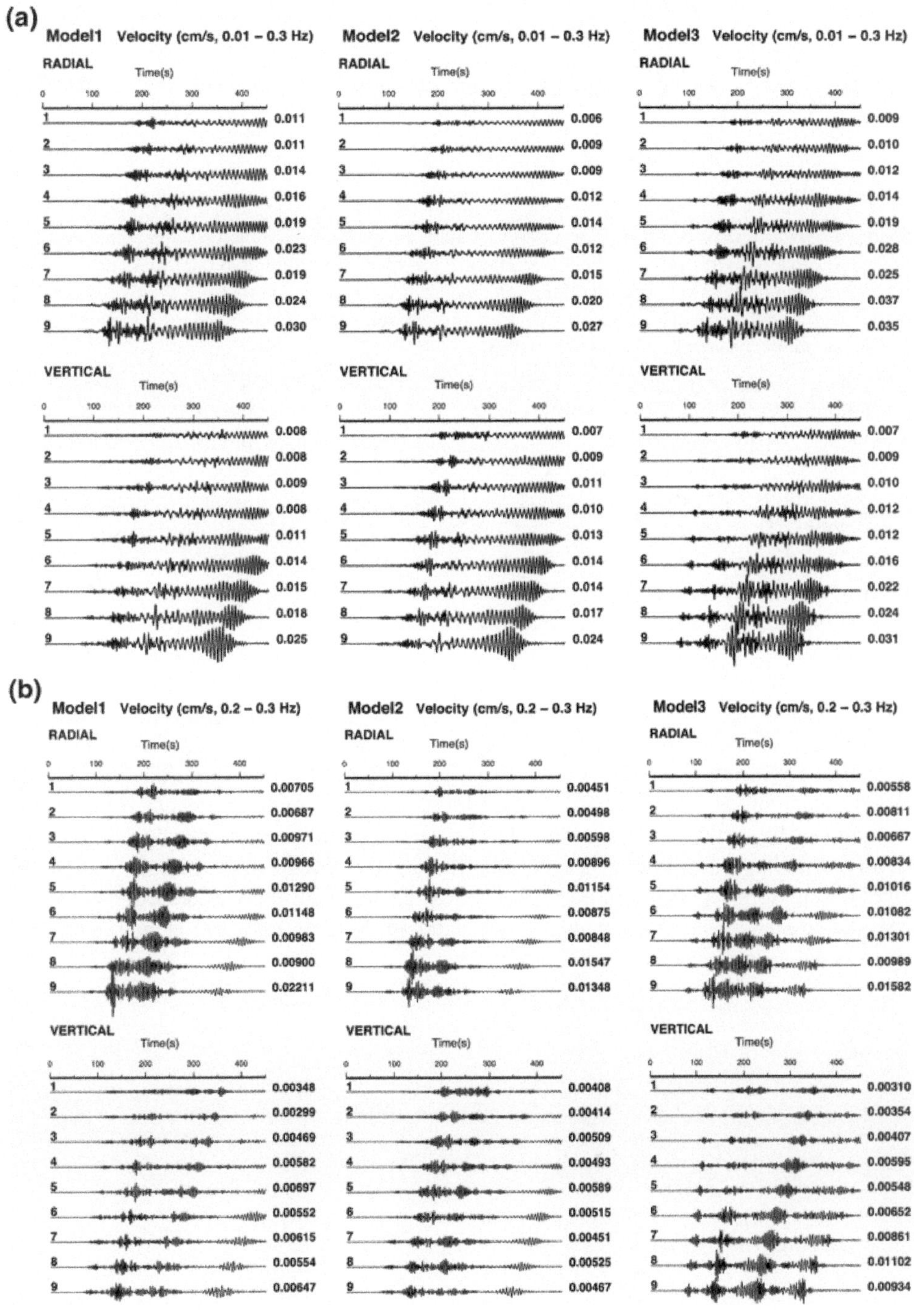

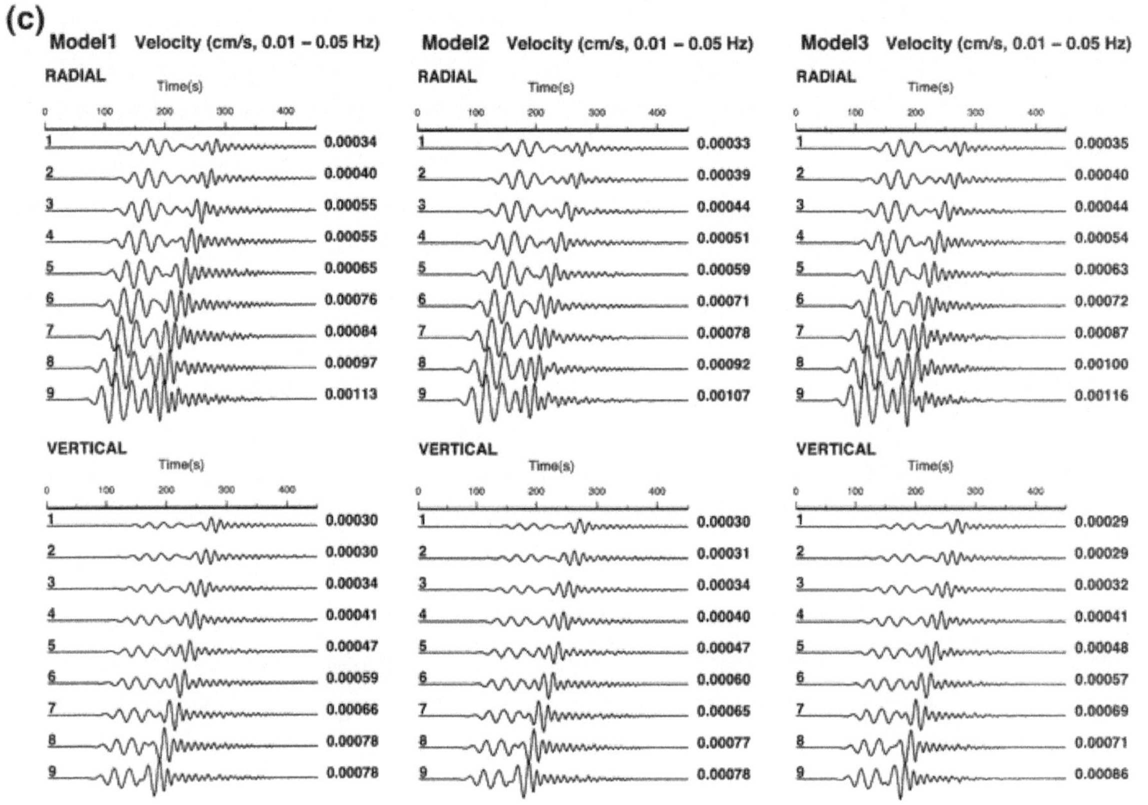

Figure 8: a Synthetic velocity seismograms computed for a double-couple point source with a thrust mechanism using Model 1 (*left panels*), Model 2 (*central panels*), and Model 3 (*right panels*). The seismograms are band-pass filtered at 0.01–0.3 Hz. The number on the left of each seismogram indicates the peak amplitude in cm/s. **b** Synthetic velocity seismograms computed for a double-couple point source with a thrust mechanism using Model 1 (*left panels*), Model 2 (*central panels*), and Model 3 (*right panels*). The seismograms are band-pass filtered at 0.2–0.3 Hz. **c** Same as Fig. 8a, but band-pass filtered at 0.01–0.05 Hz

The ground motion in both radial and vertical components is dominated by dispersive basin surface waves that are trapped in the sedimentary layers. The basin-trapped waves are seen in different wave trains with group velocities that depend on basin layer velocity. The simulations reproduce the exceptionally long duration of ground motion of more than 350 s that is observed in the recorded data. Our simulation demonstrates that due to basin wave guiding, the amplitude of surface waves and their frequency content depends on the overall velocity structure of the basin. In general, Model 3 produces the largest ground motion, and Model 2 produces the weakest ground motion. This characteristic is more pronounced in the frequency range of 0.1–0.3 Hz (see Fig. 8b). In contrast, in the period range of 20–50 s ground motion at very long periods is dominated by surface waves, and the three basin models produce very similar waveforms (Fig. 8c). This is not surprising since the three models have the same basin geometry and velocity background.

The most significant manifestation of basin-guided waves is seen in the comparison of simulated waves in the period range of 2.5–5 s (0.2–0.3 Hz). In this period range the waveforms vary among the models. The variation suggests that ground motion is sensitive to small-scale structural features. For example, compared to Model 1, Model 3, which has a sharper edge in the east side of the basin, produces larger ground motion, especially at stations 1 and 2. Stations 1 and 2 record waves that travel across the shallow part of the basin structure. The geometry of the basin edge in Model 3, and its location above the source facilitates trapping of waves that enter the basin. The basin-guided wave effect increases the amplitude of ground motion. This distance-dependent effect can also be seen in the acceleration response spectra computed for the radial component of motion, shown in Figs. 9, 10. The response spectra show a larger separation at stations 1, 2, and 3, located near the western edge of the basin. The differences among models are larger at short periods, smaller than 3 s.

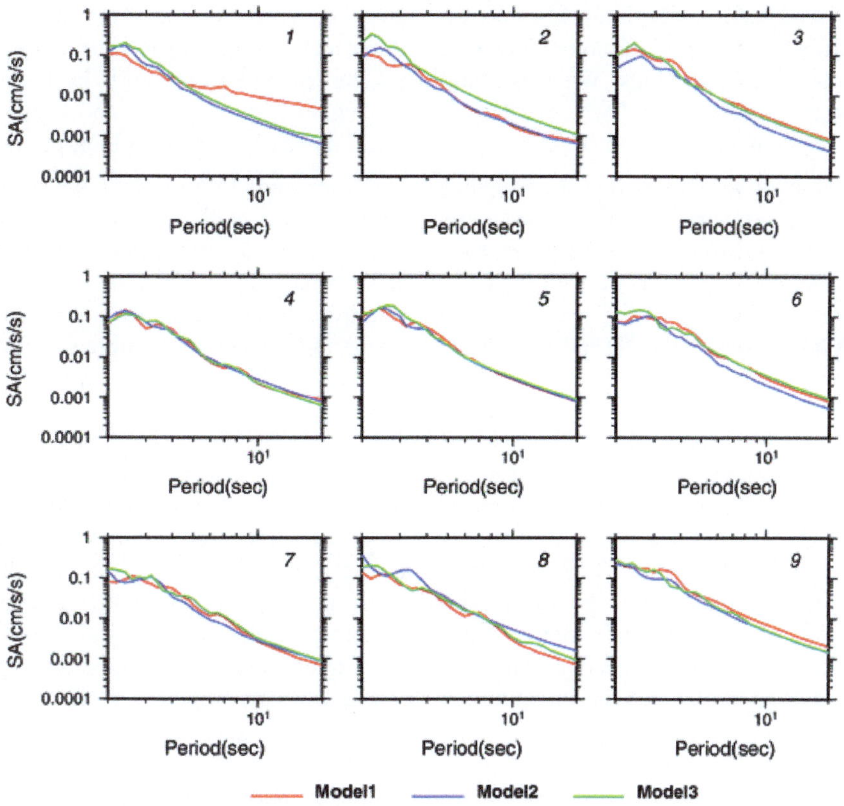

Figure 9: Comparison of acceleration response spectra of simulated ground motion for Model 1 (*red traces*), Model 2 (*blue traces*), and Model 3 (*green traces*)

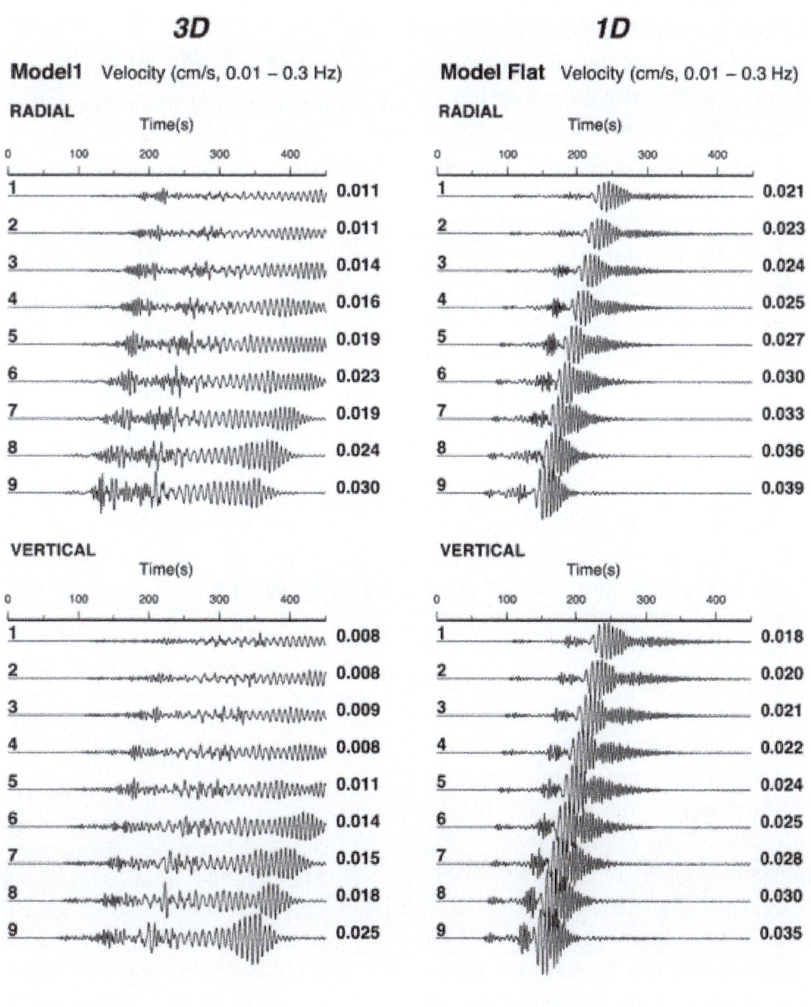

Figure 10: Comparison of synthetic velocity seismograms computed using a 3D model (Model 1, *left panels*) and a 1D model (*right panels*) for a double-couple point source with thrust focal mechanism. The seismograms are band-pass filtered at 0.01–0.3 Hz

Our simulation results suggest that the basin structure on the east side of the Gulf affects the ground motion on the west side of the Gulf, especially at short periods. We also found that, at these periods, the effect of shallow basin structure is significant in the vertical component of motion (Fig. 8b).

The effect of 3D structure on ground motion at basin sites becomes very clear when we compare synthetic seismograms computed with a 3D basin model (Model 1) and a 1D reference velocity model. The 1D-layered model was derived from the velocity structure below station 1 in Model 1. The comparison of synthetic seismograms obtained with the 3D and 1D models is shown in Fig. 10. The comparison suggests that the effect of 3D basin structure causes the increase of ground motion duration. The ground motion generated with 3D basin models is dominated by secondary surface waves with durations that depend on station location. In contrast, the 1D model creates dispersive surface waves that are confined within a relatively short time window, after the body waves. This numerical experiment demonstrates that ground motion produced with 1D regional velocity models of the Gulf basin may produce unrealistic ground motion. The seismograms will lack the energetic basin-induced surface waves with very long duration enhanced by 3D wave scattering.

CONCLUSIONS

This study is an attempt to analyze ground motion characteristics observed in the western side of the Arabian Gulf from earthquakes in the Zagros Mountains using comparisons with GMPEs and ground motion simulation. The comparisons with GMPEs showed that ground motion recorded in the Gulf region from earthquakes in the Zagros Mountains region is anomalous. The ground motion response in the period range of 5–10 s is much higher than the predicted one, with factors approaching 100, depending on the station and event locations. In contrast, at shorter periods the ground motion response is lower by at least a factor of 2. The extraordinary long duration of the seismic energy observed in the intermediate and long-period bands may have a significant impact on large structures along the Gulf shoreline. We speculate that the duration and amplitude of ground motion is due to waveguide effects in the sedimentary structure of the Gulf basin.

This hypothesis was tested using large-scale 3D waveform modeling. Sensitivity analysis of the basin induced waves and the corresponding response spectra due to shallow complexities in the basin models reveal the significant implication of basin structure in predicting strong ground motion in the Gulf region. Our high performance computer simulations suggest that a plausible explanation of the anomalous observed ground motion is the generation of basin reverberation waves that are trapped in the shallow sedimentary layers of the basin. Non-uniform structural effects on wave propagation cause the basin response to be period-dependent and highly variable. In order to reproduce the extremely long duration of the observed ground motion our simulations required a $Q_p = 200$ and $Q_s = 100$ in the top sedimentary layers of the basin. These relatively high-quality factors imply that long-period (2–10 s) waves are subject to rather low attenuation when propagating in the Gulf basin. The observation of strongly path-dependent amplitudes in the basin structure suggest that this region might be a good natural laboratory to test numerical simulation of ground motions to predict motions in the 1–10 s period band. Rapid development of infrastructure, including tall buildings, bridges, pipelines, and ports in this region will likely expose the region to seismic risk and further indicate that improved seismic hazard analysis of long-period motions is warranted.

The simulation results confirm our hypothesis that long-period waves with extremely long duration and relatively large amplitudes are caused by the geometry of the basin sedimentary layers and, to some extent, by the relatively shallow earthquake depth in the Zagros mountains, which favors the channeling of more seismic energy into the shallow sedimentary layers. Combined effects of basin edge geometry with sharp velocity contrast, and shallow earthquakes (depth <10 km) on the eastern part of the Arabian Gulf can cause large long-period ground motion on the western part of the Gulf. This study is a first step toward understanding and characterizing ground motion in the eastern part of the Gulf from earthquakes in the Zagros Mountains region. The development of well-constrained and high-resolution crustal velocity models of the region is crucial for improving the quality of earthquake ground motion prediction on a broad period range.

Both GMPEs adopted in this study may not be reliable at very long distances. Nevertheless, we used them as a reference to show the extremely large difference between recorded and predicted ground motion observed at all stations. More strong motion stations are need in order to develop adequate GMPEs, and improve strong ground motion characterization for the region.

ACKNOWLEDGMENTS

The authors would like express their gratitude to the National Plan for Science, Technology, and Innovation at King Abdulaziz City for Science and Technology (KACST), Saudi Arabia, for funding this project (Grant No. 09–INF945-02). The authors would like to thank two anonymous reviewers for their comments and suggestions which greatly improved the clarity of the manuscript. Numerical simulations were performed on the SIERRA and CAB Linux clusters operated by Livermore Computing Center. This work was performed under the auspices of the U.S. Department of Energy by Lawrence Livermore National Laboratory under Contract DE-AC52-07NA27344. Some figures in this paper were drawn using Generic Mapping Tools software package developed by Wessel and Smith (1998).

REFERENCES

1. Aagaard, B., *et al.* (2008), *Ground motion modeling of the 1906 San Francisco earthquake II: Ground motion estimates for the 1906 earthquake and scenario events*, Bull. Seismol. Soc. Am., *98*, 10121046, doi:10.1785/0120060410.
2. Al-Amri, A. M. (2013). *Seismotectonics and seismogenic source zones of the Arabian Platform*, K. Al Hosani et al. (eds.), Lithosphere Dynamics and Sedimentary Basins: The Arabian Plate and Analogues, Frontiers in Earth Sciences, doi:10.1007/978-3-642-30609-9-15, Springer-Verlag Berlin Heidelberg.
3. Al-Amri, A.M. (1999). *The crustal and upper-mantle structure of the interior Arabian platform*, Geophysical Journal International, *136*, 421–430.

4. Al-Amri, A.M. (1998). *The crustal structure of the western Arabian Platform from the spectral analysis of long-period P-wave amplitude ratios*, Tectonophysics, *290*, 271–283.

5. Al-Amri, A. L, R. Mellors, and F. L. Vernon (1999). *Broadband Seismic Noise Characteristics of the Arabian Shield*, Arabian Journal of Science and Engineering, King Fahd University of Petroleum and Minerals, *24*, 99–113.

6. Al-Amri, A.M. A. J. Rodgers, and T. A. Al-Khalifah (2008). *Improving the level of seismic hazard parameters in Saudi Arabia using earthquake location*. Arab J Geosci. *1*, 1, 1–15.

7. Al-Damegh, K., Sandvol, E., Al-Lazki, A. and Barazangi, M. (2004). *Regional wave propagation (Sn and Lg) and Pn attenuation in the Arabian plate and surrounding regions*, Geophys. J. Int., *157*, 775–795.

8. Alsinawi, S. A., 1986. *Historical Seismicity of the Arab Region*. Proceedings of the Third Arab Seismological Seminar, Riyadh Saudi Arabia, pp. 11–33.

9. Ambraseys N. N., 1988. *The seismicity of Saudi Arabia and adjacent area, Report Prepared for the Kingdom of Saudi Arabia*. ESEE, Imperial Col. Sci. Tech., London, U.K. Report 88/11. p. 294.

10. Ambraseys, N. N.; Melville, C.P. and Adams, R. D., 1994. *The Seismicity of Egypt, Arabia and the Red Sea-A historical review*. Cambridge University Press and King Abdul Aziz City for Science and Technology, p. 181.

11. Appelo, D., and N. A. Petersson (2008), *A stable finite difference method for the elastic wave equation on complex geometries with free surfaces*, Commun. Comput. Phys., *5*, 84–107.

12. Bassin, C. G. Laske, and G. Masters (2000). *The current limits of resolution in surface wave tomography in North America*, EOS Trans. AGU, *F897*, 81.

13. Boore and Atkinson (2006), *Earthquake Ground-Motion Prediction Equations for Eastern North America*. Bull. Seism. Soc. Am., *96*, no 6, 2181–2205.

14. Campbell and Bozorgnia (2008) *NGA ground motion model for the geometric mean horizontal component of PGA, PGV,PGD and 5% damped linear elastic response spectra for periods ranging from 0.01 to 10 s*. Earthquake Spectra, *24*, No. 1, 139–171.

15. Frankel, A., R. W. Clayton (1986). *Finite Difference Simulations of Seismic Scattering: Implications for the Propagation of Short-Period Seismic Waves in the Crust and Models of Crustal Heterogeneity*. JGR, Vol. *91*, B6, 6465–6489.

16. Gök, R., H. Mahdi, H. Al-Shukri and A. Rodgers (2008). *Crustal structure of Iraq from receiver functions and surface wave dispersion*, Geophys. Journal Int., Vol. *172*(3), pp. 1179–1187.

17. Johnson, P.R. (1998). *Tectonic Map of Saudi Arabia and Adjacent Areas, Ministry of Petroleum and Mineral Resources, Deputy Ministry Mineral, Resources Technical Report USGS-TR-98-3*, Jeddah, Saudi Arabia.

18. Laske, G. and T.G. Masters (1997). *A global digital map of sediment thickness*, EOS Trans. AGU, *78*, F483, http://igppweb.ucsd.edu/~gabi/sediment.html.

19. Mellors, R., Vernon, F., Camp, V., Al-Amri, A.M K Gharib, A. (1999). *Regional waveform propagation in the Saudi Arabian Peninsula and evidence for a hot upper mantle under western Arabia*. J. Geophys. Res., *104*, 20,221–20,235.

20. Mooney, W.D., G. Laske, and G. Masters (1998). *CRUST5.1: A global crustal model at 5° × 5°*, J. Geophys. Res., *103*, 727–747.

21. Nilsson, S., N. A. Petersson, B. Sjogreen, and H.-O. Kreiss (2007), *Stable difference approximations for the elastic wave equation in second order formulation*, SIAM J. Numer. Anal., *45*, 1902–1936, doi:10.1137/060663520.

22. Pasyanos, M.E., W. R. Walter, E. M. Matzel (2009). *A simultaneous multiphase approach to determine P-wave and S-wave attenuation of the Crust and upper Mantle*. Bull. Seism. Soc. Am., *99*, 3314–3325.

23. Pasyanos, M.E., T.G. Masters, G. Laske, and Z. Ma (2014). *LITHO1.0: An updated crust and lithospheric model of the Earth*, J. Geophys. Res. Solid Earth, *119*, 2153–2173, doi:10.1002/2013JB010626.

24. Petersson, N. A., and B. Sjogreen (2010a), *Reference guide to WPP version 2.0, Tech Rep. LLNL - TR - 422928*, Lawrence Livermore Natl. Lab., Livermore, Calif.

25. Petersson N.A., and B. Sjogreen (2010b). *Stable grid refinement and singular source discretization for seismic wave simulations*. Communications in Computational Physics, vol. *8*, no. 5, pp. 1074–1110.

26. Pitarka, A., K. Irikura, T. Iwata and H. Sekiguchi (1998). *Three-dimensional simulation of the near-fault ground motion for the 1995 Hyogo-ken Nanbu (Kobe), Japan, earthquake*. Bull. Seism. Soc. Am., *88*, 428–440.

27. Rodgers, A.J, A. Fowler, A.S. Al-Amri, and A. Al-Enezi (2006). *The March 11, 2002 Masafi, United Arab Emirates earthquake: Insights into the seismotectonics of the northern Oman Mountains*. Tectonophysics, *415* (2006) 57–64.

28. Rodgers, A., W. Walter, R. Mellors, A. M. S. Al-Amri and Y. S. Zhang (1999). *Lithospheric structure of the Arabian Shield and Platform from complete regional waveform modeling and surface wave group velocities*, Geophys. J. Int., *138*, 871–878.

29. Rodgers, A., J.F. Ni, and T. M. Hearn (1997). *Propagation characteristics of short-period Sn and Lg in the Middle East*, Bull. Seism. Soc. Am. *87*, 396–413.

30. Talebian, M., and J. Jackson (2004), *A reappraisal of earthquake focal mechanisms and active shortening in the Zagros mountains of Iran*, Geophys. J. Int., *156*, 506–526.

31. Tkalčiç H., M. Pasyanos, A. Rodgers, R. Gok and A. Al-Amri (2006). *A multi-step approach for joint modeling of surface wave dispersion and teleseismic receiver functions: Implications for lithospheric structure of the Arabian Peninsula*. J. Geophys. Res., *111*, B113111, doi:10.1029/2005JB004130.

32. Wessel; P, W.H.F. Smith (1998). *New improved version of generic mapping tools released*. EOS Trans. Amer. Geophys. Union *79* (47):579. doi:10.1029/98EO00426

CHAPTER 9

ACOUSTIC–GRAVITY WAVES IN THE IONOSPHERE DURING SOLAR ECLIPSE EVENTS

Petra Koucká Knížová and Zbyšek Mošna

INTRODUCTION

Terrestrial atmosphere shows a high variability over a broad range of periodicities, which mostly consists of wave-like perturbations characterized by various spatial and temporal scales. The interest for short time variability in ionospheric attributes is related to the role that ionosphere plays in the Earth's environment and space weather. Acoustic-gravity waves (AGWs), waves in the period range from sub-seconds to several hours, are sources of most of the short-time ionospheric variability and play an important role in the dynamics and energetics of atmosphere and ionosphere systems. Many different mechanisms are likely to contribute to the acoustic-gravity wave generation: for instance, excitation at high latitudes induced by geomagnetic and consequent auroral activity, meteorological phenomena, excitation in situ by the solar terminator passages and by the occurrence of solar eclipses. During solar eclipse, the lunar shadow creates a cool spot in the atmosphere that sweeps at supersonic speed across the Earth's atmosphere. The atmosphere strongly responds to the decrease in ionization flux and heating. The very sharp border between sunlit and eclipsed region, characterized by strong gradients in temperature and ionization flux, moves throughout the atmosphere and drives it into a non-equilibrium state. Acoustic-gravity waves contribute to the return to equilibrium. At thermospheric heights, the reduction in temperature causes a decrease in pressure over the totality footprint to which the neutral winds respond. Thermal cooling and downward transport of gases lead to neutral composition changes in the thermosphere that have significant influence on the resulting electron density distribution. Although the mechanisms are not well understood, several studies show direct evidence that solar eclipses induce wave-like oscillations in the acousticgravity wave domain. Many different mechanisms are likely to contribute to wave generation and enhancement at ionospheric heights. Hence, it is difficult to clearly separate or differentiate each contributing agent and to decide which part of wave field belongs to the in situ generated and which part comes from distant regions. First experimental evidence of the existence of gravity waves in the ionosphere during solar eclipse was reported by Walker et al. (1991), where waves with periods of 30–33 min were observed on ionosonde sounding virtual heights.

Ionospheric Sounding

As the solar radiation penetrates Earth's atmosphere it forms pairs of charged particles. Under a normal day-time conditions the ionization solar flux increases immediately after sunrise, reaches maximum around local noon and decreases again till sunset. Under such conditions concentration of charged particles significantly grows in the atmosphere and forms atmospheric plasma called ionosphere. Due to the composition of the neutral atmosphere together with the changing efficiency of the incoming solar radiation, ionosphere is stratified into the layers denoted D, E, F1 and F2. After sunset, electrons and ions recombine rapidly in the D, E and F1 layer. Due to slower recombination processes of atomic ions that dominate at heights approximately above 150km altitude, F2 layer remains present all the night. Special stratification Es, sporadic E layer, occurs sometimes at heights of E layer (Davies, 1990). Ionosphere significantly affects propagation of the electromagnetic waves. According to a frequency of the wave with respect to

Citation: Petra Koucká Knížová and Zbyšek Mošna (2011). Acoustic–Gravity Waves in the Ionosphere During Solar Eclipse Events, Prof. Marco G. Beghi (Ed.), InTech, DOI: 10.5772/19722.

a concentration of the ionospheric plasma, wave propagates through the medium or it is reflected. Electromagnetic waves with frequency lower than plasma frequency of the particular plasma parcel are reflected, which allows to estimate plasma frequency. Higher frequency waves propagate through plasma. An instrument called ionosonde (or digisonde) transmits electromagnetic wave of a defined frequency and detects it after reflection from the ionosphere. Typical ionosonde sounding range is 1 MHz – 20 MHz. For each sounding wave ionosonde records time of flight τ on the path transmitter - reflection point – receiver. Time of flight is simply converted into a virtual height $h_{virtual} = \frac{\tau.c}{2}$ that corresponds to wave propagation in the vacuum (c stands here for speed of light). Virtual height is equal or higher than the corresponding real height. The output of the measurement is height-frequency characteristics called an ionogram. Real height electron concentration profiles can further be inverted from ionograms using for instance programs POLAN (Titheridge, 1985) or NHPC (Huang and Reinish, 1996). Ionosphere represents inhomogeneous and anisotropic medium which leads to a wave splitting into an ordinary and extraordinary wave modes. Hence, two reflection traces are recorded by the ionosonde (as seen on ionograms in Figure 1). However, the extraordinary mode is not further used for electron concentration profile inversion. Figure 1 shows typical day-time and night-time ionograms recorded by a digisonde in the observatory Pruhonice. Together with the ionograms there are plots of the real height electron concentration derived by NHPC routine. On the day-time electron concentration profile, three ionospheric layers E, F1 and F2 are present while on the night-time profile there is only F2 layer detected by the ionosonde. Sequences of ionograms are widely used for analyses of variability of atmospheric plasma ranging from detection of rapid changes with periods of minutes to the study of long-term trends.

Basic theory of AGWs in the Earth's Atmosphere

Most of the wave-like oscillations in the atmosphere can be described/parametrized using basic acoustic-gravity wave theory in the atmosphere. Details can be found, for instance, in works of Davies (1990), Bodo et al. (2001), Hargreaves (1982), Yeh & Liu (1974) among others. Here, we show brief derivation of the dispersion relation that any wave motion of the AGW type must satisfy. In a plane-stratified, isothermal atmosphere under gravity that is constant with height, two frequency domains exist in the atmosphere where atmospheric waves can propagate, acoustic and gravity wave. Atmosphere represents compressible gas that once compressed and then released would expand and oscillate about its equilibrium state. Its oscillation frequency is known as an acoustic cut-off frequency

$$\omega_a = \frac{c}{2H} \tag{1}$$

where c is speed of sound

$$c = \sqrt{\gamma g H} \tag{2}$$

γ is the ratio of specific heats at constant pressure and constant volume, g is the gravitational acceleration, and H is the scale height. For diatomic gas $\gamma \sim 1.4$.

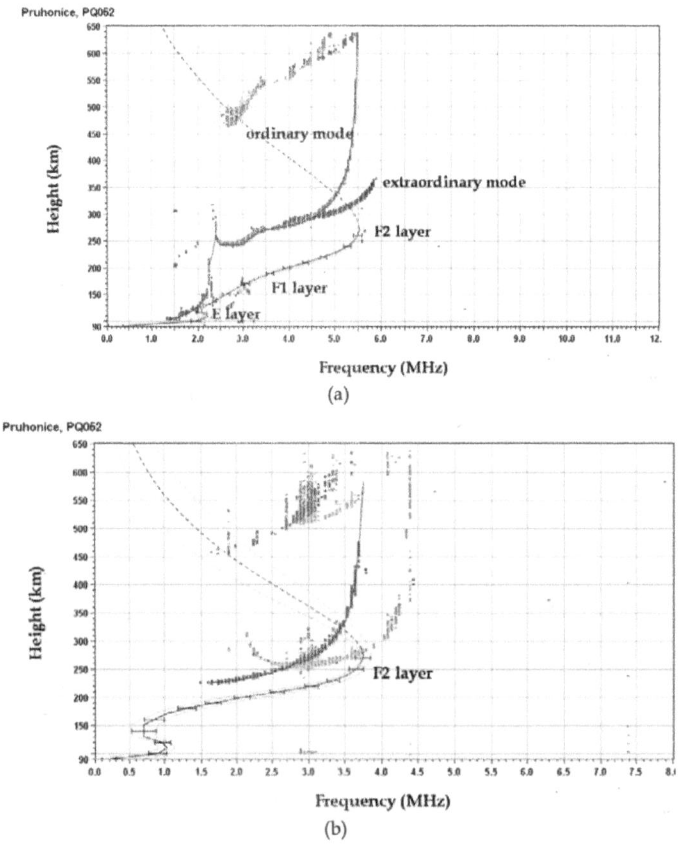

Figure 1: Typical day-time (a) and night-time ionogram (b) measured by digisonde DPS 4 in the Observatory Pruhonice. On both plots, there is real hight electron concentration (solid line with error bars) provided as obtained by the NHPC routine.

Single element of fluid, parcel of the atmosphere, at height z with density ρ which is displaced in the vertical by Δz to a place where its density changes to ρ+Δρ, remains in pressure equilibrium with its surroundings. Displacement takes place adiabatically. This is valid when the motion is so slow that sound waves with speed

$$c = \sqrt{\frac{dp}{d\rho}} \tag{3}$$

where p stands for pressure can traverse the system faster than the time-scale of interest and the motion is so fast that the entropy is preserved. The parcel is no longer in equilibrium and starts to oscillate about its equilibrium height with buoyancy frequency. The buoyancy force which acts on the parcel is balanced by inertial force (Newton's second law):

$$\rho \frac{d^2}{dt^2}(\Delta z) = -g\Delta\rho \tag{4}$$

where Δρ is the difference between internal and external densities.

Internal and external Δρ are derived as:

$$(\Delta\rho)_{internal} = \Delta p / c^2 = -\frac{g\rho}{c^2}\Delta z \tag{5}$$

which is due to compressibility of the fluid within the membrane and

$$(\Delta\rho)_{external} = -\frac{d\rho}{dz}\Delta z \tag{6}$$

is the change of background density at new position due to inhomogeneous nature of the atmosphere. Taking both the contributions of Δρ we get

$$\frac{d^2}{dt^2}(\Delta z) = \left(g\frac{d}{dz}(\ln\rho) + g^2/c^2\right)\Delta z \tag{7}$$

which can be recast into

$$\frac{d^2}{dt^2}(\Delta z) + \omega_B^2 \Delta z = 0 \tag{8}$$

Where

$$\omega_B^2 = -g\left(\frac{d}{dz}(\ln\rho) + g/c^2\right) \tag{9}$$

If $\omega_B^2 > 0$, the solution is oscillatory and the fluid parcel will oscillate with characteristic buoyancy frequency ω_B called Brunt-Vaisala frequency.

More convenient form used for atmosphere is following:

$$\omega_B^2 = (\gamma - 1)g^2/c^2 + g / c^2 dc^2 / dz \tag{10}$$

This approximation is valid in the atmosphere-ionosphere system of our interest.

In isothermal atmosphere ω_B reduces to

$$\omega_B^2 = (\gamma - 1)g^2 / c^2 \tag{11}$$

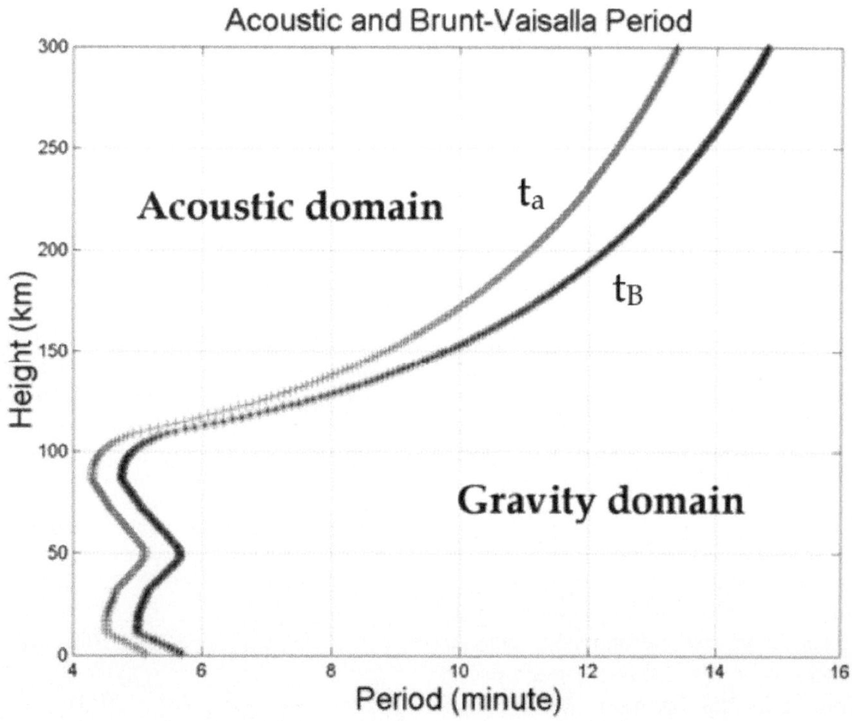

Figure 2: Height dependence of acoustic cut-off period ta and Brunt-Vaisala period tB that represent limits dividing periods into acoustic and gravity wave domains. Period domain between acoustic cut-off and Brunt-Vaisala represents region where no AGW propagates.

Wave motion in the atmosphere can be described using mass conservation (continuity equation), and equation of motion:

$$\frac{\partial \rho}{\partial t}+\rho \nabla .\vec{u}+\left(\vec{u}.\nabla\right)\rho=0 \tag{12}$$

$$\rho\left(\frac{\partial \vec{u}}{\partial t}+\left(\vec{u}.\nabla\right)\vec{u}\right)=-\nabla p+\rho g \tag{13}$$

where pressure gradients and gravity are the only forces causing the acceleration. Oscillation takes place adiabatically

$$\rho\left(\frac{\partial p}{\partial t}+\vec{u}.\nabla p\right)=\gamma p\left(\frac{\partial \rho}{\partial t}+\vec{u}.\nabla\right) \tag{14}$$

where ρ, p, γ and $\vec{u}$ are parameters of the atmosphere – density, pressure, ratio of specific heats, and velocity.

Applying the perturbation approach we are searching for wave-like solutions for the perturbation quantities. Further simplification comes from the assumption that the background state is of constant temperature T in which p_0/ρ_0 must be a constant.

$$p_0/\rho_0=c^2/\gamma \tag{15}$$

Then the system (12), (13) and (14) reduces to:

$$\frac{\partial \rho'}{\partial t}+\rho_0\vec{\nabla}.\vec{u}'-\rho_0 u_z/H=0 \tag{16}$$

$$\rho_0\frac{\partial \vec{u}}{\partial t}+\nabla p'-\rho' g=0 \tag{17}$$

$$\rho_0\left(\frac{\partial p'}{\partial t}-p_0 u_z'/H\right)=\gamma p_0\left(\frac{\partial \rho'}{\partial t}-\rho_0 u_z'/H\right) \tag{18}$$

where index 0 denotes stationary (non fluctuating) component and the apostrophe denotes perturbation. These are the basic governing equations for the gravity waves. For a nontrivial solution the following prescription of the dispersion relation must be satisfied:

$$\omega^4-\omega^2\omega_a^2-k_x^2c^2\left(\omega^2-\omega_g^2\right)-c^2\omega^2k_z^2=0 \tag{19}$$

From disperse relation, it is evident that between buoyancy frequency and acoustic cut-off frequencies one cannot have both k_x and k_z real. Figure 2 shows two period domains with border limits of acoustic cut-off period and buoyancy period. An attenuation or growth in the wave amplitude must occur in either the vertical or the horizontal directions. We suppose that there is no variation in amplitude in horizontal directions so that k_x is purely real and k_z has an imaginary component. At frequencies exceeding acoustic cut-off ω_a, expression (19) becomes simple and the waves may be termed as ACOUSTIC WAVES. At frequencies smaller than Brunt-Vaisala frequency where gravity plays an important role, the waves are called GRAVITY or INTERNAL GRAVITY WAVES. Brunt-Vaisala frequency and acoustic cut-off frequency divide the frequency spectrum into two domains in which ω_g forms the high frequency limit for one class $\omega<\omega_a$ called the acoustic waves. A gap in the frequency spectrum exists between ω_g and ω_a where no internal waves can propagate. Important approximations can be obtained under the assumption $|k_z|>>1/2H$ and $\omega << \omega_g$ then:

$$k_z^2=\left(\omega_g^2/\omega^2\right)k_x^2 \tag{20}$$

These approximations apply to much of the observed gravity waves. From (20) we see that the angle of ascent of the phase α is:

$$tg\alpha=k_z/k_x=\omega_g/\omega=\tau/\tau_g \tag{21}$$

The motions of the air parcels are, in general, ellipses in the plane of propagation and have components transverse to the direction of wave propagation. The ratio of the horizontal displacement ξ to its vertical displacement ζ is:

$$\frac{\xi}{\zeta}=\frac{\frac{ck_x}{\omega}}{\left(\frac{ck_x}{\omega}\right)^2-1}\left(\frac{ck_z}{\omega}-i\sqrt{\left(\frac{\omega_a}{\omega}\right)^2-\left(\frac{\omega_g}{\omega}\right)^2}\right) \tag{22}$$

On the frequencies just above the acoustic cutoff the air motion is essentially vertical. With acoustic waves on high frequencies the motion is radial as in sound waves. The motion is circular with horizontal propagation at a frequency $\omega_a\sqrt{2/\gamma}$. Gravity wave propagation is limited to angles between

$$\phi_{min}=\sin^{-1}\left(\frac{\omega}{\omega_g}\right),\ \phi_{max}=\pi-\sin^{-1}\left(\frac{\omega}{\omega_g}\right) \tag{23}$$

The sense of rotation of the air for gravity waves is opposite than for acoustic waves. As Φ approaches its asymptotic values the air motion becomes linear and transverse to the direction of propagation. Air parcel rotation is clockwise in case of acoustic waves while anticlockwise in case of gravity waves. Energy vector lies in the same quadrant as direction of propagation of acoustic waves. Energy flows up when phase travels down and vice versa in case of gravity waves propagation. This is important property since it accounts for the observed downward phase propagation when the source is below the level at which a disturbance is observed. The horizontal u_x and vertical u_z components of the packet velocity are obtained from disperse relation:

$$u_x=\frac{c^2k_x\left(\omega^2-\omega_g^2\right)}{\omega\left(2\omega^2-\omega_a^2-c^2k^2\right)}$$

$$u_z=\frac{c^2k_z\omega^2}{\omega\left(2\omega^2-\omega_a^2-c^2k^2\right)} \tag{24}$$

Due to coupling between neutral and charged components the initial wave-like oscillation in the neutral atmosphere induces wave-like perturbation in the ionosphere. Perturbation in the ion production is the most effective when solar ionizing rays are nearly in alignment with the initial wave front. Perturbations in the neutral atmosphere may cause perturbations in chemical processes. Presence of AGW influences the ionisation rate through changes in the local neutral density and temperature, and through changes in the ionisation radiation absorption (Hooke, 1970).

AGW IN THE IONOSPHERIC PLASMA

Acoustic-gravity waves are always present in the Earth's atmosphere. AGWs arise from many natural sources like convection, topography, wind shear, moving solar terminator, earthquakes, tsunami, etc. Increase in wave-like activity is associated also with human activity including coordinated experiments or unwilling accidents. AGWs influence on the upper atmosphere is not yet understood enough. They produce a great amount of variability and contribute to the background conditions in a specific parcel of the atmosphere. Gravity waves propagating from lower laying atmosphere have been long regarded as a very important source of the energy and momentum transfer in the upper atmosphere (Hines, 1960). The breaking of the upward propagating waves affects wind system, generates turbulence and heats the atmospheric gas. Waves that reach upper atmosphere produce travelling atmospheric disturbances (TAD) or travelling ionospheric disturbances (TID) and even form the ionospheric inhomogenities which grow and finally break into the plasma instabilities observed by radar techniques that might cause scintillation of the communication signals propagating through the ionosphere. From the observation it is evident that the thermosphere is continuously swept by the acoustic-gravity waves. Statistically, the waves show a moderate preference for southward travel, with this preference being reduced or shifted to southeastward travel during disturbed times (Oliver et al., 1997). Experimental studies show that AGW activity in the ionosphere slightly increases during dawn and dusk periods of the day (Galushko et al., 1998; Somsikov & Ganguly, 1995; Sauli et al., 2005 among others). Influence of infrasonic waves generated by ground experimental sources on the ionosphere was reported for instance by Rapoport et al. (2004). Solar eclipse represents well defined source of the AGW in the atmosphere and ionosphere systems. During solar eclipse event, solar ionization flux decreases producing well-defined cool spot in the atmosphere that moves through the Earth's atmosphere. Moving source in the atmosphere can emit both acoustic and gravity waves. Supersonic motion of the source forms wave field with bow wave. Both acoustic and gravity waves can be radiated in association with supersonic motion in the atmosphere. When the source is moving within atmosphere with subsonic velocity only gravity waves can be emitted (Kato et al., 1977).

SOLAR ECLIPSE EVENT – SIGNATURES IN THE IONOSPHERIC PLASMA

It has been proposed by Chimonas and Hines (1970) that solar eclipses can act as sources for AGWs. The lunar shadow creates a cool spot in the atmosphere that sweeps at supersonic speed across the Earth. The sharp border between sunlit and eclipsed regions, characterized by strong gradients in temperature and ionization flux, moves throughout atmosphere and drives it into a non-equilibrium state. Earth atmosphere shows variable sensitivity to the changes of ionization flux.

Experiments

Solar eclipse event represents phenomenon that can be precisely predicted, hence many observational campaigns are organised around the world. Effects of the solar eclipses on the ionospheric plasma are studied by mean of GPS techniques, radars, vertical ionospheric soundings etc. Study limitations lay mainly in the fact that there are no identical solar eclipse events. Moreover, solar eclipse induced effects are easily to be mixed with effects caused by geomagnetic field variations, diurnal changes of the ionosphere, seasonal variability of the atmosphere/ionosphere etc. In the upper atmosphere, AGWs can be observed either directly as neutral gas fluctuations or indirectly as induced ionospheric plasma variations. Despite intensive research many questions in the problem of the generation and propagation remain to be understood. Studies by Fritts and Luo (1993) suggest that perturbations generated by the eclipse induced ozone heating interruption may propagate upwards into the thermosphere–ionosphere system where they have an important influence. Temperature fluctuations and electron density changes propagate as a wave, away from the totality path, cf. Muller-Wodarg et al. (1998). By means of vertical ionospheric sounding, Liu et al. (1998) detected waves excited during solar eclipse event at F1 layer heights and their generation and/or enhancement attributed to changes of temperatures and variations of the height of the transition level for the loss coefficient and the height of the peak of electron production. Studies reported by Farges et al. (2001) suggest a longitudinal diversity of the disturbances with respect to prenoon and postnoon phases. Xinmiao et al. (2010) reported synchronous oscillations in the Es and F layer during the recovery phase of the solar eclipse. Ivanov et al. (1998) found that during solar eclipse with maximum obscuration of about 70% the F-region electron density decreased by 6-8% compared to a control day and detected travelling ionospheric disturbances. Additionally, they detected strong variations in the difference group delays with a period about 40 minutes associated with the start and end of the eclipse. Oscillations in the ionosphere, similar to gravity waves, were observed following some solar eclipse events (Chimonas and Hines, 1970; Cheng et al., 1992; Liu et al., 1998; Sauli et al., 2006). Investigation of the latitudinal dependence of NmF2 (the maximum electron density of the F2 layer) indicated that the strongest response was at middle latitudes (Le et al., 2009). The response of the sporadic-E (Es) layer also differed in each solar eclipse event. A remarkable decrease in Es layer ionization was observed during the eclipse of 20 July 1963 (Davis et al., 1964). Enhancement of Es layer ionization has also been reported and it has been suggested that it is related to internal gravity waves generated in the atmosphere during the solar eclipse (Datta, 1972).

Processes Induced by Solar Eclipse

During the solar eclipse, on the time scale shorter than day-night change, the ionosphere reconfigures itself into a state similar to that of night situation. Photochemical ionization falls heavily almost to a night-time level. With the decreasing solar flux, atmospheric temperature falls in the moon shadow creating a cool spot with well defined border. Then the increasing solar flux starts ionization processes and warms the atmosphere again to daytime level. Such changes in the ionization cause variation in the reflection heights, decrease/increase in electron concentration at all ionospheric heights, decrease/increase in the total electron content, rising/falling of the layer height. Such effects are characteristic for the processes during sunrise/sunset in the ionosphere. However, supersonic movement of the eclipsed region represents a key difference from the regular solar terminator motion at sunrise and sunset times. These changes in the neutral atmosphere and ionosphere induced by solar eclipse force the evolution of the ionospheric plasma toward a new equilibrium state. The return to equilibrium is likely accompanied by the eclipse induced wave motions excited in the atmosphere. Any moving discontinuity of gas parameters such as temperature, pressure etc. will generate transit-like waves. In the upper ionosphere, waves can be generated by a strong horizontal electron pressure gradient. Possible mechanisms contributing to the wave generation in the region of solar terminator are in detail discussed by Somsikov & Ganguly (1995).

Solar eclipse induces changes in all atmospheric regions extend from the upper atmosphere down to ground level. Despite the low magnitude of the eclipse induced effects at ground level, Jones et al. (1992) reported wave-like oscillation related to eclipse on the microbarometer pressure records. The cooling effect of the Moon's shadow may induce the powerful meridional airflow in the atmosphere, which accelerates the ionized clouds in the Es layer and forms the wind shear to raise the observed Doppler frequency shift and foEs values, respectively (Chen et al., 2010).

SOLAR ECLIPSE OBSERVED BY VERTICAL IONOSPHERIC SOUNDING IN MIDLATITUDES

Vertical sounding measurements provide local information on the electron density distribution of the bottomside ionosphere. Electron concentration in the plasma and its corresponding plasma frequency are related via following equation:

$$f_p^2 = \frac{Ne^2}{4\pi\varepsilon_0 m} \tag{25}$$

where f_p denotes plasma frequency and N, e, ε_0 and m stand for the electron concentration, the charge of electron, permittivity of free space, and the mass of the electron, respectively. This section summarizes experimental results from the midlatitude ionospheric observatory Pruhonice (50N, 15E). At the observatory, the vertical sounding measurements were performed with ionosonde IPS 42 KEL Aerospace till the end of year 2003. Then this older equipment was replaced by digisonde DPS 4. Special campaigns of rapid sequence soundings were organized in order to study in detail ionospheric behavior during partial solar eclipses of 11 August 1999, 4 January 2011 and annular solar eclipse 3 October 2005. All three analyzed events were characterized by low geomagnetic activity; hence they represent a good occasion to observe mostly solar eclipse induced effects in the ionosphere. However, inconclusive results of the solar eclipse observations rise from the fact that different solar eclipses produce different plasma motions. Indeed, the travel cone geometry and its angular effects on the magnetized plasmas are different for each eclipse. Solar eclipse of 11 August 1999 (as a total seen in place as close as 200 km from the measurement point) represents so far the event of the highest solar disc coverage observed in the Observatory Pruhonice. Figure 3 depicts sequence of raw ionograms measured during this event by IPS 42 KEL Aerospace equipment. The ionograms were recorded with the cadence of 1 minute. On the ionograms there is clearly seen that the eclipse event affects whole electron density profile. Critical frequencies in the E and F layer decrease before maximum disc occultation and then increase again. The electron density decrease in the E layer is much stronger than in the F layer due to different dominant type of the recombination. Electron density fall and increase occur simultaneously with occultation and de-occultation of the solar disc in the E and F1 layer while the F2 layer electron density reacts with slight delay. There are special structures of the spread F type developed on the profile after beginning of the solar disc occultation (clearly seen on ionograms at 9.14 UT and 9.16 UT). Shape of the F layer is affected as well. Unfortunately, effects in the F1 region cannot be discussed here in details because F1 layer is blanketed by strong sporadic E layer during part of the solar eclipse.

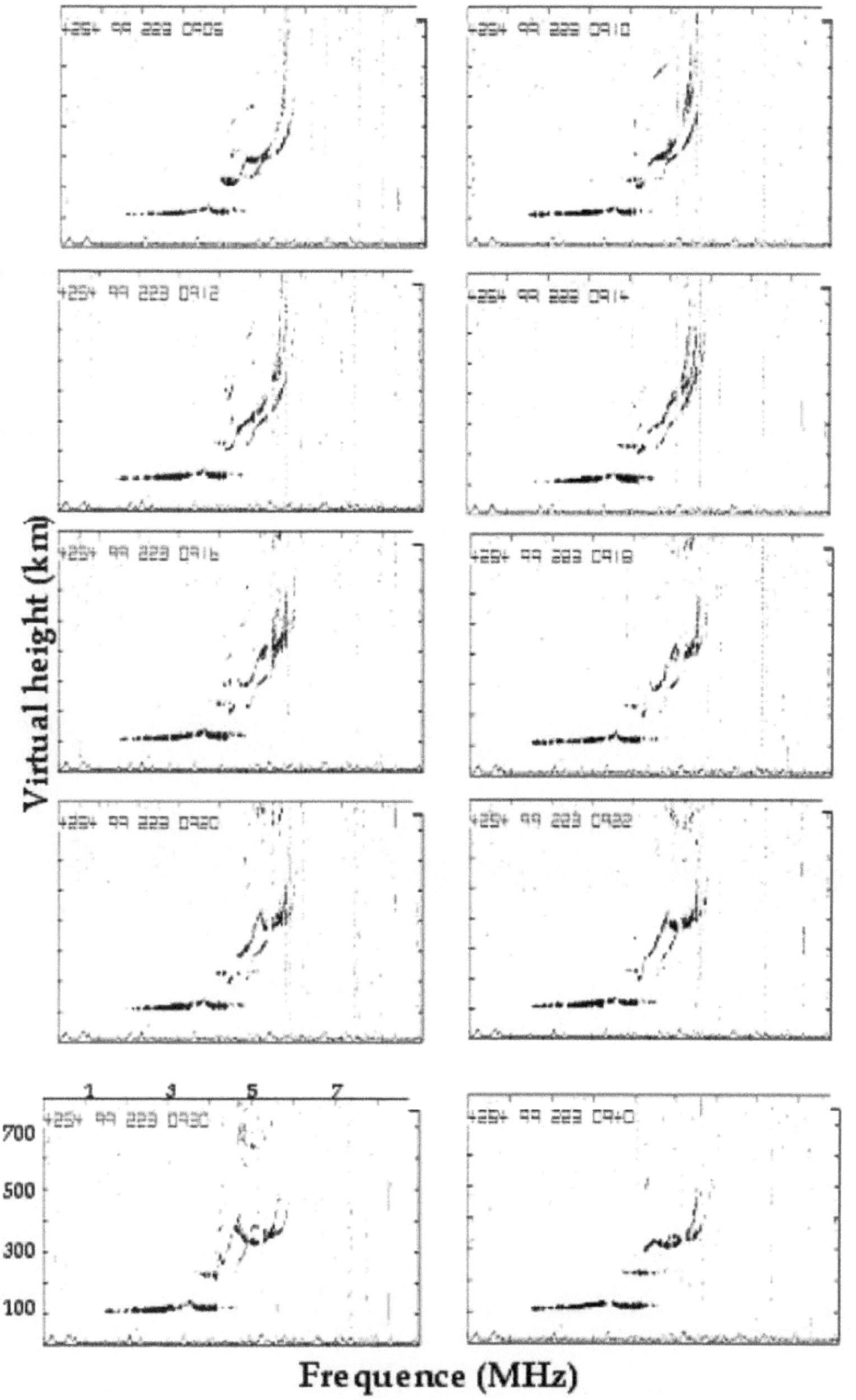

Figure 3: Sequence of raw ionograms measured by the ionosonde KEL Aerospace IPS 42 at the observatory Pruhonice. During the special campaign ionograms were recorded with oneminute resolution in order to study rapid ionospheric changes during the solar eclipse.

Detail analysis of electron concentration by mean of spectral analysis reveals that within oscillation of electron concentration there occur several clear wave-like oscillations. It has been shown by Sauli et al. (2007) that wavelet spectral analysis is very convenient approach for such wave detection. The advantage of the wavelet based analysis is identification of the structure occurrence time which helps to associate particular wave-like structure to the agent. Figure 4 shows estimated wave parameters for selected structure that is coherent through all studied heights. Parametrization of the wave-like structure is based on AGW approximation described in Section 2. From Figure 4 it is evident that wave originates at height of about 200 km and propagates upward and downward from the source region.

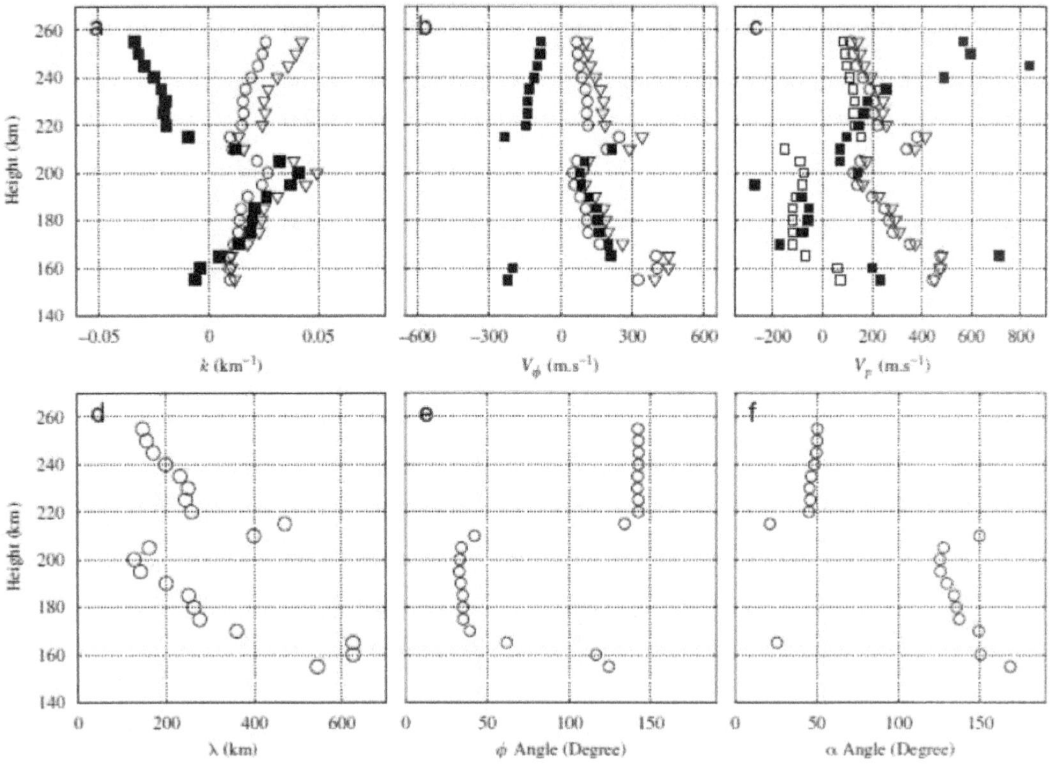

Figure 4: Parameters of acoustic-gravity wave structure detected within ionospheric plasma during solar eclipse event 11 August 1999 (Sauli et al., 2007). Panels: wave vector (a), phase velocity (b), packet velocity (c), wave number (d), energy (e) and phase (f) angles. For the vectors of first row, the '□' correspond to the measured (full squares) and computed (empty symbols) z-components, the '○' correspond to the horizontal components while the '∇ ' are related to the modulus.

Another representation of the rapid changes in the ionospheric plasma is shown on the profilogram (Figure 5) measured during solar eclipse 3 October 2005 by DPS 4. Decrease in the plasma frequency at all heights is well developed. Within plasma frequency oscillation, several wave coherent structures were found that can be attributed to the eclipse event. These structures occur in the plasma at the maximum of the eclipse and after the event. In all cases we detected a component of upward energy progression. Due to the occurrence time and low geomagnetic activity the detected wave-like oscillations in the ionospheric plasma are likely signatures of bow shock and possibly waves excited by cooling of ozone in the lower laying atmosphere. Estimated velocities for one particular structure are shown in Figure 6.

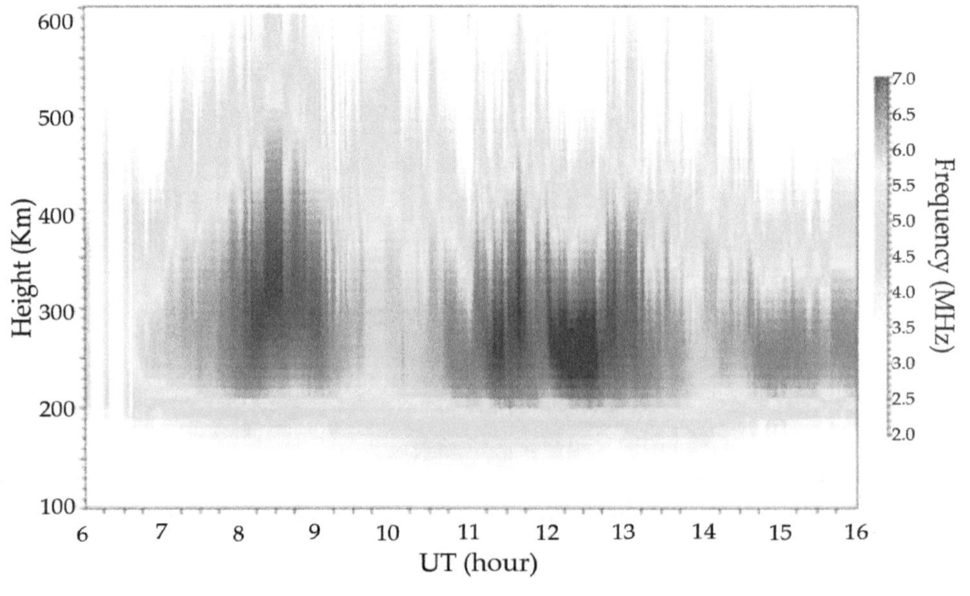

Figure 5: Profilogram (height-time-plasma frequency development) during solar eclipse 3 October 2005 as measured by DPS 4. Ionograms were measured every 2 minutes. All ionograms were manually scaled and inverted into true-height profiles using True Height Profile Inversion Tool NHPC.

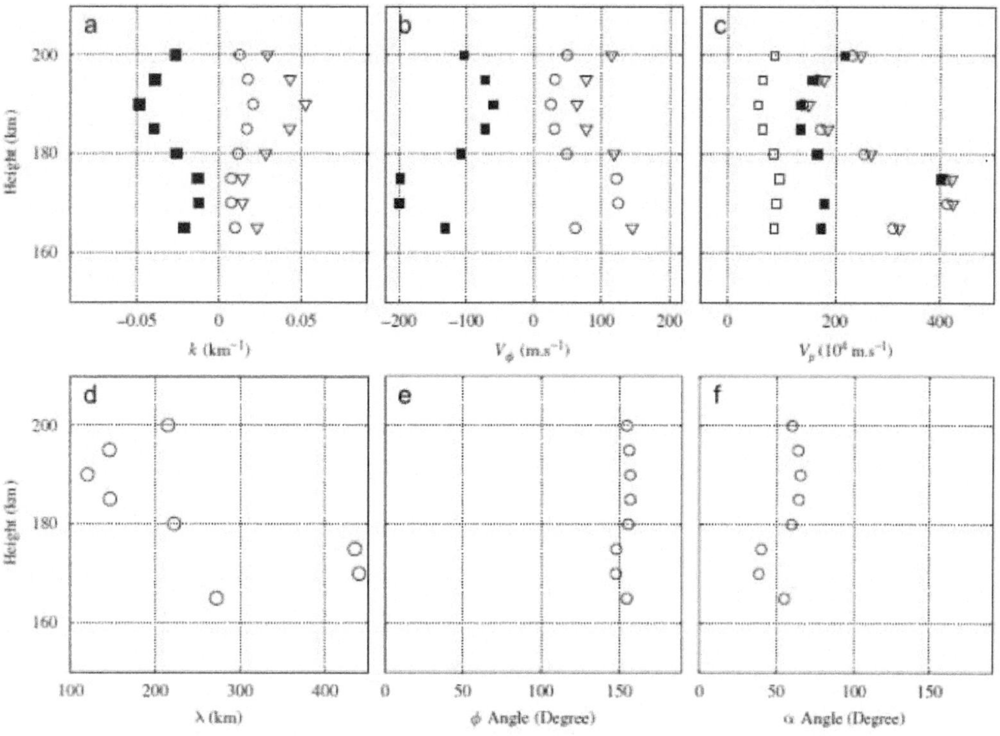

Figure 6: Parameters of acoustic-gravity wave structure detected within ionospheric plasma during solar eclipse event 3 October 2005 (Sauli et al. 2007). Panels: wave vector (a), phase velocity (b), packet velocity (c), wave number (d), energy (e) and phase (f) angles. For the vectors of first row, the '□' correspond to the measured (black) and computed (empty) zcomponents, the '○' correspond to the horizontal components while the '∇ ' are related to the modulus.

Result of the annular eclipse is significantly different from the case of the total eclipse event of 1999 where the dominant AGW activity took place at the beginning of eclipse. The atmospheric cooling and decrease in radiation flux during an annular solar eclipse is not as strong as during a total eclipse and the ionospheric response occurs with time delay.

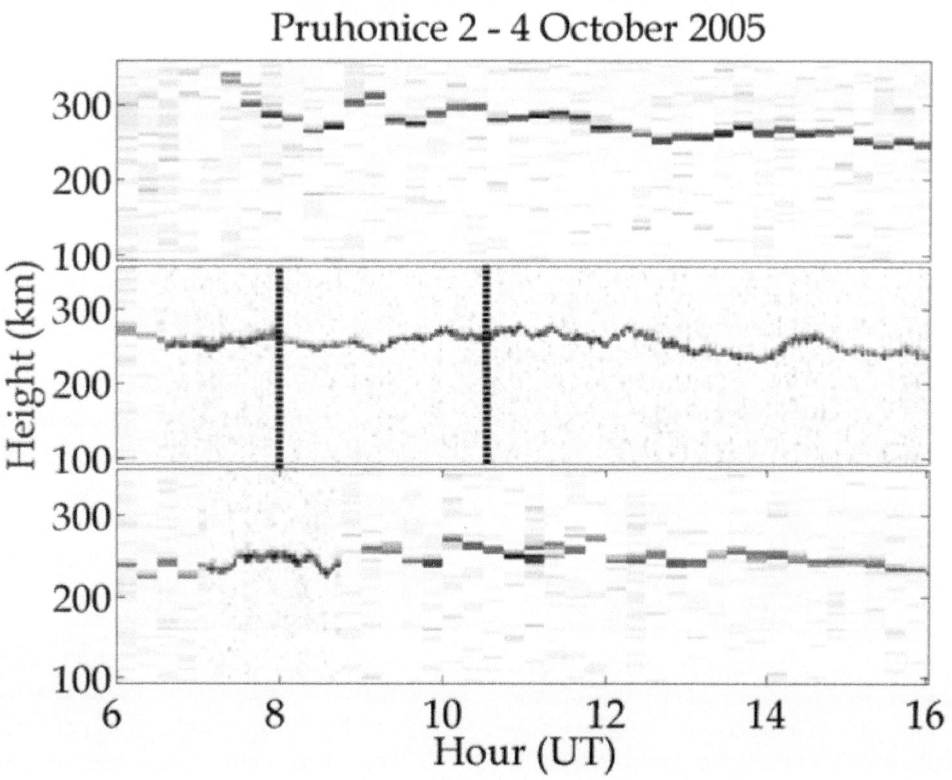

Figure 7: Virtual reflection heights of plasma frequency in the range 4.2 - 4.3 MHz derived from raw ionograms. From up to bottom: day before, day of eclipse, day after eclipse. Vertical lines in middle panel depict beginning and end of the eclipse. Time resolution is different for day of solar eclipse (2 min) and days before/after (15 min).

In Figure 7 and Figure 8, there are plots of virtual reflection height variations at single frequency during three consecutive days, day of solar eclipse event and one day before and after the event. Variation of the reflection height during eclipse event of 3 October 2005 does not differ much from the corresponding variation during reference time span day before and day after. Wave-like oscillations excited by solar eclipse are of comparable magnitude as those induced by other sources preceding and consecutive day. On the contrary, clear difference in reflection height oscillation during reference days and solar eclipse event is perfectly seen in Figure 8. Records of virtual heights at fixed frequency from January 4, 2011 present strong ionospheric response which is exhibited as periodic changes in reflection height. Sharp changes in the reflection height develop immediately after the beginning of the solar disc occultation and last till the end of eclipse event. Higher wave-like activity remains remarkable whole day. In this partial solar eclipse event, wave-like oscillations can be very probably attributed to the solar eclipse. Strong decrease in electron concentration in practically whole electron profile as well as the wave-like changes were observed during and after August 11, 1999 and January 4, 2011. Wave-like activity develops immediately after the start of the solar disc obscuration during partial solar eclipse. During annular solar eclipse, significant acoustic-gravity wave type bursts develop around and after maximum phase of the eclipse.

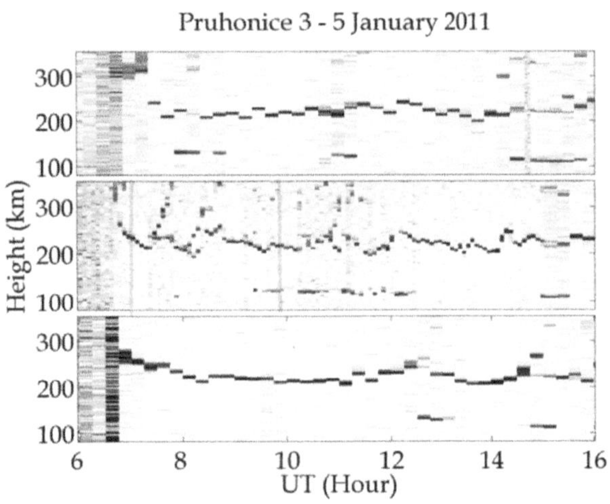

Figure 8: Virtual reflection heights of plasma frequency range 3.4 – 3.5 MHz derived from raw ionograms. From up to bottom: day before, day of eclipse, day after eclipse. Vertical lines in middle panel depict beginning and end of the eclipse. Time resolution is different for day of solar eclipse (5 min) and days before/after (15 min).

CONCLUSION

Acoustic-Gravity waves play important role in the dynamic of the upper atmosphere. Vertical ionospheric sounding represents powerful tool that allows us to monitor acousticgravity wave activity in the ionosphere. Ionospheric observation of such a strong event as solar eclipse gives us an opportunity to better understand processes of creation and dissipation of the AGW in the area of the ionosphere. Although the acoustic-gravity waves are always present in the area of our interest, sharp temporally well-defined changes of solar flux during the solar eclipse give us a possibility to define sources of AGW. It is rather uneasy to unambiguously assess causality between the solar eclipse events and the detected wave structures in the ionospheric plasma. Difficulties result from the fact that there are no two exactly identical solar eclipse events and from limitations of sounding techniques. Despite the fact that various AGW sources have been identified, many others remain to be found. Amongst irregular AGW bursts, regular increase in AGW activity were found to occur around sunrise and sunset hours, excited by Solar Terminator movement. Most of other sources (meteorological systems, geomagnetic and solar disturbances, etc.) and corresponding wave-like oscillations contribute to the irregular patterns of AGW activity observed in the ionospheric plasma.

As the solar eclipses, analyzed in the Section 5, occur sufficiently long time after the sunrise hours, one can assume that none of the reported waves are induced by solar terminator. During the analyzed sounding campaigns, no wave coming from auroral zone was expected, due to the quiet geomagnetic and solar activity. Additionally, meteorological analysis shows that meteorological systems very probably did not influence the ionosphere during studied events by means of AGW. The acoustic-gravity wave activity increases after a notably larger delay for the annular solar eclipse compared to the total solar eclipses: waves are found during the maximum phase of the eclipse only for the former while they occur during the initial phase for the latter. This discrepancy in gravity waves generation/occurrence can likely be explained by differences in the terrestrial atmosphere cooling: the border between sunlit and eclipsed region is much sharper in the case of total eclipse. Analyzing wave propagations, we observe predominantly upward propagating structures. The wave structure, that propagate upward and downward from the source region located around 200 km height, was created during an exceptional case related to the Solar eclipse of 11 August 1999.

REFERENCES

1. Altadill, D., Sole, J.G. & Apostolov, E.M. (2001). Vertical structure of a gravity wave like oscillation in the ionosphere generated by the solar eclipse of August 11, 1999. Journal of Geophysical Research, 106 (A10), 21419–21428, ISSN 0148-0227.
2. Bodo, G., Kalkofen, W., Massaglia, S. & Rossi, P. (2001). Acoustic waves in a stratified atmosphere. III. Temperature inhomogenities. Astronomy & Astrophysics, 370, pp. 1088-1091, ISSN 0004-6361.
3. Chen, G., Zhao, Z., Zhou, C., Yang, G. & Zhang, Y. (2010). Solar eclipse effects of 22 July 2009 on Sporadic-E. Annales Geophysicae, 28, 353–357, ISSN 0992-7689.
4. Cheng, K., Huang, Y.N. & Chen, S.W. (1992). Ionospheric effects of the solar eclipse of September 23, 1987, around the equatorial anomaly crest region. Journal of Geophysical Research, 97, A1, 103–111. ISSN 0148-0227.
5. Chimonas, G. & Hines, C.O. (1970). Atmospheric gravity waves induced by a solar eclipse. Journal of Geophysical Research, 75, 4, pp. 875, ISSN 0148-0227.
6. Datta, R.N. (1972). Solar-eclipse effect on sporadic-E ionization. Journal of Geophysical Research, 77, 1, 260–262.
7. Davies, K. (1990). Ionospheric radio. Peter Peregrinus

Ltd., ISBN 0 86341 186 X, London, United Kingdom.

8. Davis, J.R., Headrick, W.C. & Ahearn, J.L. (1964). A HF backscatter study of solar eclipse effects upon the ionosphere. Journal of Geophysical Research, 69 (1), 190–193. ISSN 0148-0227.
9. Farges, T., Jodogne, J.C., Bamford, R., Le Roux, Y., Gauthier, F., Vila, P.M., Altadill, D., Sole, J.G. & Miro, G. (2001). Disturbances of the western European ionosphere during the total solar eclipse of 11 August 1999 measured by a wide ionosonde and radar network. Journal of Atmospheric and Solar-Terrestrial Physics, 63, 9, pp. 915–924, ISSN 1364-6826.
10. Fritts, D.C. & Luo, Z. (1993). Gravity wave forcing in the middle atmosphere due to reduced ozone heating during a solar eclipse. Journal of Geophysical Research, 98, pp. 3011– 3021, ISSN 0148-0227.
11. Galushko, V.G., Paznukhov, V.V., Yampolski, Y.M. & Foster, J.C. (1998). Incoherent scatter radar observations of EGW/TID events generated by the moving solar terminator. Annales Geophysicae, 16, pp. 821-827, ISSN 0992-7689.
12. Hargreaves, J.K. (1982). The upper atmosphere and solar-terrestrial relations. An introduction to the aerospace environment. Van Nostrand Reinhold, ISBN 0 521 32748 2, Cambridge, United Kingdom.
13. Hines, C.O. (1960). Internal atmospheric gravity waves at ionospheric heights. Canadian Journal of Physics, 38, pp. 1441–1481, ISSN 1208-6045.
14. Hooke, W.H. (1968). Ionospheric irregularities produced by internal atmospheric gravity waves. Journal of Atmospheric and Solar-Terrestrial Physics, 30, pp. 795-829, ISSN 1364- 6826.
15. Huang X. & Reinish B.W (1996). Vertical electron density profiles from the digisonde network. Advances in Space Research., 18, pp. 121129, ISSN 0273-1177.
16. Ivanov, V.A., Ryabova, N.V., Shumaev, V.V., Uryadov, V.P., Nosov, V.E., Brinko, I.G. & Mozerov, N.S. (1998). Effects of the solar eclipse of 22 July 1990 at mid-latitude path of HF propagation. of Atmospheric and Solar-Terrestrial Physics, 60, 10, pp. 1013-1016, ISSN 1364-6826.
17. Jones, B.W., Miseldine, G.J. & Lambourne, R.J.A. (1992). A possible atmospheric-pressure wave from the total solar eclipse of 22 July 1990. Journal of Atmospheric and Terrestrial Physics, 54, 2, pp.113-115, ISSN 1364-6826.
18. Kato, S., Kawakami, T. & St. John, D. (1977). Theory of gravity wave emission from moving sources in the upper atmosphere. Journal of Atmospheric and Terrestrial Physics, 39, pp. 581–588, ISSN 1364-6826.
19. Le, H., Liu, L., Yue, X., Wan, W. & Ning, B. (2009). Latitudinal dependence of the ionospheric response to solar eclipses. Journal of Geophysical Research, 114, A07308, ISSN 0148-0227.
20. Liu, J.Y., Hsiao, C.C., Tsai, L.C., Liu, C.H., Kuo, F.S., Lue, H.Y. & Huang, C.M. (1998). Vertical phase and group velocities of internal gravity waves derived from ionograms during the solar eclipe of 24 October 1995. Journal of Atmospheric and Solar-Terrestrial Physics, 60, pp. 1679–1686, ISSN 1364-6826.
21. Muller-Wodarg, I.C.F., Aylward, A.D. & Lockwood, M. (1998). Effects of a mid-latitude solar eclipse on the thermosphere and ionosphere - a modeling study. Geophysical Reseach Letters, 25, 20, pp. 3787–3790, ISSN 0094-8276.
22. Oliver, W.L., Otsuka, Y., Sato, M., Takami, T. & Fukao, S. (1997). A climatology of F region gravity wave propagation over the middle and upper atmosphere radar. Journal of Geophysical Research, 102, A7, pp.14,499-14,512, ISSN 0148-0227.
23. Rapoport, V.O., Bespalov, P.A., Mityakov, N.A., Parrot, M. & Ryzhov, N.A. (2004). Feasibility study of ionospheric perturbations triggered by monochromatic infrasonic waves emitted with a ground-based experiment. Journal of Atmospheric and Solar-Terrestrial Physics, 66, pp. 1011-1017, ISSN 1364-6826.
24. Reinisch, B.W., Huang, X., Galkin, I.A., Paznukhov, V. & Kozlov, A. (2005). Recent advances in real-time analysis of ionograms and ionospheric drift measurements with digisondes, Journal of Atmospheric and Solar-Terrestrial Physics 67, pp. 1054-1062, ISSN 1364-6826.
25. Sauli, P., Abry, P., Boska, P. & Duchayne, L. (2006). Wavelet characterisation of ionospheric acoustic and gravity waves occuring during solar eclipse of August 11, 1999. Journal of Atmospheric and Solar—Terrestrial Physics, 68, pp. 586-598, ISSN 1364-6826.
26. Sauli, P., Roux, S.G., Abry, P. & Boska, J. (2007). Acoustic–gravity waves during solar eclipses: Detection and characterization using wavelet transforms. Journal of Atmospheric and Solar-Terrestrial Physics, 69, pp. 2465–2484, ISSN 1364-6826.
27. Sauli, P., Abry, P., Altadill, D. & Boska, P. (2005). Detection of the wave-like structures in the F-region electron density: two station measurement. Studia Geophysica & Geodetica, 50, pp. 131-146, ISSN 0039-3169.
28. Somsikov, V.M. & Ganguly, B. (1995). On the formation of atmospheric inhomogenities in the solar terminator region. Journal of Atmospheric and Solar—Terrestrial Physics, 57, 12, pp. 1513-1523, ISSN 1364-6826.
29. Titheridge J.E., 1985. Ionogram Analysis with the Generalised Program POLAN. UAG Report-93, 1985 (http://www.ips.gov.au/IPSHosted/INAG/uag_93/uag_93.html).
30. Walker, G.O., Li, T.Y.Y., Wong, Y.W., Kikuchi, T. & Huang, Y.N. (1991). Ionospheric and Geomagnetic effects of the solar eclipse of 18 march 1988 in East-Asia. Journal of Atmospheric and Solar—Terrestrial Physics, 53, 1–2, pp. 25–37, ISSN 1364-6826.
31. Xinmiao, Z., Zhengyu, Z., Yuannong, Z. & Chen, Z. (2010). Observations of the ionosphere in the equatorial anomaly region using WISS during the total solar eclipse of 22 July 2009. Journal of Atmospheric and Solar—Terrestrial Physics, 72, pp. 869–875, 1364-6826.
32. Yeh, K.C. & Liu, C.H. (1974). Acoustic-gravity waves in the upper atmosphere. Reviews of Geophysics and Space Physics, 12, 2, pp. 193-216, ISSN 8755-1209.

CHAPTER 10

ASYMMETRIC LONGITUDINAL OSCILLATIONS OF A PIKELNER'S MODEL PROMINENCE

J. Kraskiewicz´,K. Murawski, A. Solov'ev, A.K. Srivastava

Keywords: MHD Sun: magnetic fields, Sun: corona, Sun: prominences

INTRODUCTION

Prominences are dense and cold solar coronal magnetic structures, which are highly complex, as shown for instance by their long and thin threads (Tandeberg-Hanssen, 1974). The average prominence temperature is about two hundred times lower and the mass density is approximately two hundred times higher than the ambient coronal values. Prominences are linked to the underlying photosphere by several foot-points and lie along the polarity inversion line. The length of prominences is within 50 − 500 Mm, their height is 10 − 100 Mm, and their average thickness is 15 Mm (Priest, 1989). Solar prominences can be classified into two groups: i) active prominences and ii) quiescent prominences (Zirin, 1988). Active prominences reveal life-times of no more than a few days, undergoing dramatic changes in plasma motions and magnetic activity. They are often associated with solar flares. Quiescent prominences can live for months, forming themselves over a magnetic neutral line that separates the regions of opposite magnetic polarities on the photosphere. The first quiescent prominence models were devised about 60 years ago (Menzel, 1951). Two classical models are commonly accepted, which are known as the Kippenhahn and Schl¨uter (1957), and Kuperus and Raadu (1974) models. In these models the cool and dense plasma is maintained against gravity by the magnetic tension at the local dip of the magnetic arcade. The distribution of magnetic polarity in this arcade prominence is the same as in the underlying photosphere, i.e. direct polarity. Kuperus and Raadu (1974) proposed a model in which the prominence, taking the form of a magnetic flux rope, is maintained in a vertical current-sheet with open magnetic-field lines. Below this current-sheet the X-point is present, and the prominence exhibits inverse polarity.

Currently, there are two well-accepted 3D prominence models with inverse polarity: the sheared arcade and the flux-rope magnetic structures (Labrosse et al., 2010, and references cited therein). While these classical models reveal the equilibrium magnetic configurations of quiescent prominences, a recent trend is emerging with the observational and theoretical reports that led to the foundation of plasma and wave dynamics within such stable magnetic configurations of the prominence system. (Luna and Karpen 2012; Luna, D´ıaz and Karpen 2012) have found that the observed largeamplitude longitudinal prominence oscillations (> 20 km s^{-1}) are driven by the projected gravity along the flux tubes and are strongly influenced by the curvature of the dips of the magnetic field in which the prominence threads reside. These oscillations reported by both Luna and Karpen (2012) and Luna, D´ıaz and Karpen (2012) for slow magnetoacoustic–gravity modes, are the symmetric longitudinal oscillations. Terradas et al. (2013) have studied the excitation of fast as well as slow antisymmetric magnetoacoustic–gravity modes of a prominence; they also dealt with magnetoststatic (MHS) equilibrium and performed linear MHD normal mode analysis. Small-amplitude longitudinal (slow) and fast transverse prominence oscillations either as a collective motion or as an individual thread (≈ 2 − 3 km s^{-1}) are well observed and modeled in the solar atmosphere (Arregui, Oliver and Ballester, 2012, and references cited therein). Terradas, Oliver and Ballester (2001) have reported that slow magnetoacoustic– gravity longitudinal oscillations of the quiescent prominences can be damped by the radiation and Newtonian cooling. Longitudinal prominence oscillations

Citation: Kraśkiewicz, J., Murawski, K., Solov'ev, A., & Srivastava, A. K. (2016). On the Asymmetric Longitudinal Oscillations of a Pikelner's Model Prominence. Solar Physics, 291(2), 429-444. DOI: http://dx.doi.org/10.1007/s11207-016-0850-0.

were observed and modeled by Li and Zhang (2012), Zhang et al. (2012), Shen et al. (2014), Bi et al. (2014), and Chen, Harra and Fang (2014). While a number of generalizations to the seminal models of Kippenhan and Schl¨uter (1957) and Kuperus and Raadu (1974) were constructed in the past, we propose an arcade model of normal polarity as sketched by Pikelner (1971).

Pikelner (1971) only sketched the model but he did not support it by any analytical expression. Our aim is to devise an analytical arcade model of a prominence by presenting the stringent mathematical expressions for the equilibrium prominence quantities. Our devised analytical model of prominence is significant to the bounded prominence structures that exhibit collective wave motions as well as the plasma dynamics. Additionally, our goal is to implement this analytical model into the FLASH numerical code (Lee, 2013) and develop a numerical model. This numerical model allows us to study a number of plasma phenomena. As a particular application of this model, in the present paper, we simulate magnetoacoustic–gravity waves in the prominence that result from a localized pressure pulse. This is the first article describing in detail our newly developed analytical model and its one numerical experiment on long-period prominence oscillations to shed light on their driving physical mechanism. The derived results are applicable to understand the physical processes of the prominence during evolved non-linear oscillations and are thereby useful for prominence seismology and related observations. This article is organized as follows: Section 2 describes the analytical model of a quiescent prominence that is based on Pikelner's model. The results of numerical simulations are outlined in Secion 3. In the last section we present the discussion and conclusions.

ANALYTICAL MODEL OF QUIESCENT PROMINENCE

MHD Equations

We consider a gravitationally stratified and magnetically confined plasma, which is described by ideal two-dimensional (2D) magnetohydrodynamic (MHD) equations:

$$\frac{\partial \varrho}{\partial t} + \nabla \cdot (\varrho \boldsymbol{V}) = 0, \tag{1}$$

$$\varrho \frac{\partial \boldsymbol{V}}{\partial t} + \varrho (\boldsymbol{V} \cdot \nabla) \boldsymbol{V} = -\nabla p + \frac{1}{\mu}(\nabla \times \boldsymbol{B}) \times \boldsymbol{B} + \varrho \boldsymbol{g}, \tag{2}$$

$$\frac{\partial \boldsymbol{B}}{\partial t} = \nabla \times (\boldsymbol{V} \times \boldsymbol{B}), \tag{3}$$

$$\nabla \cdot \boldsymbol{B} = 0, \tag{4}$$

$$\frac{\partial p}{\partial t} + \boldsymbol{V} \cdot \nabla p = -\gamma p \nabla \cdot \boldsymbol{V}, \tag{5}$$

$$p = \frac{k_B}{m} \varrho T. \tag{6}$$

Here ϱ is the mass density, p a gas pressure, $V = [V_x, V_y, 0]$, $B = [B_x, B_y, 0]$, and $g = [0, -g, 0]$ represent the plasma velocity, the magnetic field and gravitational acceleration, respectively. The value of g is equal to 274 m s^{-2}. In addition, T is the plasma temperature, $\gamma = 5/3$ is the adiabatic index, μ is the magnetic permeability of the plasma, and m is a particle mass that is specified by mean molecular weight of 0.6. Although this value is valid for fully ionized plasma of the corona we believe that its larger value will not qualitatively change the response of the system. We assumed that z is an invariant coordinate ($\partial/\partial z = 0$) and we set the z-components of velocity and magnetic field equal to zero. This assumption removes Alfv´en waves from the system in which magnetoacoustic– gravity waves are able to propagate.

Equilibrium Configuration

We assume that the solar atmosphere is in static equilibrium $V_e = 0$). It follows from Equations (1)–(5) that in such model this equilibrium is described by

$$-\nabla p_e + \frac{1}{\mu}(\nabla \times \boldsymbol{B}_e) \times \boldsymbol{B}_e + \varrho_e \boldsymbol{g} = \boldsymbol{0}, \tag{7}$$

$$\nabla \cdot \boldsymbol{B}_e = 0. \tag{8}$$

Here the subscript $_e$ corresponds to the equilibrium quantities.

As a result of Equation (8), the magnetic field, B_e, can be expressed with the use of the magnetic-flux function, A(x, y), as

$$\boldsymbol{B}_e(x, y) = \nabla \times (A\hat{\boldsymbol{z}}) \tag{9}$$

with $\hat{z}$ being a unit vector along the z-direction. As a result, the x- and ycomponents of the magnetic field are

$$B_{ex} = \frac{\partial A}{\partial y}, \quad B_{ey} = -\frac{\partial A}{\partial x}. \tag{10}$$

Under the above assumptions, Equations (6) and (7) are reduced to the following expressions (Low, 1975, 1980):

$$\nabla^2 A(x, y) = -\mu \frac{\partial p_e(A, y)}{\partial A}, \tag{11}$$

$$\varrho_e(x, y) g = -\frac{\partial p_e(A, y)}{\partial y}. \tag{12}$$

Now, we can specify the RHS of Equation (11) and thereafter proceed to solve this equation for the magnetic field, however, this is a formidable nonlinear Dirichlet problem to solve (Low, 1975, 1980). Here we adopt the approach proposed by (Low, 1980, 1981, 1982) who reversed the mathematical problem by specifying the magnetic field first and then deriving the equilibrium conditions for the mass density and the gas pressure. By defining the magnetic flux function, A(x, y), we can in general integrate Equation (11) to find formula for the equilibrium gas pressure and then using Equation (12) calculate the mass density. Following this idea, we integrate

Equation (11) regarding y-coordinate as a fixed parameter (Solov'ev, 2010). Then, a variation of A is

$$dA = \frac{\partial A}{\partial x}dx + \frac{\partial A}{\partial y}dy + \frac{\partial A}{\partial z}dz = \frac{\partial A}{\partial x}dx, \tag{13}$$

as a result of $\partial A/\partial z = 0$ and $dy = 0$.

Using this expression in Equation (11) we obtain

$$\mu p_e = -\int \left(\frac{\partial^2 A}{\partial x^2} + \frac{\partial^2 A}{\partial y^2} \right) \frac{\partial A}{\partial x} dx + \mu C(y), \tag{14}$$

where C(y) is an integration function. As the integration over the first integrand can be performed explicitly, therefore, we obtain

$$\mu p_e = -\frac{1}{2}\left(\frac{\partial A}{\partial x}\right)^2 - \int \frac{\partial^2 A}{\partial y^2}\frac{\partial A}{\partial x}dx + \mu C(y). \tag{15}$$

As the magnetic field tends to zero at $x \longrightarrow \pm\infty$ and $p_e = p_h(y)$ we find

$$C(y) = p_h(y), \tag{16}$$

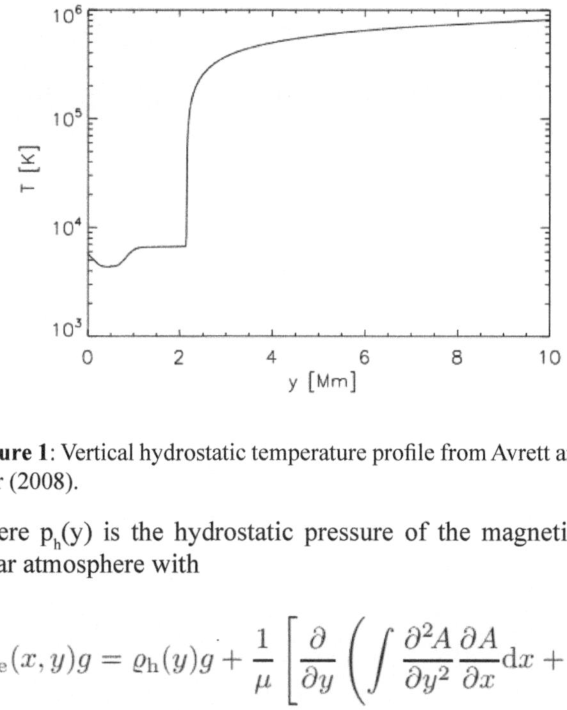

Figure 1: Vertical hydrostatic temperature profile from Avrett and Loeser (2008).

where $p_h(y)$ is the hydrostatic pressure of the magnetic-free solar atmosphere with

$$-\frac{\partial p_h}{\partial y} = \varrho_h(y)g, \tag{17}$$

which is specified by a temperature profile $T_e(y)$. In this case we adopt the temperature model of Avrett and Loeser (2008). Note that temperature reaches its minimum of 4300 K at approx0.6 Mm (Figure 1). At higher altitudes, the temperature rises with y. At the transition region that is located at $y \approx 2.1$ Mm, the temperature shows abrupt growth of about 0.8 MK in the solar corona at y = 10 Mm. The temperature profile determines uniquely the hydrostatic mass density and a gas pressure, which fall off with y (not shown here). For a recent implementation of $p_h(y)$ see Murawski et al. (2013). From Equation (15) we get

$$p_e(x,y) = p_h(y) - \frac{1}{\mu}\left(\int \frac{\partial^2 A}{\partial y^2}\frac{\partial A}{\partial x}dx + \frac{1}{2}\left(\frac{\partial A}{\partial x}\right)^2 \right). \tag{18}$$

From Equation (12) it follows that in order to find the distribution of the mass density, $\varrho_e(x,y)$, we need to calculate $\partial p_e(A, y)/\partial y$. In order to execute this action, we perform the following analysis. Let A(x, y) be a function of independent variables, (x, y). Then we can express any differentiable function, S(x, y), as S(y, A(x, y)) and the following equation is derived:

$$\frac{\partial S(x,y)}{\partial y} = \frac{\partial S(y,A)}{\partial y} + \frac{\partial A}{\partial y}\frac{\partial S(y,A)}{\partial A}. \tag{19}$$

Then

$$\frac{\partial S(y,A)}{\partial y} = \frac{\partial S(x,y)}{\partial y} - \frac{\partial A}{\partial y}\frac{\partial S(y,A)}{\partial A}. \tag{20}$$

Replacing S(y, A) by $p_e(y, A)$ and utilizing Equation (11) we find

$$\frac{\partial p_e(y,A)}{\partial y} = \frac{\partial p_e(x,y)}{\partial y} - \frac{\partial A}{\partial y}\nabla^2 A, \tag{21}$$

and then expressing $p_e(y, A)$ by Equation (18) we find the formula for the equilibrium mass density,

$$\varrho_e(x,y)g = \varrho_h(y)g + \frac{1}{\mu}\left[\frac{\partial}{\partial y}\left(\int \frac{\partial^2 A}{\partial y^2}\frac{\partial A}{\partial x}dx + \frac{1}{2}\left(\frac{\partial A}{\partial x}\right)^2 \right) - \frac{\partial A}{\partial y}\nabla^2 A \right]. \tag{22}$$

Note that p_e and ϱ_e are specified by Equation (18) and Equation (22), respectively. These equations are general in nature. Specific forms of p_e and ϱ_e are obtained after the estimation of A(x, y), which is a free function. This function must be chosen by some physical reasons. In the following part of this article we present this function for a solar prominence.

Pikelner's Prominence Model

For the solar prominence model of Pikelner (1971) we adopt the following expression for the flux function:

$$A(x,y) = \frac{B_0}{k}(1 + a_1 k^2 x^2)\exp[-k^2(a_2(y - y_{\rm ref})^2 + x^2)], \tag{23}$$

where a_1 and a_2 are some positive constants, k denotes inverse length, which determines the spatial scale of the structure, and B_0 is the magnetic-field strength at the reference point, ($x = 0$, $y = y_{ref}$). We choose and hold fixed $y_{ref} = 10$ Mm, $k = 1/50$ $(Mm)^{-1}$, and $B_0 \approx 6$ Gauss. This choice of A(x, y) leads to magnetic-field lines that are characteristic for a prominence and the equilibrium mass density and gas pressure can be given by relatively simple expressions. The magnetic field resulting from Equations (23) and (10) is illustrated in Figure 2. The dip in the magnetic-field vectors is discernible along the vertical line x = 0, and this dip grows with a_1 as magnetic lines are more tilted for a large value of a1 Figure 2).

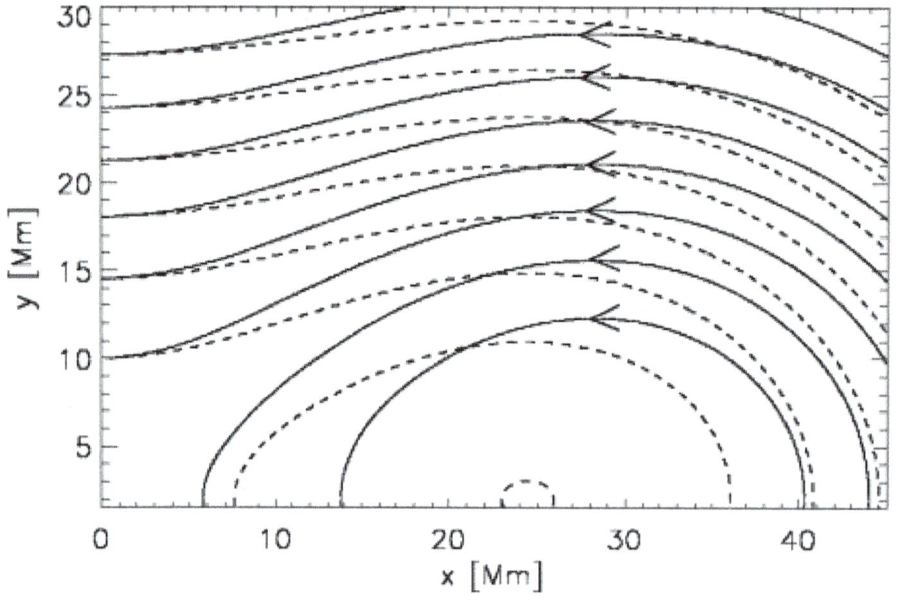

Figure 2: Spatial profiles of magnetic-field lines in Pikelner's prominence model for $a_2 = 1.5$, $a_1 = 1.6$ (dashed lines) and $a_1 = 2.0$ (solid lines). Note that only the right-hand ($x \geq 0$) side of the prominence is displayed.

Having specified A(x, y) by Equation (23), with the use of Equations (18) and (22), we express the equilibrium gas pressure and the mass density for the Pikelner's model as

$$p_e(x,y) = p_h(y) - \\ - 0.5\left[(p_3x^4 + p_4x^2 + p_7)y^2 + p_9x^4x^2 + p_1x^4 + p_2x^2 - 2a_2^2\right] \times \\ \times \exp[p_{10}(y - y_{ref})^2 + p_{11}x^2]B_0, \tag{24}$$

$$\varrho_e(x,y) = \varrho_h(y) + \\ + 4\,(p_5x^4 + p_6x^2 + p_8)(y - y_{ref})\exp[p_{10}(y - y_{ref})^2 + p_{11}x^2]B_0/g, \tag{25}$$

where

$$p_1 = k^4(-2a_1^2a_2 - 8a_1^2 + 8a_1), \quad p_2 = k^2(-4a_1a_2 + 4a_1^2 - 8a_1 + 4),$$
$$p_3 = 4k^6a_1^2a_2^2, \quad p_4 = 8k^4a_1a_2^2,$$
$$p_5 = k^6(a_1^2a_2^2 - a_1^2a_2), \quad p_6 = k^4[2a_1a_2^2 + (-a_1^2 - 2a_1)a_2],$$
$$p_7 = 4k^2a_2^2, \quad p_8 = k^2[a_2^2 + (a_1 - 1)a_2],$$
$$p_9 = 4k^6a_1^2, \quad p_{10} = -2k^2a_2, \; p_{11} = -2k^2.$$

The prominence equilibrium mass density and temperature profiles that result from Equations (24), (25), and (6) are displayed in Figure 3. Note that the prominence plasma occupies the dense (at x = 0, y ≈ 20 Mm) and cold (at y ≈ 35 Mm) region. While the classical models of a prominence exhibit a single cold region that is centrally located, Pikelner's model reveals the central cold region and two side, cold regions (Figure 3 c and d). These side regions result from faster fall-off in gas pressure than in mass density.

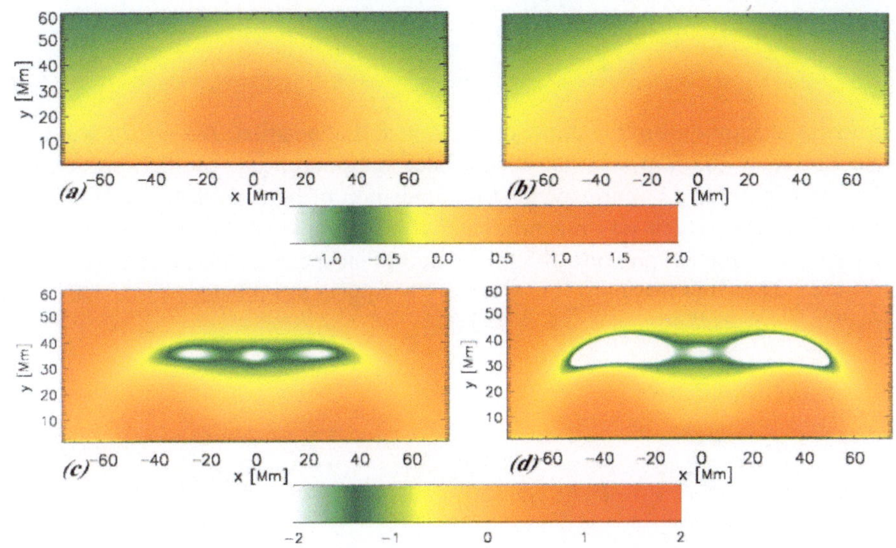

Figure 3: Spatial profiles of log(ϱ_e (x, y)) for $a_2 = 1.5$, $a_1 = 1.6$ (a) and $a_1 = 2.0$ (b). Temperature profiles log(T_e(x, y)) for $a_1 = 1.6$ (c) and $a_1 = 2.0$ (d).

RESULTS OF NUMERICAL SIMULATIONS

Equations (1) – (6) are solved numerically using the FLASH code (Lee and Deane, 2009; Lee, 2013). This code implements a third-order, unsplit Godunov solver with various slope limiters and Riemann solvers (e.g. T´oth, 2000), as well as adaptive mesh refinement (MacNeice et al., 1999). We use the minmod slope limiter and the Roe Riemann solver. For all cases considered we set the

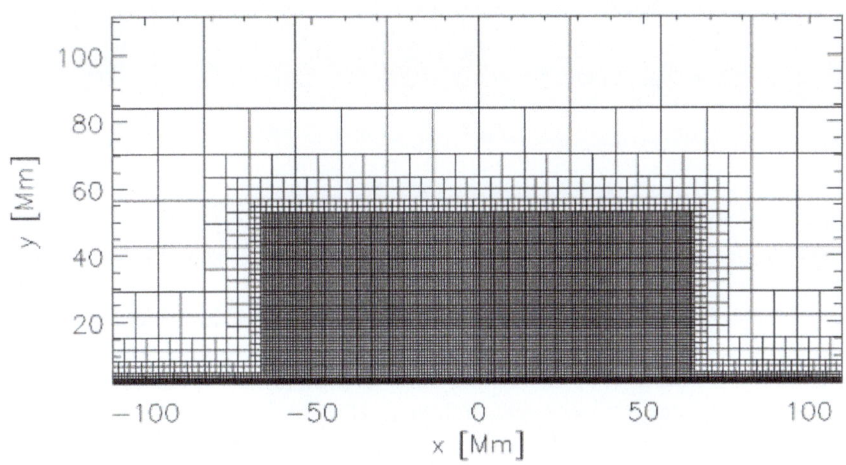

Figure 4: Block system used in the simulation studies.

simulation box as (−110, 110) Mm×(1, 111) Mm and impose boundary conditions by fixing in time all plasma quantities at all four boundaries to their equilibrium values. In all of our studies we use a static grid with a minimum (maximum) level of refinement set to 3 (9). The whole computational region is covered by a set of blocks with different grid cell sizes, which are organized in a hierarchical fashion using a tree data structure Figure (4). The blocks at the first/top level of the tree consist of the largest cells, while their children have a factor of two smaller cells and are called to be refined. As a result, at level 3 (9) the block size contains cells by the factor of 2^2 (2^8) smaller than the top level cells. In this way, we attain the effective finest spatial resolution of about 20 km below the transition region, y = 2.1 Mm. The total number of grid cells implemented at t = 0 seconds in the model is almost 106. The typical computation time is 48 hours and the computations were performed with the 32 CPUs.

Perturbations in the Pikelner's Prominence

We initially perturb the above described equilibrium impulsively by adding a Gaussian pulse in a gas pressure, viz

$$p(x, y, t = 0) = p_e(x, y)\left[1 + A_p \exp\left(-\frac{x^2 + (y - y_0)^2}{w^2}\right)\right] . \tag{26}$$

Here the symbol A_p denotes the amplitude of the initial pulse, y_0 its initial position, and w is its width. The relative amplitude of the initial pressure pulse, i.e., $A_p = -0.95$ means that we consider large-amplitude oscillations. We set and hold fixed $A_p = -0.95$, w = 4 Mm, and $y_0 = 35$ Mm, where y_0 corresponds to the position of bottom of the magnetic dip. The negative value of A_p mimics rapid cooling of plasma, which was recently studied by Murawski, Zaqarashvili and Nakariakov (2011). The sudden, spatially localized decrease in the gas pressure around the null point sucks the plasma in, creating a region of the density enhancement around the X-point. The compression of mass density is subject to buoyancy force, and its influence is expressed by changing the blob shape.

Dynamics of the Perturbed Plasma

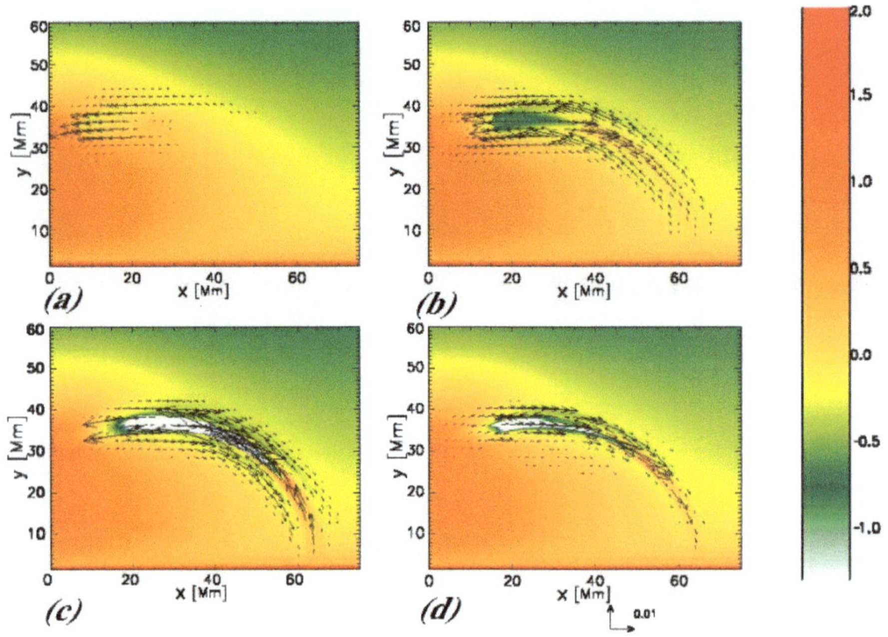

Figure 5: Spatial profiles of log $(\varrho(x, y))$ for $a_1 = 1.6$ and $a_2 = 1.5$ at t = 1000 seconds (a), t = 2500 seconds (b), t = 4000 seconds (c), and t = 5500 seconds (d). The vectors denote the velocity of the perturbed plasma. The only $x \geq 0$ Mm is displayed.

Figure 5 displays log $(\varrho_e(x, y, t))$ at four consecutive moments of time. As a result of the initial perturbation magnetoacoustic–gravity waves, which propagate in the system are essentially excited within the Pikelner's prominence. However, these waves quickly leave the prominence domain. Later the under-pressure that settles at the launching place results in slow magnetoacoustic–gravity waves, which propagate essentially along the magnetic-field lines (Figure 5 b).

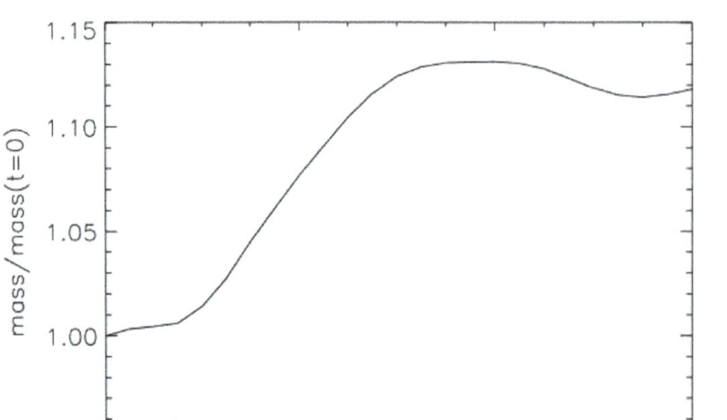

Figure 6: The relative mass of the prominence vs. time.

This under-pressure region brought some extra plasma from the ambient region into the launching place (Figure 5 c and d). Figure 6 illustrates the relative mass, Mr, of the prominence. This mass is evaluated within the region $|x| \le 20$ Mm, 20 Mm $\le y \le 45$ Mm. Indeed, according to our expectations, this mass grows in time; at t = 0 s, $M_r = 1$ while at t = 4000 s, M_r reaches a magnitude of 1.13. Figure 7 shows temperature profiles at four instants of time. We see that plasma at the top of the Pikelner's arcade becomes colder than in the ambient medium. As we are considering adiabatic equations, the cooling plasma results from the inflow of plasma, which is caused by the initial depression in the gas pressure there. As a result of this depression plasma is attracted by the pressure gradient force into the launching place. Note that we do not include any dissipation mechanism in our model, such as radiation or thermal conduction. The initial negative pressure imbalance produces an increase of the mass density at the dip of the model prominence because it produces an inflow there mainly from chromospheric plasma, which is seen in Figure 8 after the first 250 seconds of the simulation.

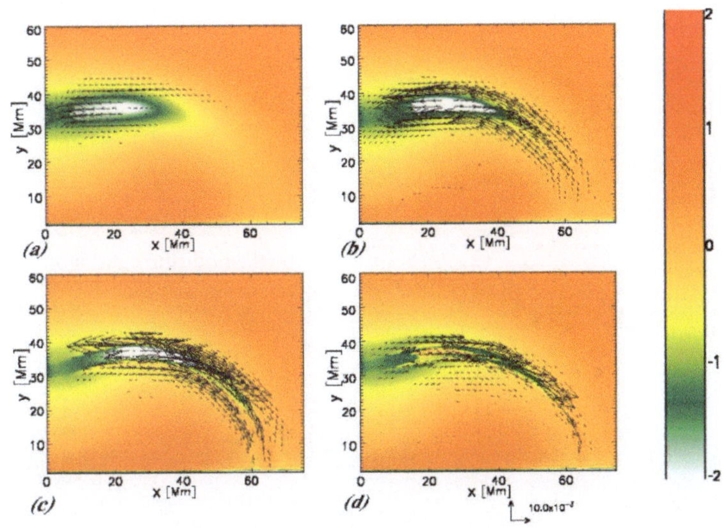

Figure 7: Spatial profiles of log(T (x, y)) for $a_1 = 1.6$ and $a_2 = 1.5$ at t = 1000 seconds (a), t = 2500 seconds (b), t = 4000 seconds (c), and t = 5500 seconds (d). The velocity vectors are denoted by arrows.

As the dip is a potential well, some part of the inflowing plasma is trapped there and the mass density grows there. A similar physical explanation can be applied to the temperature field of the prominence. The plasma flows from the foot-points of the prominence along its field lines, which is colder than the plasma in the dip at t = 0 seconds. Figure 9 displays the temporal signatures that are drawn by collecting wave signals in the mass density at the point (x = 0, y = 35) Mm, which corresponds to the location of the prominence dip. These time-signatures reveal an initial phase with strong mass-density variations. These initial phase lasts until t ≈ 1500 seconds. Later on, the oscillations are of lower amplitude indicating a strong attenuation. The attenuation may result from energy leakage through the footpoints of the prominence that are located at the transition region, and also from the dip in magnetic-field lines. We will discuss some estimations to support the inherent wave-leakage process to dissipate the magnetoacoustic oscillations in the later part of the manuscript.

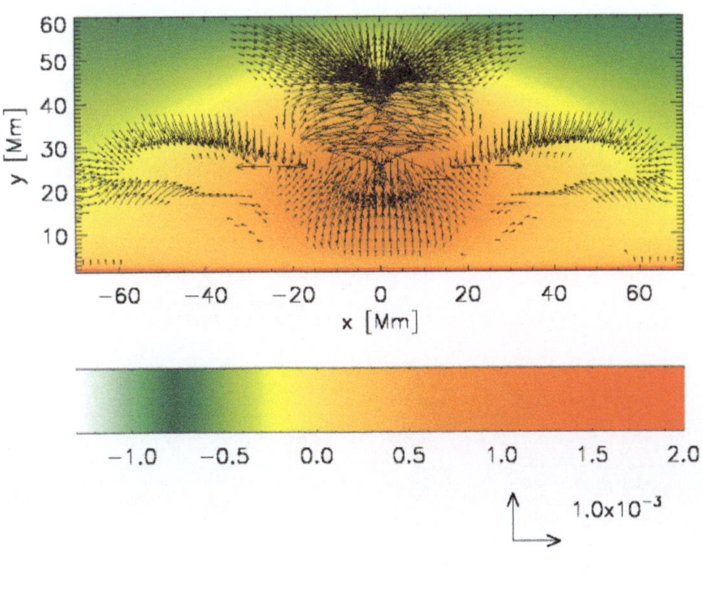

Figure 8: Spatial profiles of log $(\varrho(x, y))$ for $a_1 = 1.6$ at t = 250 seconds. The vectors denote the velocity of the perturbed plasma.

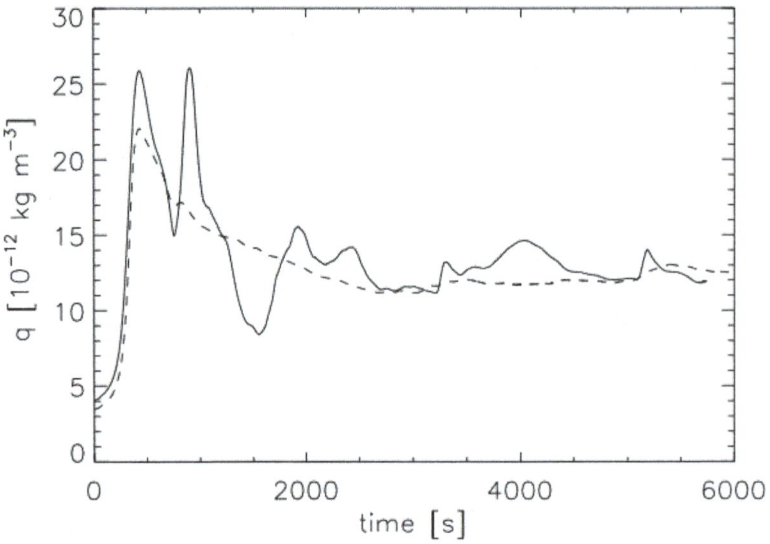

Figure 9: Time-signatures of $\varrho(x = 0\,\mathrm{Mm}, y = 35\,\mathrm{Mm})$ vs. time for $a_1 = 1.6$ (dashed line) and $a_1 = 2.0$ (solid line).

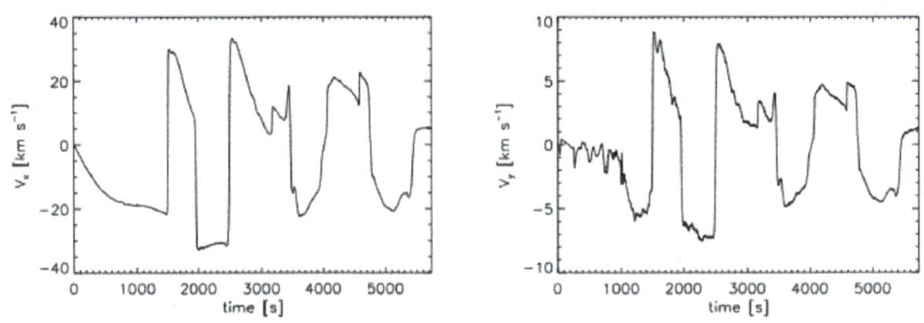

Figure 10: Time-signatures of velocity V_x(x = 20 Mm, y = 35 Mm) km s^{-1} (the left panel) and V_y(x = 20 Mm, y = 35 Mm) km s^{-1} (the right panel) vs time for $a_1 = 2.0$.

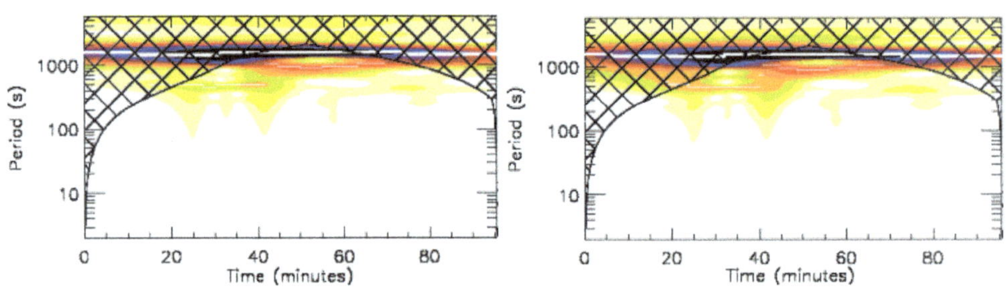

Figure 11: The wavelet spectrum of velocity V_x(x = 20 Mm, y = 35 Mm) km s^{-1} (the left panel) and V_y(x = 20 Mm, y = 35 Mm) km s^{-1} (the right panel) for $a_1 = 2.0$.

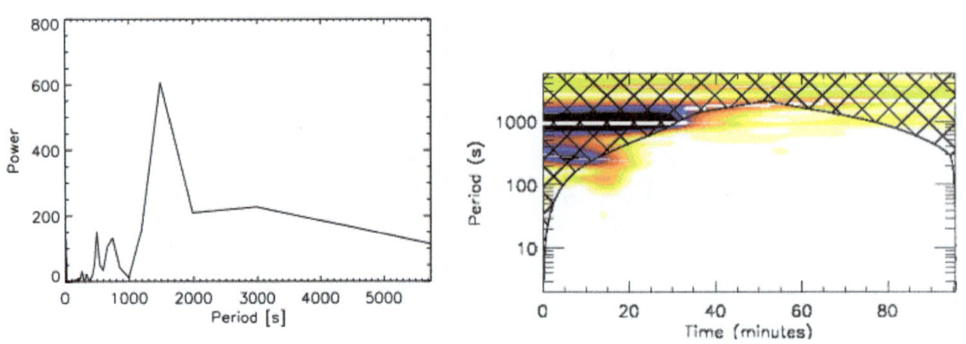

Figure 12: The Fast Fourier Transform (FFT) spectrum of time-signatures of mass density (the left panel) and the wavelet analysis of the density (the right panel) for $a_1 = 2.0$.

The temporal evolution of the mass density exhibits a clear non-linear behavior because the initial mass density, i.e. the equilibrium value, is smaller than the ending value, which remains almost constant.

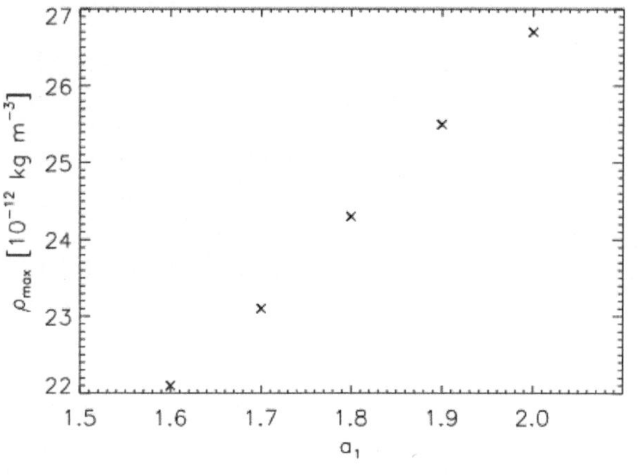

Figure 13: Maximum of the mass density taken from the left panel of Figure 9 vs. a_1.

The velocity field also oscillates according to the mass-density perturbations. Figure 10 shows the velocity oscillation near the maximum of the prominence at (x = 20, y = 35) Mm. The velocity-field perturbations are aligned with the magnetic-field lines indicating that the motions are associated with slow modes. These oscillations for the case of $a_1 = 2.0$, which represent the slow magnetoacoustic– gravity oscillations along the arcade, are analyzed by the Fast Fourier Transform (FFT) method. They exhibit the main period of 1577 seconds as is seen in Figure 11 and 12. Figure 13 shows that the maximum of the mass density, ϱ_{max} grows with a_1. Note that a larger value of a1 corresponds to a larger depth of the dips in the magnetic-field lines; the prominence with a larger depth of the dips, once perturbed and departed from the equilibrium, exhibits a larger amplitude of the slow magnetoacoustic–gravity oscillations. As a result, more plasma can be trapped in the dips and the mass density becomes larger there, which explains the growing trend exhibited in Figure 13. From Figure 9 we infer that a larger depth of the dips in the magnetic-field lines of a prominence leads to more oscillations in the mass density. Compare the solid line drawn for the larger depth in the dip with the dashed line, which corresponds to the case of the shallower dip.

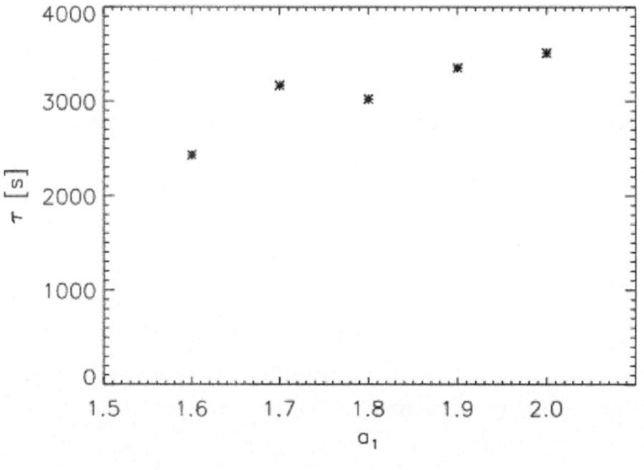

Figure 14: Variation of attenuation time τ vs. a_1.

The decay of the slow magnetoacoustic–gravity oscillations clearly decreases with the depth of the magnetic dip of a prominence. Figure 9 shows the fast attenuation and almost no oscillatory motion in the case of a shallower depth of the magnetic dip (dashed line). Figure 14 illustrates how attenuation time, τ, grows with a_1. This means that a prominence with a larger dip (larger a_1) has a longer attenuation time (i.e., longer decay) and vice-versa. Terradas et al. (2013) also found the larger oscillation wave periods for heavy prominences, while lower periods waves for light prominences. Our model deals with the prominence and its ambient medium within the framework of the analytical technique of Solov'ev (2010) in which all forces acting on the plasma are taken into account. The oscillations excited in the system are not pure gravity or gas pressure, which is a different aspect compared to the previous model. The results in our article are associated with the antisymmetric longitudinal oscillations, which is a case study of our developed prominence model, and they match the results obtained by Terradas et al. (2013). However, it should be noted that the present model is based on nonlinear full MHD, while the model by Terradas et al. (2013) deals with magnetoststatic equilibrium and linear MHD normal-mode analysis. Therefore, in the context of the model description, these two cases have weak relevance. As stated above, the most likely dissipation mechanism of the excited oscillations is a wave leakage. This can result from energy leakage through the foot-points of the prominence, which are located at the transition region and from the energy leakage through the dip in magnetic- field lines also. Figure 15 shows the ratio of kinetic energy of the plasma just under the foot-points to the total kinetic energy above the foot-points, i.e., above the transition region. We see that this ratio grows with time, which indicates the leakage of energy from the prominence. Also in Figure 16, we see some leakage of energy under and above the prominence in the form of side streams of fast-moving plasma. Therefore, we quantify the presence of wave-leakage process as the dissipative agent for the evolved antisymmetric magnetoacoustic–gravity oscillations in Pikelner's model prominence. We infer from Figure 16 that maximal pulse velocity is about 50 km s^{-1}. This magnitude of the flow matches the typical observational data. For a large amplitude of the initial pulse, nonlinear effects would become more important. However, we have verified by running the appropriate cases that the general scenario of the system evolution remains similar.

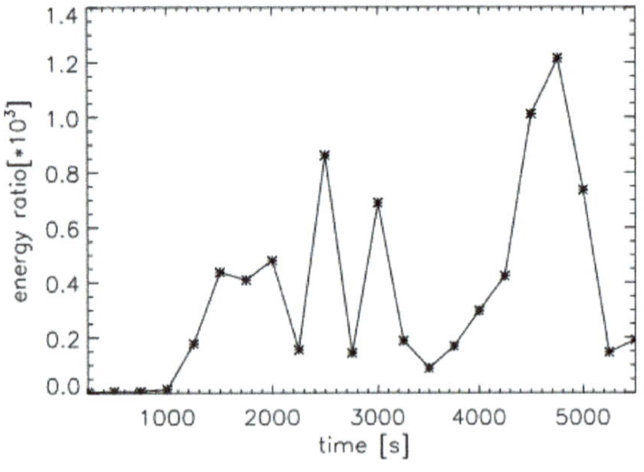

Figure 15: The ratio of energy under the foot-points to total energy.

SUMMARY

In this article we adopted the analytical methods of Solov'ev (2010) to derive Pikelner's model (Pikelner, 1971) of the normal polarity solar prominence. We implemented our analytical model into the publicly available FLASH code (Lee and Deane, 2009) and demonstrated the feasibility of fluid simulations in obtaining quantitative features in weakly magnetized and gravitationally stratified prominence plasma.

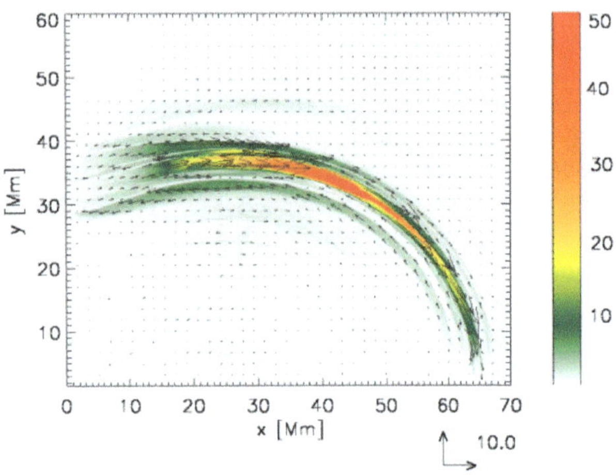

Figure 16: Total velocity in km s^{-1} (colour maps and vectors) for a_1 = 1.6 at time t = 6000 seconds.

In particular, we focused on perturbation of the developed normal polarity prominence model and its subsequent evolution. This perturbation was materialized by launching the initial pulse in gas pressure, which excited fast and slow magnetoacoustic–gravity waves. Fast waves were present at the initial stage of the prominence evolution while slow waves resulted later on and they lead to accumulation of plasma at the launching place which correspond to location of the magnetic dip. The parametric studies that we performed revealed that this accumulation varies with the shallowness of the dip; for a shallower dip there is less accumulated plasma and smaller oscillations at the magnetic dip. Such large-amplitude oscillations consist of the motions with the observed velocities greater than 20 km s^{-1} (Arregui, Oliver and Ballester, 2012), which is also evident in our model results. Large-amplitude longitudinal oscillations can be excited by impulsive events, e.g., microflares due to impulsive heating (e.g., Zhang et al., 2012). Therefore in our model, we consider the gas pressure perturbation as a physical initial trigger mechanism of the oscillations that excite longitudinal fast and slow magnetoacoustic–gravity waves. Our new analytical prominence model with realistic temperature distribution shows reasonable physical behavior of the typical slow acoustic oscillations in such quiescent prominences, which also matches the numerical results of Terredas et al. (2013). However, it should be noted that Terradas et al. (2013) deal with magnetoststatic (MHS) equilibrium and linear MHD normal mode analysis, while our model is based on non-linear full MHD equations. The general damping mechanism may most likely be the radiative cooling as invoked by many analytical and numerical investigations (e.g., Terradas, Oliver and Ballester, 2001; Terradas et al., 2013) related to the slow acoustic oscillations of the quiescent prominences. It should be noted that the fast dissipation of large-amplitude prominence oscillations is clearly evident in the observational data (e.g., Terradas et al. 2002; Arregui, Oliver and Ballester, 2012, and references cited therein). In the present case, we specifically found and quantified that the wave-leakage process is the dissipative agent for the antisymmetric magnetoacoustic–gravity oscillations in Pikelner's model prominence. Since in our model we do not invoke any non-adiabatic thermodynamical effects, e.g., radiative cooling, for the dissipation of such oscillations, even then the wave-leakage turns out to be a very effective mechanism for the dissipation of asymmetric oscillations. The study of the relative significance of various dissipative agents on the magnetoacoustic– gravity mode oscillations, e.g., wave-leakage, radiative cooling, etc. may be an important task that we will study in a future project.

In conclusion, we have tested our new analytical prominence model numerically, and excited the slow magnetoacoustic–gravity oscillations along its magnetic dip by perturbing a gas pressure within the prominence. Our model will be further used in a more detailed study of prominence dynamics and to constrain its oscillations to exploit prominence seismology in order to deduce local plasma conditions.

ACKNOWLEDGEMENTS

This work was supported by the project from the Polish National Foundation (NCN) Grant no. 2014/15/B/ST9/00106. The work has also been supported by a Marie Curie International Research Staff Exchange Scheme Fellowship within the 7th European Community Framework Program (K. Murawski, A. Solov'ev and J. Kra´skiewicz.). In addition, A. Solov'ev thanks the Russian Scientific Foundation for the support in the frame of project No 15-12-20001. The software used in this work was in part developed by the DOEsupported ASCI/Alliance Center for Astrophysical Thermonuclear Flashes at the University of Chicago. The visualizations of the simulation variables have been carried out using the IDL (Interactive Data Language) software package. Numerical simulations were performed on the Solaris cluster at Institute of Mathematics of M. Curie-Sk lodowska University in Lublin, Poland.

REFERENCES

1. Arregui, I., Oliver, R., Ballester, J.L.: 2012, Living Rev. Solar Phys. 9, 2. DOI:10.12942/lrsp- 2012-2

2. Avrett, E.H., Loeser, R.: 2008, Astrophys. J. Suppl. Sec. 175, 229. DOI:10.1086/523671

3. Bi, Y., Jiang Y., Yang J., Hong J., Li H., Yang D., Yang B.: 2014, Astrophys. J. 790, 100. DOI:http://dx.doi.org/10.1088/0004-637X/790/2/100

4. Chen, P.F., Harra L.K., Fang C.: 2014, Astrophys. J. 784, 50. DOI:10.1088/0004- 637X/784/1/50 Kuperus, M., Raadu, M.A.: 1974, Astron. Astrophys. 31, 189.

5. Kippenhahn, R., Schl¨uter, R.: 1957, Zaitsch. Astroph. 43, 36.

6. Labrosse, N., Heinzel, P., Vial, J.C., Kucera, T., Parenti, S., Gunar, S., Schmieder, B., Kilper, G.: 2010, Space Sci. Rev. 151, 243. DOI:10.1007/s11214-010-9630-6

7. Lee, D.: 2013, J. Comp. Phys. 243, 269. DOI:10.1016/j.jcp.2013.02.049

8. Lee, D., Deane, A.E.: 2009, J. Comp. Phys. 228, 952. DOI:10.1016/j.jcp.2008.08.026 Li, T., Zhang, J. 2012.: Astrophys. J. Lett. 760 (1), L10.

9. Low, B.C.: 1975, Astrophys. J. 197 251. DOI:10.1086/153508 Low, B.C.: 1980, Solar Phys. 65, 147. DOI:10.1007/BF00151389

10. Low, B.C.: 1981, Astrophys. J. 246, 538. DOI:10.1086/158954

11. Low, B.C.: 1982, Solar Phys. 75, 119. DOI:10.1007/BF00153465

12. Luna, M., Karpen, J.: 2012, Astrophys. J. Lett. 750, 1. DOI:10.1088/2041-8205/750/1/L1

13. Luna, M., D´ıaz, A.J., Karpen, J.: 2012, Astrophys. J. 757, 98. DOI:10.1088/0004- 637X/757/1/98

14. MacNeice, P., Olson, K.M., Mobarry, C., de Fainchtein, R., Packer, C.: 2000, Comp. Phys. Comm. 126, 330.

CHAPTER 11

INVESTIGATING ALFVÉNIC WAVE PROPAGATION IN CORONAL OPEN-FIELD REGIONS

R.J. Morton, S. Tomczyk & R. Pinto

Keyword: Physical sciences, Astronomy Fluids and plasma physics

INTRODUCTION

It is now well known that the fast solar wind originates from within open magnetic field regions in the Sun's atmosphere, for example, coronal holes, with wind acceleration beginning in the transition region and low corona1·2·3·4·5. However, the mechanism(s) responsible for heating the coronal plasma and accelerating the solar wind within the open fields is still unclear. A host of different physical models exist to explain this phenomena6 and one broad category of models relies on the propagation, reflection and dissipation of Alfvén waves7·8·9·10·11. Alfvén waves are perfect for the transport of energy as they are incompressible, making it difficult for their energy to be dissipated without the presence of large gradients in the Alfvén speed. This property should allow them to propagate out into the extended solar atmosphere and deposit their energy and momentum beyond the critical point and the point of temperature maximum, an important feature needed to explain the acceleration of fast solar winds9·12·13. These Alfvén waves would likely be generated towards the foot points of magnetic flux tubes, which are rooted in the photosphere and extend out into interplanetary space. In theory, after traversing the lower solar atmosphere, the waves reach the corona where they propagate and are partially reflected by the gradual variation in Alfvén speed, enabling the development of magnetohydrodynamic (MHD) wave turbulence[7,8,9,10,11,14] and the subsequent dissipation of the energy transported by the waves.

Citation: Morton, R. J., Tomczyk, S., & Pinto, R. (2015). Investigating Alfvenic wave propagation in coronal open-field regions. Nature communications, 6. DOI: http://dx.doi.org/10.1038/ncomms8813.

The textbook Alfvén wave, first described by Hannes Alfvén, requires the existence of an infinite, homogenous plasma. In this scenario, other perturbations of the plasma can also be categorized neatly as slow and fast MHD waves. The distinct characteristics of Alfvén waves are that they are the sole propagator of vorticity perturbations and magnetic tension is the dominant restoring force, in addition to incompressibility. Conversely, slow and fast MHD waves are characterized by compressibility and do not perturb vorticity[15]. However, these characteristics are textbook ideas of MHD waves. In highly inhomogeneous plasmas, such as the solar atmosphere, MHD wave modes have mixed properties and cannot be so precisely categorized. The fine-scale structure in the solar atmosphere is frequently modelled as a collection of over-dense, magnetic waveguides. In such a setting, the wave energy is confined predominantly to the regions of enhanced density. Two wave modes in the system share characteristics reminiscent of the textbook Alfvén wave, namely the fundamental modes of the torsional Alfvén wave and the fast kink wave[15]. The torsional mode causes an incompressible rotation of magnetic surfaces. The fast kink wave generally has mixed properties. It is identified by a non-axisymmetric displacement of the waveguide (or a swaying motion) where magnetic tension is the dominant restoring force. The fast kink wave is also highly incompressible, especially in the limit where the wavelength is much larger than the transverse scale of the density inhomogeneity that defines the waveguide[16]. The kink mode also transports vorticity and shares a remarkable similarity with the surface Alfvén wave[15]. From theses considerations, we follow Goossens et al. (ref. 15, 16) and use the adjective Alfvénic to describe wave modes with characteristics similar to the textbook Alfvén waves. At present, it is unknown whether the models of Alfvén wave turbulence remain valid if the Alfvénic waves of inhomogeneous plasmas were used instead. The common characteristics shared by the waves could suggest that central features of the wave turbulence models may survive.

The existence of Alfvénic fluctuations far into the solar wind (at 1 AU) has been known for a number of years[17] and their presence in the corona have been anticipated from the measurements of non-thermal line widths[18,19,20,21,22,23,24,25]. Recently, imaging and spectroscopic observations confirmed the presence of Alfvénic waves throughout the solar atmosphere [26,27,28,29,30,31], with fast kink modes appearing omnipresent along fine-scale magnetic structures. The exact relationship between the motions encapsulated by the non-thermal width measurements and the wave motions seen in images is still unclear, typified by the apparent disparate measured amplitudes (and, hence, energetics). This is likely someway due to the complex dependence of non-thermal line widths on the resolution of the instrument, line of sight integration and the addition of different sources of motion, that is, flows and the multitude of MHD wave modes[32,33,34].

Nevertheless, the associated flux of energy estimated to be carried by the Alfvénic waves at the coronal base appears to be sufficient to meet the requirements for heating the coronal plasma and/or solar wind acceleration. Although, again, there are still some remaining issues that need to be clarified with respect to the energy flux[31,35,36,37] (also, see the Results section).

The presence of Alfvénic waves in the chromosphere and corona does not, however, provide sufficient support for acceleration of the solar wind via Alfvénic waves. For Alfvénic waves to be a viable option for wind acceleration, the presence of counter-propagating Alfvénic waves, over an extended frequency range, is crucial for MHD turbulence to occur in the open-field regions. Counter-propagation allows for the nonlinear interaction of the waves, which leads to a turbulent cascade. To date, only measurements of wave amplitudes and periods exist in such regions[18,19,20,21,22,23,24,25,30,31]. Further, current coronal observations of Alfvénic waves place only relatively simple constraints on the models, that is, non-thermal widths have provided frequency-integrated velocity amplitudes. More rigorous constraints can be placed on the models if insights are obtained into the evolution of the Alfvénic waves as they propagate from the corona out into the solar wind, along with measurements of key plasma parameters, for example, density, magnetic field strength.

In the following, these issues are explored through the analysis of both spectroscopic and imaging data from ground- and space-based observatories. We investigate whether the conditions in an open magnetic field region are suitable for MHD turbulence to develop, search for indications of the waves origin and whether the observed coronal Alfvénic waves are connected to those found far out in the solar wind. It is envisioned that the results will help constrain the variety of Alfvénic wave models. This is supported by an exploitation of the observed waves to probe some of the key plasma parameters of the open magnetic field region.

RESULTS

The Coronal Multi-Channel Polarimeter

The Coronal Multi-Channel Polarimeter (CoMP)[38] is used to analyse Doppler velocity fluctuations and their propagation in an open-field region located at the northern pole of the Sun. CoMP was the first instrument to confirm the ubiquity of Alfvénic waves in the corona29and, still, has an unequalled ability to observe coronal velocity fluctuations, over a large field of view at high temporal cadence.

The Sun, on 27 March 2012, had two dominant open-field regions with the southern one more readily identifiable. The southern region is marked by the presence of a coronal hole that has extremely low emission in Fe XIII images of the corona obtained with CoMP (Fig. 1). This lack of emission means that the data preparation technique (see Methods) fails low in the coronal hole and as such there is no line centre intensity measurements (or Doppler velocities/widths). The region for which velocity measurements could be made is highlighted by the contour on the CoMP image in Fig. 1a. The northern open-field region is less obvious but there is reduced emission at the pole compared with the rest of the corona. The lack of an obvious coronal hole is also apparent in coronal Fe IX and XII emission from higher resolution, full disk images provided by the Solar Dynamics Observatory (SDO) Atmospheric Imaging Assembly (AIA), although radially orientated, field-aligned fine-scale magnetic structures are visible (Fig. 1c,d). The relationship of this fine-scale magnetic structure to plumes and inter-plumes is unclear from the current data set. We are able to say that they are much narrower than the classic plumes, which are reported to span ~30 Mm (40 arcseconds) on average.

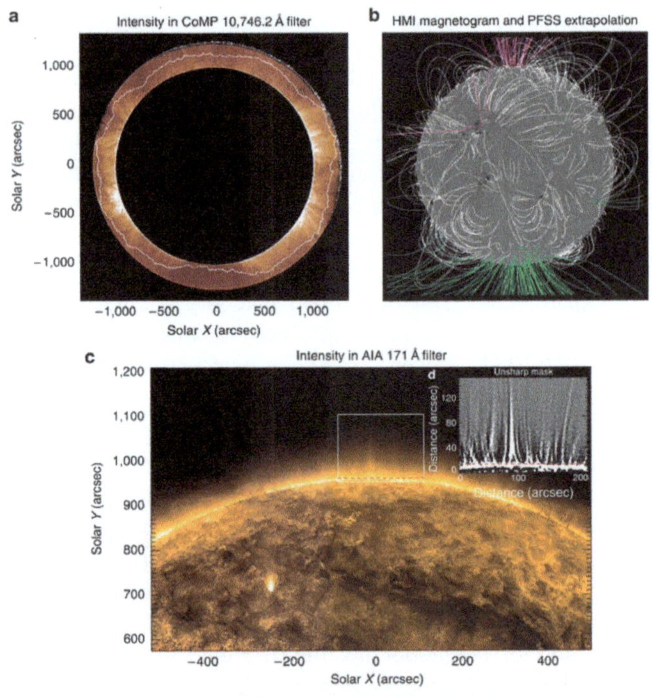

Figure 1: The solar corona (**a**) An intensity image of the corona as seen in the infrared Fe XIII emission line, where the disk of the Sun has been occulted. The image is enhanced with a radial gradient filter for visual purposes only. The white contour in the corona demarcates the region for which Doppler velocities could be measured. (**b**) A magnetogram, based on Helioseismic and Magnetic Imager (HMI) data and evolved with the surface-flux assimilation model, demonstrates the photospheric magnetic field and is over-plotted with magnetic field lines from the PFSS potential field extrapolation. The purple and green lines show open-field regions of negative and positive polarity, while the white lines show closed field regions. (**c**) An

enlarged image of the polar region as seen with SDO/AIA in the Fe IX coronal emission line. Fine-scale structures exist at the limb and are visible as elongated intensity enhancements. (**d**) An unsharp masked cut-out of a portion of the open-field region (box in**c**) clearly reveals this fine-scale magnetic structure extending radially away from the surface. The red dashed lines in **c** and **d** shows the position of the slit used to generate the time–distance diagram in Fig. 5a.

To confirm the presence of open fields in this region, we use the evolving surface-flux assimilation model and potential field source surface (PFSS) extrapolation tool39 (the extrapolation is shown in Fig. 1b). The southern coronal hole is seen in the extrapolation as a large open-field region that has some presence on the solar disk, while the northern open-field region is restricted to high latitudes. There is a direct correspondence between the locations of weak emission at low coronal heights in the CoMP data and the open-field lines in the PFSS extrapolation. We note that the visible structure in the region of interest remains practically unchanged throughout the observations, indicating little or no large-scale changes in the magnetic topology of the region.

THE EXISTENCE OF COUNTER-PROPAGATING ALFVÉNIC WAVES

We focus on the northern open-field region and examine the Doppler velocities in this part of the corona. It is evident from the data that there is a predominantly outward propagating Doppler signal in this region. To analyse the fluctuations in the Doppler velocities, we use two different methods to extract the data in the form of velocity time–distance diagrams (Fig. 2). The two independent methods are used to validate the analysis techniques and results of each other. For the first method, the crude assumption is made that the direction of potential wave propagation is strictly in the North–South direction. Individual Doppler velocity time–distance diagrams are generated from vertical strips of pixels in the boxed region (shown in Fig. 2a). The presence of the outward propagating Doppler velocity signals is visible in these diagrams and so is the periodic nature of these features, implying that the observed signals correspond to MHD waves.

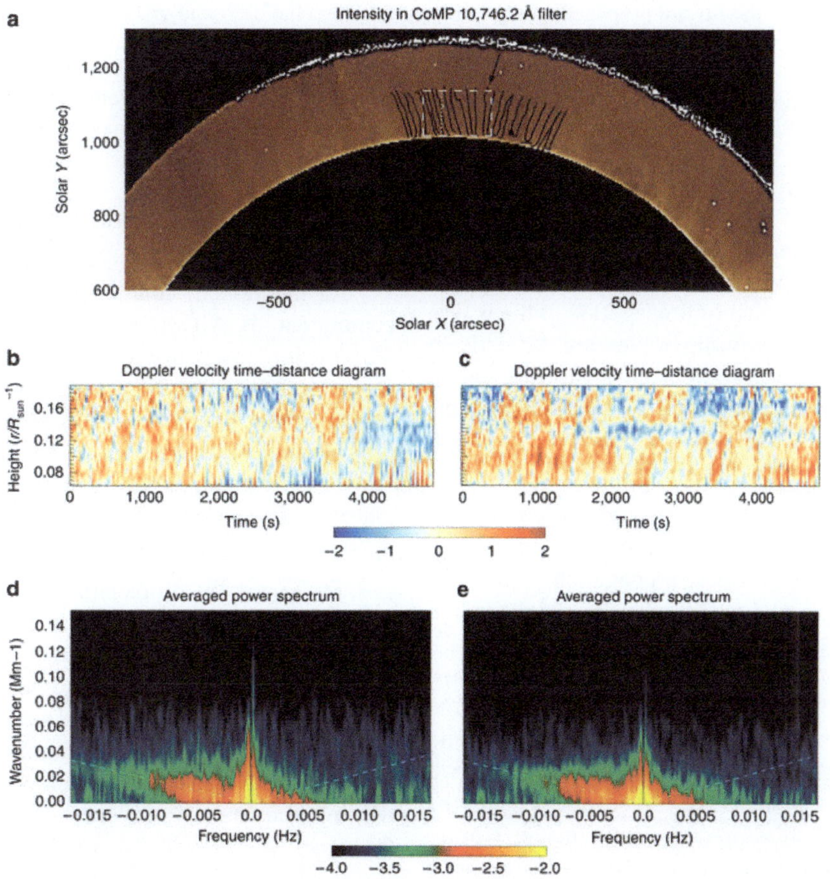

Figure 2: Propagating Alfvénic waves in the open-field region (**a**) A close-up CoMP intensity image of the northern polar region that demonstrates the locations of the box (dashed white lines) and the tracks (black lines) used in the two methods for isolating propagating waves in the open-field region. The predominantly upwardly propagating waves can be seen in the Doppler velocity time–distance diagrams from the box (**b**) and from the track (**c**) methods, where the scale bar indicates the velocity (km s^{-1}). The locations of these Doppler velocity time–distance diagrams are highlighted in **a** by the white dotted line (**b**) and black arrow (**c**). The time–distance diagrams are used to derive the corresponding frequency-wavenumber power plots for the box (**d**) and track (**e**) methods. A pronounced ridge of power in the negative frequency domain is clearly visible corresponding to the dominance of outward propagating waves. Crucially, the power plots also reveal the existence of inwardly propagating waves, but with reduced power. The scale bar indicates the power ($km^2 s^{-2}$) to the log base 10. The over-plotted white dashed lines show the calculated average inward and outward wave propagation velocities from the Doppler velocity time–distance diagrams.

The power spectrum for the Doppler velocity time–distance diagrams is determined for a range of frequencies (f) and wavenumbers (k) in the vertical direction, and the outward and inward propagating signals are decomposed. To analyse the waves, both an f–k diagram (Fig. 2) and wavenumber-integrated power spectra are calculated (Fig. 3). A prominent ridge in the left half of the f–k diagram corresponding to negative frequencies confirms the dominance of outward propagating waves. No clear ridge exists for the inward propagating waves although there is significant power in the low frequency part of the spectrum. Such ridges have also been found in f–k diagrams for large quiescent coronal loops40.

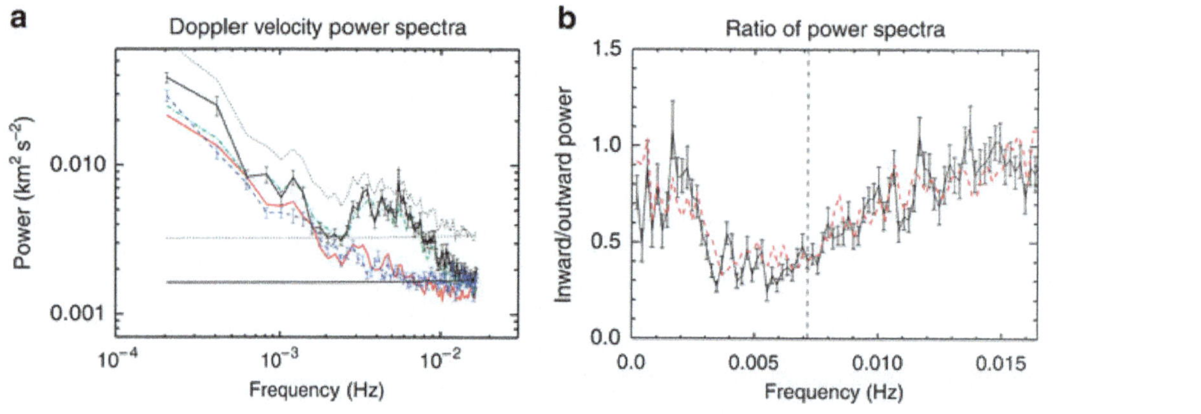

Figure 3: Average frequency power spectra and ratios of counter-propagating waves. The predominance of power in outward propagating waves (box method—black solid and track method—green dash-dot) is clear in the wavenumber-integrated power spectra (**a**). The spectra also provide insight into the power of the inward propagating waves (box method—blue dash and track method—red dash-triple dot). The total power spectrum is also shown for reference (dots). The horizontal dotted line is the estimate for the data noise associated with the total power spectra, while the solid horizontal line is the data noise split between inward and outward components. (**b**) The ratio of inward to outward power is displayed as a function of frequency and reveals a frequency-dependent trend suggestive of partial reflection of the waves in the corona. The vertical dashed line highlights the frequency after which noise begins to dominate the inward power spectra. The error bars in both plots show the s.e.m.

The second method is a more refined version of the first, providing thef–k diagram, power spectra and, in addition, the propagation velocities of the waves (see Methods). The method explicitly determines the direction of propagation of the waves and assuming that these propagating waves are Alfvénic in nature, the direction of propagation defines the orientation of the magnetic field in the plane of the sky. Building up this picture of orientations over the entire atmosphere, the local direction of the magnetic field is determined.

Starting at a particular pixel, the direction of the magnetic field in the plane of sky is followed and tracks are built that are aligned with the direction of wave propagation. The tracks are almost vertical with respect to the solar disk, highlighting the radial structure of the magnetic field (black lines in Fig. 2). The Doppler velocity time series at each point along the track is then used to build Doppler velocity time–distance maps and, as with the preceding method, the f–k diagram (Fig. 2) and the k-integrated power spectra (Fig. 3) are obtained. The two methods show qualitatively similar results suggesting that the wave propagation direction is predominantly in the North–South direction, that is, radial for this small open-field region.

An estimate of the propagation speed of the waves is obtained from these Doppler velocity time–distance diagrams with a weighted average giving the typical speeds and standard deviation for outward propagation as 444±2 km s^{-1} and inward propagation as 365±76 km s^{-1}. The propagation speed of Alfvénic waves is given by $c_p \approx \omega/k$ and this relationship is used to overplot the outward and inward velocities on the f–k diagrams in Fig. 2. Good agreement between the location of the ridge and the estimated outward propagation speed is clear, giving confidence in the values. These speeds are significantly greater than the sound speed for a 1.6 MK plasma $c_s \sim 150$ km s^{-1} and, hence, the observed waves are likely Alfvénic.

As mentioned, counter-propagating waves are a crucial component of Alfvén wave heating/acceleration models if MHD wave turbulence is to develop[14,41], generating the small scales necessary for the dissipation of Alfvénic wave energy through their interaction. The power spectra obtained here provide evidence for the presence of both outward and inward propagating Alfvénic waves in open-field regions, suggesting that the Alfvénic waves are at least partially reflected in the open magnetic field regions of corona, or further out.

The inwardly propagating waves can be seen to have significantly less power, both in the f–k diagrams (Fig. 2c,d) and the wavenumber-integrated power spectra (Fig. 3a). In Fig. 3b, the ratio of inward to outward wave power is given as a function of frequency. The ratio reveals a frequency dependence of the power ratio, potentially indicative of wave reflection in the corona. The results suggest (for f<6 mHz) that the low frequency waves are more readily reflected than high frequency waves, which is to be expected for coronal Alfvén waves7,9. However, the ratio begins to increase after f~6–7 mHz. The increase in the ratio for these higher frequencies is likely related to the inward power being dominated by noise for f>6 mHz. The noise in the data has a flat power spectrum and the continued decrease in the outward wave power with frequency >6 mHz leads to the increase in the ratio. The ratio flattens off for f>10 mHz, a sign that the outward power spectra also becomes dominated by the noise. The influence of noise is confirmed upon estimating the contribution of the data noise to the power spectra.

Connecting the Coronal Waves To The Solar Wind

The power spectra of the outward and inward propagating waves are found to have a slope of ~1/f (see Table 1 for exact values) and an enhancement of power at ~3–5 mHz. The functional 1/f form of the coronal power spectra is also found in magnetic and velocity fluctuation power spectra observed in solar wind streams in the heliosphere at 0.3 AU (64 $R_\odot$)[42,43]. The similarity between the slopes of the coronal and wind power spectra lends support to the ideas presented in ref. [43], which suggest that the shape of the wind spectrum is already set in the low solar atmosphere, at least in the low corona and transported outwards.

Table 1: Properties of the $10^a f^b$ **fit to the outward (Out) and inward (In) power spectra shown in** Fig. 3, for the box (M1) and track (M2) methods.

	Frequency range (mHz)	a	b	χ^2_ν
Out M1	0.2-2	−5.0 ± 0.1	−0.92 ± 0.03	4.3
	4-10	−4.8 ± 0.1	−1.04 ± 0.05	3.8
In M1	0.2-2	−5.3 ± 0.1	−0.98 ± 0.03	3.8
Out M2	0.2-2	−4.4 ± 0.1	−0.7 ± 0.04	0.9
	4-10	−5.0 ± 0.1	−1.1 ± 0.05	1.4
In M2	0.2-2	−5.1 ± 0.1	−0.89 ± 0.04	2.7

Interestingly, the presence of an enhancement in power is observed at~ 1–2 mHz in some solar wind spectra42, similar to the enhancement in the coronal power spectra at frequencies of 3–5 mHz. This may imply that the features seen in solar wind spectra are potential remnants of the functional form of the power spectra set in the low solar atmosphere (<1.3 $R_\odot$). The frequency shift of this feature between the two spectra, that is, coronal and at 0.3 AU, may be explained by the stretching of the perpendicular wavenumber scale with height owing to the radial expansion of the magnetic field combined with solar wind acceleration and magnetic field line rotation44.

Evidence *p*-Modes Play a Role in Alfvénic Wave Generation

The enhancement of power seen in the coronal power spectra is of interest in its own right. The frequency range occupied by the power enhancement is coincidental with the frequency range of peak power ofp-modes and has also been observed in CoMP power spectra for large quiescent coronal loops[40]. This leads us to suspect that p-modes play an important role in the generation of Alfvénic waves in open-field regions, as well as in the quiescent Sun.

Theoretical considerations support the idea that p-modes can be guided into the corona by the magnetic field. In particular, mode conversion would play a dominant role in generating coronal Alfvénic waves[45]. At some height in the lower solar atmosphere, dependent upon the local pressure and magnetic field, there will exist a canopy where the gas pressure and magnetic pressure are approximately equal in magnitude ($\beta = 8\pi p/B^2 \approx 1$). At the canopy, a transformation from slow to fast magnetoacoustic waves can occur, with the efficiency of this transformation dependent upon the angle between the magnetic field and the direction of wave propagation. This phenomenon has been well studied in a variety of magnetic field configurations and plasma conditions[46,47,48,49]. Although slow magnetoacoustic modes above this canopy are expected to steepen and shock owing to the rise in temperature in the upper chromosphere, the fast magnetoacoustic waves are reflected due to the steep gradient in Alfvén speed at the transition region. In the case when the gravity, magnetic field and wave vector of the fast magnetoacoustic wave are not co-planar, then there is a coupling of the fast wave to the Alfvén wave[45,50,51,52]. This coupling enables wave energy to cross the transition region and propagate into the corona as an Alfvén wave. Finally, regions of strong magnetic field provide the necessary conditions for p-modes to leak out from the interior into the atmosphere[53,54,55], where they can then be converted via various methods of mode coupling. This is of particular importance as the open coronal fields are generally thought to be rooted in the kilo-Gauss faculae that form the magnetic network in the photosphere[56,57]. It has been estimated that a sufficient amount of energy can be converted from p-modes to coronal Alfvén waves to explain the approximate energy content of observed Alfvénic motions in the corona[45].

We note that the discussed theoretical work[45,50,51,52] has only been undertaken for a homogenous atmosphere, that is, the excitation of classical Alfvén waves. It is still unknown whether p-modes could excite Alfvénic waves in a corona with a cross-field density structuring. However, it is clear from the theory that vorticity perturbations must be excited in the corona by the mode conversion at the transition region; otherwise Alfvén waves would not be produced. In the case of a structured corona, any generated vorticity at the transition region must also be transported and the Alfvénic modes would seemingly be the best choice for this. The observed enhanced power of the coronal Alfvénic waves coincident with the frequency range of the dominant p-mode power may support this but further investigation is required for the confirmation of this hypothesis.

Probing Open-Field Regions with Solar Magneto-Seismology

Moving away from the power spectra, another interesting aspect observed is that the propagation velocities of the outward waves is greater than the inward. Under the assumption that there exists a background outflow, in this case most likely the solar wind, then a simplified relationship describing the Alfvénic waves can be obtained (see Methods),

$$\omega = \left(U \pm c_{ph}\right)k_z. \quad (1)$$

Here, c_{ph} is the phase speed of the wave, U is the flow speed and z is the direction parallel to the magnetic field. The propagation speed of the outward (plus) and inward (minus) waves is modified by the presence of the flow. Hence, outward and inward wave speed estimates obtained earlier from the Doppler velocity time–distance diagrams suggest the presence of an outward moving flow with an average speed U=31±7 km s^{-1}. This value is comparable to the spectroscopic estimate of outflows in open-field regions obtained from the transition region and low coronal emission lines58.

This basic process is taken further. Following the same principles outlined for creating Doppler velocity time–distance diagrams, estimates for the outward and inward propagation velocities throughout the corona are obtained (Fig. 4a,b, see Methods for details). Subsequently, a flow map for the northern polar region is derived (Fig. 4c), which reveals a scene that is dominated by outflows, as should be expected if the open magnetic fields are a source region for the solar wind. The flow is seen to accelerate as a function of height (Fig. 4d), tending towards a value of 60–70 km s^{-1} at 1.2 $R_\odot$. The observed acceleration of the wind above ~1.09 $R_\odot$ is consistent with predictions of wind acceleration in the low corona found from solutions of one-dimensional hydrodynamic solar wind models (ref. [59]—wind speed from the model is shown in Fig. 4d). Interestingly, the wind speed values from the model at ~1.4 $R_\odot$ (not shown in the figure) are comparable to previously measured outflow speeds in open-field regions at the corresponding height[4]. Note, the measured variation of average flow speed from inflow to outflow below 1.09 $R_\odot$ is thought to be an artefact of erroneously high-estimated inward propagation speeds, hence inflow speeds, near the occulting disk (see Methods).

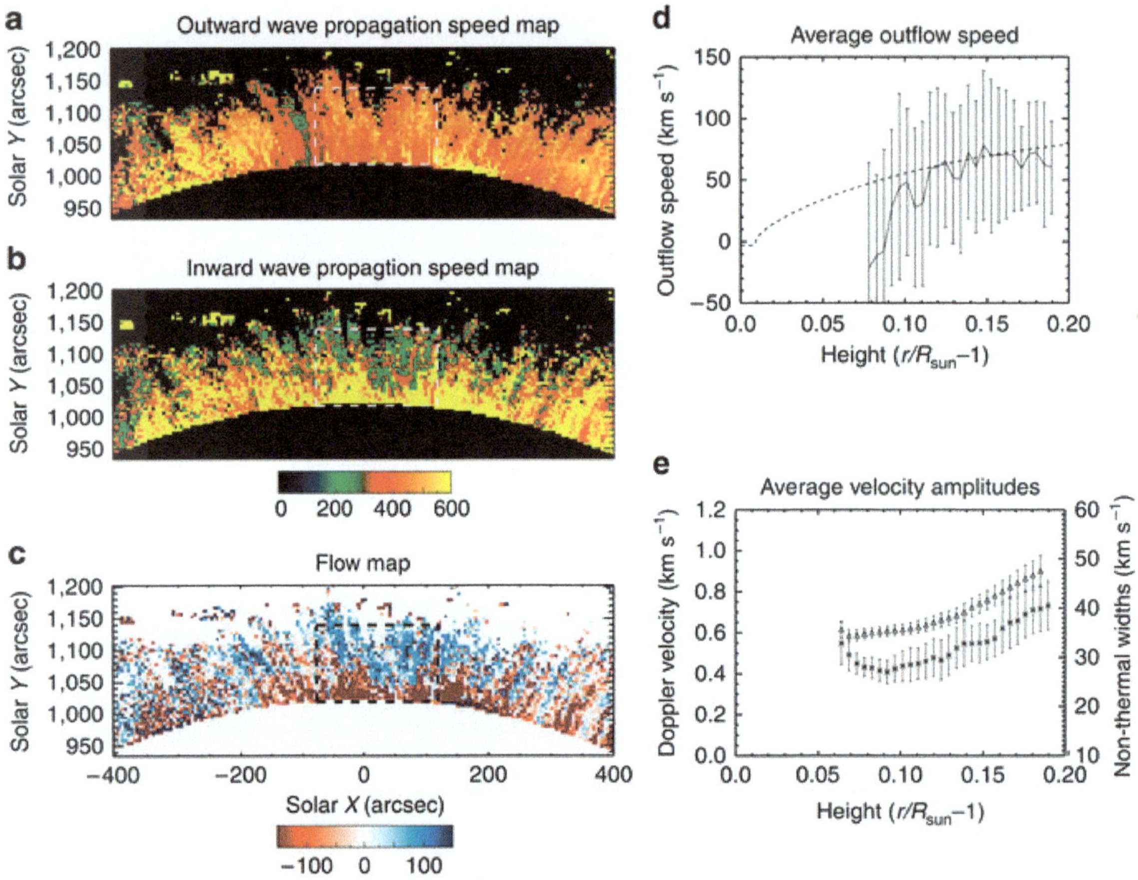

Figure 4: Measurements of the flow speed along the open fields. The measured propagation speed for the outward (**a**) and inward (**b**) Alfvénic waves can be used to determine the flow speed of plasma along the magnetic field (**c**). The black and white boxes in the propagation and flow maps mark the region that is used for determining the average values in the following panels. The scale bars indicate propagation speed (km s^{-1}) and flow speed (km s^{-1}). The open-field region is dominated by outflowing plasma (**d**), which is likely the beginnings of the solar wind and is comparable to outflows obtained from hydrodynamic wind models (dashed line). The error bars show the standard deviation of the average flow. (**e**) The RMS Doppler velocity (stars) and non-thermal widths (triangles) are shown as a function of height revealing that the Alfvénic wave amplitude increases as they propagate away from the limb. The error bars show the s.d. of the respective quantities.

The measured properties of the Alfvénic waves, that is, amplitude and propagation speed, are further exploited to gain estimates for the gradients of magnetic field strength and density in the open-field region (see Methods). These gradients are shown in combination with electron density measurements available from CoMP, an analytic model of electron densities in an open magnetic field region from measurements in white light7 and the magnetic field estimated from the PFSS extrapolation model (Fig. 5a–c).

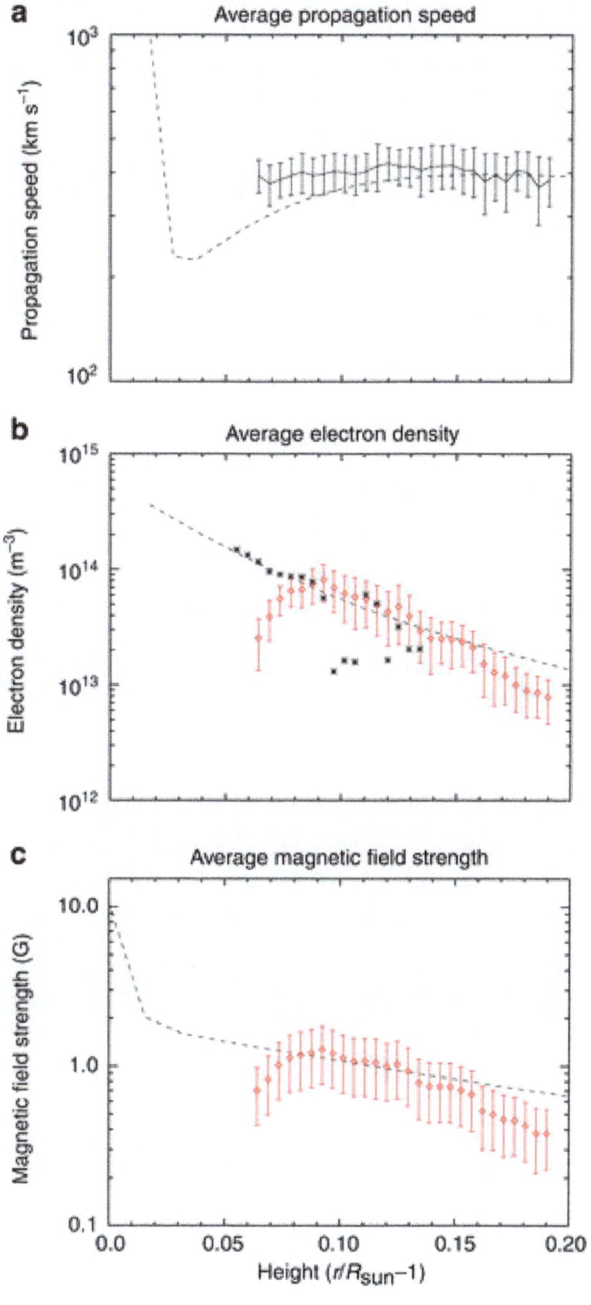

Figure 5: Physical properties of the open-field region. Multiple complementary techniques are used to estimate the physical properties of the magnetized plasma. (**a**) The measured outward propagation speed (solid line), averaged over the boxed region in Fig. 4a, is compared with the estimated propagation speed calculated from the analytic density profile (dashed line in **b**) and PFSS magnetic field (dashed line in **c**). (**b**) The electron density estimates from CoMP (stars) is shown and is compared with the density profiles obtained from magneto-seismology of solar atmosphere (red diamonds) and an analytic density profile (dashed line—ref. 7). (**c**) The magnetic field strength profile from the magneto-seismology (red diamonds) is compared with the magnetic field strength estimated from the PFSS potential field extrapolation (dashed line). The error bars in all plots are the s.d. of the respective quantities.

The estimated electron density from CoMP is found to follow the expected trend from the analytic model and this is supported by the density gradient obtained from the magneto-seismological inversion. In addition, the magnetic field gradients from the magneto-seismology show good agreement with that predicted by the PFSS. The quality of the agreement between the different measures for the plasma parameters suggests that the physical properties of this particular open magnetic field region are well constrained by this set of observations.

Energetics of the Waves

Now, combining CoMP and SDO observations, we are able to provide insight and constraints on the energetics of the observed Alfvénic waves. Using SDO, we measure the swaying motions of the fine-scale structures (Fig. 1d) at a height of ~1.01 $R_{\odot}$ (Fig. 6), which is interpreted in terms of the MHD kink wave mode[28,31,60]. It is likely to be these swaying motions that correspond to the periodic fluctuations of the CoMP Doppler velocities33. The observational signature of torsional motions is oppositely directed Doppler velocities on either side of the fine structure. Owing to large spatial resolution of CoMP and its inability to resolve the finest structure in the corona, these oppositely directed velocities will average to zero and have a negligible contribution to the measured Doppler velocities. The kink motions are observed to be periodic, with the distribution of periods peaking at~300 s (3.3 mHz)—coinciding with the frequency range of the enhanced power in the CoMP observations (Fig. 3a). The measured velocity amplitudes have an average value of 14 km s^{-1}. This value is significantly larger than CoMP Doppler velocity amplitudes (<1 km s^{-1}—Fig. 4e), although the poor spatial resolution of CoMP is known to lead to a significant underestimation of the Doppler velocities due to bulk plasma motions[33,34]. Conversely, the measurements from SDO data are almost a third of the value of the non-thermal velocities (Fig. 4e) that CoMP measures (and other instruments[20,21,23,24,25])—although velocity amplitudes up to 40 km s^{-1} are found in the SDO measurements (Fig. 6). The reasons for the larger amplitudes inferred from the non-thermal widths are still unclear. Should unresolved torsional Alfvénic motions be present along fine structure, it would then be expected that they would contribute to non-thermal line broadening[61]. In addition, unresolved kink motions with small displacement amplitudes (so as to be also unobservable with SDO) would also contribute.

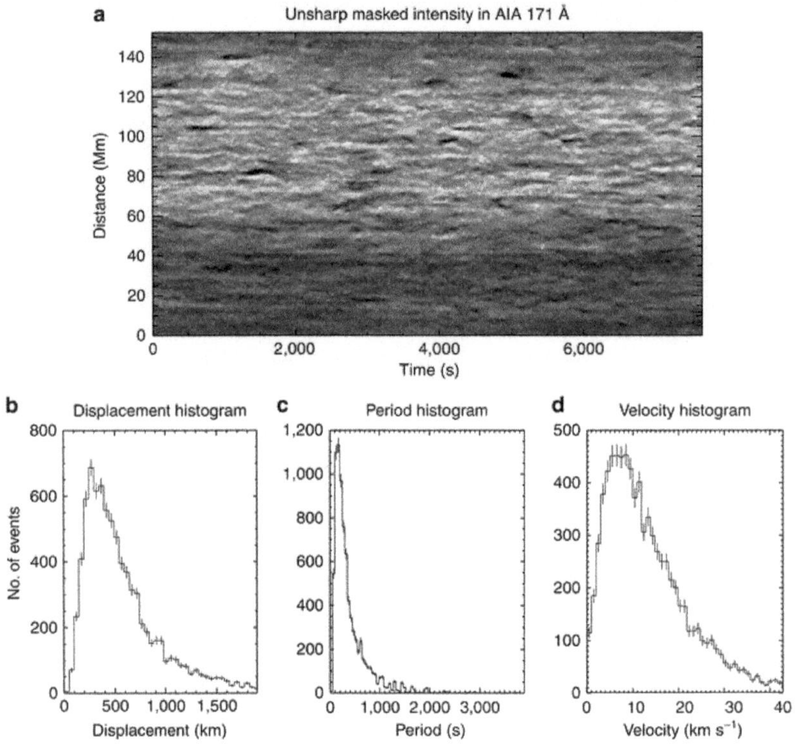

Figure 6: Direct observations of Alfvénic waves with SDO. An example time–distance diagram (location shown in Fig. 1c,d) reveals that the fine-scale magnetic structure in the open-field region sways periodically (**a**), and this motion is interpreted from MHD wave theory as the kink mode. These Alfvénic waves observed with SDO contribute to both the CoMP Doppler velocities and non-thermal widths. The wave motion is measured directly (**b**–**d**) and the mean values with s.e.m. are 590±5 km for displacement amplitude, 470±6 s for the period and 14.7±0.2 km s^{-1} for the velocity amplitude. Error bars show the sample s.d. for each bin in the histogram.

The upper limit on the energy flux of propagating Alfvénic waves[37] is given by $F \approx 2<\rho>v_{rms}^2 c_{ph}\delta$, we estimate the contribution of the kink MHD waves is on the order of $\sim 50\delta$ W m^{-2}, where v_{rms} is the root mean square velocity amplitude and $<\rho>=9.5 \times 10^{-13}$ kg m^{-3} is the average density at 1.01 $R_\odot$ extrapolated from CoMP electron density measurements. A phase speed of 250 km s^{-1} is also used30. The factor δcorresponds to the filling factor of the waveguides in the observed region. The value of δ is unclear although it likely <50% (see Fig. 1). The total wave energy including both kink and torsional motions may be best calculated by the non-thermal widths, giving a total value of~300δ W m^{-2} at 1.06 $R_\odot$ ($<\rho>=2.7 \times 10^{-13}$ kg m^{-3}, $c_{ph}=450$ km s^{-1})—which can meet the energy flux requirements for accelerating the solar wind assuming $\delta \sim 50\%$ (ref. 62). However, this number should be treated with some caution as it is uncertain exactly what physical phenomena are contributing to the non-thermal broadening[25,33].

DISCUSSION

The initial study presented here demonstrates the potential for using Doppler velocity time series of the corona to understand Alfvénic wave propagation through the solar atmosphere and connecting the dynamics of the corona to the solar wind. The results reveal the existence of both outward and inward propagating Alfvénic waves in open magnetic field regions—a crucial element required if MHD wave turbulence is to develop. We remind the reader that it is presently unknown whether the details of the nonlinear interaction of fast kink waves are similar to those of Alfvén waves. Although the kink waves are Alfvénic, the turbulent cascade of vorticity to small scales through their nonlinear interaction seems entirely plausible. Alternatively, the kink waves are subject to mode conversion via resonant absorption[15] and the energy could be transferred to torsional Alfvén waves before it is cascaded to higher frequencies.

Further, the ratio of outward to inward wave power demonstrates a frequency dependence that is suggestive of the waves being partially reflected in the corona, that is, a decreasing proportion of inward power to outward power with increasing frequency, which is also a feature in the models of Alfvénic wave propagation.

In addition to this, the power spectra reveal insights into the generation of the waves and their propagation. The slope of the power spectra for the Alfvénic waves shows a 1/f functional form that is comparable to previously measured solar wind spectra at 0.3 AU. This implies that the waves propagate outwards from the corona into the solar wind and at least partially maintain the shape of the power spectra[44]. Moreover, the observations reveal the presence of enhanced power centred on the frequencies routinely associated with peak p-mode power. Current theoretical models suggest that the acoustic waves generated in the solar interior can generate coronal Alfvénic waves through mode conversion[45,50]. We again highlight that these models use a homogenous corona, which raises questions about whether these results are directly applicable to the observations presented here. However, the current models suggest that vorticity is generated in the corona and the fast kink modes are one of the key transporters of vorticity in an inhomogeneous plasma[15]. This may advocate that p-modes could play a dominant role in generating coronal Alfvénic waves in the frequency range neighbouring 3 mHz. The role of acoustic waves is excluded from certain surface to heliosphere solar wind acceleration models dependent on Alfvén waves due to a simplifying assumption of incompressibility throughout the system (note that not all Alfvén wave models assume incompressibility). The indication that p-modes could play an important role in wave generation suggests that this assumption would neglect a fundamental physical process.

Currently, the estimates of energy flux from observable Alfvénic (that is, kink) waves in open-field regions advocate that there is insufficient energy to meet the demands for solar wind acceleration[31]. Although, CoMP measurements of the 10,747 Å line profile (and a host of other coronal emission line profile measurements from different instruments) reveal that the velocity amplitudes from non-thermal widths imply otherwise. This would indicate that only by including contributions from both resolved and unresolved (and both kink and torsional) Alfvénic waves can the energy flux requirement for solar wind acceleration be potentially met exclusively by waves. Further work is clearly needed to deconstruct the motions that comprise the non-thermal widths and provide a clear estimate for the energy flux of all Alfvénic waves.

This study also demonstrates the power of magneto-seismology of the solar atmosphere to probe the corona. Using the measured properties of the waves, evidence for the origins of the solar wind are seen in flow maps and significant constraints can be placed on other physical properties of the open magnetic field region, that is, magnetic field strength, density and the gradients of the two quantities. It is hoped that the measurements of the plasma parameters, along with the measured wave properties, for example, amplitude and power spectra, should provide a significant test scenario for models of solar wind acceleration, especially those which focus on MHD Alfvénic wave turbulence.

METHODS

Observations and Data Reduction

The data used here were obtained with the CoMP38 instrument on 27 March 2012 at 18:51:02 UT to 20:13:02 UT. The details of the acquisition and reduction of the data are fully described in refs 29, 38, 40 and we make use of the final data product that has a cadence of 30 s and a pixel size of 4″.46 (3,234 km). The final data product is intensities at three infrared wavelengths (10,745.0 Å—I_1, 10,746.2 Å—I_2 and 10,747.4 Å—I_3), which are positions centred on the Fe XIII emission line at 10,747 Å. The filter has a bandpass of 1.3 Å. Following ref. 63, for each pixel in the CoMP field of view in each time frame, we calculate the central intensity, Doppler shift and Doppler width of the line profile using an analytic fit of a Gaussian to the intensity values at each wavelength position.

We also make use of SDO AIA64 and the Helioseismic and Magnetic Imager65 data, which are taken from 18:00 UT to 20:10 UT and processed following standard procedures.

Velocity Time–Distance Maps

Here, we describe the two complementary methods for analysing the wave power spectra.

For the first method, a box is defined over the northern region, which is 200″ wide and 125″ tall (shown in Fig. 2). We make the crude assumption that, in the boxed region, the direction of potential wave propagation is strictly in the North–South direction. Displayed in Fig. 2 is an example of Doppler velocity time–distance diagram from a strip of pixels within the box, orientated in the vertical direction and one pixel wide. The presence of the outward propagating Doppler velocity signals is visible and so is the periodic nature of these features.

For each pixel in the box, the time-averaged value of the velocity is subtracted from the time series. Then, for each vertical strip in the box, the time–distance diagram is subjected to a two-dimensional Hamming window before a two-dimensional Fast Fourier Transform (FFT) is performed and the power for a range of frequencies and wavenumbers in the vertical direction is calculated. The outward propagating motions are decomposed from the inward propagating motions. Averaging the outward and inward power over the different strips, the power is integrated over wavenumber (k) and the outward and inward power spectra are determined (Fig. 3). The spectra are fit with a power law of the form $10^a f^b$. The outward spectra are fit in two frequency ranges while, the inward wave power spectra are not fit for the high frequency ranges due to the power spectra being predominantly noise in this regime. The results for the fits are given in Table 1.

It is highly instructive to examine the averaged f–k diagram obtainable from the time–distance diagrams. To improve the frequency/wavenumber resolution in the f–k diagrams, the mean subtracted and windowed time–distance diagrams are padded with zeros before the FFT is performed. The results are shown in Fig. 2d.

The second method used is a more refined version of the first, providing the f–k power spectra and, in addition, the propagation velocities of the waves. This technique was demonstrated in ref. 40 and involves the tracking of wave packets via a coherence approach. In brief, for each pixel in the CoMP data, the time series is filtered at 3.5 mHz and the coherence of this filtered time series with the filtered time series in neighbouring pixels is calculated. This provides islands of high coherence, typically elongated in a particular direction that corresponds to the angle of wave propagation. Assuming that these propagating waves are Alfvénic, the direction of propagation defines the orientation of the magnetic field in the plane of the sky. It has been demonstrated that this technique provides magnetic field orientations that are in excellent agreement with the CoMP magnetic field measurements from the linear polarization of the 10,747 Å emission line40, and, hence, supporting the assumption of Alfvénic waves.

Starting at a particular pixel, we follow the direction of the magnetic field in the plane of sky using the calculated wave angles. Doing so allows tracks to be built that are aligned with the direction of wave propagation. We select 36 of these tracks that lie in the northern open-field region and these tracks are over-plotted on an intensity image inFig. 2. The velocity time series at each point along the track is then extracted via cubic interpolation. An example of one of the time–distance diagrams from a track is shown in Fig. 2c. As with the first method, these Doppler velocity time–distance diagrams are used to create an average f–k diagram (Fig. 2e) and the k-integrated power spectra (Fig. 3a). Note that the power spectra and ratios in Fig. 3 are calculated from the original Doppler velocity time–distance diagrams, zero padding was not used to create these figures.

Note that the power in the f–k plots in Fig. 2 has been clipped between the given ranges. This leads to no loss of useful information as the signal is progressively noisier for larger k values and is done purely for aesthetic reasons.

Propagation Speed Calculation

An estimate of the propagation speed of the waves is obtained from the Doppler velocity time–distance diagrams. The procedure is based on the technique described in ref. 40, so we give a brief overview of the technique and additions made to the process.

For each pixel, a time–distance diagram is created from tracks that are built using the wave propagation angles. Each time–distance diagram is subject to an FFT and then decomposed into inward and outward parts, both of which are then subject to an inverse FFT. The following is applied to both the inward and outward time series separately. The time series central pixel along the track is used as the 'master time series' and is cross-correlated with the time series of the other pixels along the track to find values for the phase lag between the series. The lag values between the different time-series are fit with a linear function, which is then used to align the time series to the 'master time series'. The aligned series are then averaged to create a higher signal-to-noise 'master time series'. This new 'master time series' is then cross-correlated with the time series of the other pixels along the track to get improved estimates on the lag values. A sub-resolution estimate of the phase lag is obtained by fitting the cross-correlation peak with a parabola. This process still leaves a number of high error or outlier lag data points that can negatively influence the result. To remove these points, two steps are applied. First, a null hypothesis test is performed. The value of the cross-correlation of any of the two series is tested to see whether the two series are uncorrelated and each series is not auto-correlated. For two such series with populations that are normally distributed, the sample cross-correlations would have zero mean and a variance of 1/N, N being the number of samples. For a 99.7% confidence interval, this assumes that the null hypothesis is incorrect if the cross-correlation value is $>2.58/\sqrt{N}$. Lag data points with cross-correlation values below the cut-off are discarded. Having removed these data points, a linear fit to the lag data points is performed. From the residuals to the fit, the residual sample standard deviation is calculated. Here, the second cut is applied. Assuming the residuals are sampled from a normal distribution, a 99.7% significance level is calculated usingt-distribution statistics for a two-tailed test. Any lag data point that lies outside this significance range is deemed an outlier and also removed. For the number of lag data points, this condition has a much stricter rejection criterion than, for example, Chauvenet's criterion. These two steps appear to be a highly robust way to remove erroneous data points. A linear function is then fit to the remaining lag data points and the propagation speed of the wave can be obtained from the gradient, with the error on the gradient calculated using the sample standard deviation.

For the propagation speed maps (Fig. 4), the time–distance diagrams used have varying spatial lengths depending on the position of the pixel under consideration. The standard length is 25 spatial pixels but can be shorter, for example, for pixels near the occulting disk.

Note that the inward propagation speeds suffer from great uncertainty in the lowest part of the corona. Typically this results in unphysically large values for the propagation speed that are well in excess of the outward propagation speed and the estimated Alfvén speed using the PFSS estimates for magnetic field and density measurements from CoMP (Fig. 5a). It can be seen in the average power spectra (Fig. 3) that the amplitude of the inward propagating waves is comparable to the noise level for the higher frequencies. This has the greatest influence in the lower corona where amplitudes of waves are smaller (Fig. 4a–e). The amplitude of both inward and outward waves grow with height above the limb (Fig. 4d) and at a certain point the inward wave amplitude becomes large enough to overcome the noise and provide reliable measurements. In addition, the time–distance diagrams for pixels of the lower corona will typically have less data points spatially, as mentioned, which leads to greater uncertainties in the results. For this reason, the propagation speeds are clipped at 600 km s^{-1} and this predominantly affects the inward wave speed (Fig. 4b).

Density Measurements

CoMP is able to measure both the 10,798 Å and the 10,747 Å Fe XIII emission lines, the ratio of which is known to be sensitive to the electron number density, n_e, in the corona66. At 20:29 UT, after the main 10,747 Å sequence was taken, CoMP was set to take scans at both 10,798 Å and 10,747 Å. The scans are used to calculate the central intensity for each wavelength and the intensity ratio is calculated for a number of consecutive frames and averaged. Then, using the CHIANTI database v7.0 (ref. 67), electron density versus intensity ratio curves are calculated for a range of heights above the photosphere taking into account the strong influence of photo-excitation on the formation of the two lines. The curves are then used to calculate the electron number density in the open-field region, which can be converted to coronal mass density using $\rho=\mu m_p n_e$, where μ is the mean atomic weight (1.3 for coronal abundances) and m_p is the proton mass, 1.67×10^{-27} kg. The electron density measurements as a function of height (stars) are shown in Fig. 5b and compared with a scaled analytic density profile (dashed line—ref. 7). The measurements only cover a relatively short section of the corona due to weak emission in the corresponding region in 10,798 Å. As the emission decreases in strength with height, the associated errors increase and the number of pixels with significant signal decreases. Both these factors are likely to be responsible for the increased deviation in measured density from the analytic profile with height.

Estimating Quantities from Magneto-Seismology

In the main text, we discussed the process of estimating the flow speed in the corona via magneto-seismology of the solar atmosphere by exploiting the measured propagation speeds of the waves. This relationship can be obtained from the following equation:

$$\rho_i\left(\frac{\partial}{\partial t}+U_i\frac{\partial}{\partial z}\right)^2\eta+\rho_e\left(\frac{\partial}{\partial t}+U_e\frac{\partial}{\partial z}\right)^2\eta=\frac{2B^2}{\mu_0}\frac{\partial^2\eta}{\partial z^2}. \quad (2)$$

Here, η is the perturbation of the displacement, μ_0 is the magnetic permeability of free space, and the subscript i and e correspond to quantities inside and external to the waveguide, respectively. The magnetic field inside and external to the waveguide is assumed equal. This equation describes the kink wave in a thin over-dense waveguide in a low-beta plasma68. The condition for the waveguide to be thin is satisfied if the tube radius is much less than the wavelength. This is satisfied for the current observations where $\lambda \sim 100$ Mm. Looking for plane wave solutions and under the assumption that $U_i \approx U_e$, it is straightforward to arrive at equation (1), with c_{ph} equal to the kink speed. The square of the kink speed is given by $c_k^2=B^2/(\mu_0\langle\rho\rangle)$, where $\langle\rho\rangle=(\rho_i+\rho_e)/2$ is the average density.

Further information about coronal plasma properties can also be obtained via magneto-seismology of the solar atmosphere, if we exploit measurements of the wave amplitude and propagation speed together. In particular, gradients in the density and magnetic field strength can be estimated[69,70]. The results can then be compared with estimates from complementary techniques, that is, the magnetic field from PFSS extrapolations; density from CoMP line ratios and the analytic model from ref. 7.

MHD wave theory reveals that the relationship between the measureable quantities and the quantities to be estimated is given by,

$$v = C\sqrt{\frac{c_k}{\omega}\frac{1}{\sqrt{B}}}, \quad (3)$$

where v is the velocity perturbation amplitude of the wave, ω is the angular frequency, C is a constant and B is the magnetic field strength. Due to the presence of the unknown C, it is only possible to find the gradients of the density and magnetic field strength. The magnitude of the quantities can only be known if we have initial values for both the quantities. Hence, the gradient of each quantity is normalized and can be scaled for comparison with the magnetic field and density estimates measured by other means.

To exploit equation (3), the propagation speed is averaged horizontally over the boxed region shown in Figs 2 and 4 and plotted as a function of height in Fig. 5a.

In addition to the propagation speed, the Doppler velocity amplitude is also required as a function of height. Each velocity time series in the boxed regions is smoothed using boxcar function of length 3. The root mean square (RMS) value of the removed signal due to smoothing is found to match the estimated RMS of the noise, implying that the smoothing isolates the real velocity signal and suppresses the noise. Taking the RMS value of the smoothed velocity signal and averaging, we obtain the Doppler velocity displayed in Fig. 4e (stars).

Now, the variation in velocity amplitude for Alfvénic waves is expected to be due to changes in density alone and is given by $v \propto n_e^{-1/4}$ (assuming electron and ion number densities are equal and wave damping is negligible). Using this expression, the density gradient inferred from the RMS Doppler velocity amplitude (red diamonds) shows good agreement with the analytic density profile (Fig. 5b). Note that the density gradient determined from the Doppler velocity has been scaled for comparison with the analytic density profile and the density values obtained from the CoMP line ratios.

Using the measured Doppler velocity amplitude and propagation speed, the gradient in magnetic field is obtained using equation (3) and shown in Fig. 5c (red diamonds). We can compare this to the average magnetic field strength in the open-field region estimated from the SDO/Helioseismic and Magnetic Imager magnetograms via the PFSS extrapolation (dashed line). As we are unable to provide an exact boundary for the open-field region, the PFSS magnetic field is examined when averaged over latitudes >80 degrees and >85 degrees, both results incorporate the open magnetic flux region seen in Fig. 1. The results from both latitude boundaries show minor differences, mainly near the solar surface and are practically identical in the corona. It can be seen that there is good agreement between the two measures for the gradients of the magnetic field. Note that, again the magneto-seismological profile has been scaled to allow comparison with the PFSS profile.

The PFSS and magneto-seismology results suggest that the magnetic field strength decreases only by ~0.5 Gauss from 1.05 to 1.2 $R_\odot$. However, the magnetic field strength in the corona is weak, which means that this corresponds to a substantial decrease of 30–50%.

Finally, to complete the comparison between the different measures, the PFSS magnetic field and analytic density profile are combined to provide an estimate for the averaged Alfvén speed (dashed line—Fig. 5a). As noted in the main text, the agreement between magneto-seismology results and the PFSS and line ratio measurements provides re-assurance that the plasma properties of this open magnetic field region are well constrained.

Non-Thermal Widths

The non-thermal widths of the spectral lines are often used as an indicator of Alfvénic wave activity[20,21,22,23,24,25], so we also derive them to provide a comparison with the Doppler shift velocities. The non-thermal widths are also calculated from the Fe XIII emission line profile. The peak formation temperature of Fe XIII is ~1.6 MK, so it is expected that the line profile has a thermal width $\sigma_T=\sqrt{(2k_bT_e/m_{ion})}=21$ km s^{-1}, where k_b is the Boltzmann constant, T_e is the electron temperature (assumed equal to the ion formation temperature) and m_{ion} is the mass of the ion. In addition, the line will be broadened by an instrumental width that has a value of 21 km s^{-1}. The thermal and instrumental widths can be removed from the measured Doppler width to provide the non-thermal width. As with Doppler velocity, the non-thermal width is averaged horizontally over the boxed region and plotted as a function of height in Fig. 4e. The gradients in the amplitude of both the Doppler velocities and non-thermal widths are found to be approximately equal, demonstrating that both are measures of Alfvénic wave activity in the corona.

Wave energy Flux

SDO/AIA is used to measure the swaying motions of the magnetic field, which can be interpreted as MHD kink waves. The magnetic field is outlined by density enhancements in the corona seen in the open-field region (Fig. 1d). The measurements are performed at a height of ~1.01 $R_\odot$ and follow the methods in ref. 31. The results are shown as histograms in Fig. 6. A median value for the sample population of each quantity, that is, displacement amplitude, period and velocity amplitude, is calculated. The distributions for each of these quantities show approximately log-normal behaviour; hence, the mean value of the log-normal distribution is calculated to give the average values. These values are given in Table 2.

Table 2: Measured wave properties from SDO/AIA data

	Median	Mean from log-normal
Displacement (km)	470	590 ± 5
Period (s)	268	414 ± 5
Velocity (km s^{-1})	11.2	14.7 ± 0.2

To estimate the wave energy flux, we require an estimate for the mass density in the corona. This is determined from extrapolating from the CoMP electron density measurements back to 1.01 $R_\odot$ using the scaled analytic function7. The value of the mass density used is then $\rho=9.5\times10^{-13}$ kg m^{-3}. In addition, the phase speed at 1.01 $R_\odot$ is taken from the estimate of average Alfvén speed in ref. 30.

The wave energy flux from the non-thermal widths uses the density and value for non-thermal widths at a height of 1.06 $R_\odot$.

ACKNOWLEDGEMENTS

R.J.M. is grateful to Northumbria University for the award of the Anniversary Fellowship and the Leverhulme Trust for the award of an Early Career Fellowship. R.J.M. also thanks the Higher Education Funding Council for England and the High Altitude Observatory for financial assistance that enabled this work and acknowledges IDL support provided by STFC. The National Center for Atmospheric Research is sponsored by the National Science Foundation. R.P. was supported by the FP7 project #606692 (HELCATS).

REFERENCES

1. Krieger, A. S., Timothy, A. F. & Roelof, E. C. A coronal hole and its identification as the source of a high velocity solar wind stream. Sol. Phys. 29, 505–525 (1973).
2. Hassler, D. M. et al. Solar wind outflow and the chromospheric magnetic network. Science 283, 810–813 (1999).
3. Peter, H. & Judge, P. G. On the Doppler shifts of solar ultraviolet emission lines. Astrophys. J. 522, 1148–1166 (1999).
4. Giordano, S. et al. Identification of the coronal sources of the fast solar wind. Astrophys. J. 531, L79–L82 (2000).
5. Tu, C. Y. et al. Solar wind origin in coronal funnels. Science 308, 519–523 (2005).
6. Cranmer, S. R. Self-consistent models of the solar wind. Space Sci. Rev. 172, 145–156 (2012).
7. Cranmer, S. R. & van Ballegooijen, A. A. On the generation, propagation, and reflection of Alfvén waves from the solar photosphere to the distant heliosphere. Astrophys. J. Suppl. 156, 265–293 (2005).
8. Suzuki, T. K. & Inutsuka, S.-i. Making the corona and the fast solar wind: a self-consistent simulation for the low-frequency Alfvén waves from the photosphere to 0.3 AU. Astrophys. J. 632, L49–L52 (2005).
9. Verdini, A. et al. A turbulence-driven model for heating and acceleration of the fast Wind in coronal holes. Astrophys. J. 708, L116–L120 (2010).
10. Perez, J. C. & Chandran, B. D. G. Direct numerical simulations of reflection-driven, reduced magnetohydrodynamic turbulence from the sun to the Alfvén critical point. Astrophys. J. 776, 124 (2013).
11. van der Holst, B. et al. Alfvén wave solar model (AWSoM): coronal heating. Astrophys. J. 782, 81 (2014).
12. Leer, E. & Holzer, T. E. Constraints on the solar coronal temperature in regions of open magnetic field. Sol. Phys. 63, 143–156 (1979).

13. Leer, E. & Holzer, T. E. Energy addition in the solar wind. J. Geophys. Res. 85, 4681–4688 (1980).
14. Matthaeus, W. H. et al. Coronal heating by magnetohydrodynamic turbulence driven by reflected low-frequency waves. Astrophys. J. 523, L93–L96 (1999).
15. Goossens, M. et al. Surface Alfvén waves in solar flux tubes. Astrophys. J. 753, 111 (2012).
16. Goossens, M. et al. On the nature of kink MHD waves in magnetic flux tubes. Astron. Astrophys. 503, 213 (2009).
17. Belcher, J. W. & Davis, L. Jr. Large-amplitude Alfvén waves in the interplanetary medium-2. J. Geophys. Res. 76, 3534–3563 (1971).
18. Hassler, D. M. et al. Line broadening of Mg X 609 and 625 A coronal emission lines observed above the solar limb. Astrophys. J. 348, L77–L80 (1990).
19. Seely, J. F. et al. Turbulent velocities and ion temperatures in the solar corona obtained from SUMER line widths. Astrophys. J. 484, L87–L90 (1997).
20. Doyle, J. G., Banerjee, D. & Perez, M. E. Coronal line-width variations. Sol. Phys. 181, 91–101 (1998).
21. Banerjee, D., Teriaca, L., Doyle, J. G. & Wilhelm, K. Broadening of SI VIII lines observed in the solar polar coronal holes. Astron. Astrophys. 339, 208–214 (1998).
22. Moran, T. G. Interpretation of coronal off-limb spectral line width measurements. Astron. Astrophys. 374, L9–L11 (2001).
23. Banerjee, D., Pèrez-Suàrez, D. & Doyle, J. G. Signatures of Alfvén waves in the polar coronal holes as seen by EIS/Hinode. Astron. Astrophys. 501, L15–L18 (2009).
24. Bemporad, A. & Abbo, L. Spectroscopic signature of Alfvén waves damping in a polar coronal hole up to 0.4 solar radii. Astrophys. J. 751, 110 (2012).
25. Chae, J., Schühle, U. & Lemaire, P. SUMER measurements of non-thermal motions: constraints on coronal heating mechanisms. Astrophys. J. 505, 957–973 (1998).
26. De Pontieu, B. et al. Chromospheric Alfvénic waves strong enough to power the solar wind. Science 318, 1574–1577 (2007).
27. Morton, R. J. et al. Observations of ubiquitous compressive waves in the sun's chromosphere. Nat. Commun. 3, 1315 (2012).
28. Morton, R. J. & McLaughlin, J. A. High-resolution observations of active region moss and its dynamics. Astrophys. J. 789, 105 (2014).
29. Tomczyk, S. et al. Alfvén waves in the solar corona. Science 317, 1192–1196 (2007).
30. McIntosh, S. W. et al. Alfvénic waves with sufficient energy to power the quiet solar corona and fast solar wind. Nature 475, 477–480 (2011).
31. Thurgood, J. O., Morton, R. J. & McLaughlin, J. A. First direct measurements of transverse waves in solar polar plumes using SDO/AIA. Astrophys. J. 790, L2 (2014).
32. Goossens, M. et al. The transverse and rotational motions of magnetohydrodynamic kink waves in the solar atmosphere. Astrophys. J. 788, 9 (2014).
33. McIntosh, S. W. & De Pontieu, B. Estimating the 'dark' energy content of the solar corona. Astrophys. J. 761, 138 (2012).
34. De Moortel, I. & Pascoe, D. J. The effects of line-of-sight integration on multi-strand coronal loop oscillations. Astrophys. J. 746, 31 (2012).
35. Morton, R. J. & McLaughlin, J. A. Hi-C and AIA observations of transverse magnetohydrodynamic waves in active regions. Astron. Astrophys. 553, 10 (2013).
36. Goossens, M. et al. Energy content and propagation in transverse solar atmospheric waves. Astrophys. J. 768, 191 (2013).
37. Van Doorsselaere, T. et al. Energy propagation by transverse waves in multiple flux tube systems using filling factors. Astrophys. J. 795, 18 (2014).
38. Tomczyk, S. et al. An instrument to measure coronal emission line polarization. Sol. Phys. 247, 411–428 (2008).
39. Schrijver, C. J. & De Rosa, M. L. Photospheric and heliospheric magnetic fields. Sol. Phys. 212, 165–200 (2003).
40. Tomczyk, S. & McIntosh, S. W. Time-distance seismology of the solar corona with CoMP. Astrophys. J. 697, 1384–1391 (2009).
41. Verdini, A. & Velli, M. Alfvén waves and turbulence in the solar atmosphere and solar wind. Astrophys. J. 662, 669–676 (2007).
42. Roberts, D. A. Demonstrations that the solar wind is not accelerated by waves or turbulence. Astrophys. J. 711, 1044–1050 (2010).
43. Roberts, D. A. Evolution of the spectrum of solar wind velocity fluctuations from 0.3 to 5 AU. J. Geophys. Res. 115, A12101 (2010).
44. Verdini, A., Grappin, R., Pinto, R. & Velli, M. On the Origin of the 1/f spectrum in the solar wind magnetic field. Astrophys. J. 750, L33 (2012).
45. Hansen, S. C. & Cally, P. S. Benchmarking fast-to-Alfvén mode conversion in a cold MHD plasma. II. how to get Alfvén waves through the solar transition region. Astrophys. J. 751, 31 (2012).
46. Bogdan, T. J. et al. Waves in the magnetized solar atmosphere. II. waves from localized sources in magnetic flux concentrations. Astrophys. J. 599, 626–660 (2003).
47. Khomenko, E., Collados, M. & Felipe, T. Nonlinear numerical simulations of magneto-acoustic wave propagation in small-scale flux tubes. Sol. Phys. 251, 589–611 (2008).
48. Vigeesh, G., Hasan, S. S. & Steiner, O. Wave propagation and energy transport in the magnetic network of the sun. Astron. Astrophys. 508, 951–962 (2009).
49. Fedun, V., Shelyag, S. & Erdélyi, R. Numerical modeling of footpoint-driven magneto-acoustic wave propagation in a localized solar flux tube. Astrophys. J. 727, 17 (2011).
50. Cally, P. S. & Goossens, M. Three-dimensional MHD wave propagation and conversion to Alfvén waves near the solar surface. I. direct numerical solution. Sol. Phys. 251, 251–265 (2008).

51. Cally, P. S. & Hansen, S. C. Benchmarking fast-to-Alfvén mode conversion in a cold magnetohydrodynamic plasma. Astrophys. J. 738, 119 (2011).
52. Khomenko, E. & Cally, P. S. Numerical simulations of conversion to Alfvén waves in sunspots. Astrophys. J. 746, 68 (2012).
53. Schunker, H. & Cally, P. S. Magnetic field inclination and atmospheric oscillations above solar active regions. Mon. Not. R. Astron. Soc 372, 551–564 (2006).
54. Jain, R., Gascoyne, A. & Hindman, B. W. Interaction of p modes with a collection of thin magnetic tubes. Mon. Not. R. Astron. Soc 415, 1276–1279 (2011).
55. Gascoyne, A., Jain, R. & Hindman, B. W. Energy loss of solar p modes due to the excitation of magnetic sausage tube waves: importance of coupling the upper atmosphere. Astrophys. J. 789, 109 (2014).
56. Dowdy, J. F. Jr., Rabin, D. & Moore, R. L. On the magnetic structure of the quiet transition region. Sol. Phys. 105, 35–45 (1986).
57. Peter, H. On the nature of the transition region from the chromosphere to the corona of the Sun. Astron. Astrophys. 374, 1108–1120 (2001).
58. Tian, H. et al. The nascent fast solar wind observed by the EUV imaging spectrometer on board hinode. Astrophys. J. 709, L88–L93 (2010).
59. Pinto, R. et al. Time-dependent hydrodynamical simulations of slow solar wind, coronal inflows, and polar plumes. Astron. Astrophys. 497, 537–543 (2009).
60. Van Doorsselaere, T. et al. Seismological demonstration of perpendicular density structuring in the solar corona. Astron. Astrophys. 491, L9–L12 (2008).
61. Zaqarashvili, T. V. Observation of coronal loop torsional oscillation. Astron. Astrophys. 399, L15–L18 (2003).
62. Hansteen, V. H. & Leer, E. Coronal heating, densities, and temperatures and solar wind acceleration. J. Geophys. Res. 100, 21577–21594 (1995).
63. Tian, H. et al. Observations of coronal mass ejections with the coronal multichannel polarimeter. Sol. Phys. 288, 637–650 (2013).
64. Lemen, J. R. et al. The atmospheric imaging assembly (AIA) on the solar dynamics observatory (SDO). Sol. Phys. 275, 17–40 (2012).
65. Scherrer, P. H. et al. The helioseismic and magnetic imager (HMI) investigation for the solar dynamics observatory (SDO). Sol. Phys. 275, 207–227 (2012).
66. Flower, D. R. & Pineau des Forets, G. Excitation of the Fe XIII spectrum in the solar corona. Astron. Astrophys. 24, 181 (1973).
67. Landi, E. et al. CHIANTI—An atomic database for emission lines. XII. version 7 of the database. Astrophys. J. 744, 99 (2012).
68. Ruderman, M. S. The effects of flows on transverse oscillations of coronal loops. Sol. Phys. 267, 377–391 (2010).
69. Verth, G., Goossens, M. & He, J.-S. Magnetoseismological determination of magnetic field and plasma density height variation in a solar spicule. Astrophys. J. 733, L15 (2011).
70. Morton, R. J. Magneto-seismological insights into the penumbral chromosphere and evidence for wave damping in spicules. Astron. Astrophys. 566, A90 (2014).

CHAPTER 12

QUANTUM EMULATION OF GRAVITATIONAL WAVES

Ivan Fernandez-Corbaton, Mauro Cirio, Alexander B'use, Lucas Lamata, Enrique Solano and Gabriel Molina-Terriza

INTRODUCTION

Gravitational waves are commonly described as ripples in the fabric of spacetime that travel at the speed of light. According to general relativity, their direct detection on Earth will only be possible for waves created when spacetime is stirred by the movement of very massive astrophysical objects, like black holes and neutron stars (Sec. 36.3 in Ref. [1]). We have indirect evidence of the existence of gravitational waves[2,3], but they have not yet been detected directly. The considerable efforts placed in developing and building detectors[4] should allow the direct observation of these elusive objects in the near future[5]. The analysis of gravitational radiation should provide a completely new source of information about our universe. For example, while we will never have access to electromagnetic information older than the Cosmic Microwave Background, we may one day be able to study gravitational waves produced at earlier epochs, much closer to the Big Bang[6].

From the theoretical side, gravitational waves are modeled as propagating perturbations of the spacetime metric. A wide class of approximate gravitational wave solutions can be studied by means of Isaacson's high frequency limit[8,9]. This limit is analogous to the geometric-optics approximation of Maxwell's equations. The analogy between gravitation and electromagnetism goes well beyond this particular mathematical approach. Several different analogies are collected under the name gravitoelectromagnetism. We refer the reader for to Sec. II.3 in Ref. [9] a concise review with numerous references. The analogies range from the similarity between Coulomb's law of electricity and Newton's law of gravitation, through the use of electric like and magnetic like vectors to approximately describe general relativity, to the non-perturbative covariant formulation of Maxwell-like forms of gravitational tensors and their dynamical equations[11].

In this article, we develop an equivalence between the propagation of gravitational waves in a curved spacetime background and the propagation of a restricted set of electromagnetic tensor waves in the same background. The factorization of a circularly polarized gravitational wave as the tensor product of two circularly polarized electromagnetic waves, which is not possible for linear polarizations, is central to this equivalence. Its physical meaning becomes clear when treating polarization by means of the helicity operator and its eigenstates. As far as we know, this equivalence has not been reported before.

After establishing the equivalence, we use it to propose a path for the emulation of the propagation of gravitational waves in a curved spacetime background by means of electromagnetic setups in a laboratory. To achieve the proposed equivalence for all regimes, it is necessary to use a genuine quantum system, i.e. entangled photons. Therefore, we propose the simulation of a purely classical system with a genuinely quantum setup. Conformal invariance is used to shrink the astronomical sizes down to laboratory scales, and suitable metamaterial media to mimic spacetime curvature. Up to our knowledge, this is a new emulation approach, but we point out that there has been an intensive research on quantum simulations of gravitational phenomena as black holes and gravitational waves in Bose-Einstein condensates[12,13,14]. In parallel, several proposals for detecting gravitational waves using cold atoms have been made[15,16].

Citation: Fernandez-Corbaton I, Cirio M, Büse A, Lamata L, Solano E, Molina-Terriza G. Quantum Emulation of Gravitational Waves. Scientific Reports. 2015;5:11538. doi:10.1038/srep11538.

PHYSICS OF GRAVITATIONAL WAVES

Let us first of all outline the gravitational wave physics that we want to capture. We adopt the conventions of Einstein's summation over repeated indexes, a flat metric $\eta_{\mu\nu}=\mathrm{diag}(1,-1,-1,-1)$, and bold symbols to denote spatial three-vectors.

The first step for studying the propagation of gravitational waves is to decompose the total metric tensor $g_{\mu\nu}$ into a slowly varying background $\gamma_{\mu\nu}$ and a rapidly varying wave $h_{\mu\nu}$ (see Ref. [1] and Sec. 35.13 14 in Ref. [7]):

$$g_{\mu\nu} = \gamma_{\mu\nu} + h_{\mu\nu}. \tag{1}$$

The gravitational wave $h_{\mu\nu}$ is assumed to be a small perturbation on top of the background $\gamma_{\mu\nu}$, and the wavelengths in $h_{\mu\nu}$ are assumed to be much smaller than any inhomogeneity scale of the background. More explicitly, if λ_h is the largest wavelength contained in $h_{\mu\nu}$ and L_γ the characteristic distance over which the background changes, we will assume throughout the paper that

$$\frac{\lambda_h}{L_\gamma} \ll 1, \tag{2}$$

which, in particular, limits the maximum curvature of the background which is of order L_γ^{-2}.

Under these restrictions, the propagation of $h_{\mu\nu}$ in empty space follows the wave equation (Eq. 5.12 in Ref.[7]):

$$\Delta_{\{\gamma\}} h_{\mu\nu} = 0, \tag{3}$$

where $\Delta_{\{\gamma\}}$ is the d'Alembertian operator in the spacetime geometry determined by $\gamma_{\mu\nu}$. In flat spacetime it reduces to the common form of the wave operator $\Delta_{\{\eta\}} = \partial_t^2/c - \Sigma_{i=1}^3 \partial_{x_i}^2$, where c is the speed of light which we will set to 1 from now on. Equation (3) already assumes the choice of a Lorenz and traceless gauge: $D_\nu h^{\mu\nu}=0$, $\gamma_{\mu\nu}h^{\mu\nu}=0$, where D_ν denotes the covariant derivative (Eqs. 5.8,5.9 in Ref. [7]). Exhausting the last remaining gauge freedom, we set $h_{\mu 0}=0$ and turn $h_{\mu\nu}$ into a purely spatial symmetric tensor (the symmetry $h_{\mu\nu}=h_{\nu\mu}$ is a gauge independent property of any metric tensor). This collection of gauge choices is known as the transverse traceless gauge, and selects the gauge invariant part of the wave (Sec. 35.4 in Ref.[1]), thereby isolating its physical degrees of freedom. We will work in the transverse traceless gauge throughout the article.

Exploiting the assumption that $\gamma_{\mu\nu}$ varies much slower than $h_{\mu\nu}$, approximate solutions of Eq. (3) are obtained with the ansatz[1]:

$$h_{\mu\nu}(x_\alpha) = B_{\mu\nu}(x_\alpha)\exp(i\phi(x_\alpha)), \tag{4}$$

where $B_{\mu\nu}$ changes slowly with the spacetime coordinates x_α, and ϕ has a possibly large first derivative $p_\mu(x_\alpha)=\partial_\mu\phi(x_\alpha)$, which also changes slowly with x_α, and no significant derivatives of higher orders. In this complex notation, the real part of $h_{\mu\nu}$ is used for the metric tensor. Substituting Eq. (4) into Eq. (3) leads to the important results that the p_μ are null vectors ($p_\mu(x_\alpha)p^\mu(x_\alpha)=0$), gravitational waves are transverse ($p_\mu(x_\alpha)B^{\mu\nu}(x_\alpha)=0$), and that their propagation can be understood as the parallel transport of both p_μ and $B_{\mu\nu}$ along the null geodesics $dx^\mu/dl=p^\mu$. Indexes are lowered and raised using $\gamma_{\mu\nu}$.

Identical results can be reached for the propagation of electromagnetic waves in curved spacetime (Box 22.4 in Ref. [1]), except that instead of the symmetric tensor $h_{\mu\nu}$, they apply to the electromagnetic four-vector potential A_μ. We will now show that the analogy with electromagnetism is exact when, instead of considering one instance of the electromagnetic field (A_μ), one considers a subspace of the tensor products of two instances of it ($A_\mu A_\nu$). As we will show later, the states in this subspace can be physically realized by two-photon states.

RELATIONSHIP BETWEEN GRAVITATIONAL WAVES AND ELECTROMAGNETIC WAVES IN FLAT SPACETIME

We start by considering solutions of Eq. (3) in a flat background, i.e. $\gamma_{\mu\nu}=\eta_{\mu\nu}$. We will study the generalization to curved backgrounds in the next section. In a flat metric, plane waves constitute a complete basis of the space of propagating solutions. Plane waves are characterized by a null four-momentum and a constant polarization tensor. For each four-momentum p_μ, there are only two independent polarization tensors that meet the constraints of the transverse traceless gauge. For $p_\mu=(\omega,0,0,\omega)$, the tensors that are typically chosen are (Sec. 6 in Ref. [16]):

$$s = \hat{\mathbf{x}}\otimes\hat{\mathbf{x}} - \hat{\mathbf{y}}\otimes\hat{\mathbf{y}}, \text{ and } m = \hat{\mathbf{x}}\otimes\hat{\mathbf{y}} + \hat{\mathbf{y}}\otimes\hat{\mathbf{x}}, \tag{5}$$

where $\otimes$ denotes the tensor product. The tensors in Eq. (5) are called the "plus" and "cross" linear polarizations. Another possible choice is the combinations $s\pm im$, which are referred to as circular polarizations. There is a key difference between these two sets of tensors: The linear polarization tensors sand m cannot be written as the tensor product of a vector with itself $\hat{\mathbf{a}}\otimes\hat{\mathbf{a}}$, but the $s\pm im$ combinations can

$$\begin{aligned} s+im &= \hat{\mathbf{x}}\otimes\hat{\mathbf{x}} - \hat{\mathbf{y}}\otimes\hat{\mathbf{y}} + i(\hat{\mathbf{x}}\otimes\hat{\mathbf{y}} + \hat{\mathbf{y}}\otimes\hat{\mathbf{x}}) = \\ &(\hat{\mathbf{x}}+i\hat{\mathbf{y}})\otimes(\hat{\mathbf{x}}+i\hat{\mathbf{y}}), \\ s-im &= \hat{\mathbf{x}}\otimes\hat{\mathbf{x}} - \hat{\mathbf{y}}\otimes\hat{\mathbf{y}} - i(\hat{\mathbf{x}}\otimes\hat{\mathbf{y}} + \hat{\mathbf{y}}\otimes\hat{\mathbf{x}}) = \\ &(\hat{\mathbf{x}}-i\hat{\mathbf{y}})\otimes(\hat{\mathbf{x}}-i\hat{\mathbf{y}}). \end{aligned} \tag{6}$$

This factorization has notable implications. We can use Eq. (6) to factorize the complete expressions of the corresponding gravitational plane waves into a tensor product of two vector plane waves

$$[(\hat{\mathbf{x}}+i\hat{\mathbf{y}})\otimes(\hat{\mathbf{x}}+i\hat{\mathbf{y}})]\exp(i\omega(z-t)) =$$

$$\left[(\hat{\mathbf{x}}+i\hat{\mathbf{y}})\exp(i\kappa^{+}\omega(z-t))\right]\otimes\left[(\hat{\mathbf{x}}+i\hat{\mathbf{y}})\exp(i(1-\kappa^{+})\omega(z-t))\right], \tag{7}$$

$$[(\hat{\mathbf{x}}-i\hat{\mathbf{y}})\otimes(\hat{\mathbf{x}}-i\hat{\mathbf{y}})]\exp(i\omega(z-t)) =$$
$$\left[(\hat{\mathbf{x}}-i\hat{\mathbf{y}})\exp(i\kappa^{-}\omega(z-t))\right]\otimes\left[(\hat{\mathbf{x}}-i\hat{\mathbf{y}})\exp(i(1-\kappa^{-})\omega(z-t))\right], \tag{8}$$

where $\kappa^{\pm}$ are scalars whose allowed values and significance will be discussed later. The vectorial waves have polarization vectors $(\hat{\mathbf{x}}\pm i\hat{\mathbf{y}})$, transverse to their four-momentum, which is proportional to that of the gravitational wave and hence also of null length. We recognize the expressions of the constituent vectorial waves in Eqs. (7) and (8) as the spatial part of the four-vector potential of circularly polarized electromagnetic plane waves in the transverse gauge ($\nabla\cdot\mathbf{A}=0$). In this gauge, $\mathbf{A}$ collects the transverse degrees of freedom of the electromagnetic field. These physically relevant components are radiative, and are said to belong to the "free field". On the other hand, the non-radiative longitudinal degrees of freedom of the field can always be attached to the sources (I.B.5 in Ref. [17], Chap. XXI.S22 in Ref. [18]). In the transverse gauge the longitudinal components are contained in A_0, the time component of the vector potential. By setting $A_0=0$, the resulting electromagnetic gauge has the same physical significance as the transverse traceless gauge in the gravitational case. This is seen by noting that with $A_0=0$, the transverse gauge is equivalent to the Lorenz gauge ($\partial_\mu A^\mu=0$), and that the zero trace and purely spatial conditions eliminate the non-radiative and longitudinal degrees of freedom of the gravitational field (Sec. 35.4 in Ref. [1]).

Up to this point, we have just formally factorized two particular gravitational plane waves as tensor products of electromagnetic plane waves. We will now use a constructive procedure to show that there exists a formal equivalence between each gravitational plane wave and a set of physically valid entangled photon states.

Let us denote by $|\mathbf{p}\,\lambda\rangle$ an electromagnetic plane wave with momentum $\mathbf{p}$ and helicity $\lambda=\pm1$. The helicity operator is the projection of the angular momentum operator vector in the direction of the linear momentum operator vector $\Lambda=\mathbf{J}\cdot\mathbf{P}/|\mathbf{P}|$ (Chap. 8.4.1 in Ref. [19]). Its eigenstates have the same polarization handedness in all their momentum components. Any valid electromagnetic tensor wave can be written as a linear superposition of states of the kind:

$$|\mathbf{p}\;\lambda\rangle\otimes|\bar{\mathbf{p}}\;\bar{\lambda}\rangle+|\bar{\mathbf{p}}\;\bar{\lambda}\rangle\otimes|\mathbf{p}\;\lambda\rangle, \tag{9}$$

where the correct exchange symmetry for bosonic fields is enforced in an obvious way.

From the point of view of field theory, an object with zero mass and allowed helicity eigenvalues $\lambda=\pm1$ is associated with the electromagnetic field, while an object with zero mass and allowed helicity eigenvalues $\lambda=\pm2$ is associated with the gravitational field (Chap. 2.5 in Ref. [20]). Consequently, for states of the kind (9) to be equivalent to gravitational waves, they must have their associated characteristics, i.e., zero mass and allowed helicity eigenvalues equal to ±2.

Let us first apply the mass squared operator to Eq. (9):

$$-\left({}_2P_0^2-\sum_{i=1}^{3}{}_2P_i^2\right)\left(|\mathbf{p}\;\lambda\rangle\otimes|\bar{\mathbf{p}}\;\bar{\lambda}\rangle+|\bar{\mathbf{p}}\;\bar{\lambda}\rangle\otimes|\mathbf{p}\;\lambda\rangle\right), \tag{10}$$

where ${}_2P_\mu$ is the generator of translations along direction μ in the tensor product space. Its expression as a function of P_μ, the momentum operator for a single constitutive space, is: ${}_2P_\mu=P_\mu\otimes I+I\otimes P_\mu$, where I is the identity operator. Using that

$$P_0|\mathbf{p}\;\lambda\rangle=\omega|\mathbf{p}\;\lambda\rangle,\;P_0|\bar{\mathbf{p}}\;\bar{\lambda}\rangle=\bar{\omega}|\bar{\mathbf{p}}\;\bar{\lambda}\rangle,$$
$$P_i|\mathbf{p}\;\lambda\rangle=p_i|\mathbf{p}\;\lambda\rangle,\;P_i|\bar{\mathbf{p}}\;\bar{\lambda}\rangle=\bar{p}_i|\bar{\mathbf{p}}\;\bar{\lambda}\rangle,$$
$$\omega^2=\sum_{i=1}^{3}p_i^2,\;\bar{\omega}^2=\sum_{i=1}^{3}\bar{p}_i^2, \tag{11}$$

we find that the result of (10) is:

$$-\left(\omega^2+\bar{\omega}^2+2\omega\bar{\omega}-\sum_{i=1}^{3}\left(p_i^2+\bar{p}_i^2+2p_i\bar{p}_i\right)\right)\left(|\mathbf{p}\;\lambda\rangle\otimes|\bar{\mathbf{p}}\;\bar{\lambda}\rangle+|\bar{\mathbf{p}}\;\bar{\lambda}\rangle\otimes|\mathbf{p}\;\lambda\rangle\right)=$$
$$-2\left(\omega\bar{\omega}-\sum_{i=1}^{3}p_i\bar{p}_i\right)\left(|\mathbf{p}\;\lambda\rangle\otimes|\bar{\mathbf{p}}\;\bar{\lambda}\rangle+|\bar{\mathbf{p}}\;\bar{\lambda}\rangle\otimes|\mathbf{p}\;\lambda\rangle\right). \tag{12}$$

Since $\omega=|\mathbf{p}|$ and $\bar{\omega}=|\bar{\mathbf{p}}|$, the term between brackets in the last line of Eq. (12) can be written as $|\mathbf{p}||\bar{\mathbf{p}}|-\mathbf{p}\cdot\bar{\mathbf{p}}$. To get mass squared equal to zero we then need

$$|\mathbf{p}||\bar{\mathbf{p}}|=\mathbf{p}\cdot\bar{\mathbf{p}}, \tag{13}$$

which is only met if $\bar{\mathbf{P}} = \alpha\mathbf{p}$ for real $\alpha > 0$. Therefore, for a state of the kind (9) to have zero mass, the two momenta $\mathbf{p}$ and $\bar{\mathbf{P}}$ must be parallel.

Let us now look at the polarization of this kind of states using the helicity basis. Given a $\mathbf{p}$ and a $\bar{\mathbf{P}}$ that fulfill the zero mass condition, the possible states of the kind (9) are:

$$\begin{aligned} &|\mathbf{p}\,+\rangle \otimes |\alpha\mathbf{p}\,+\rangle + |\alpha\mathbf{p}\,+\rangle \otimes |\mathbf{p}\,+\rangle, \\ &|\mathbf{p}\,-\rangle \otimes |\alpha\mathbf{p}\,-\rangle + |\alpha\mathbf{p}\,-\rangle \otimes |\mathbf{p}\,-\rangle, \\ &|\mathbf{p}\,+\rangle \otimes |\alpha\mathbf{p}\,-\rangle + |\alpha\mathbf{p}\,-\rangle \otimes |\mathbf{p}\,+\rangle, \\ &|\mathbf{p}\,-\rangle \otimes |\alpha\mathbf{p}\,+\rangle + |\alpha\mathbf{p}\,+\rangle \otimes |\mathbf{p}\,-\rangle. \end{aligned} \tag{14}$$

The application of the helicity operator ${}_2\Lambda = \Lambda \otimes I + I \otimes \Lambda$ to these states shows that the four of them are eigenstates of helicity with eigenvalues +2, −2, 0 and 0, respectively. The last two do not correspond to a gravitational wave and must be discarded for our purposes.

Applying the ${}_2P_\mu$ operators shows that the states in Eq. (14) have four-momentum eigenvalues equal to $(1+\alpha)p_\mu$. At this point we can affirm that, given an arbitrary gravitational plane wave of helicity λ^g and four-momentum q_μ, the following set of electromagnetic tensor waves is equivalent to the gravitational wave:

$$|\kappa\mathbf{q}\ \frac{\lambda^g}{2}\rangle \otimes |(1-\kappa)\mathbf{q}\ \frac{\lambda^g}{2}\rangle + |(1-\kappa)\mathbf{q}\ \frac{\lambda^g}{2}\rangle \otimes |\kappa\mathbf{q}\ \frac{\lambda^g}{2}\rangle, \tag{15}$$

for any real $\kappa \in (0,1)$. The allowed values of κ follow from the massless condition $\kappa\mathbf{q} = \alpha(1-\kappa)\mathbf{q}$, with real $\alpha > 0$. Figure 1 illustrates the results of the derivation.

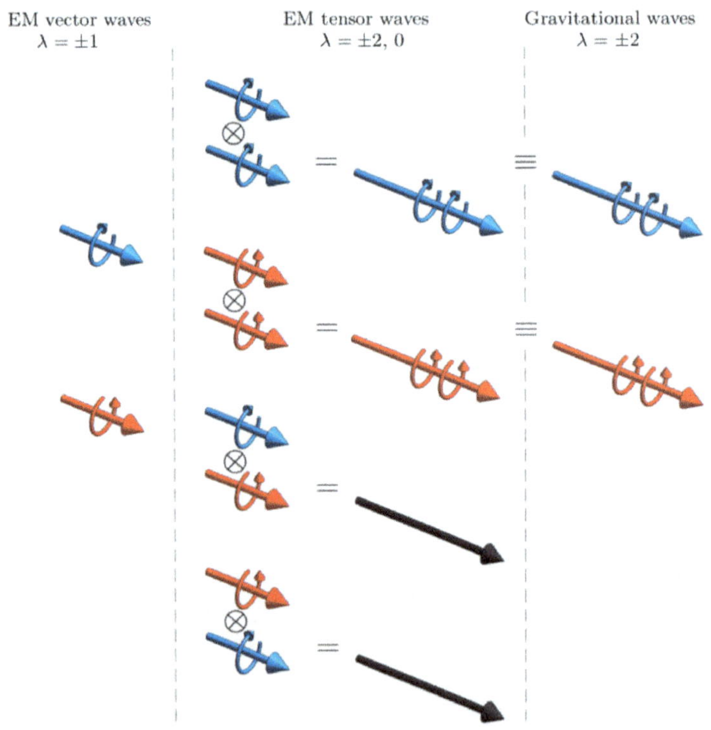

Figure 1: Equivalence between electromagnetic tensor product states and gravitational waves. The different tensor products of electromagnetic plane waves of well-defined helicity equal to ±1 (in the left panel) result in electromagnetic tensor waves of helicity ±2 and 0 (middle panel), see Eq. (14). Only the first two have a gravitational wave equivalent (right panel). The momenta of the two electromagnetic plane waves do not need to be equal: Parallel momenta is enough to ensure that the tensor product will have a gravitational wave equivalent.

These results are consistent with and get reinforced by group theoretical arguments involving the decomposition of tensor products of two identical massless representations of the Poincare group. The reduction of the direct product of two massless representations with the same helicity λ results in both massless and massive irreducible representations[22]. All of the massless representations have helicity equal to 2λ, and their multiplicity is infinite and uncountable, following the range of a real and positive constant (Eq. 7.1 in Ref. [21]), which corresponds exactly to our parameter α.

Since any gravitational wave solution of Eq. (3) in flat spacetime can be written as a linear combination of gravitational plane waves of well-defined helicity, we conclude that any gravitational wave in a flat background is equivalent to a sum of electromagnetic tensor waves. Each electromagnetic tensor wave is the tensor product of two plane waves of the same helicity and parallel momentum.

Equations (7) and (8) can now be interpreted in light of these results. By re-writing the right hand side of Eq.(7) in an exchange symmetric form (similarly for Eq. (8))

$$\frac{1}{2}(\hat{\mathbf{x}}+i\hat{\mathbf{y}})\exp(i\kappa^{+}\omega(z-t))\otimes(\hat{\mathbf{x}}+i\hat{\mathbf{y}})\exp(i(1-\kappa^{+})\omega(z-t))$$
$$+\frac{1}{2}(\hat{\mathbf{x}}+i\hat{\mathbf{y}})\exp(i(1-\kappa^{+})\omega(z-t))\otimes(\hat{\mathbf{x}}+i\hat{\mathbf{y}})\exp(i\kappa^{+}\omega(z-t)), \quad (16)$$

we obtain a representation of the state

$$|\kappa^{+}\omega\hat{\mathbf{z}}\,+\rangle|(1-\kappa^{+})\omega\hat{\mathbf{z}}+\rangle+|(1-\kappa^{+})\omega\hat{\mathbf{z}}\,+\rangle|\kappa^{+}\omega\hat{\mathbf{z}}+\rangle, \quad (17)$$

restricted to equal position and time coordinates (z and t) in both factors of the tensor products. This restriction is needed to fully equate measurements on two-body objects to measurements on single-body objects.

GENERALIZATION TO CURVED SPACETIME BACKGROUNDS

We will now show that these flat spacetime results also hold in the case of gravitational waves propagating in a non-flat background, provided that the waves have a small amplitude and vary rapidly with respect to the background metric $\gamma_{\mu\nu}$, as indicated by Eq. (2). With these restrictions, we go back to the approximate solution of Eq. (4), $h_{\mu\nu}(x_\alpha)=B_{\mu\nu}(x_\alpha)\exp(i\phi(x_\alpha))$, and start by exploiting the conditions that the polarizability tensor must meet. In the chosen gauge, $B_{\mu\nu}(x_\alpha)$ is transverse to the phase gradient four-vector $p_\mu(x_\alpha)=\partial_\mu\phi(x_\alpha)$, and purely spatial. This leaves a space which can be expanded using tensor products of two linearly independent spatial vectors. Since the tensor must also be symmetric, we can write:

$$\begin{aligned}B_{\mu\nu}(x_\alpha) &= c_1(x_\alpha)\left[\hat{\mathbf{e}}_1(x_\alpha)\otimes\hat{\mathbf{e}}_1(x_\alpha)\right]\\ &+c_2(x_\alpha)\left[\hat{\mathbf{e}}_2(x_\alpha)\otimes\hat{\mathbf{e}}_2(x_\alpha)\right]\\ &+c_3(x_\alpha)\left[\hat{\mathbf{e}}_1(x_\alpha)\otimes\hat{\mathbf{e}}_2(x_\alpha)+\hat{\mathbf{e}}_2(x_\alpha)\otimes\hat{\mathbf{e}}_1(x_\alpha)\right],\end{aligned} \quad (18)$$

where $c_i(x_\alpha)$ are complex scalars. We also know that gravitational waves have only two gauge invariant degrees of freedom, and that the transverse traceless gauge isolates them. Therefore, expression (18) has one degree of freedom too many. In order to eliminate it, we will impose that the two degrees of freedom correspond to helicity eigenstates with eigenvalues ±2. This demand is consistent with the fact that the equations of general relativity in vacuum are invariant under the gravitational version of the duality transformations (Sec. 3 of Ref. [10]). It follows that their solutions can be classified by the eigenvalues of the generator of duality transformations: Helicity[23,24]. We hence require the existence of waves $h^{\pm}_{\mu\nu}(x_\alpha)$ that are invariant under rotations by an arbitrary angle θ along the spatial momentum direction ($R_p(\theta)$), and pick up a factor of $\exp(\mp i2\theta)$. In the geometric optics approximation, the spatial momentum direction is given by the spatial part of $p_\mu(x_\alpha)$. The requirement of definite helicity equal to ±2 reads then:

$$R_{\hat{\mathbf{p}}(x_\alpha)}(\theta)h^{\pm}_{\mu\nu}(x_\alpha)=\exp(\mp i2\theta)h^{\pm}_{\mu\nu}(x_\alpha). \quad (19)$$

Our requirement is met by assigning the first two degrees of freedom in Eq. (18) to the two helicities (±2) and discarding the third one. The required transformation properties of the tensors are then:

$$\begin{aligned}&R_{\hat{\mathbf{p}}(x_\alpha)}(\theta)\left[\hat{\mathbf{e}}_\pm(x_\alpha)\otimes\hat{\mathbf{e}}_\pm(x_\alpha)\right]=\\ &\left[R_{\hat{\mathbf{p}}(x_\alpha)}(\theta)\hat{\mathbf{e}}_\pm(x_\alpha)\right]\otimes\left[R_{\hat{\mathbf{p}}(x_\alpha)}(\theta)\hat{\mathbf{e}}_\pm(x_\alpha)\right]=\\ &\exp(\mp i2\theta)\left[\hat{\mathbf{e}}_\pm(x_\alpha)\otimes\hat{\mathbf{e}}_\pm(x_\alpha)\right],\end{aligned} \quad (20)$$

which means that the two vectors $\hat{\mathbf{e}}_\pm(x_\alpha)$ must meet

$$R_{\hat{\mathbf{p}}(x_\alpha)}(\theta)\hat{\mathbf{e}}_\pm(x_\alpha)=\exp(\mp i\theta)\hat{\mathbf{e}}_\pm(x_\alpha). \quad (21)$$

With this choice of $\hat{\mathbf{e}}_1=\hat{\mathbf{e}}_+$ and $\hat{\mathbf{e}}_2=\hat{\mathbf{e}}_-$, the discarded degree of freedom is, as in the flat background case, a helicity eigenstate with zero eigenvalue. With these results and the insight from the flat spacetime case we can write:

$$\begin{aligned}x_\alpha) &= B_{\mu\nu}(x_\alpha)\exp(i\phi(x_\alpha))=\\ &\frac{c_+(x_\alpha)}{2}\left[\hat{\mathbf{e}}_+(x_\alpha)\exp(i(1-\kappa^+)\phi(x_\alpha))\otimes\hat{\mathbf{e}}_+(x_\alpha)\exp(i\kappa^+\phi(x_\alpha))\right]+\\ &\frac{c_+(x_\alpha)}{2}\left[\hat{\mathbf{e}}_+(x_\alpha)\exp(i\kappa^+\phi(x_\alpha))\otimes\hat{\mathbf{e}}_+(x_\alpha)\exp(i(1-\kappa^+)\phi(x_\alpha))\right]+\\ &\frac{c_-(x_\alpha)}{2}\left[\hat{\mathbf{e}}_-(x_\alpha)\exp(i(1-\kappa^-)\phi(x_\alpha))\otimes\hat{\mathbf{e}}_-(x_\alpha)\exp(i\kappa^-\phi(x_\alpha))\right]+\\ &\frac{c_-(x_\alpha)}{2}\left[\hat{\mathbf{e}}_-(x_\alpha)\exp(i\kappa^-\phi(x_\alpha))\otimes\hat{\mathbf{e}}_-(x_\alpha)\exp(i(1-\kappa^-)\phi(x_\alpha))\right]=h^+_{\mu\nu}(x_\alpha)+h^-_{\mu\nu}(\end{aligned} \quad (22)$$

According to Eq. (21) the constituent vector waves in Eq. (22) are helicity eigenstates with eigenvalues ±1. According to Eq. (22), their four-momenta have zero length because they are proportional to $p_\mu(x_\alpha)$. We recognize them again as electromagnetic waves, this time in the geometric optics approximation. We note that the decomposition of the polarization tensor into two components of definite helicity can be also achieved by taking the gauge invariant part of the null vierbein decomposition of $B_{\mu\nu}$ that appears in Eq. (5.17) of Ref. [24].

Let us now consider the propagation of the waves in Eq. (22). We recall that in the geometric optics approximation, propagation is accomplished by parallel transporting both the phase gradient four-vector p_μ and the polarization tensor along the null geodesics $dx^\mu/dl = p^\mu$. All the tensors in Eq. (22) have the same total phase attached to them. They will be transported along the same path. This null geodesic ray must also be the path of parallel transport for the constitutive vector waves, otherwise the factors would split up and break the required structure. This imposes $\kappa^\pm \in (0,1)$, since if either κ or $(1-\kappa)$ were negative, the corresponding factor would propagate backwards along the geodesic. In the case of flat spacetime, we obtained the same allowed range of values for $\kappa^\pm$ from requiring massless states. The requisite of a consistent null geodesic ray is the massless condition in geometric optics. When higher order terms are added to the geometric optics solution, the two helicity components do not follow the same path (Sec. II.3.5 in Ref. [9]). Nevertheless, since the vectorial factors in each helicity should still propagate together, the same restriction $\kappa^\pm \in (0,1)$ is needed.

The exact value of $\kappa^\pm$ does not change the null ray or the helicity. Its meaning is hence the same as in the flat spacetime case: It collects the set of electromagnetic tensor waves which are equivalent to a single gravitational wave.

Let us now investigate the effects of propagation in the factorized polarization tensor. We write the parallel transport equation for the total polarization tensor in (22) by means of the covariant derivatives D_μ:

$$0 = p^\sigma D_\sigma \left[c_+ \hat{\mathbf{e}}_+ \otimes \hat{\mathbf{e}}_+ + c_- \hat{\mathbf{e}}_- \otimes \hat{\mathbf{e}}_-\right]. \tag{23}$$

The first order linear differential equation (23) with the initial condition at some initial point $x_\alpha = x_\beta$ has a unique solution at each point of the path. We now make the guess that the solution can be obtained by first parallel transporting both $\hat{\mathbf{e}}_\pm$ and then taking their tensor products. This guess turns out to be correct. To verify it, we expand the covariant derivatives in (23) across the tensor products:

$$\begin{aligned} 0 = &c_+(x_\beta)\left(\left[p^\sigma D_\sigma\left(\hat{\mathbf{e}}_+\right)\right] \otimes \hat{\mathbf{e}}_+ + \hat{\mathbf{e}}_+ \otimes \left[p^\sigma D_\sigma\left(\hat{\mathbf{e}}_+\right)\right]\right) + \\ &c_-(x_\beta)\left(\left[p^\sigma D_\sigma\left(\hat{\mathbf{e}}_-\right)\right] \otimes \hat{\mathbf{e}}_- + \hat{\mathbf{e}}_- \otimes \left[p^\sigma D_\sigma\left(\hat{\mathbf{e}}_-\right)\right]\right). \end{aligned} \tag{24}$$

Note that the covariant derivatives do not act on the scalars $c_\pm(x_\beta)$, which are fixed initial conditions. Our guess assumes that the parallel transport equations for the polarization vectors $\hat{\mathbf{e}}_\pm$ are met, i.e:

$$p^\sigma D_\sigma\left(\hat{\mathbf{e}}_\pm\right) = 0, \tag{25}$$

which is clearly a solution of Eq. (24).

We can hence conclude that, in order to propagate the tensorial gravitational wave in Eq. (22), we may as well take the constitutive electromagnetic waves in the tensor product factors, propagate them in the curved background by the rules of geometric optics, take the tensor products of the results and add both tensor products weighted by $c_\pm(x_\beta)$. The two helicity components of an electromagnetic vector wave do not mix during parallel transport. They pick up phases with the same absolute value and different sign[26,27]. The same is true for gravitational waves. The factorization allows us to recover the known result that the phase for the gravitational case is exactly twice that of the electromagnetic wave (Sec. IV in Ref. [24]).

All the wave (Sec. IV in Ref. [24]) above arguments are valid for any initial x_β and $p_\mu(x_\beta)$, so they apply in all spacetime to each of the terms in a sum of an arbitrary number of solutions of the type $B_{\mu\nu}\exp(i\phi)$:

$$h_{\mu\nu} = \sum_n B_{\mu\nu}(n)\exp(i\phi(n)). \tag{26}$$

This finishes the proof of the equivalence: Any gravitational wave of the kind (26) propagating in curved spacetime is equivalent to a sum of electromagnetic tensor product waves propagating in the same spacetime. Each term of the sum is the tensor product of two electromagnetic waves of the same helicity and parallel momentum. These results hold under the assumptions of small amplitude and fast variation of the gravitational wave with respect to the background metric (see Eq. (2)).

We finalize this discussion by noting that the expression of the stress-energy tensor of the waves does not depend on whether they are gravitational or electromagnetic (see Eq. (35.77j) in Ref. [1] and Secs. 4 and 6 in Ref. [8]). Therefore, the portion of the background curvature due to the waves themselves is also equivalent in both cases, and extends the equivalence to the back action of the waves onto the background.

Throughout the analysis, we have assumed propagation in empty space. The different couplings of the electromagnetic and gravitational fields to material particles, for example to electrically charged particles, indicates that the equivalence will not hold in the presence of matter.

QUANTUM EMULATION OF GRAVITATIONAL WAVES WITH ENTANGLED PHOTON STATES

We now use these results to propose a path for the quantum emulation of the propagation of gravitational waves in a laboratory by means of experimental electromagnetic setups. Besides the equivalence that we have shown, we will need two more ingredients: The conformal invariance of electromagnetism, and the fact that Maxwell's equations in an empty but curved spacetime are identical to Maxwell's equations in a material medium in flat spacetime.

Let us assume that we are interested in emulating a given scenario where gravitational waves from one or more sources propagate through curved spacetime regions (see the upper panel of Fig. 2). The wavelengths of the detectable gravitational field are typically very large, larger than 10^5 meters[7], as are the features of the curved background. Using the results contained in this article, we can map the scenario of interest to one where the gravitational waves are substituted by electromagnetic tensor waves. Now, the conformal invariance will allow us to shrink the problem. One of the symmetry transformations allowed by conformal invariance is a pointwise scaling of spacetime: $\bar{x}_\alpha = s(x_\alpha)x_\alpha$. We will use a constant $s(x_\alpha)=s$. Let us say that we want to study the solutions of Maxwell's equations in a spacetime with metric $\gamma_{\mu\nu}(x_\alpha)$. The invariance under scalings means that we can equivalently study the solutions of the system obtained after the change $\bar{x}_\alpha = sx_\alpha$, whose metric is $s^{-2}\gamma_{\mu\nu}(\bar{x}_\alpha/s)$. In terms of the electromagnetic field tensors, the solutions of the original and scaled system are related as $\bar{F}_{\mu\nu}(\bar{x}_\alpha) = s^{-2}F_{\mu\nu}(\bar{x}_\alpha/s)$. Features of size L in the original spacetime geometry have size sL in the scaled system, and the same relationship holds for the electromagnetic wavelengths in the two systems. After shrinking the problem to a suitable size, we are left with the task of creating a spacetime curvature $s^{-2}\gamma_{\mu\nu}$ in the laboratory. Fortunately, we do not need to tackle this problem directly: The propagation of electromagnetic waves in a given empty but curved spacetime is equivalent to their propagation in a particular material medium in flat spacetime[28]. The constitutive relations of the material are a function of the metric and, in general, describe a continuous, inhomogeneous and anisotropic medium. Provided that one can fabricate such media, it is possible to emulate the propagation and interference of an arbitrary number of gravitational waves traveling through the spacetime geometry created by an arbitrary number of cosmological objects by means of experimental electromagnetic setups. Figure 2 illustrates this quantum emulation. It is important to note that the spacetime curvature caused by the gravitational waves themselves is not dynamically produced in this emulation framework. That is, the part of $\gamma_{\mu\nu}$ attributable to them must be small enough so that it can be ignored.

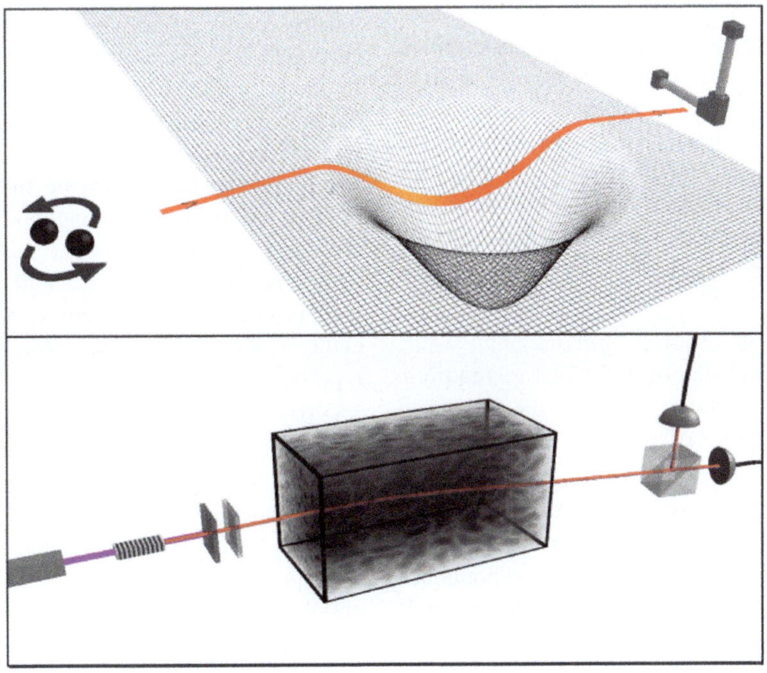

Figure 2: Schematic comparison of a gravitational wave propagating through curved spacetime (upper panel) with the equivalent situation of two-photon states (lower panel) propagating through a metamaterial. The propagation of the gravitational wave is equivalent to the propagation of a subset of two-photon states. Conformal invariance is used to shrink the astronomical sizes down to laboratory scales, and a metamaterial media to mimic the specific spacetime curvature. The source for the two-photon states is a non-linear crystal, for example a periodically poled potassium titanyl phosphate (ppKTP), pumped by a narrow band laser. An interference filter after the crystal removes the pump light and could restrict the frequencies for the down-converted photons. A quarter waveplate would finally transform the state to the helicity basis. In order to emulate an arbitrary gravitational wave polarization, superpositions of photon pairs with both helicities are required. This can be achieved by combining two sources with orthogonal linear polarizations before the quarter waveplate.

A technical challenge for this quantum emulation framework is the fabrication of the appropriate inhomogeneous and anisotropic media that mimic the desired spacetime geometry. Their design and fabrication is actively being investigated by the metamaterials community. These media implement transformation devices like for example cloaks and perfect lenses[29,30,31]. Actually, the helicity preservation inherent to a general curved spacetime (Sec. 11 in Ref. [31]) can be exploited to provide design guidelines for transformation devices[33].

Having solved the scale problem and the propagation in a curved spacetime, we now turn to the generation of electromagnetic fields with the desired properties: That they can be decomposed as sums of tensor products of two electromagnetic plane waves of the same helicity, whose momentum four-vectors are parallel and proportional to one another. A system which already fulfills most of these properties is a spontaneous parametric down-conversion (SPDC) source of momentum correlated photons. Entangled photon sources based on SPDC generated photons have been regularly used in quantum optics experiments in the last 40 years. In these quantum light sources, a quadratic nonlinear crystal is pumped by a high frequency beam of light, typically in the ultraviolet or violet regime. The nonlinear crystal then mediates the low efficiency process of down-converting some of the pump photons into two photons at a lower frequency (typically in the infra-red). The possibility of engineering the phase matching of nonlinear crystals has allowed for a wide range of engineered quantum states of light.

In particular, the use of collinear sources, i.e. sources where the generated photons are emitted along the propagation direction of the pump beam, allows for a tight control of the momentum and polarization correlations of the down-converted photons. Collinear sources based on Type 0 periodically poled potassium titanyl phosphate (ppKTP) can generate entangled photons which are momentum correlated and have the same linear polarization. Their polarization can then be transformed to the required helicity basis with a quarter waveplate. The momentum correlation is achieved by using a pump beam with a width which is much smaller than $(S/k_p)^{1/2}$, where S is the length of the crystal and k_p the wavenumber of the pump beam. In the two-photon state, the terms exhibiting correlations between non-parallel momenta cannot be completely suppressed, but can be made arbitrarily small by either increasing the length of the crystal or using a tighter pump focus[33].

In order to emulate an arbitrary gravitational wave, superpositions of photon pairs with both helicities are required. This can be achieved by combining two of the described sources with orthogonal linear polarizations before the quarter waveplate.

We note that the parameter $\kappa \in (0,1)$ in Eq. (15), which ties together all the electromagnetic tensor states that are equivalent to a single gravitational wave, has a physical correspondence in the source of momentum correlated photons. It corresponds to the split of the frequency of the pump w_p, which we assume monochromatic, into the frequencies of the emitted photons (ω_s, ω_i). Since energy conservation enforces $\omega_p = \omega_s + \omega_i$, we may write $\omega_s = \kappa\omega_p$ and $\omega_i = (1-\kappa)\omega_p$ for $\kappa \in (0,1)$, which makes the correspondence clear. Should the metamaterial exhibit a significant frequency dependence, $\kappa^{\pm} = 1/2$ can be chosen by means of a narrow-band interference filter selecting the frequency degenerate down-converted photon. This completes the design of a quantum source of light, capable of emulating a gravitational wave.

For the detection of the field after it has passed through the metamaterial, we need to keep in mind that we are emulating single-body objects by means of two-body objects. The measurements should rely on detecting both photons "at the same place and at the same time". This may be done by counting coincidences, where both photons of a pair arrive at the detector within a very small time interval. Measurements that are unique to a two-particle state, like quantum interference, have no correspondence in the gravitational wave case, which places them outside the emulation framework.

The applicability of the emulation scheme to a particular background metric is determined by the applicability of the theoretical results contained in the previous section: The amplitudes of the gravitational waves must be much smaller than those of the background, and, the waves must vary much faster than the background (seeEq.(2)). For a given wavelength, this last condition limits the maximum allowable curvature. When the curvature increases, e.g. while approaching a singularity, the applicability reduces to correspondingly smaller wavelengths.

DISCUSSION

In conclusion, we have shown that the propagation of gravitational waves in a curved spacetime background is equivalent to the propagation of a restricted set of electromagnetic tensor product waves. The defining characteristics of gravitational waves, zero mass and possible helicities equal to ±2, select the appropriate subspace of electromagnetic states. They turn out to be the tensor products of two electromagnetic waves of the same helicity and parallel momentum. Therefore, any linearly or elliptically polarized gravitational wave can be expanded as the sum of two electromagnetic tensor waves, one for each circular polarization handedness. The consideration of helicity eigenstates of both the gravitational and the electromagnetic waves has been crucial: A linearly polarized gravitational wave cannot be formally factorized as a tensor product of two electromagnetic waves, while a circularly polarized gravitational wave can. Our analysis is restricted to gravitational waves of small amplitude that vary rapidly with respect to the spacetime background and propagate in empty space. As far as we know, this equivalence has not been reported before.

We have shown that the implementation of the equivalence needs genuine quantum correlations between two photon states. As an application of these results, we have proposed a path for the quantum emulation of the propagation of gravitational waves by means of experimental electromagnetic setups. First, the equivalence derived in this article is applied to map the gravitational waves into electromagnetic tensor waves, then the conformal invariance of electromagnetism is used to shrink the problem from large wavelengths and large sized spacetime features down to suitable sizes, and finally, the scaled down curved spacetime background is emulated in the laboratory by an appropriate material medium. Precisely this type of media are already the object of research in the field of metamaterials.

To finish, we suggest that the decomposition of gravitational waves into tensor products of electromagnetic waves could be used to reduce the complexity of simulating gravitational waves in a computer. If the scenario of interest meets the specified restrictions, setting $\kappa^{\pm} = 1/2$ in Eq. (22), one can first simulate the propagation of two electromagnetic vector waves, one for each helicity, and then sum the tensor products of each resulting vector wave with itself.

Future expansions of this framework could include its extension outside the geometrical optics approximation in order to study the interaction of gravitational waves with strongly curved spacetimes, and the use of nonlinear optical metamaterials capable of reproducing the back action of the gravitational wave on the background metric.

ACKNOWLEDGMENTS

This work was funded by the Center of Excellence for Engineered Quantum Systems and by the Basque Government IT472- 10 Grant, Spanish MINECO FIS2012-36673-C03-02, Ramón y Cajal Grant RYC-2012-11391, UPV/EHU UFI 11/55, UPV/EHU EHUA14/04, CCQED, PROMISCE and SCALEQIT European projects. G.M.-T is also funded by the Future Fellowship program (FF). I.F.C wishes to acknowledge Ms Magda Felo for drawing Fig. 1.

REFERENCES

1. Misner, C. W., Thorne, K. S. & Wheeler, J. A. Gravitation (W.H. Freeman, San Francisco, 1973).
2. Weisberg, J. M., Taylor, J. H. & Fowler, L. A. Gravitational waves from an orbiting pulsar. Sci. Am. 245, 74–82 (1981).
3. Taylor, J. H. & Weisberg, J. M. A new test of general relativity - gravitational radiation and the binary pulsar PSR 1913+16. Astrophys. J. 253, 908 (1982).
4. Barish, B. C. & Weiss, R. LIGO and the detection of gravitational waves. Phys. Today 52, 44–50 (1999).
5. Ju, L., Blair, D. G. & Zhao, C. Detection of gravitational waves. Rep. Prog. Phys. 63, 1317 (2000).
6. Thorne, K. S. Gravitational radiation – a new window onto the universe. arXiv:gr-qc/9704042 (1997).
7. Isaacson, R. A. Gravitational radiation in the limit of high frequency. I. the linear approximation and geometrical optics. Phys. Rev. 166, 1263–1271 (1968).
8. Isaacson, R. A. Gravitational radiation in the limit of high frequency. II. nonlinear terms and the effective stress tensor. Phys. Rev. 166, 1272–1280 (1968).
9. Iorio, L. The Measurement of Gravitomagnetism: A Challenging Enterprise (Nova Publishers, 2007).
10. Maartens, R. & Bassett, B. A. Gravitoelectromagnetism. Class. Quantum Grav. 15, 705 (1998).
11. Garay, L. J., Anglin, J. R., Cirac, J. I. & Zoller, P. Sonic analog of gravitational black holes in Bose-Einstein condensates. Phys. Rev. Lett. 85, 4643–4647 (2000).
12. Lahav, O. et al. Realization of a sonic black hole analog in a bose-einstein condensate. Phys. Rev. Lett. 105, 240401 (2010).
13. Bravo, T., Sab´ın, C. & Fuentes, I. Analog quantum simulation of gravitational waves in a bose-einstein condensate. arXiv:1406.3572
14. cond-mat, physics:gr-qc, physics:quant-ph. (2014). arXiv: 1406.3572.
15. Graham, P. W., Hogan, J. M., Kasevich, M. A. & Rajendran, S. New method for gravitational wave detection with atomic sensors. Phys. Rev. Lett. 110, 171102 (2013).
16. Sab´ın, C., Bruschi, D. E., Ahmadi, M. & Fuentes, I. Phonon creation by gravitational waves. arXiv:1402.7009
17. cond-mat, physics:gr-qc, physics:quant-ph. (2014). arXiv: 1402.7009.
18. Carroll, S. M. Lecture notes on general relativity. arXiv:gr-qc/9712019 (1997).
19. Cohen-Tannoudji, C., Dupont-Roc, J. & Grynberg, G. Photons and atoms: introduction to quantum electrodynamics (Wiley, New York, NY, 1989). Trans. of : Photons et atomes. (Paris, InterEditions, 1987).
20. Messiah, A. Quantum Mechanics (Dover, 1999).
21. Tung, W.-K. Group Theory in Physics (World Scientific, 1985).
22. Weinberg, S. The Quantum Theory of Fields (Volume 1) (Cambridge University Press, 1995), 1 edn.
23. Lomont, J. S. Decomposition of direct products of representations of the inhomogeneous Lorentz group. J. Math. Phys. 1, 237–243 (2004).
24. Calkin, M. G. An invariance property of the free electromagnetic field. Am. J. Phys. 33, 958 (1965).
25. Zwanziger, D. Quantum field theory of particles with both electric and magnetic charges. Phys. Rev. 176, 1489–1495 (1968).
26. Ramos, J. & Mashhoon, B. Helicity-rotation-gravity coupling for gravitational waves. Phys. Rev. D 73, 084003 (2006).
27. Brodutch, A. & Terno, D. R. Polarization rotation, reference frames, and Mach's principle. Phys. Rev. D 84, 121501 (2011).
28. Brodutch, A., Demarie, T. F. & Terno, D. R. Photon polarization and geometric phase in general relativity. Phys. Rev. D 84, 104043 (2011).
29. Plebanski, J. Electromagnetic waves in gravitational fields. Phys. Rev. 118, 1396–1408 (1960).
30. Leonhardt, U. Optical conformal mapping. Science 312, 1777–1780 (2006).
31. Wegener, M. & Linden, S. Shaping optical space with metamaterials. Phys. Today 63, 32–36 (2010).
32. Soukoulis, C. M. & Wegener, M. Past achievements and future challenges in the development of three-dimensional photonic metamaterials. Nat. Photonics 5, 523–530 (2011).
33. Bialynicki-Birula, I. Photon wave function. In Wolf, E. (Ed.) Prog. Optics, Vol. 36, 245–294 (Elsevier, 1996).
34. Fernandez-Corbaton, I. & Molina-Terriza, G. Role of duality symmetry in transformation optics. Phys. Rev. B 88, 085111 (2013).
35. Leach, J. et al. Quantum correlations in position, momentum, and intermediate bases for a full optical field of view. Phys. Rev. A 85, 013827 (2012).

CHAPTER 13

ELECTROMAGNETIC WAVES AND THEIR APPLICATION TO CHARGED PARTICLE ACCELERATION

Hitendra K. Malik

INTRODUCTION

A wave is a disturbance that propagates through space and time, usually with the transference of energy from one point to another without permanent displacement of particles of the medium. The particles under this situation only oscillate about their equilibrium positions. If the particles oscillate in the direction of wave propagation, then the wave is called longitudinal wave. However, if these oscillations take place in perpendicular direction with the direction of wave propagation, the wave is said to be transverse in nature. Electromagnetic (EM) waves are transverse in nature. In electromagnetic waves such as light waves, it is the changes in electric field and magnetic field that represent the wave disturbance. The propagation of the wave is described by the passage of a waveform through the medium with a certain velocity called the phase (or wave) velocity. However, the energy is transferred at the group velocity of the waves making the waveform. Electromagnetic radiation is a form of energy exhibiting wave like behavior as it travels through the space. The electromagnetic radiation is classified based on the frequency of its wave. Figure 1 shows the electromagnetic spectrum that consists of radio waves, microwaves, infrared (IR) radiation, visible light, ultraviolet (UV) radiation, X-rays and gamma rays. T-rays shown in the spectrum represent the terahertz (THz) radiations. This region of frequency (10^{11}Hz to ~10^{13} Hz) had remained the last unexplored region between long wavelength and visible electromagnetic radiation for a long time due to the lack of efficient emitters and receptors. Interestingly this region of the THz rays demarcates the regions of most fascinating subjects of *electronics* and*photonics*.

PROPAGATION OF ELECTROMAGNETIC WAVES

In order to study the propagation of wave, we first let the sinusoidal variation of oscillating quantities as $\sim e^{i(\vec{k}\cdot\vec{r}-\omega t)}$ that are associated with the wave. Here $\vec{k}$ is the wave vector that tells about the direction of wave propagation and also gives the wavelength of oscillations as $\lambda=2\pi/k$ and ω is the angular frequency of the oscillations that gives rise to the time period $T=2\pi/\omega$. The relation between ω and k is called the dispersion relation, based on which the wave propagation is investigated in a medium. The propagation of electromagnetic waves in any medium can be understood based on the fundamental equations of electromagnetic wave theory, i.e. the Maxwell's equations, which were established by James Clerk Maxwell in 1873 and experimentally verified by Heinrich Hertz in 1888. These Maxwell's equations are

$$\vec{\nabla}\cdot\vec{D}=\rho \tag{1}$$

$$\vec{\nabla}\cdot\vec{B}=0 \tag{2}$$

$$\vec{\nabla}\times\vec{E}=-\frac{\partial\vec{B}}{\partial t} \tag{3}$$

$$\vec{\nabla}\times\vec{H}=\vec{J}+\frac{\partial\vec{D}}{\partial t} \tag{4}$$

Citation: Hitendra K. Malik (2013). Electromagnetic Waves and Their Application to Charged Particle Acceleration, DOI: 10.5772/52246.S

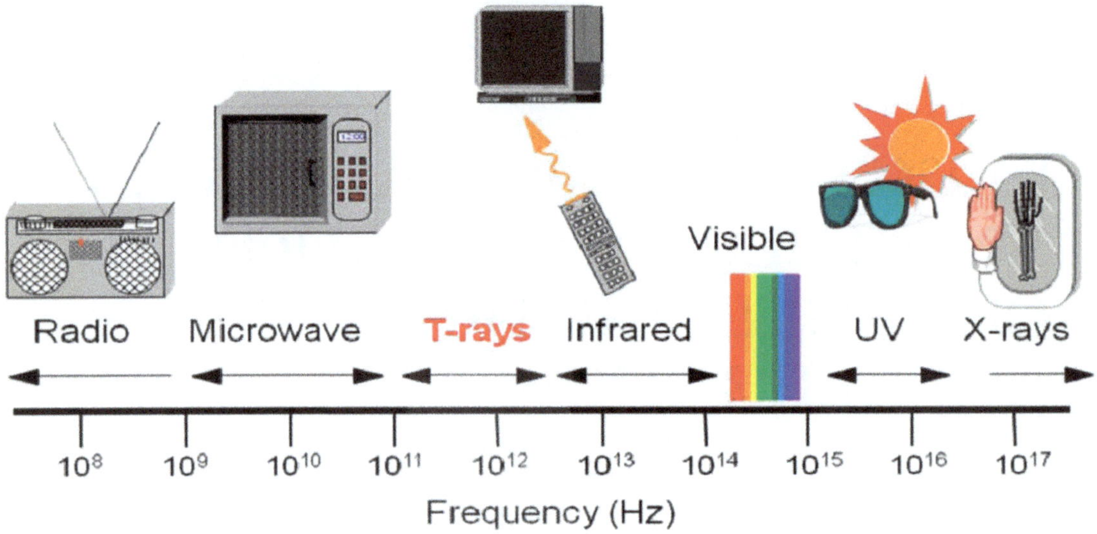

Figure 1: Electromagnetic spectrum.

Here D⃗, B⃗, E⃗, H⃗, J⃗ and ρ are respectively the electric displacement (C/m^2), magnetic flux density (Wb/m^2), electric field strength (V/m), magnetic field strength (A/m), electric current density (A/m^2) and electric charge density (C/m^3), which are the real functions of position and time. For an isotropic medium

D⃗ =εE⃗, i.e. the vector E⃗ is parallel to D⃗, and B⃗ =μH⃗ B→=μH→, i.e. the vector H⃗ is parallel to B⃗. Here ε is the electric permittivity of the medium that tells about the polarization of the medium and is determined by the electrical properties of the medium. μ is the permeability of the medium that tells about the magnetization of the medium and is determined by the magnetic properties of the medium. Hence, the properties of the medium associated with the Maxwell's equations affect the electromagnetic wave propagation. Below we discuss the propagation of electromagnetic waves in different media.

EM Wave Propagation in Free Space / Vacuum

For free space or vacuum ε=ε0, $\mu=\mu_0$, J⃗ =0J→=0 and ρ=0. If we put these values in Eqs.(1) – (4) and take the curl of Eq.(3), we get

$$\overline{\nabla}(\vec{\nabla}\cdot\vec{E})-\nabla^2\vec{E}=-\mu_0\frac{\partial}{\partial t}(\overline{\nabla}\times\overline{H}) \quad (5)$$

The use of Eqs.(4) and (1) in Eq.(5) yields the following de-coupled equation in E⃗

$$\nabla^2\overline{E}-\mu_0\varepsilon_0\frac{\partial^2\overline{E}}{\partial t^2}=0 \quad (6)$$

Similarly the wave equation for the field H⃗ is obtained as

$$\nabla^2\overline{H}-\mu_0\varepsilon_0\frac{\partial^2\overline{H}}{\partial t^2}=0 \quad (7)$$

Equations (6) and (7) describe respectively the propagation of the fields E⃗ and H⃗ in free space. For the plane uniform wave, following are the solutions of these second order homogeneous differential equations

$$\overline{E}(\vec{r},t)=\overline{E}_0e^{i(\vec{k}\cdot\vec{r}-\omega t)} \text{ and } \overline{H}(\vec{r},t)=\overline{H}_0e^{i(\vec{k}\cdot\vec{r}-\omega t)} \quad (8)$$

The above solutions should satisfy the respective wave equations. For example, when we put the solution for

E⃗ in Eq.(6) and replace ∇⃗ by ik⃗ and ∂/∂t by -iω, we get for harmonic wave with single frequency

$$(k^2-\omega^2\mu_0\varepsilon_0)\overline{E}=0 \quad (9)$$

Since E⃗ cannot be zero for the existence of wave, the wave equation will be satisfied only if

$$k^2-\omega^2\mu_0\varepsilon_0=0 \quad (10)$$

This is the dispersion relation of the electromagnetic wave in free space or vacuum. The ratio of ω and k gives rise to the phase velocity (say v) of the wave, i.e.

$$v=\frac{\omega}{k}=\frac{1}{\sqrt{\mu_0\varepsilon_0}}=3\times10^8 \text{ m/sec}=c, \text{the speed of light.}$$

Hence, it is clear that the electromagnetic wave propagates with the speed of light in free space.

In addition, we can examine the nature of the electromagnetic wave based on the directions of the fields

E⃗ ,H⃗ and the wave vector k⃗. The use of solution (8) in Eq.(3) yields

$$\vec{k}\times\overline{E}=\omega\mu_0\overline{H} \quad (11)$$

Similarly we obtain from Eq.(4)

$$\vec{k}\times\overline{\mathrm{H}}=\omega\varepsilon_0\overline{\mathrm{E}} \tag{12}$$

Equation (11) says that the field vector H⃗ is perpendicular to both k⃗ and E⃗ vectors. Also the vector E⃗ is perpendicular to both k⃗ and H⃗ vectors [see Eq.(12)]. When we combine both the equations (11) and (12), it is inferred that the vectors E⃗ H⃗ and k⃗ form a set of orthogonal vectors such that the cross product of E⃗ and

H⃗ is always in the direction of k⃗. For this reason, the energy associated with the electromagnetic waves is carried in the direction of wave propagation. On the other hand, Eq.(1) reveals that

k⃗·E⃗ =0 whereas Eq.(2) yields k⃗ ·H⃗ =0. It means the oscillations of the electric field E⃗ are perpendicular to the direction of wave propagation; the same is the case with the magnetic field. Hence, it is evident that the electromagnetic waves are transverse in nature.

EM Wave Propagation in a Dielectric

In an isotropic dielectric medium, the current density J⃗ and volume charge density ρ are zero. Also the vectors

D⃗ and B⃗ are defined as D⃗ = ε_0E⃗ +P⃗ ≡εE⃗ D→=ε_0E→+P→≡εE→ and B⃗ =μ_0H⃗ +μ_0M⃗ ≡μH⃗ B→=μ0H for the isotropic linear dielectric medium, which is polarizable and magnetic. Here the vectors P⃗ and M⃗ give respectively the measure of polarization and magnetization of the medium. Nonetheless, for the dielectric medium it would be sufficient to remember that ε_0 and μ_0 of free space have been replaced with ε and μ. Hence, for the dielectric medium the Maxwell's equations (1) – (4) take the form

$$\vec{\nabla}\cdot\overline{\mathrm{E}}=0 \tag{13}$$

$$\vec{\nabla}\cdot\overline{\mathrm{H}}=0 \tag{14}$$

$$\vec{\nabla}\times\overline{\mathrm{E}}=-\mu\frac{\partial\overline{\mathrm{H}}}{\partial t} \tag{15}$$

$$\vec{\nabla}\times\overline{\mathrm{H}}=-\varepsilon\frac{\partial\overline{\mathrm{E}}}{\partial t} \tag{16}$$

Following the similar procedure as done in the case of free space, the wave equations for the fields

E⃗ and H⃗ are obtained as

$$\nabla^2\vec{E}=\mu\varepsilon\frac{\partial^2\vec{E}}{\partial t^2},\ \nabla^2\vec{H}=\mu\varepsilon\frac{\partial^2\vec{H}}{\partial t^2} \tag{17}$$

A comparison of these wave equations with Eqs.(6) and (7) reveals that the phase velocity v of the wave in a linear dielectric medium is

$$v=\frac{1}{\sqrt{\mu\varepsilon}}=\frac{1}{\sqrt{\mu_0\mu_r\varepsilon_0\varepsilon_r}}=\frac{c}{\sqrt{\mu_r\varepsilon_r}} \tag{18}$$

From the above equation, it is clear that the propagation velocity of an electromagnetic wave in a dielectric medium is less than that in free space. Also the refractive index, say n, can be evaluated as $n=\frac{c}{v}=\sqrt{\mu_r\varepsilon_r}$. Since for a non-magnetic dielectric medium μr≈1, the refractive index is simply given by square root of the relative permittivity, i.e. n≈ε_r. This is also true for most materials as for them μ≈μ_0 and hence μr≈1.

EM WAVE PROPAGATION IN A CONDUCTOR

We consider a conducting linear and isotropic medium whose permeability is μ, permittivity is ε and the conductivity is σ. In the cases of vacuum and dielectrics or insulators, the conductivity is zero and hence the current density J⃗ was neglected in the Maxwell's equations. Moreover, the free charge density ρ was taken to be zero in these cases. In the case of conductors, the flow of charge however is not independently controlled and the current density in general cannot be neglected. Since any free charge supplied to a conductor gets dissipated, we can rather take ρ=0. This can be seen based on the continuity equation $\frac{\partial\rho}{\partial t}+\vec{\nabla}\cdot\vec{J}=0$. The use of Ohm's law J⃗ =σE⃗ and Gauss law of electricity $\vec{\nabla}\cdot\vec{D}=\rho$ in this equation leads $\frac{1}{\rho}\frac{\partial\rho}{\partial t}=-\frac{\sigma}{\varepsilon}$, the integration of which gives $\rho(t)=\rho(0)e^{-\frac{\sigma}{\varepsilon}t}$ together with ρ(0) as the initial free charge density. This relation shows that if we put some free charge on a conductor, it will flow out to the edges in a characteristic time τf≡εσ. For a perfect conductor this characteristic time τf=0 as σ=∞, and for a good conductor τf will be much less than the other relevant times, for example 1ω in an oscillatory system, i.e. τf<<1ω. Under this situation, we can write the Maxwell's equations as

$$\vec{\nabla}\cdot\overline{\mathrm{E}}=0 \tag{19}$$

$$\vec{\nabla}\cdot\overline{\mathrm{H}}=0 \tag{20}$$

$$\vec{\nabla}\times\overline{\mathrm{E}}=-\frac{\partial\overline{\mathrm{B}}}{\partial t}=-\mu\frac{\partial\overline{\mathrm{H}}}{\partial t} \tag{21}$$

$$\vec{\nabla}\times\overline{\mathrm{H}}=\sigma\overline{\mathrm{E}}+\varepsilon\frac{\partial\overline{\mathrm{E}}}{\partial t} \tag{22}$$

Taking the curl of Eq.(21) and then making use of Eqs. (19) and (22), we get the following electromagnetic wave equation for the field E⃗ in a conductor

$$\nabla^2 \bar{E} = \mu\sigma \frac{\partial \bar{E}}{\partial t} + \mu\varepsilon \frac{\partial^2 \bar{E}}{\partial t^2} \tag{23}$$

Similarly the wave equation for the field $\vec{H}$ is obtained as

$$\nabla^2 \bar{H} = \mu\sigma \frac{\partial \bar{H}}{\partial t} + \mu\varepsilon \frac{\partial^2 \bar{H}}{\partial t^2} \tag{24}$$

In one-dimension (along z-axis) the wave equations are written as

$$\frac{\partial^2 \vec{E}}{\partial z^2} = \mu\sigma \frac{\partial \vec{E}}{\partial t} + \mu\varepsilon \frac{\partial^2 \vec{E}}{\partial t^2} \tag{25}$$

$$\frac{\partial^2 \vec{H}}{\partial z^2} = \mu\sigma \frac{\partial \vec{H}}{\partial t} + \mu\varepsilon \frac{\partial^2 \vec{H}}{\partial t^2} \tag{26}$$

If we compare Eq.(23) with Eq.(6), we notice that an additional term $\mu\sigma\partial\vec{E}/\partial t$ appears in the wave equation for the $\vec{E}$ field; the same is the case with Eq.(24) and an additional term $\mu\sigma\partial\vec{H}/\partial t$ appears. Hence, these wave equations are called modified wave equations for the electromagnetic field in a conductor. Owing to the inclusion of conductivity σ, both the additional terms are called the dissipative terms as these allow the current to flow through the medium.

We can assume the following plane wave solution (in one-dimension) to the wave equations (25) and (26)

$$\frac{\partial^2 \vec{H}}{\partial z^2} = \mu\sigma \frac{\partial \vec{H}}{\partial t} + \mu\varepsilon \frac{\partial^2 \vec{H}}{\partial t^2} \tag{27}$$

Putting the above solution of $\vec{E}$ in Eq.(25) or of $\vec{H}$ in Eq.(26) we get

$$k^2 = \mu\varepsilon\omega^2 + i\mu\sigma\omega \tag{28}$$

This relation shows that the wave vector is a complex quantity, say. With this the fields $\vec{E}$ and $\vec{H}$ become

$$\vec{E}(z,t) = \vec{E}_0 e^{-k_i z} e^{i(k_r z - \omega t)} \text{ and } \vec{H}(z,t) = \vec{H}_0 e^{-k_i z} e^{i(k_r z - \omega t)} \tag{29}$$

It is evident from the above expressions that when the electromagnetic wave propagates through a conductor, its amplitude decreases and hence the attenuation of the wave takes place. The distance through which the amplitude is reduced by a factor of 1/e is called skin depth (say δ). The skind depth is decided by the imaginary part of the wave vector, i.e. ki, as it can be seen that

$$\delta \equiv \frac{1}{|k_i|} \tag{30}$$

The real part of the wave vector determines the wavelength, phase velocity, and the refractive index in the usual manner, i.e. $\lambda = \frac{2\pi}{k_r}$, $v = \frac{\omega}{k_r}$ and $n = \frac{c}{v} = \frac{ck_r}{\omega}$. Putting $k = k_r + i_{ki}$ in Eq.(28) we obtain

$$k_r = \omega\sqrt{\frac{\mu\varepsilon}{2}}\left\{\sqrt{1+\left(\frac{\sigma}{\omega\varepsilon}\right)^2}+1\right\} \text{ and } k_i = \omega\sqrt{\frac{\mu\varepsilon}{2}}\left\{\sqrt{1+\left(\frac{\sigma}{\omega\varepsilon}\right)^2}-1\right\} \tag{31}$$

It is evident that the propagation of the wave and the skin depth depend on the properties of the conductor and the frequency of the wave. Based on the expression of k_i, this can be seen that the skin depth for the electromagnetic waves having high frequencies is smaller. Since the skin depth is a measure of how far the wave penetrates into the conductor, the high frequency waves are found to penetrate less into the conductor. For example, in the case of copper, the skin depth of approximately 6 cm is obtained at the frequency of 1 Hz and it decreases to about 2 mm if the frequency is increased to 1 KHz. The skin depth causes the effective resistance of the conductor to increase at higher frequencies where the skin depth is smaller, thus reducing the effective cross-section of the conductor. If we talk in general about the skin depth, it is the tendency of an alternating electric current to distribute itself within a conductor with the largest current density near the surface of the conductor and decreased density at greater depths. Under this situation, the electric current flows mainly at the skin of the conductor. Hence, the word *skin* comes into picture.

During the wave propagation in conductors, unlike the cases of vacuum and dielectrics, the electric field and magnetic field vectors do not remain in phase. This can be seen as follows. Taking the direction of electric field

$\vec{E}$ along the *x*-axis, we write it as $\vec{E}(z,t) = \hat{x}E_0 e^{-k_i z} e^{i(k_r z - \omega t)}$. Using this in Eq.(21), we get $\vec{H}(z,t) = \hat{y}\frac{kE_0}{\omega\mu} e^{-k_i z} e^{i(k_r z - \omega t)}$. Clearly the amplitude of field $\vec{H}$ contains k, which is a complex quantity and can be expressed in terms of its magnitude (say k′) and phase (say θk) as $k = k' e^{i\theta k}$. Here $k' = \sqrt{k_r^2 + k_i^2} = \omega\sqrt{\mu\varepsilon\sqrt{\left\{1+\left(\frac{\sigma}{\varepsilon\omega}\right)^2\right\}}}$. With this the field can be written as $\vec{H}(z,t) = \hat{y}\frac{k'E_0}{\omega\mu} e^{-k_i z} e^{i(k_r z - \omega t + \theta_k)}$. A comparison of this expression with $\vec{E}(z,t) = \hat{x}E_0 e^{-k_i z} e^{i(k_r z - \omega t)}$ clearly infers that $\theta_H - \theta_E = \theta_k$, where θ_H is the phase of the magnetic field and θ_E is the phase of the electric field. Hence, the magnetic field lags behind the electric field during the electromagnetic wave propagation in a conductor.

EM WAVES AND PLASMA INTERACTION

Our aim is to disucss the electromagnetic waves and plasma interaction in view of the particle acceleration. Hence, now we introduce the plasma as a new medium, which is sometimes referred to as the fourth state of the matter.

Plasma: Fourth State of Matter

Everybody is well aware of three states of the matter, i.e. the solid, liquid and gas. In solids, the atoms are packed very close to each other and are fixed at definite positions. These are connected with each other by the interatomic forces. The atoms of solids start oscillating about their equilibrium positions when we supply energy to them, and as a result the interatomic forces become weaker and the atoms are separated significantly. This way the solid takes the form of liquid, where the atoms or molecules override. The liquid has a specific volume but does not have precise shape. So it changes shape according to the shape of the container in which it is kept. If we further supply energy to the atoms, the interatomic forces become insignificant, the atoms get separated and start moving freely. Under this situation, the liquid takes the form of gas. In gas, the atoms are not connected with each other and hence can move in any direction. The gas neither has precise shape nor the fixed volume. It takes the shape and volume of the container in which it is kept. If more energy is supplied to the atoms (or molecules) of a gas, the electrons from the outermost level of the atoms get easily detached and hence the atoms become ionized. As a result, we are left with the collection of ions, electrons and some neutrals (unionized atoms). This collection of charged and neutral particles is referred to as plasma. This is sometimes called the fourth state of matter, as it is found in natural conditions. For example, the gases near the sun are always in ionized state that qualifies for plasma. The species of the plasma being charged are connected with each other by the electromagnetic forces. This can be understood as follows. Since the charges separated with each other set up the electric field, the plasma species produce the electric field. However, the separation of charges of plasma is not fixed (as the species do not remain stationary). So this electric field is time varying field, which will generate magnetic field according to the Maxwell's fourth equation. On the other hand, the motion of charges generates current and hence the magnetic field. In view of this, the plasma species produce time varying magnetic field that will induce electric field according to the Maxwell's third equation. Thus, it can be said that the plasma species are connected with each other by the electromagnetic fields. In view of almost equal number of ions and electrons in the plasma, the plasma as a whole is neutral. However, the plasma is quasineutral, as we cannot neglect the internal forces at the same time. Moreover, if we attempt to disturb a part of the plasma, the whole body of the plasma gets perturbed due to the connection of all the species with each other. This property is known as collective behaviour of the plasma. Therefore, an ionized gas can qualify for plasma state, if it is quasineutral and it shows collective behaviour.

Another interesting property of the plasma is its ability to shield out the field that is applied on it. This can be better understood, for example, when we insert the electrodes of a battery into the plasma. Then the positive (negative) electrode attracts the electrons (ions) whose number is decided by the charge carried by the electrode. So an electron cloud is developed around the electrode that shields / cancels the external field. The thickness of this electron cloud is known as Debye length (say, $_{\mathrm{De}}$). Since the electrons are light species compared with ions, the shielding is generally accomplished by the electrons only. It is clear that the field exists within the cloud or the Debye sphere (sphere with the radius $_{\mathrm{De}}$). Now imagine if the Debye length is much less than the dimension (L) of the plasma. Then the bulk of the plasma will remain neutral. Therefore, $_{\mathrm{De}} << L$ is the required condition for the quasineutrality. In aadtion, if the number of electrons in the Debye sphere (say, N_{De}) is much larger than unity, i.e. $N_{De} >> 1$, then the condition of collective behaviour will be fulfilled. Any distance in the plasma system is measured in terms of Debye length λ_{De} and the time is measured in terms of reciprocal of plasma frequency (say f_{pe}). The plasma frequency is nothing but the natural frequency of the plasma, the same as all the materials have their natural frequencies. Actually this is the frequency of oscillations made by the electrons about their equilibrium positions. The plasma frequency f_{pe} and the Debye length $_{\mathrm{De}}$ in SI system of units are given by

$$f_{pe} = \frac{1}{2\pi}\left(\frac{n_0 e^2}{\varepsilon_0 m_e}\right)^{1/2} \text{ and } \lambda_{De} = \left(\frac{\varepsilon_0 k T_e}{n_0 e^2}\right)^{1/2}$$

In these expressions, n_0 is the plasma density, which is the common density of ions (n_i) and electrons (n_e), i.e. $n_0 = n_i = n_e$, T_e is the electron temperature, k is the Boltzmann constant ($k = 1.38\times10^{-23}$ J/K), $e$is the electronic charge and m_e is the electron mass. In plasmas, generally we do not talk about the temperature of the ions and electrons, but we specifically focus on their energies. It means the temperature is written in terms of the energy. For example, 1eV energy of the electron would be equal to its thermal energy kT_e (for two-dimensional system, and in general in plasma physics). So

$$1\text{eV} = kT_e$$

$$\text{or} \quad 1.6\times10^{-19}(J) = 1.38\times10^{-23}T_e(\text{J/K})$$

$$\text{or} \quad T_e = 11{,}600 \text{ K.}$$

It means 1eV energy is equivalent to 11,600 K temperature. The electron temperature in laboratory plasmas generally varies from 1 eV to 5 eV. For a plasma with number density of $10^{18}/\text{m}^3$ and temperature 2 eV, the Debye length comes out to be of the order of μm and the plasma frequency of the order of GHz (10^9 Hz).

Hence, it is clear that only the electrons would be able to respond to the high frequency field of the electromagnetic waves, for example, microwaves or lasers. As mentioned, our aim is to develop an understanding for the electromagnetic waves and plasma interaction for their possible applications to the particle acceleration. Below we discuss about this topic in greater detail and summarize the research conducted in this direction. At first we talk about the phenomena that may be realized during the interaction of electromagnetic waves and plasmas.

Some Basic Phenomena

According to linear theory, only the electromagnetic wave of frequency ω higher than plasma frequency ω_{pe} can propagate through the plasma. The wave whose frequency ω is below ω_{pe} gets reflected and the one with $\omega=\omega_{pe}$ gets absorbed resonantly in collisionless plasma. The plasma itself can support several types of electrostatic and electromagnetic waves such as electron plasma wave, ion acoustic wave, electron electromagnetic waves, etc. The interaction of electromagnetic wave with plasma can take place through the exciation of such waves and in this process the exchange of energy can be possible between the electromagnetic wave and plasma species. If the amplitude of wave is much higher than its nonlinear interaction with other collective modes in plasma, plasma instabilities are dominant. On the other hand, the wave can also decay by Landau damping and if plasma is underdense ($\omega > \omega_{pe}$) then the wave can decay in electrostatic wave and some other electromagnetic waves, resulting in parametric instabilities including Raman scattering, Brillouin scattering, etc. In case of large amplitude wave, the effect of ponderomotive force also comes into picture. This is very important phenomenon in view of harmonic generation, beat wave excitation, wakefield excitation for particle acceleration, self-focusing of laser beam, filamentation of laser beam, etc.

In the theory for resonance absorption, wave propagation in the resonance layer is described either by electron-ion collisions and thermal dispersion or by nonlinear effects like wave breaking, etc. [1 – 3]. The anomalous absorption of electromagnetic waves on a surface of an inhomogeneous unmagnetized plasma was theoretically predicted by Gildenburg [4]. Later this phenomenon was confirmed experimentally, out of which some experiments have shown that large amount of power can be absorbed by magnetized plasma at the electron and ion cyclotron frequencies. A usual way of coupling transverse waves into a plasma for the purpose of such resonant absorption has been to use a magnetic beach as suggested by Stix [5]. Breizman et al. [6] have presented a self-consistent theory of the rf-wave propagation and ion motion through the resonance. An important ingredient of the problem is the ion flow along the magnetic field. The flow velocity limits the time the ions spend at the resonance, which in turn limits the ion energy gain. A feature that makes the problem nonlinear is that the flow accelerates under the effect of

$\nabla^{\rightarrow}$ B force and rf-pressure. This acceleration can produce a steep reduction in the plasma density at the resonance, resulting in partial reflection of the incident wave. The propagation and collisionless absorption of electromagnetic waves propagating in nonuniformly magnetized plasmas with regions of cyclotron resonance were computed by Kuckes [7]. He considered the particle dynamics associated with motion in a nonuniform magnetic field near cyclotron resonance explicitly and predicted the complete wave absorption above a critical plasma density.

The nonlinear behaviour of the large-amplitude plasma wave and the effect of an inhomogeneous plasma on its growth and saturation in a collisionless plasma due to the beating of two laser beams with frequencies much above the plasma frequency $_{pe}$ has been considered taking into account the modulation of the Lorentz force by the large-amplitude plasma wave as well as the temporal variation of its phase [8]. In this case, a novel parametric instability as a result of the modulation of the Lorentz force by the large-amplitude plasma wave is found when the beat frequency is twice the plasma frequency. The high phase velocity electron plasma wave excited by collinear optical mixing has been detected directly [9], where the frequency, wave number, spatial extent, saturation time, and peak amplitude were all measured experimentally and found to be in reasonable agreement with the theoretical expectations. The resonant excitation of an electron plasma wave and its effects on the density profile steepening have been theoretically studied by using a modified, warm-capacitor model [10], where the scaling laws characterizing the process were established and the wave structure and density profile were self-consistently determined.

Chang et al. [11] have observed experimentally the parametric excitation of ion acoustic waves and cyclotron harmonic waves by a high frequency electric field with frequencies near the harmonics of the cyclotron frequency. They have verified both the wave vector and the frequency selection rules. Parametric excitation of longitudinal oscillations of plasma was studied by Kitsenko et al. [12] in a weak alternating electric field with frequency ω_0 close to that of electron-ion hydrodynamic longitudinal oscillations of cold plasma, $\omega_1 = \omega_{LH}\sqrt{\left[1+(m_i/m_e)\cos^2\theta\right]}$, where

ω_{LH} is the lower hybrid resonance frequency. In their study, the angle θ between the direction of propagation of the oscillations and the magnetic field was close to $\pi/2$ and it was shown that oscillations can be excited in the plasma with frequencies much less than ω_0, if the drift velocity of the particles in the steady external magnetic field and the alternating electric field is less than the thermal velocity of the ions.Optical investigations have been reported of the interaction of 0.3 TW, 250 fs Ti: sapphire laser pulses with underdense plasmas created from high density gas jet targets [13]. Time resolved shadowgraphy using a 2ω probe pulse, images of the transmitted radiation and images of 1ω and 2ω side radiations were presented for various gases. Their experimental results and analysis based on a simple numerical Gaussian beam model showed that ionization-induced refraction dominates the interaction process for all gases except hydrogen. The numerical modeling has also shown that for a given laser power there exists only a narrow density range in which self-focusing can be expected to occur. On the other hand, it has been observed that the nonlinear frequency shift of a strong electromagnetic wave in a plasma due to weak relativistic effects and the

v⃗ ×B⃗ v→×B→

force can cause modulation and self-focusing instabilities [14]. Kaw et al. [15] have shown that an electromagnetic wave interacting with a plasma is subject to instabilities that leads to light filamentation. Numerical studies of beam filamentation in laser produced plasma have been presented by Nickolas et al. [16] based on a parabolic wave equation, known as the Schroedinger equation, coupled with thermal transport equations for both the ions and electrons in two-dimensions. Also the results of a numerical code have been described which models the relativistic self-focusing of high intensity laser beams in plasmas by the nonlinear relativistic dependence of the optical constants on laser intensity [17]. Here rapid relativistic self-focussing down to a beam diameter of one micron in a distance of the order of the original beam diameter was observed. They also observed the production of GeV ions moving against the laser light.

Particle Acceleration

Particle accelerators are among the largest machines built by humans. In the conventional linear accelerators (LINACs), the acceleration gradients are however limited to some tens of MeV/m. Since the energy gain of particles is the product of such acceleration gradient and the acceleration distance, we need to extend only the acceleration distances in order to reach high energies. That's why these tools for high energy physics are becoming larger and larger, and increasingly more expensive. For the first time, it was realized by Tajima and Dawson [18] that a laser beam propagating in a plasma can excite electron plasma wave, which being longitudinal can be used to accelerate electrons. To understand the underline principle for plasma based acceleration, consider the limits of conventional particle accelerators based on rf-waves propagating in corrugated metallic cavities. They are limited first by the availability of high peak power drivers and ultimately by electrical breakdown of the metal structure. These factors correspond to linear accelerating gradient of 20 – 100 MeV/m. Plasmas though are not limited by breakdown as they are already ionized and indeed can support electric fields of the order 10 – 100 GeV/m. Consequently, with regard to the energy gain of particles in accelerators, a plasma accelerator can cut down significantly the acceleration distance to boost particles from rest to several MeV over a short distance (less than the millimeter range) and still provide high quality electron beam. Thus, plasma based particle accelerators opened a new and exciting field of extreme gradient (beyond 1TV/m). There has been a tremendous progress in recent years, due to the advances in technology, particularly by the development of compact terawatt laser systems based on the technique known as chirped-pulse amplification (CPA).

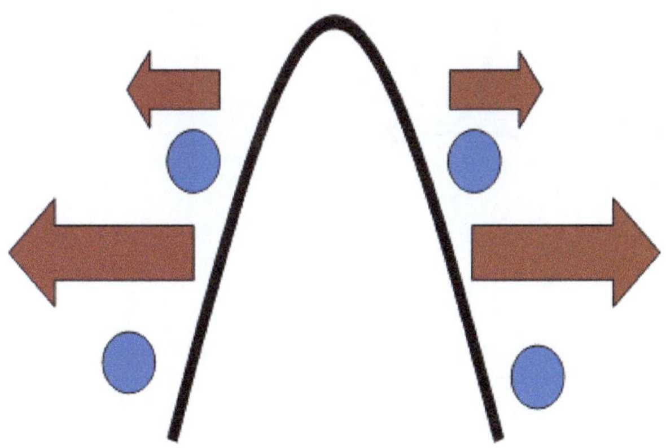

Figure 2: Laser intensity profile: Ponderomotive force.

Excitation of Langmuir Waves: Wakefield Generation

Although electric fields of the order of 1 TeV/m are readily achievable these days at the focus of a laser beam, these fields in vacuum cannot be used directly for the purpose of particle acceleration. This is because they are transverse and oscillatory in nature. However, if laser light can be used to excite Langmuir waves in plasma, these waves being longitudinal can be used to accelerate charged particles.

The motion of the electron in the presence of electric field is governed by the Lorentz force. In the case of high, nonuniform electromagnetic (or purely electric) field, the expression for Lorentz force has a second order term, which is proportional to the laser intensity gradient. This second order force term is known as the ponderomotive force, given by

$$\vec{F}_{pm} = \left\{e^2\left(1+\alpha^2\right)/4m_e\omega^2\right\}\vec{\nabla}E^2 \tag{32}$$

Here ω is the frequency of laser having the electric field E and α is the ellipticity of the laser light, which is equal to zero for the linearly polarized light and is unity for the circularly polarized light. The above expression is for ponderomotive force on a single electron. However, for the plasma the ponderomotive force on the electrons is defined for unit volume as per the following relation

$$\vec{F}_{pm} = -\left(\omega_{pe}/\omega\right)^2\left\{\left(1+\alpha^2\right)/2c\right\}\vec{\nabla}I \tag{33}$$

Thus, any spatial variation of the laser intensity I will act to push the electrons / ions from the region of higher intensity to the region of lower intensity through the ponderomotive force (Fig.2). This displacement of electrons creates large amplitude plasma wave, which is called the wake. The field corresponding to this wake, i.e. the wakefield, can reach up to 100 GeV/m provided there is a resonance between the plasma frequency $_{pe}$ and the ponderomotive force. The concept of wakefield acceleration can be understood based on the following example. When a speed-boat travels in water, it produces two types of waves viz. bow waves and wakefield waves. The bow waves are conical waves having tip at the front end of the boat. These are produced because the velocity of the boat exceeds that of the water waves. The wakefield waves are waves set up at the back (or wake) of the boat, which travels with the velocity equal to the velocity of the boat. According to the principle of the Landau damping, a floating ball dropped in the wakefield wave of the boat will get accelerated to the velocity of the boat if its initial velocity is slightly less than that of the boat. This is exactly the principle of wakefield acceleration.

Below we discuss a few methods that are used to excite plasma wave and hence the wakefield.

Laser beat wave accelerator (LBWA)

In the LBWA method, the plasma wave is excited by beating two optical waves of slightly different frequencies. Two laser waves of frequencies ω_1 and ω_2, having polarization in the same direction, traveling in preformed plasma of uniform density n_0 (corresponding plasma frequency ω_{pe}) will beat at a frequency $\Delta\omega=\omega_1-\omega_2$. If this frequency difference is exactly equal to the plasma frequency (i.e. $\Delta\omega=\omega_{pe}$), then strong Langmuir wave will be excited in the plasma by the longitudinal ponderomotive force of the beat wave. Since the beat wave moves with the laser pulse, the plasma wave will also move with a phase velocity equal to the group velocity (near light velocity) of the laser pulses. Then a properly placed bunch of electrons with a velocity slightly lesser than the laser group velocity will get accelerated by wave-to-particle energy transfer. However, in this process there is a problem of detuning of resonance condition, which is attributed to the modified plasma frequency $\left(\omega_{pe} \propto \sqrt{1/m_e}\right)$ due to the change in electron mass because of their reltivistic speeds in very large amplitude of the wakefield.

Laser wakefield accelerator (LWFA)

In beat wave acceleration scheme, it is necessary to have plasma of uniform density along with strict requirement on plasma density to exactly match with the beat wave frequency and clamping of field due to relativistic effects. Hence, laser wakefield acceleration shceme was proposed in which all the above problems are absent. For LWFA, one uses a short pulse of very high intensity. When such a high intensity laser pulse is incident on a gas, it ionizes the gas. The laser light propagates in this plasma with a velocity equal to the group velocity (vg) in plasma, which is nearly equal to the velocity of the light. The short laser pulse duration τ has a strong intensity variation in time and correspondingly in space. This leads to a strong longitudinal pondermotive force. The wavelength of this pondermotive force, and that of the density perturbation caused by it, is of the order of 2cτ. If this is made equal to the plasma wavelength (defined as $\lambda p=2\pi c/\omega_{pe}$), then high amplitude wakefields are produced due to resonance (Fig.3). Similar to the case of the boat, the laser wakefield moves with the pulse at a velocity equal to the group velocity of the laser pulse. Under this situation, a correctly injected bunch of electrons can be accelerated by the longitudinal field of the plasma waves (Fig.3), where an electron bunch is injected in the plasma wave midway between every two alternate plasma wave peaks. If the plasma wave itself moves with a phase velocity v_p and the electron beam moves with a velocity v_b, then the beam will be forced by the plasma wave to travel with a velocity equal to that of the plasma wave. This is because, if $v_b<v_p$, the electron bunch at point C (Fig.4) will start trailing from the midpoint and will experience a positive force due to electron bunch at point A. This will accelerate it in the +z direction till it attains a velocity equal to v_p

. If $v_b>v_p$, then it will start drifting towards point B and the bunch at B will repel it backward till it slows down to a velocity equal to v_p. If the electron beam has a velocity much different from that of the plasma wave, it will cross the repulsive barriers at point A or B and its velocity will keep oscillating about its mean velocity. In other words, such a beam of electrons will not have a net exchange of energy with the plasma wave. Hence, a beam of electrons traveling with a velocity slightly less than that of the plasma wave will get accelerated. Moreover, if the phase velocity of the plasma wave is relativistic, then the slight gain in velocity corresponds to a large gain in the energy.

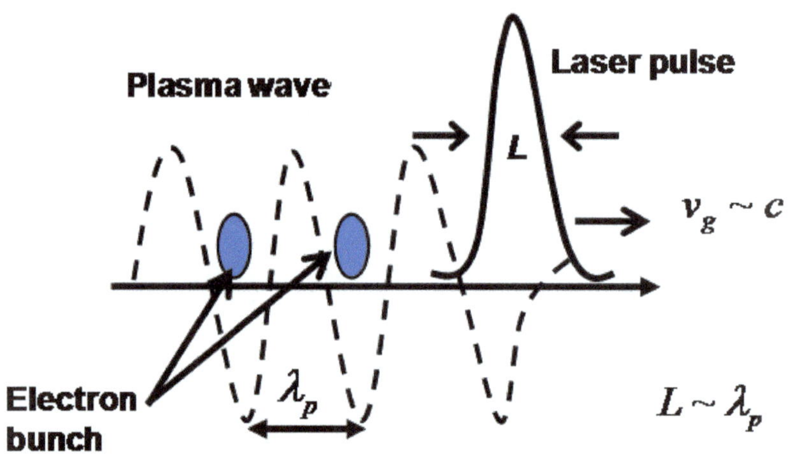

Figure 3: Schematic of LWFA.

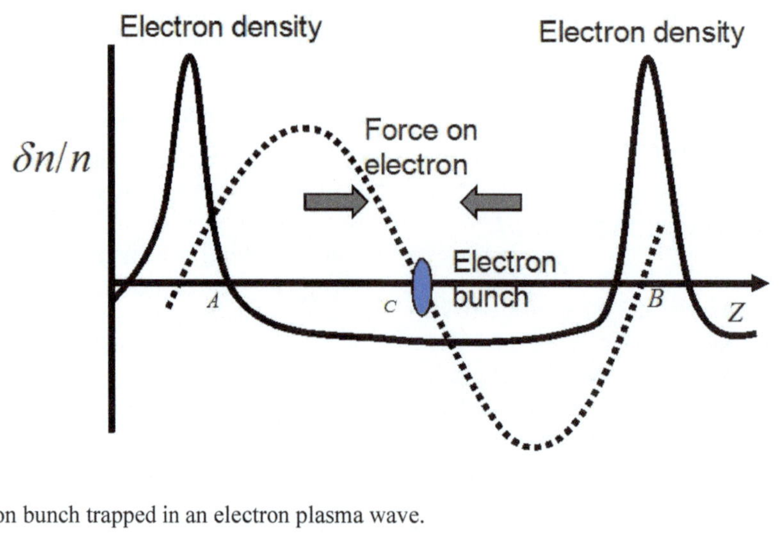

Figure 4: Force on an electron bunch trapped in an electron plasma wave.

Self modulated LWFA (SM-LWFA)

In this scheme, the electron plasma wave is excited resonantly by the modulation of the laser pulse envelope. This occurs for a laser pulse having length (L) few times longer than the plasma wavelength (λp) and pulse power larger than the power (critical power Pc) required to self-focus the laser beam. Owing to the finite pulse shape, a small plasma wave is excited non-resonantly, which results in growth of forward Raman scattering (FRS) instability. The FRS wave and the laser wave beat at the plasma frequency, which gives rise to an enhancement of the electron plasma wave. Thus, there exists an oscillating density perturbation within the pulse envelope. The laser pulse therefore sees a refractive index that is alternately peaked and dented at interval of λp/2. As the phase velocity of the laser wave depends on the density, the modulation in density gives rise to redistribution of the photon flux within the laser pulse, which leads to modulations in the envelope with a period of λp. This modulation gives rise to strong ponderomotive force with wavelength exactly equal to the plasma wavelength (as in LWFA). This strongly enhances the plasma wave amplitude. This effect grows in time, thereby transforming the initial laser pulse envelope into a train of shorter laser pulses with width of λ_n or duration proportional to $1/\omega_{pe}$ Since $\lambda_p \sim \sqrt{\frac{1}{n_0}}$ and $P_c \sim \frac{1}{n_0}$, the conditions $L > \lambda_p$ and $P > P_c$ for fixed laser parameters can usually be satisfied by operating at sufficiently high plasma density. Figure 5 shows the self-modulated scheme of laser wakefield acceleration. The advantages of the self-modulated LWFA over the standard LWFA are the simplicity and enhanced acceleration. Simplicity is that a preformed density channel and pulse tailoring are not required for the matching condition of $L \sim \lambda_p$. Enhanced acceleration is achieved for the following reasons. First, the SM-LWFA operates at a higher density, which leads to a larger wakefield ($E \sim \sqrt{n_0}$). Second, the wakefield is resonantly excited by a series of pulses as opposed to a single pulse in the standard LWFA, relativistic optical guiding allows the modulated pulse structure to propagate for several Rayleigh lengths. This extends the acceleration distance and hence the large energy gain is achieved in this scheme.

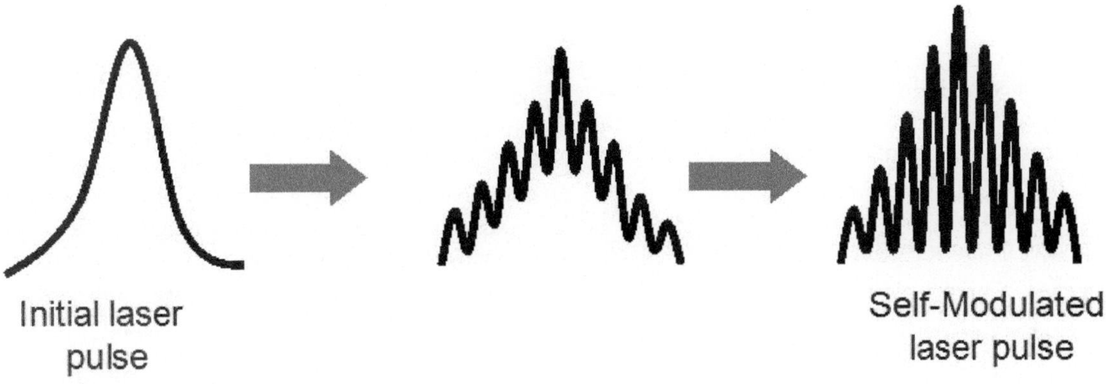

Figure 5: The self-modulated laser wakefield acceleration scheme.

So far we have seen that when a short laser pulse propagates through underdense plasma, a large amplitude plasma wave is excited in the wake of the laser pulse by the ponderomotive force associated with the temporal profile of the pulse. For tightly focused pulses $k_p \times w_0 \leq 1$, where k_p and w_0 are the plasma wave vector and the beam size at the waist, respectively), both longitudinal and radial components of the ponderomotive force generate a density perturbation, whereas in loosely focusing geometry $(k_p \times w_0 >> 1)$, only a longitudinal electron plasma wave is generated. The amplitude of the wave is maximum when $\omega_{pe} \times \tau \sim 1$ where t is the pulse duration and ω_{pe} is the plasma frequency.

Plasma Wakefield Accelerator (PWFA)

In a plasma wakefield accelerator (PWFA), the electron plasma wave is driven by one or more electron beams. Effectively the wakefield can be excited by a relativistic electron beam. This can be achieved if the electron beam terminates in a time shorter than the plasma period $1/\omega_{pe}$. In such a scheme, the ratio of energy gain to the drive beam energy (called transformation ratio) is limited to £ 2 for a symmetric driving beam in the linear regime. However, it can be increased by using an asymmetric drive beam.

Studies on Particle Acceleration

The researchers all over the world have made various attempts to accelerate the charged particles using wakefield and other mechanisms. Below we summarize the work done using lasers, microwaves and electron bunches.

Acceleration by Wakefield

The investigations on the excitation of wakefield began with the pioneering work of Chen et al. [19], and the first experimental evidence was reported by Rosenzweig and coworkers [20, 21] followed by Nakajima et al. [22]. The wakefield generation has been widely studied experimentally, analytically and using simulations [23 – 29]. Nishida et al. [30] have successfully excited wakefield in the ion wave regime with long pulse duration by employing a variety of driving bunch shapes. Later, Aossey et al. [31] observed such type of wakefield in three-component plasma also. On the other hand, efforts have been made related to wakefield excitation by relativistic electron bunch [29], [32], and coupling of longitudinal and transverse motion of accelerated electrons in laser wakefield [25]. Lotov [24] has analytically studied the laser wakefield acceleration in narrow plasma filled channels. Analytical investigations on wakefield acceleration using a dielectric lined waveguide structure showed the acceleration gradient for electrons or positrons in the range of 50 – 100 MV/m for a few nC driving bunches [33]. In another wakefield accelerator, a peak acceleration gradient of 155 MeV/m was predicted for a 2 nC rectangular drive bunch [34]. Jing et al. [35] have found transverse wakefield of about 0.13 MeV/mnC (0.2 MeV/mnC) due to X-dipole modes (Y-dipole modes) in an X-band structure generated by an electron bunch in dipole-mode wakefield in a waveguide accelerating structure. Short microwave pulses have also been used in some experiments to excite a nonlinear large amplitude ion wave at resonance absorption region [36]. This has also been suggested that the wakefield of an ultra short laser pulse can be amplified by a second laser pulse copropagating behind with duration of a few plasma wavelengths or longer [37]. Malik [38] has analytically investigated the wakefield in waveguide generated by the different types of microwave pulses with moderate intensities.

For the purpose of efficient acceleration, it is necessary to excite the wakefield of a large amplitude along with its speed nearly equal to the speed of light. The wakefield is reported to be enhanced by the nonlinearities in response of plasma to ponderomotive force of a long smooth laser pulse of relativistic intensity whose pulse length is much larger than the half of the plasma wavelength [39]. The amplitude of the laser wakefield has also been found to increase by the ionization processes of the gases at comparatively higher laser peak intensities [40]. A capillary tube can be used as a waveguide in order to enhance the interaction length [41]. Tapered plasma channels have been proposed for the enhancement of interaction length to achieve greater acceleration [42]. However, in such interactions, when the plasma wave acquires sufficiently large amplitude it becomes susceptible to instability, which is also an important issue in nonlinear plasma physics [43 – 53] in addition to other types of waves, structures and instabilities [54 – 63] including the laser produced plasmas [64] that may support different types of growing waves under the effect of high magnetic field [65].

Acceleration Using Lasers

McKinstrie and Startsev [66] have proposed that a laser field can accelerate the pre-accelerated electron significantly. However, they neglected the effect of longitudinal field of the laser pulse. On the basis of 3-D particle-in-cell simulations for the ion acceleration from a foil irradiated by a laser pulse, Pukhov [67] has shown that at the front side the laser ponderomotive force pushes electrons inward and creates the electric field by charge separation, which drags the ions. Yu et al. [68] considered the electron acceleration from the interaction of an intense short pulse laser with low density plasma and the optimum condition for the acceleration in the wake was obtained. They showed that the electron acceleration within the pulse dominates as the pulse becomes sufficiently short. By using 2-D particle-in-cell simulation, Suk [69] has studied the electron acceleration based on self-trapping by plasma wake. Sentoku et al. [70] examined experimentally the interaction of short laser pulse with dense plasma target for the proton acceleration and found that the peak proton energy increases in inverse proportion to the target thickness. Singh and Tripathi [71] have studied the laser induced electron acceleration in a tapered magnetic wiggler where the IFEL resonance condition was maintained for longer duration. With regard to the importance of polarization effects, Kado et al. [72] have observed strongly collimated proton beam from Tantalum targets when irradiated with circularly polarized laser pulses. With the help of radially polarized ultra relativistic laser pulses, Karmakar and Pukhov [73] have shown that collimated attosecond GeV electron bunches can be produced by ionization of high-Z material. They also compared the results with the case of Gaussian laser pulses and found that the radially polarized laser pulses are superior both in the maximum energy gain and in the quality of the produced electron beams. Xu et al. [74] made a comparison between circularly polarized (CP) and linearly polarized (LP) fields with regard to the laser driven electron acceleration in vacuum and found that the CP field can give rise to greater acceleration efficiency.

Acceleration using microwaves

The researchers have made efforts to use microwave field for the particle acceleration [35, 36, 75 – 85]. In microwave plasma interaction experiments, electron acceleration has been realized via the $\vec{v}_p\times\vec{B}$ process [75, 76] and that of resonance absorption during wave particle interaction [77]. In the $\vec{v}_p\times\vec{B}$ process, where $\vec{v}$ v_p is the phase velocity of the wave, an electrostatic wave (e.g. electron plasma wave) propagates in a direction perpendicular to a magnetic field $\vec{B}$. Here an electron that is trapped in the wave trough gets accelerated in the $\vec{v}_p\times\vec{B}$ direction. In these experiments, the electrons could be accelerated up to 400 eV. In another experiment, a nonlinear large amplitude ion wave was excited by using short microwave pulses at the resonance absorption region [36], where a strong electron wave was found to be excited after shut-off of the incident microwave pulse and high energy electrons got emitted and accelerated by the electron wave wakefield. Hirshfield et al. [80] have proposed a cyclotron autoresonance accelerator using rf gyroresonant acceleration, where the resonance for a TE_{11} mode was maintained along a waveguide by the applied magnetic field and group velocity axial tapers, and the maximum energy achieved by the electron beam in this process was up to 2.82 MeV. Yoder et al. [86] have measured the energy gain from a microwave inverse free electron laser accelerator including the energy change as a function of relative injection phase of the electron bunches. In this accelerator, the effective accelerating gradient was achieved as 0.43 MV/m and the gain for a 6 MeV electron bunch was observed about 360 keV. Carlsten [81] has done modal analysis and gain calculation for a sheet electron beam in a ridged waveguide slow wave structure. Kumar and Malik [87] have discussed the importance of obliquely applied magnetic field to an electron acceleration and obtained that the larger acceleration is possible when the condition $\omega_{pe} > \omega_c$ (ω_{pe} is the electron plasma frequency and ω_c is the electron cyclotron frequency) is achieved in the plasma filled waveguide. Also it was proposed to use the field of superposed mode in waveguide for the effective electron acceleration [88].

CASE STUDY: WAKEFIELD BY LASERS AND MICROWAVES

Here we take an example of wakefield excitation by short pulse lasers in an infinite plasma [38] and by the microwave pulses in a rectangular waveguide [89, 90].

Wakefield by Different Types of Laser Pulses

A laser pulse with frequency ω (= 2πf), intensity I_0 (corresponding field E_0) and pulse duration τ (= f_p^{-1}= $2\pi/\omega_{pe}$) is considered to propagate in a homogeneous plasma of density n_0 and excite wakefield E_x(corresponding potential f) behind it. The ions (density n_i) in the plasma are taken to be immobile on the time scale of the interest and the plasma response to the electromagnetic field is given by the following cold and collisionless electron fluid equations

$$\partial n_e/\partial t + \vec{\nabla}\cdot(n_e\vec{v}) = 0 \tag{34}$$

$$d\vec{p}/dt = -e(\vec{E} + \vec{v}\times\vec{B}) \quad (35)]$$

$$\vec{\nabla}\times\vec{E} = -\partial\vec{B}/\partial t \tag{36}$$

$$\vec{\nabla}\times\vec{B} = \mu_0\vec{j} + (1/c^2)\partial\vec{E}/\partial t \text{ with } \vec{j} = -n_e e\vec{v} \tag{37}$$

$$\vec{\nabla}\cdot\vec{B} = 0 \tag{38}$$

$$\varepsilon_0\vec{\nabla}\cdot\vec{E} = e(n_i - n_e) \tag{39}$$

With the help of above equations one can easily obtain the following dispersion relation for the laser propagation in the plasma $\omega^2 = c^2k^2 + \omega_{pe}^2$, from which the group velocity of the laser is found as $v_g = c\sqrt{(1-\omega_{pe}^2/\omega^2)}$. Clearly the group velocity depends on the plasma density and it can be adjusted as per the requirement.

We consider one-dimensional weakly relativistic case for the nonevolving system, i.e. when all the quantities depend only on $\xi = x - v_g t$, and take the electron density $n_e = n_0 + n_e'$ together with n_e' as the density perturbation due to the laser pulse and n_0 as the unperturbed density in a homogeneous plasma where $\partial n_0/\partial\xi = 0$. Then the fluid equations are integrated under the condition that the oscillating quantities vanish as $|\xi| \to \infty$ and also when the perturbations are not so great

$(n_e'/n_0 << 1)$. This yields

$$\frac{\partial^2\phi}{\partial\xi^2} + \left(\frac{\omega_{pe}}{v_g}\right)^2\phi - \left(\frac{c^2 e}{2m_e v_g^{\,4}}\right)E^2 = 0 \tag{40}$$

This is the general equation for the wake potential f which can directly use different envelopes of E, i.e. different shapes of the laser pulses. Here we concentrate on three types of the shapes, namely Gaussian-like (GL) pulse, rectangular-triangular (RT) pulse and rectangular-Gaussian (RG) pulse, as shown inFig.6.

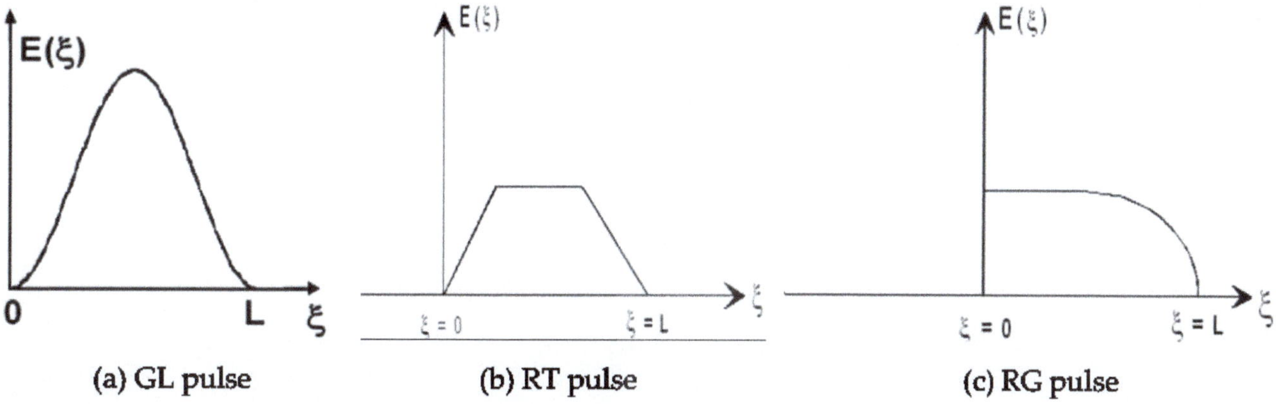

Figure 6: Different shapes of the laser pulses with pulse length L (duration τ).

The wakefield E_{GL} for the case of GL pulse is obtained as $E_{GL}(\xi) = a_1\cos(k_pL/2)\cos k_p(\xi - L/2)$, where $a_1 = (ec^2E_0^2\pi L/m_e v_g^4)[1/(L^2\alpha^2 - \pi^2)]$ and $k_p = \omega_{pe}/v_g = 2\pi/\lambda_p$ together with λ_p, as the plasma wavelength, which is described by the group velocity of the laser pulse in the plasma. The density perturbations behind the pulse are obtained as $(n_e'/n_0)_{GL} = (ea_1/2m_e v_g^2\alpha)\{2\cos(k_pL/2)\sin k_p(\xi - L/2) + (L^2\alpha^2 - \pi^2)/\alpha\pi L\}$.

The wakefield E_{RT} for the case of RT pulse is obtained as $E_{RT}(\xi) = a_2a_3\cos k_p(\xi - L/2)$, where $a_2 = (ec^2E_0^2/2m_e v_g^4\alpha)$ and

$$a_3 = \left\{ \begin{matrix} \left[c_1 - 1 - \left(\frac{2\pi c_1}{L\alpha} \right) \sin(k_p L/4) + \frac{(1-c_2)}{2} \cos(k_p L) + \frac{(1-c_2)}{2} \cos(2k_p L/3) \right] \sin(k_p L/2) \\ - \left[\frac{2\pi c_1}{L\alpha} \cos(k_p L/4) + \frac{(1-c_2)}{2} \sin(k_p L) + \frac{(1-c_2)}{2} \sin(2k_p L/3) \right] \cos(k_p L/2) \end{matrix} \right\}$$

together with

$$c_1 = L^2\alpha^2 / (L^2\alpha^2 - 4\pi^2) \text{ and } c_2 = L^2\alpha^2 / (L^2\alpha^2 - 9\pi^2).$$

The density perturbations behind the pulse are obtained as

$$(n_e'/n_0)_{RT} = (ea_2/m_e v_g^2 \alpha)[1 + a_3 \sin k_p (\xi - L/2)].$$

For the case of RG pulse the wakefield $_{ERG}$ and density perturbations $(n_e'/n_0)_{RG}$ are calculated as

$$E_{RG}(\xi) = -a_4 a_5 \cos k_p (\xi - L/2),$$

$$(n_e'/n_0)_{RG} = -(ea_4 a_5 / m_e v_g^2 \alpha)\{a_5 \sin k_p (\xi - L/2) - (L^2\alpha^2 - \pi^2)\}.$$

The constants a4 and a5 in these expressions are given by

$$a_4 = ec^2 E_0^2 / \{2 m_e v_g^4 \alpha (L^2\alpha^2 - \pi^2)\}$$

$$a_5 = \left[\begin{matrix} \{2(L^2\alpha^2 - \pi^2)\sin^2(k_p L/4) + L^2\alpha^2 \cos(k_p L/2) - \pi\alpha L \sin(k_p L)\} \sin(k_p L/2) \\ - \{\pi^2 \sin(k_p L/2) + \pi\alpha L \cos(k_p L)\} \cos(k_p L/2) \end{matrix} \right].$$

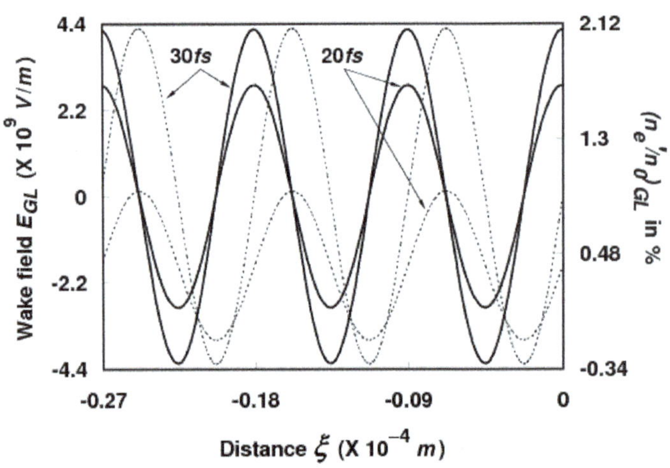

Figure 7: Wakefield E_{GL} and density perturbations $(n_e'/n_0)_{GL}$ behind the Gaussian-like pulse up to distance 3L for different pulse durations (τ = 20 ps, 30 ps) when the laser intensity is 3×10^{18} W/m² and laser frequency is 1.6 PHz.

In Fig. 7 we show the variation of wakefield E_{GL} by solid line graphs and of density perturbations $(n_e'/n_0)_{GL}$ by dotted line graphs behind the laser pulse ($\xi<0$) up to the distance ξ = 3L when f =1.6 PHz and $I_0 = 3\times10^{18}$ W/m². The wakefield of the strength of 4.24×10^9 V/m is excited by the GL pulse and the density perturbations n_e' = 0.0208 times of the unperturbed density n_0 (= 1.37×10^{25}/m³) for the pulse duration of 30 fs. A comparison of the graphs marked with 30 f_s and 20 f_s reveals that the effect of pulse duration is to increase the wakefield as well as the density perturbations. This may be attributed to the decreased plasma density n_0 for the larger pulse duration as we considered $\tau = f_{pe}^{-1} = 2\pi/\omega_{pe}$ as a condition for the wakefield excitation. For the fixed laser intensity, larger perturbations are realized in relatively lower density plasma and hence the enhanced field is obtained. Similar effects are observed for the cases of RT (Fig.8) and RG (Fig.9) pulses. Here the wakefield of the strength of 4.98×10^{9} V/m (4.28×10^{9}V/m) and density perturbations of 0.023 (0.0209) times of the unperturbed density are obtained in case of RT (RG) pulse of the same duration of 30 fs. This can also be seen that the pulses of higher intensity produce relatively larger wakefields and the density perturbations. However, a very weak effect of the laser frequency is noticed on the wakefields. In the present study of three pulses infers that the rectangular-triangular (RT) pulse is more suitable for the purpose of wakefield excitation in a homogeneous plasma.

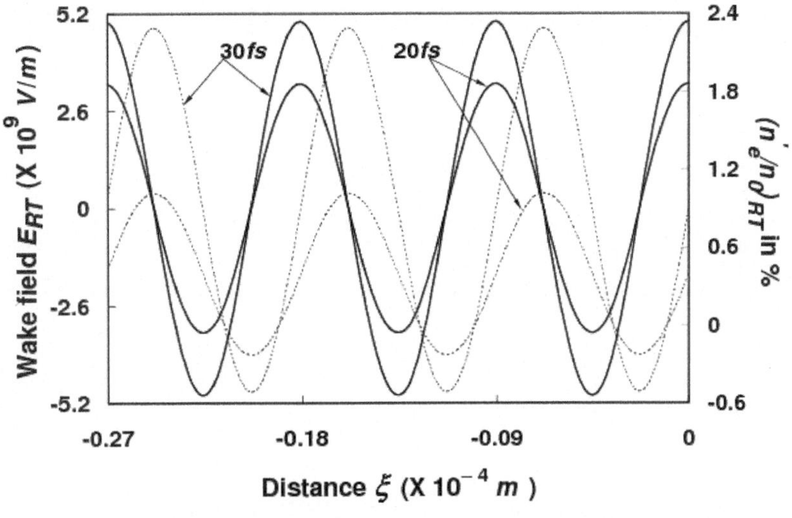

Figure 8. Wakefield E_{RT} and density perturbations $(n'_e/n_0)_{RT}$ behind the rectangular-triangular pulse up to distance 3L for different pulse durations (τ = 20 ps, 30 ps) and other parameters the same as in Fig.7.

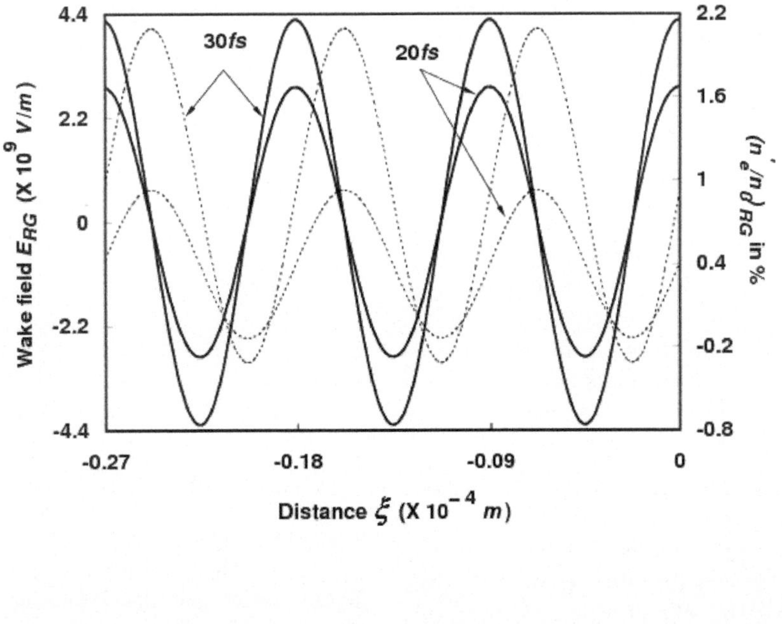

Figure 9.

Wakefield E_{RG} and density perturbation $(n'_e/n_0)_{RG}$ behind the rectangular-Gaussian pulse up to distance 3L for different pulse durations (τ = 20 ps, 30 ps) and other parameters the same as in Fig.7.

Calculation of Electron Energy Gain

In order to calculate the energy gain achieved by the electron, we proceed with the momentum equation $\frac{dp}{dt}=-eE(\xi)$ and the relativistic factor γ relation $\gamma=\sqrt{1+p^2/m_e^2c^2}$. Here E is either E_{GL}, E_{RT} or E_{RG}. We introduce a similarity variable $\eta=k_p(\xi-L/2)$ that represents the phase of the wakefield as seen by the electron. For one-dimensional motion of the electron and $\xi=x-v_gt$, we obtain $\frac{d\gamma}{dt}=-\frac{peE(\eta)}{\gamma m_e^2c^2}$ and $\frac{d\eta}{dt}=k_p\left[c(1-1/\gamma^2)^{\frac{1}{2}}-v_g\right]$ from the above relations. Dividing $d\gamma/dt$ by $d\eta/dt$ by and integrating the resultant equation by taking $dx/dt=v_x=c(1-1/\gamma^2)^{1/2}$ and $\beta_r=v_g/c$ we get

$$\gamma-\beta_r(\gamma^2-1)^{\frac{1}{2}}=-\frac{e}{m_ec^2k_p}\int E(\eta)d\eta \quad (41)$$

This is the general equation that describes the electron acceleration in the wakefield E(η). By using the expressions of wakefield E for different shapes of the laser pulses we can determine the corresponding relativistic factor (or the energy gain).

For the case of GL pulse, the integration of the resultant equation with the initial value of γ as γ_0 at ξ = 0 yields

$$\gamma-\gamma_0-\beta_r\left[(\gamma^2-1)^{\frac{1}{2}}-(\gamma_0^2-1)^{\frac{1}{2}}\right]=-\frac{e^2E_0^2\pi L}{m_e^2v_g^4k_p(L^2\alpha^2-\pi^2)}\cos(k_pL/2)\left\{\sin\eta+\sin\left(\frac{k_pL}{2}\right)\right\}.$$

. Without loss of generality we can assume $\gamma^2,\gamma_0^2>>1$. Hence,

$$\Delta\gamma_{GL}=\gamma-\gamma_0=-\frac{e^2E_0^2\pi L}{m_e^2v_g^4k_p(1-\beta_r)\left[L^2\alpha^2-\pi^2\right]}\cos\left(\frac{k_pL}{2}\right)\left\{\sin\eta+\sin\left(\frac{k_pL}{2}\right)\right\}.$$

. Therefore, the energy gain obtained by the electron during its acceleration in the wakefield excited by the GL pulse can be given by

$$\Delta W_{GL}=m_ec^2\Delta\gamma_{GL}.$$

Similarly the electron energy gain in the case of RT and RG pulses are obtained as

$$\Delta W_{RT}=-\frac{ea_2a_3}{k_p(1-\beta_r)}\left\{\sin\eta+\sin\left(\frac{k_pL}{2}\right)\right\} \text{ and } \Delta W_{RG}=\frac{ea_4a_5}{k_p(1-\beta_r)}\left\{\sin\eta+\sin\left(\frac{k_pL}{2}\right)\right\}.$$

Now we examine the effects of pulse duration (plasma density), laser intensity and laser frequency on the electron acceleration for different shapes of the laser pulses and make a comparative study.

Effect of Pulse Duration τ

We have already seen that the wakefield gets enhanced with the increased pulse duration τ for all the shapes of the laser pulses. Therefore, it is obvious that the electron will gain larger energy in the wakefield, which is excited by the pulses of longer durations. The same has been portrayed in Fig.10 for the laser intensity of 3×10^{18} W/m² and its frequency as 1.6 PHz. A comparison of the three graphs infers that the energy gains follow the trend

$W_{RT}>W_{RG}>W_{GL}$. The increased gain for the longer pulse durations is attributed to the enhanced plasma wavelength λ_p. Since λ_p is independent of the pulse shapes, the electron gets larger energy for the increasing τ for all types of the pulses irrespective of their shapes. This can also be seen from this figure that the change in energy gain is faster when the pulses of longer durations are employed for the wakefield excitation.

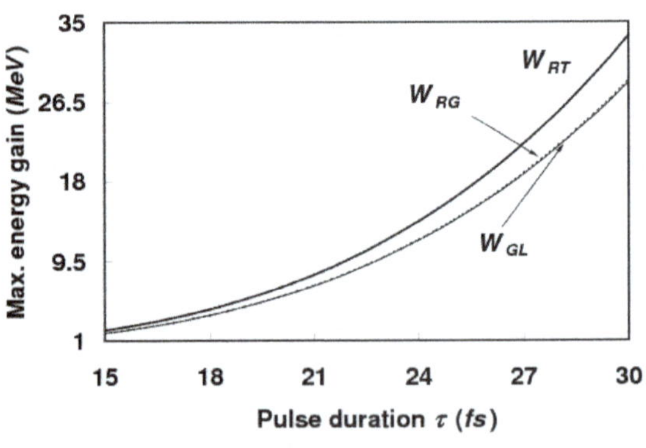

Figure 10: Dependence of maximum energy gain of electron on the laser pulse duration for the same parameters as in Fig.2. W_{RT} is the gain in case of rectangular-triangular pulse, W_{RG} is for rectangular-Gaussian pulse and W_{GL} is for Gaussian-like pulse.

Effect of Laser Frequency f

The effect of laser frequency f on the maximum energy gain attained by an electron is shown in Fig.11, from where it is evident that the gain is larger in case of RT pulse. Moreover, the slopes of the graphs reveal that the effect of laser frequency is more significant in the case of RT pulse in comparison with RG and GL pulses. Since the wakefield and plasma wavelength show weak dependence on the frequency f, it is worth clarifying the main factor that leads to significant increase in the electron energy gain with f. Actually a slight change in v_g due to f causes a greater change in the factor $(1-\beta_r)$ appearing in the denominator of the energy gain expressions. Since v_g increases for the larger frequencies, the gain gets larger with the increasing laser frequencies.

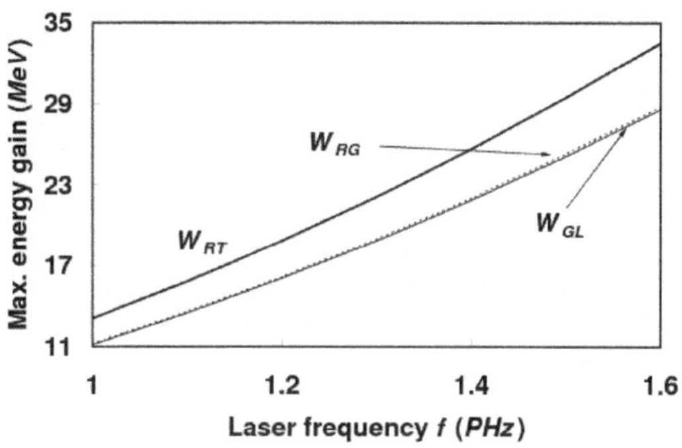

Figure 11: Dependence of maximum energy gain of electron on the laser frequency when the laser intensity is 3×10^{18} W/m^2 and pulse duration is 30 fs. W_{RT}, W_{RG} and W_{GL} have the same meaning as in Fig.10.

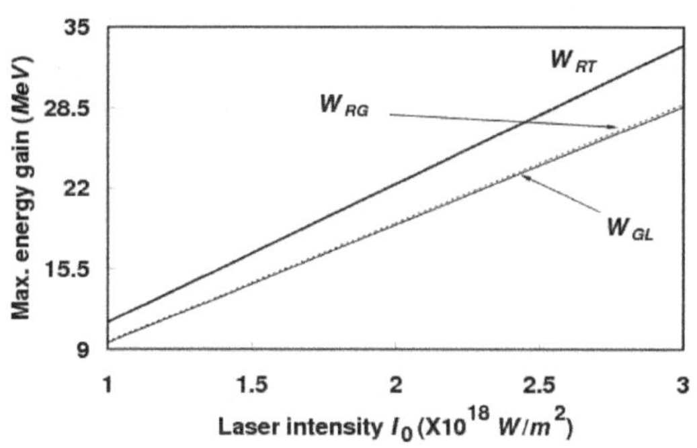

Figure 12: Variation of maximum energy gain of electron with the laser intensity when the pulse duration is 30 fs and laser frequency is 1.6 PHz. W_{RT}, W_{RG} and W_{GL} have the same meaning as in Fig.10.

Effect of Laser Intensity I_0

The expressions of wakefield for all types of the pulses show that the wake amplitude is directly proportional to

E_0^2, i.e. the intensity of the laser. It means higher intensity pulses will excite larger amplitude wakefield owing to the larger density perturbations in the plasma. However, we cannot indefinitely increase the amplitude of the wakefield because there is a limit on the maximum field that a plasma can support. Figure 12 shows that the maximum energy gain is increased from 9.5 MeV to 33.5 MeV, when the laser intensity is raised from 1×10^{18} W/m^2 to 3×10^{18} W/m^2 in case of RT pulse. A comparison of all the graphs shows that the RT pulse supersedes and gives the best results. Also the difference in energy gain becomes more and more significant when the intensity of the pulses is increased. The better results, i.e. higher amplitude wakefield and larger energy gain, obtained in case of RT pulse having smooth/fast rising time are consistent with the observations of Bulanov et al. [91] where he observed regular wakefields by a pulse with sharp steepening of its leading front.

Wakefield by Different Microwave Pulses in Waveguides

Here we present some results on wakefield excitation in a waveguide by different shapes of the microwave pulses, i.e. GL, RG and RT pulses.

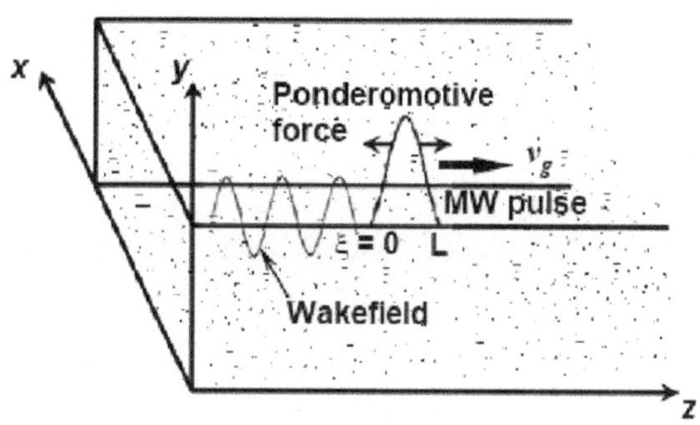

Figure 13: Schematic of wakefield generation in plasma filled rectangular waveguide by microwave pulse. Here

$\xi = z - v_g t$ and L is the pulse width.

We consider that a microwave pulse of pulse duration τ at a frequency f propagates in a plasma filled b×h rectangular waveguide. This pulse resonantly excites the wakefield (corresponding potential f) in the waveguide under the action of ponderomotive force (Fig.13), when the pulse duration τ matches with the inverse of the plasma frequency, i.e. $\tau = f_{pe}^{-1}$. The electric and magnetic fields associated with the microwave are represented by $\vec{E}$ and

$\vec{B}$. We use the Maxwell's equations and obtain the group velocity of the microwave pulse as $v_g = c\left(1-\omega_p^2/\omega^2 - \pi^2c^2/b^2\omega^2\right)^{1/2}$ that coincides with the phase velocity of the wakefield.

For the rectangular waveguide, we take the distribution of the microwave field as

$$\vec{E} = \hat{y}E(\xi)\sin\left(\frac{\pi x}{b}\right) \text{ and } \vec{B} = \hat{x}B_x(\xi)\sin\left(\frac{\pi x}{b}\right) + \hat{z}B_z(\xi)\cos\left(\frac{\pi x}{b}\right).$$

Using these relations in the basic fluid equations, we integrate them under the conditions that all the oscillating quantities tend to zero as $\xi\to\infty$ under the weakly nonlinear theory. With this we get the following equation

$$\frac{\partial^2\phi}{\partial\xi^2}+\frac{e}{2m_e v_g{}^2}\left(\frac{\partial\phi}{\partial\xi}\right)^2+\frac{\omega_p{}^2}{v_g^2}\phi-\left(\frac{c^2 e}{m_e v_g{}^2}\right)\sin\left(\frac{\pi x}{b}\right)F(\xi)=0, \quad (42)$$

Where

$$F(\xi)=\frac{1}{2}\left[\left(\frac{1}{v_g{}^2}-\frac{1}{c^2}\right)E^2(\xi)-B_z{}^2(\xi)\right].$$

This equation can be viewed as the equation governing the forced (driven) harmonic oscillator. Here the last term is the force term that evolves due to the microwave field (ponderomotive force) and drives the wake in the plasma. The third term is proportional to f and hence its coefficient determines the natural frequency of the wake. The second term is the damping term through which the nonlinearity enters the system as it is proportional to square of

$\partial\phi/\partial\xi$ (nonlinear term). Thus, the wake with potential f is evolved in the plasma as a combined contribution of each term of Eq.(42).

We can use the information related to the shape of the pulse through the last term of Eq.(42) via the coefficient

$F(\xi)$. Using the fourth-order Runge-Kutta method we simulate this equation for the above-mentioned three types of the pulse shapes. Here we take different profiles for the electric field of the pulse keeping in mind its shape and calculate $B_z(\xi)$ with the help of Maxwell's equation. The relation between $B_z(\xi)$ and $E(\xi)$ thereby comes out to be

$$B_z(\xi)=\frac{1}{v_g}\frac{\pi}{b}\int E(\xi)d\xi .$$

Results on Wakefield in the Waveguide

As mentioned, we solve Eq.(42) numerically and obtain the potential f from which we look for the wakefield amplitude for the mentioned three types of the pulse shapes. Figures 14 – 16 show the profile of wakefield generated by GL pulse, RG pulse and RT pulse, respectively. It can be easily seen that the amplitude of the wakefield is the largest in the case of RT pulse and is the least for the case of GL pulse; in other words, the wakefield amplitude follows the trend $E_{RT}>E_{RG}>E_{GL}$.

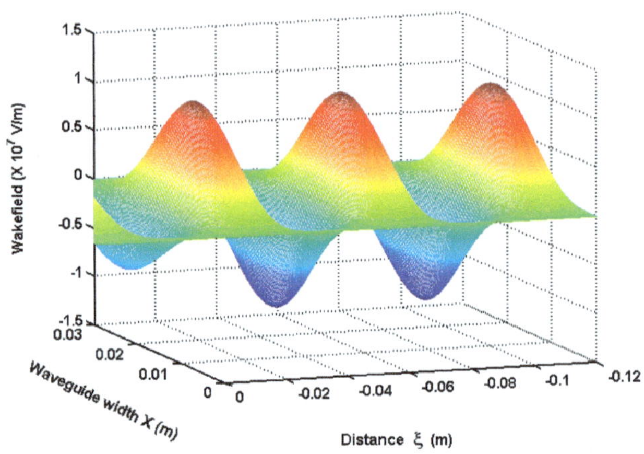

Figure 14: Variation of wakefield generated by microwave GL pulse in a waveguide for microwave intensity

$I=2GW/m^2$, frequency f=30GHz, pulse duration τ=2ns, plasma density $n_0=4.5\times10^{17}m^{-3}$and waveguide width b=0.03m

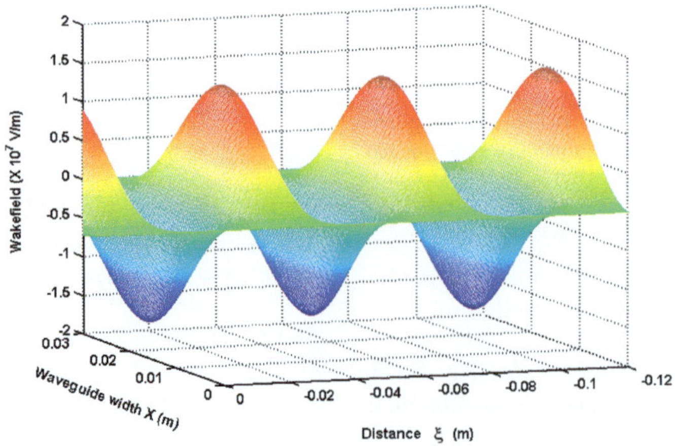

Figure 15: Variation of wakefield generated by microwave RG pulse in a waveguide for the same parameters as in Fig.14.

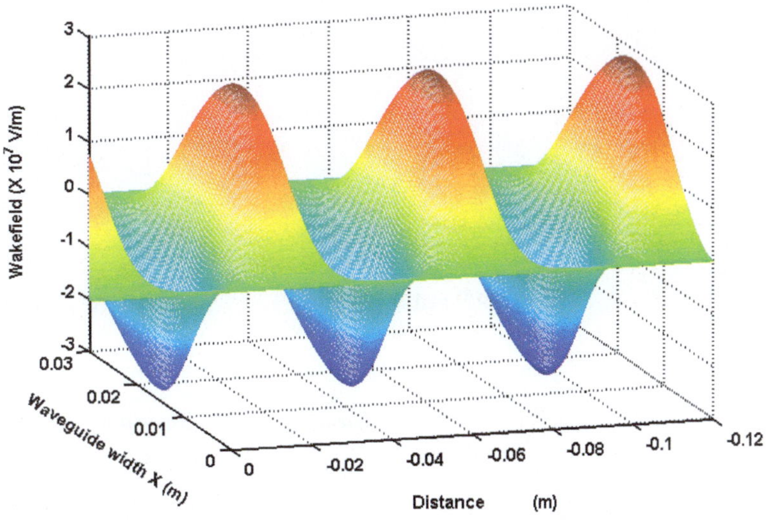

Figure 16: Variation of wakefield generated by microwave RT pulse in a waveguide for the same parameters as in Fig.14.

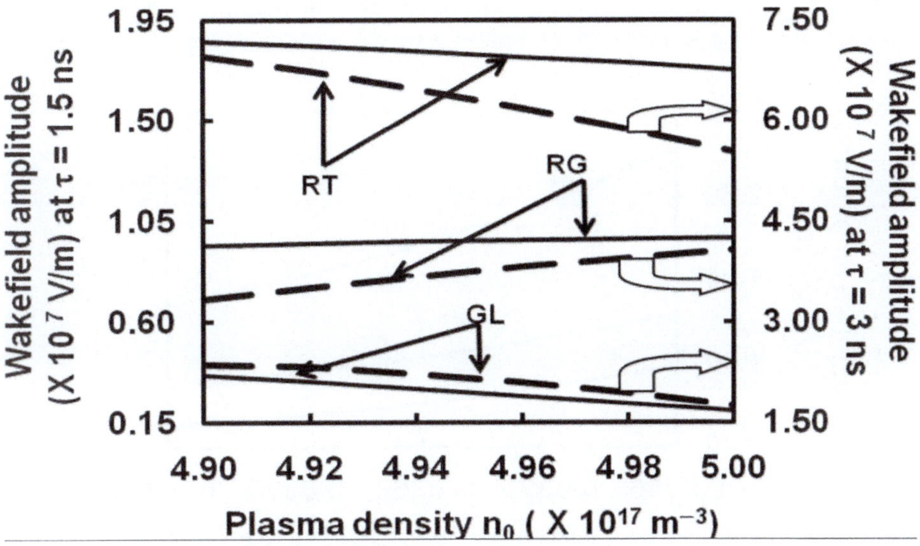

Figure 17: Dependence of wakefield amplitude on plasma density n_0 for two different microwave pulse durations τ=1.5ns (solid line graphs, left axis) and τ=3.0ns (dashed line graphs, right axis), when the other parameters are the same as in Fig.14.

Dependence of the wakefield amplitude on the plasma density for two different pulse durations is shown in Fig.17, from where it is found that the amplitude is increased for the higher plasma density in the waveguide for the case of RG pulse. However, the opposite trend is realized for the other types of the pulses (GL and RT pulses); the wakefield amplitude remains the largest in case of the RT pulse. A comparison of slopes of the graphs yields that the RT pulse shows stronger dependence on the plasma density in comparison with the other types of the pulses. With regard to the effect of pulse duration, we notice that the larger wakefield is obtained for the case of longer pulse durations; this is true for all types of the pulses.

In Fig.18, the effects of microwave frequency and its intensity are studied on the wakefield amplitude, where it is seen that an increase in the frequency leads to an enhancement in the wakefield amplitude for the cases of RG pulse and GL pulse; opposite is true for the RT pulse, which also shows a strong dependence (slope 0.097 at I=2×109W/m^2) on the frequency as compared with the other pulses. With regard to the effect of microwave intensity, we observe that the larger wakefield is obtained for the higher microwave intensity. This is further evident that the amplitude is modified at a faster rate in the case of RT pulse in comparison with the other pulses. Generally, we can conclude that the RT pulse is most sensitive to microwave frequency and intensity.

It is worth noticing from Figs. 14 – 18 that tens of MV/m wakefield is attained with the use of moderate intensity microwave pulses. Therefore, in view of the effect of microwave intensity (Fig.18), it is expected that the wakefield of the order of GV/m can be generated if the microwave pulses of intensity ~ TW/m^2 are available. Since the wakefield of this order is generally obtained by ultra high intensity lasers in usual wakefield generation schemes, the present mechanism of exciting wakefield in the waveguide by microwave pulses seems to be more effective and feasible as it can reduce the cost of accelerator and also it will provide an additional controlling parameter (the waveguide width).

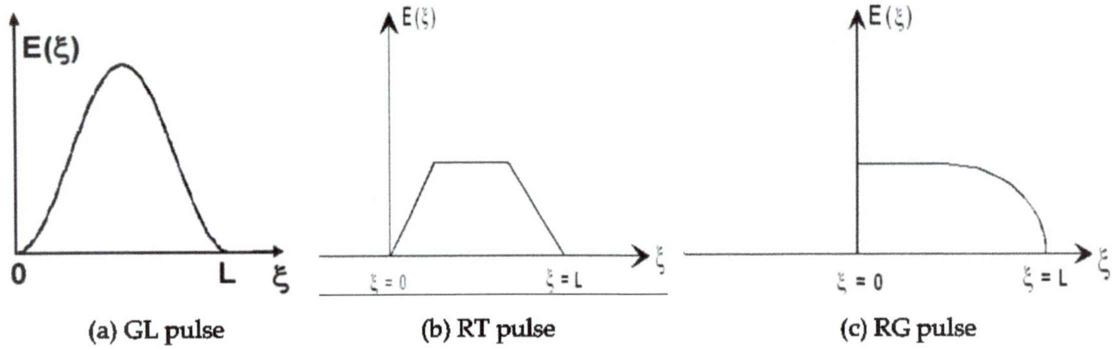

Figure 18: Variation of wakefield amplitude with microwave frequency f for two different microwave intensities

$I=2\times10^9 W/m^2$ (solid line graphs, left axis) and $I=1\times10^9 W/m^2$ (dashed line graphs, right axis), when the other parameters are the same as in Fig. 14.

Variation of the wakefield amplitude with the waveguide width is shown in Fig.19, where it is observed that the amplitude is decreased with the increase of waveguide width; same result was obtained in an analytical calculation [38]. It means the larger wakefield can be obtained for the case of plasma filled narrower waveguide. A comparison of the slopes of the graphs reveals that the wakefield amplitude changes at a faster rate in the case of RT pulse. Therefore, the RT pulse is found to be more sensitive to the waveguide width. Thus, we can conclude that a plasma filled narrower waveguide is best suited for an effective wakefield excitation and the significant particle acceleration if the RT pulse is used.

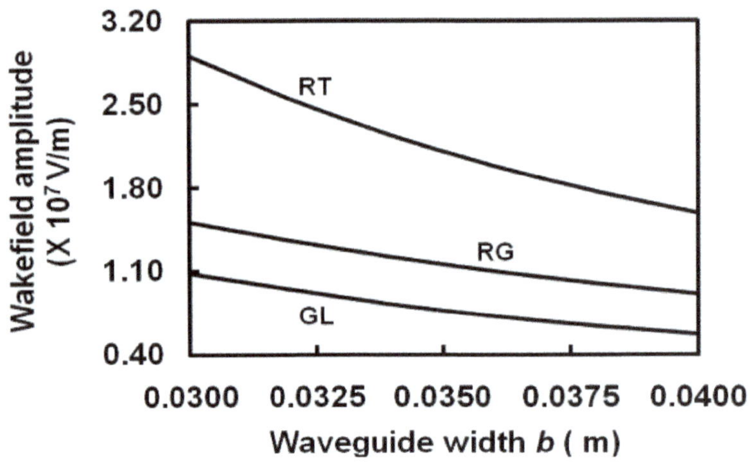

Figure 19: Variation of wakefield amplitude with waveguide width. The values of intensity, frequency, pulse duration and plasma density are given in the caption of Fig. 14.

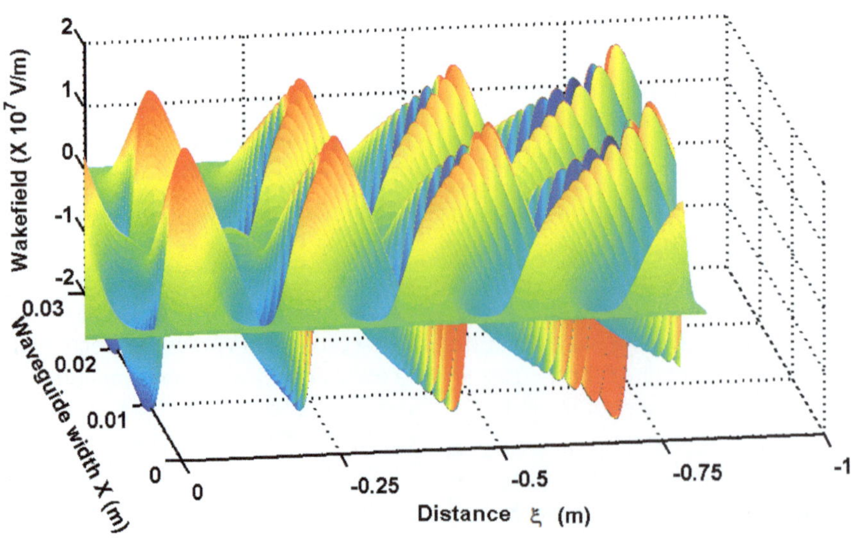

Figure 20: Profile of wakefield EW in a plasma filled waveguide near cutoff conditions when f=6.76GHz, τ=0.9ns, fc=5.11GHz, $I=1\times10^9 W/m^2$ and b=0.03m.

Wakefield near Cutoff Conditions

This has already been explored that the wakefield of larger amplitude is obtained for the smaller waveguide width and the longer pulse duration. For fixed microwave frequency, the cutoff frequency f_c gets higher under the effect of decreased width b and plasma density n_0. These effects can be viewed as if the microwave frequency is brought near the cutoff frequency f_c. Therefore, it is of much importance to investigate the wakefield structure near cutoff conditions, i.e. when f is near f_c. These results are presented in Figs.20 and 21 for f=6.76GHz whereas fc=5.11GHz. Fig.20 shows that the amplitude of wake wave gets increased under this situation as we move away from the microwave pulse, i.e. for decreasing values of ξ. This is further noticed that various peaks develop along the waveguide width during the growth of wakefield amplitude and it becomes unstable. Thus, it is plausible that some instability develops near the cutoff conditions. In order to further investigate this effect, we show in Fig.21 the maximum distance by which this growth occurs. Here we observe that the amplitude gets terminated around ξ=3.7L in the plasma and the field breaks down.

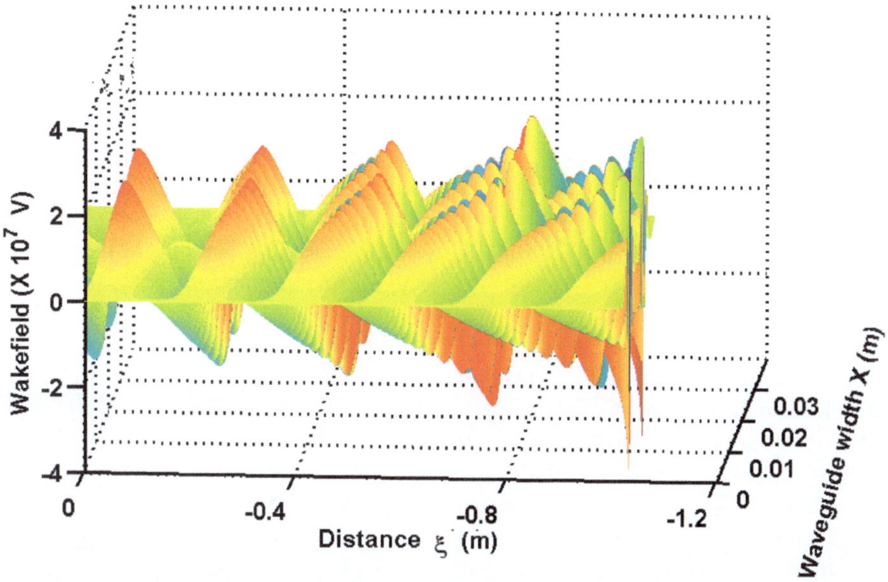

Figure 21: Profile of wakefield corresponding to Fig.20, showing its cutoff around maximum distance of ξ=3.7L from the microwave pulse.

These results suggest that the microwave of higher frequency should be employed for avoiding any instability in the system and the wakefield can be effectively used for the purpose of particle acceleration. The present mechanism of wakefield generation can be realized experimentally if we use wider waveguide filled with higher plasma density. The high density plasma can be produced in the waveguide under the action of Electron Cyclotron Resonance (ECR). However, under such situation short microwave pulses would be more effective in order to resonantly excite the plasma wake wave.

CONCLUDING REMARKS

The electromagnetic waves were classified based on their frequency and a small region (10^{11}Hz to $\sim 10^{13}$ Hz) that remained the last unexplored region was introduced as the THz rays. While explaining the propagation of EM waves, it was shown that their propagation velocity depends on the properties of the medium, and unlike the cases of vacuum and dielectrics, their electric field and magnetic field vectors do not remain in phase in the case of conductors. Very fascinating phenomenon of skin depth was discussed in the conductors where the wave vector was found to become a complex quantity and its imaginary part led to the attenuation of the wave. It was mentioned that the EM waves can propagate through the plasma medium if their frequency is larger than the plasma frequency. In the case of plasmas, another interesting phenomenon of wakefield excitation by the laser or microwave pulses was talked about in detail. Two case studies were conducted using the laser pulses and microwave pulses with different envelopes. It was shown that moderate intensity microwave pulses can also generate the wakefield effectively and accelerate the particles to sufficiently large energies. Moreover, the high cost laser systems can be replaced with microwave systems if the microwave pulses can be tailored properly.

APPLICATIONS AND FUTURE PROSPECTS

The electromagnetic wave and plasma interaction has diverse applications in different fields such as nuclear fusion, particle acceleration, heating of ionospheric and laboratory plasmas by radio waves etc. along with controlled fusion applications to ITER (International Thermonuclear Experimental Reactor), frequency upshifting, resonance absorption, laser focusing and defocusing, material processing, generation of X-ray, THz and microwave radiations, higher order harmonic generation, laser filamentation etc. With the inclusion of plasma, the performance of some devices such as backward wave oscillator (BWO), travelling wave tube (TWT) amplifiers, gyrotrons and other microwave tubes have been found to increase.

The use of laser plasma accelerators has been made in radioisotope production through (γ,n) reactions with laser accelerated electron bunches in the range tens of MeV [92, 93]. The short pulse nature and high charge of the accelerated bunches also has applications in the production of coherent THz radiation, which is achieved when femtosecond electron bunches cross the plasma vacuum boundary and emit transition radiation [94 – 96]. The generation of THz radiation has interesting applications in nonlinear THz spectroscopy, material characterization, imaging, topography, remote sensing, chemical and security identification [97, 98] etc. Another application of accelecerated electron beams / bunches is in the generation of femtosecond X-ray pulses produced by the betatron radiation emitted when the electron beam propagates through the plasma. By making an array of nanoholes on an alumina target, X-ray emission from laser produced plasma can be greatly enhanced even in soft X-ray energy regions (< 0.25 keV). The enhancement increases as the ionization level of Al becomes higher and the X-ray wavelength becomes shorter. Over 50 fold enhancement was obtained at a soft X-ray wavelength around 6 nm, which corresponds to the emission from $Al^{8+,9+}$ ions. X-ray pulse duration was 17 ps, which is much shorter than that obtained by using the prepulse technique [99]. Towards the generation of other types of electromagnetic radiation, Tripathi and Liu [82] have proposed a dielectric-lined waveguide for the free-electron laser emission in millimeter wavelength band. Farokhi et al. [83] have presented a linear theory for a free electron laser with a three-dimensional helical wiggler and axial magnetic field in the collective regime in a configuration consisting of an annular electron beam propagating inside a cylindrical waveguide. For the generation of high power (140 MW) subnanosecond (75 ps) microwave pulses in the range of 38 – 150 GHz, Yalandin et al. [84] have done experiments on coherent stimulated radiation from intense, subnanosecond electron bunches moving through a periodic waveguide and interacting with a backward propagating TM_{01} wave. Hayashi et al. [85] have also designed a two-stage ferroelectric electron gun and a peak power of 5.9 MW microwave radiation was observed when a 100A 450 kV electron beam was used.

Research directed towards the development of high power electromagnetic radiation sources accounts for much of the current interest in the plasma filled waveguides. Plasma filled waveguides may also be used for the transportation of electromagnetic energy and charged particles, and in the basic study of plasma phenomena. In spite of such extensive work over the years, the understanding of the physics of wave plasma interaction is still an active area of research, which also finds additional applications in plasma based focused ion beams (FIB), plasma sources for negative ion beams for neutral beam injection, rf-based plasma thrusters, etc. With regard to the particle acceleration, we have carried out analytical and numerical studies on the wakefield excitation by different types of the pulses in a rectangular waveguide filled with homogeneous plasma. Our analyses reveal that moderate intensity (~ 10^9 W/m^2) microwave pulses can produce up to 100 MV/m wakefield in the waveguide if the nanosecond pulses are used. Since the amplitude of resonantly excited wakefield changes at a faster rate with the waveguide width, pulse duration and microwave intensity and it is larger for the smaller waveguide width, longer pulse duration and the higher microwave intensity in the case of rectangular triangular pulse, the significant wakefield can be excited in the waveguide and effective particle acceleration can be achieved with the use of RT pulses for which the parameters can be optimized using the present studies. This theoretical work on the contribution of different microwave and laser pulses for the purpose of particle acceleration and the THz generation [100, 101] shall induce experimentalists to develop rectangular, rectangular-Gaussian, rectangular-triangular, sawtooth and triagular pulses of appropriate lengths for accomplishing various experiments on wakefield generation, particle acceleration, and the THz generation. Through such efforts our researchers would be able to benfit the society more via the medical, scientific, and technological applications of the subject electromagnetic wave and plasma interaction.

REFERENCES

1. V. L. Ginzburg, The propagation of electromagnetic waves in plasmas, Pengamon, New York (1970).
2. N. G. Denisov, Zh. Eksp. Teor. Fiz. 31, 609 (1956); Engi. Transi. Sov. Phys. JETP 4, 544 (1957).
3. P. Koch and J. Albritton, Phys. Rev. Left. 32, 1420 (1974); J. Aibritton and P. Koch, Phys. Fluids 18, 1136 (1975).
4. V. B. Gildenburg, Sov. Phys. JETP 18, 1359 (1964).
5. T. H. Stix, Phys. Fluids 1, 308 (1958).
6. B. N. Breizman and A. V. Arefiev, Phys. Plasmas 8, 907 (2001);
7. A. F. Kuckes, Plasma Phys. 10, 367 (1968)
8. M. N. Rosenbluth and C. S. Liu, Phys. Rev. Lett. 29, 701 (1972).
9. C. E. Clayton, C. Joshi, C. Darrow and D. Umstadter, Phys. Rev. Lett. 54, 2343 (1985).
10. W. Yu and Z. -Z. Xu, Phys. Rev. A. 36, 285 (1987).
11. R. P. H. Chang, M. Porkolab and B. Grek, Phys. Rev. Lett. 28, 206 (1972)
12. A. B. Kitsenkov, I. Panchenko and K. N. Stepanov, Plasma Phys. 16 1109 (1974).

CHAPTER 14

SURFACE ACOUSTIC WAVES AND NANO–ELECTROMECHANICAL SYSTEMS

Dustin J. Kreft and Robert H. Blick

INTRODUCTION

Surface acoustic waves (SAW) follow the industrial trend of reducing the size, enhancing the speed, while enhancing the efficiency of energy coupling. Integrating this with microelectromechanical systems (MEMS) and nano-electromechanical systems (NEMS) offers a wide variety of applications such as touch screens, gas and biological sensors, and embedded RFID devices. With modern lithographic techniques, allowing the fabrication of smaller SAW devices, we now use SAWs to probe the mechanical interactions of nano structures. In particular, SAWs can be used to actuate NEMS which gives rise to many interesting phenomena including anomalous acoustoelectric currents, shock waves in suspended devices, and few electron transport, to only name a few, (Beil et al., 2008; Talyanskii et al., 1997). Today, SAWs are also used to generate a quantized current for use as a current standard. In practice two counter-propagating SAWs are used to observe a quantized acoustoelectric current. This leads to population and depopulation of discrete states (Kataoka et al., 2007). In the following we want to give an overview of the state of the art of applying SAWs to nanomechanical devices. We will also give a brief introduction to recent nanoelectromechanical systems with integrated low-dimensional electron gases, which have the potential to reveal insights into quantized acoustoelectric states.

FABRICATION

The focus for generating SAWs in this chapter will involve the fabrication of interdigitated transducers (IDT). An IDT is simple in concept but can be very involved when fine tuning a structure for engineering applications. Such topics as electronic impedance matching to RF lines, effects of bulk waves in contrast to SAWs, and increasing bandwidth will not be covered; though, this is simply a shortened list of things to consider when designing a proper IDT for engineering applications, they do fall outside the scope of this chapter. Nevertheless, another fabrication step we will consider is the use of acoustic waveguides for acoustic impedance matching of IDTs to nanomechanical devices.

Interdigitated Transducer Design

The main equation to consider when designing an IDT is:

$$v = \lambda f \tag{1}$$

Where v is the velocity of sound in the material, λ is the SAW wavelength or pitch of the IDT, and f is the frequency of the propagating SAW. The pitch of the IDT fingers is the same as the SAW wavelength, λ, which will propagate across the sample. Fig. 1 shows a simple IDT schematic along with a scanning electron micrograph (SEM) image and optical image.

Citation: Dustin J. Kreft and Robert H. Blick (2011). Surface Acoustic Waves and Nano–Electromechanical Systems, - From Microdevices to Helioseismology, Prof. Marco G. Beghi (Ed.), InTech, DOI: 10.5772/19927.

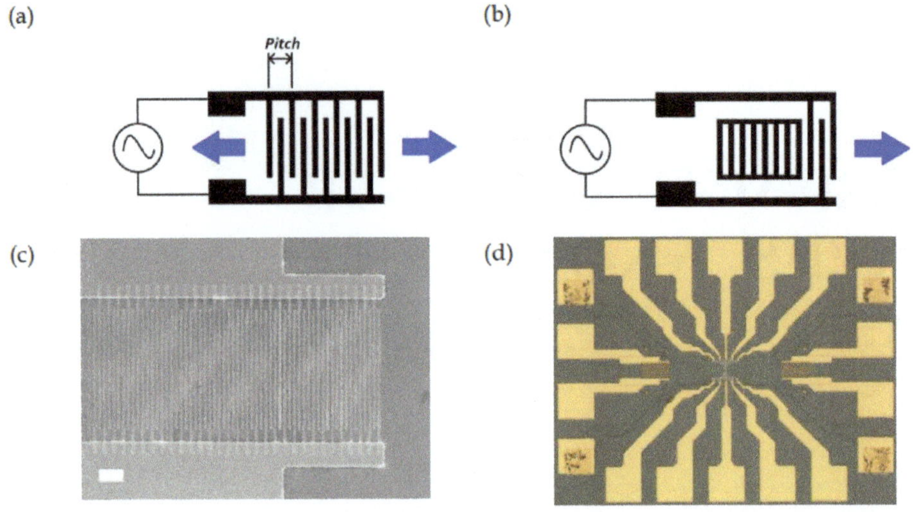

Figure 1: Schematics and scanning electron microscope images of IDTs. (a) Schematic of bidirectional IDT, (b) schematic of unidirectional IDT through the use of a reflection grating, (c) SEM image of a bidirectional IDT on GaAs with a pitch of 4 μm and a center frequency of ~715 MHz. The scale bar is 10 μm. (d) Optical image of device with an IDT on both ends and a nanomechanical device placed in the center

In Fig. 1 the arrows indicate the SAW propagation directions. It can be easily seen that the SAW will propagate in both directions away from the IDT. A singly propagating SAW direction can be achieved by placing a reflection grating on one end of the SAW. The reflection grating will have the same geometry as the IDT; that is, the same finger spacing and width. The grating distance from the IDT is ~ λ/2. The IDT fingers are typically chosen with evenly spaced fingers, where the spacing between the fingers is equal to the finger width. The finger width and finger spacing is λ/4 in this scenario and gives a metallization ratio of $\eta = 0.5$, which will generate only odd harmonics with no response of the third harmonic (Campbell, 1998). The bandwidth of an IDT is defined as $BW\% = f / N_{Pairs}$. Where f, again, is the center frequency and N_{Pairs} is the number of finger pairs of the IDT. The IDT in Fig. 1a has five pairs of fingers and is symmetric about a center point along the axis of SAW propagation. It is typical to fabricate two IDTs, one on each end of the device, so that their propagating waves can interfere either constructively or destructively across the center region of the device; this is typically the region where a Quantum Point Contact (QPC) or other structures resides. This can be seen in Fig. 1d where in the center a double quantum dot (DQD) is placed and the IDTs are placed to the left and right.

Acoustic Waveguide Design

Another detail of fabrication is the use of waveguides for the SAW. A waveguide for SAWs is a pattern that is fabricated onto the device so the SAWs can be guided into a certain region allowing a stronger SAW signal or amplitude; it is analogous to a coplanar waveguide for RF signals. When fabricating IDTs, and a device as a whole, using waveguides does not have much of a use in the areas of RFID, cellular delay lines, sensors, and other non-region specific devices. However, this chapter will focus on NEMS and the use of SAWs to interact with these devices. Since the NEMS device is orders of magnitude smaller than the IDT aperture very little SAW power can interact with the NEMS structure; it can be beneficial to include a waveguide to focus the SAW onto the nanostructure. Waveguides can be modeled using standard acoustic, or horn, waveguides with some minor tuning. There are several shapes which can be used and each shape offers its own benefits regarding a particular application. We will start by first mentioning the base equation for hyperbolic horns used in speaker systems, these horns can also be referred to as hyperbolicexponential horns or hypex horns (Salmon, 1946).

$$A = A_t[\mathrm{Cosh}(kx) + T\cdot \mathrm{Sinh}(kx)]^2 \tag{2}$$

$$z = \frac{\rho_0^{\,v}}{A_t}\left(\sqrt{1-\left(\frac{f_c}{f}\right)^2} + i\frac{f_c T}{f}\right)\left(1-\frac{f_c^2(1-T^2)}{f^2}\right)^{-1} \tag{3}$$

Eq. 2 gives the wave front area expansion of the wave as propagating through the waveguide or horn. A_t is the area of the throat, or base, of the waveguide, T is a factor describing the shape of waveguide; for T = 1 the waveguide becomes exponential in shape, and becomes canonical as $T \rightarrow \infty$, k is given as $(2\pi f_c)/v$; where v is the velocity of sound, and f_0 is the cutoff frequency of the waveguide. For hyperbolic and exponential waveguides there is no transmission below the cutoff frequency, fc, since at this point the impedance is purely reactive. Eq. 3 gives the impedance of the waveguide; all variables are defined previously with the addition of ρ_0 which is the density of the material. Fig. 2 contains three SEM images of an exponential waveguide with a tube waveguide in the center. A tube waveguide is simply a small or narrow SAW delay line. The material is a GaAs/AlGaAs heterostructure containing a two-dimensional electron gas (2DEG) and a sacrificial layer. We use a 2DEG because the high electron mobility makes this an ideal candidate for nanostructures, such as quantum dot (QD) and DQD systems. The waveguide was defined using electron-beam lithography to open up an etch region through the PMMA. The material was then wet etched with H_3PO_4:H_2O_2:H_20 into the sacrificial layer. Then diluted HCl was used to remove the sacrificial layer which resulted in a suspended region in the center and around the waveguide. This suspended region is visible as a darkened shadow indicating a height difference due to strain relaxation of the material. The suspended region forms the actual nanomechanical device. Since this device contains a DQD in the center region of the structure it is beneficial to use a waveguide to ensure a higher SAW power density coupled into that region. Since the IDT aperture of this device is 50 μm and the DQD region is roughly 600 nm at the widest QPCs.

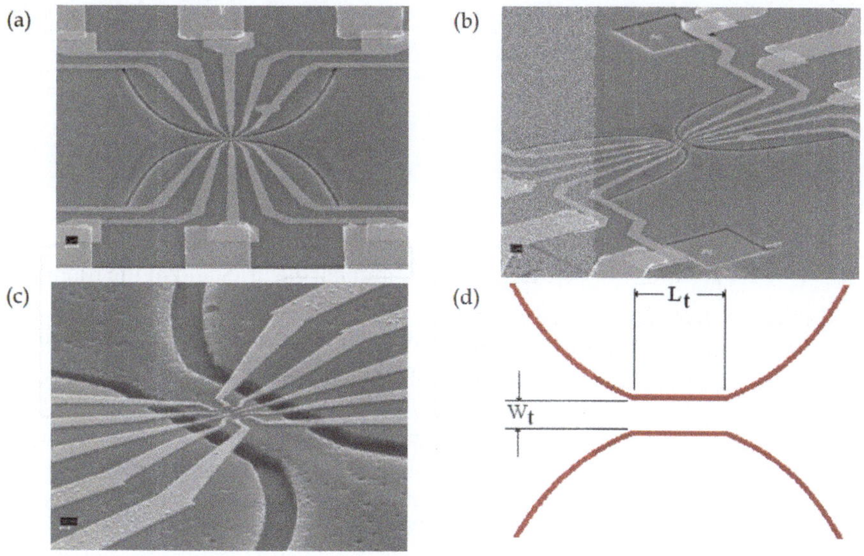

Figure 2: SEM images of an exponential waveguide with a cutoff frequency of 75 MHz and throat width of ~1.5 μm. The darker region around the waveguide shows that the area is suspended, which was achieved by an HCl etch to remove the sacrificial layer below the 2DEG, (a), (b) and (c). (d) Schematic of waveguide, center region is the throat (pipe waveguide) with width W_t and length L_t. The outer curves are the exponential portion of the waveguide.

SURFACE ACOUSTIC WAVES IN TWO DIMENSIONAL ELECTRON GASES

Surface Acoustic Wave Basics

When applying an RF signal to the IDTs an alternating electric potential is seen by the electrodes, this will create an electric field distribution within the piezoelectric substrate, see Fig. 3a. This field distribution, in turn, will cause a mechanical deformation of the material through the inverse piezoelectric effect. As stated before, this mechanical deformation will propagate away from the IDTs and continue to move along the surface of the device, see Fig. 3d. Here we neglect other forms of waves being produced such as bulk waves. Once the RF signal of the IDT is coupled into the piezoelectric substrate a Rayleigh-wave is produced. The elliptical wave propagating (Fig. 3b, d) along the surface can be described quite accurately by a traveling wave:

$$U = |U| e^{i(\omega t - kz)} e^{-k|y|} \tag{4}$$

The electric potential being created by the propagating wave can also be described in a like manner:

$$\Phi = |\Phi| e^{i(\omega t - kz)} e^{-k|y|} \tag{5}$$

In both Eqs. 4 and 5, ω is the angular frequency and k is the phase constant. The SAW penetrates into the depth of the material by about one wavelength, ~λ. This value is different in the suspended region. Hence, the actuation of nanomechanical resonators is considerably enhanced. Another property of the SAW is that the electric field created from the induced electric potential does not terminate at the surface of the material but can extend beyond by λ.

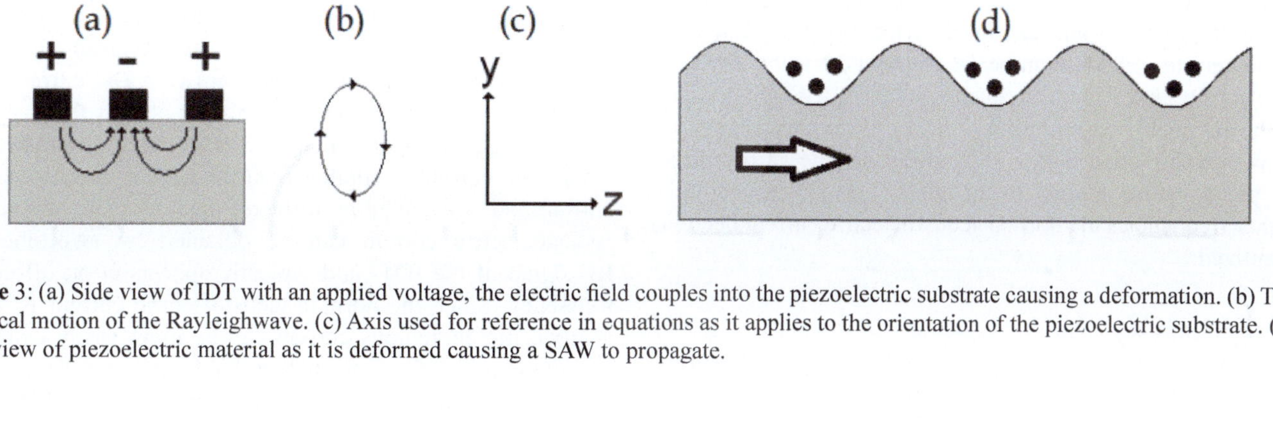

Figure 3: (a) Side view of IDT with an applied voltage, the electric field couples into the piezoelectric substrate causing a deformation. (b) The elliptical motion of the Rayleighwave. (c) Axis used for reference in equations as it applies to the orientation of the piezoelectric substrate. (d) Side view of piezoelectric material as it is deformed causing a SAW to propagate.

The set of base equations used for describing SAW phenomena are listed.

$$E_i = -\frac{\partial \Phi}{\partial x_i} \tag{6}$$

$$T_{ij} = c_{ijkl} S_{kl} + e_{nij} E_n \tag{7}$$

$$S_{kl} = \frac{1}{2}\frac{\partial u_k}{\partial x_l} + \frac{\partial u_l}{\partial x_k} \tag{8}$$

Eq. 6 is the electric field intensity that is produced from the deformed piezoelectric material from the SAW. Eq. 7 is the piezoelectric mechanical stress and Eq. 8 is the linear strain displacement.

Attenuation

SAW attenuation can be described by the following equations (Wixworth et al., 1989):

$$\Gamma = k \frac{K^2_{eff}}{2} \frac{\sigma_s / \sigma_M}{1 + (\sigma_s / \sigma_M)^2} \tag{9}$$

$$\frac{\Delta v}{v} = k \frac{K^2_{eff}}{2} \frac{1}{1 + (\sigma_s / \sigma_M)^2} \tag{10}$$

Here the attenuation occurs because part of the longitudinal electric field of the propagating wave couples into the electrons of the 2DEG. This not only causes a current to flow but pulls power from the SAW due to ohmic losses. This attenuation is described by Eq. 9. A SAW velocity shift is also observed due to the piezoelectric stiffening of the substrate, see Eq. 10 (Wixworth et al., 1989). Below are the recreated graphs from (Wixworth et al., 1989) to show the relationship of the attenuation and sound velocity shift due to a change in conductivity.

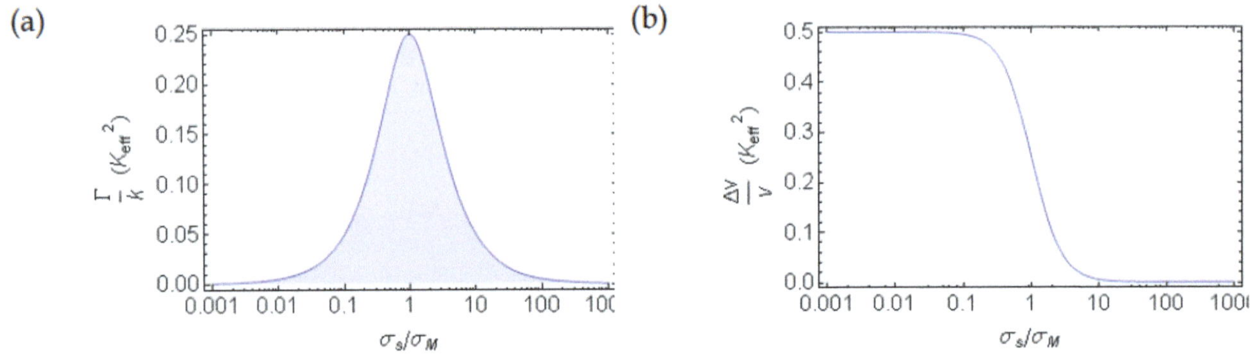

Figure 4: (a) The SAW attenuation in units of K_{eff}^2, see Eq. 9. (b) The change in SAW velocity in units of K_{eff}^2, see Eq. 10.

Where k is the SAW wave vector, K^2_{eff} is the effective piezoelectric coupling coefficient, σ_s is the 2DEG sheet conductivity, and $\sigma_M = v0(\varepsilon_1 + \varepsilon_2)$. Again, where v0 is the sound velocity and ε_1 and ε_2 are the dielectric constants of the piezoelectric substrate and half space above it.

Acoustoelectric Current

As the SAW propagates across the material it creates two types of currents, one is the normal acoustoelectric current and the other being the anomalous acoustoelectric current. The normal acoustoelectric current is created by electrons being "dragged" across the material and can be described by Eq. 11 where I is the current, n is the number of electrons, e is the charge of an electron, and f is the frequency of the SAW or RF signal to the IDT. This current always flows in the direction of the SAW and is produced as a DC current despite an oscillating RF signal being applied to the IDTs. Eq. 11 shows that at higher frequencies the normal acoustoelectric current becomes quantized.

$$I = nef \tag{11}$$

$$j(z) = \frac{a\omega}{2\pi} \int_0^{2\pi/\omega} \sigma_{zz}(z,t) E_z(z,t) dt \tag{12}$$

Once this deformation occurs the energy bands in the material bend as well causing the electrons to fall into the created quantum wells and are dragged along with the SAW, see Fig. 3d. As the frequency increases the wavelength, and pitch of the IDT, decreases causing fewer energy states to be available within the wells. This idea is also implemented to generate QDs within the SAW, (Barnes et al., 2000). The anomalous acoustoelectric current is produced from the deformation of the material and flows as a DC current. The difference is that the anomalous current always flows in one direction regardless of which IDT, left or right, produced the SAW. The current is smaller than the normal acoustoelectric current and is detected by different methods. The anomalous acoustoelectric current can be obtained by sweeping the RF signal of the IDTs and typically appears at an off-center frequency. Since the normal acoustoelectric current is much smaller, the anomalous current can be more easily detected, as shown in Fig. 5b. One such method for direct detection of the anomalous acoustoelectric current is to apply an RF signal to both the left and right IDTs, while phase locking the two signals (Beil et al., 2008). By phase locking the IDT signals a standing SAW can be created. Thus, a surface deformation is produced without a net direction of propagation. As seen in Fig. 5a the phase is shifted between the two IDTs as the current is measured. At a phase difference of 180° the current tends towards zero.

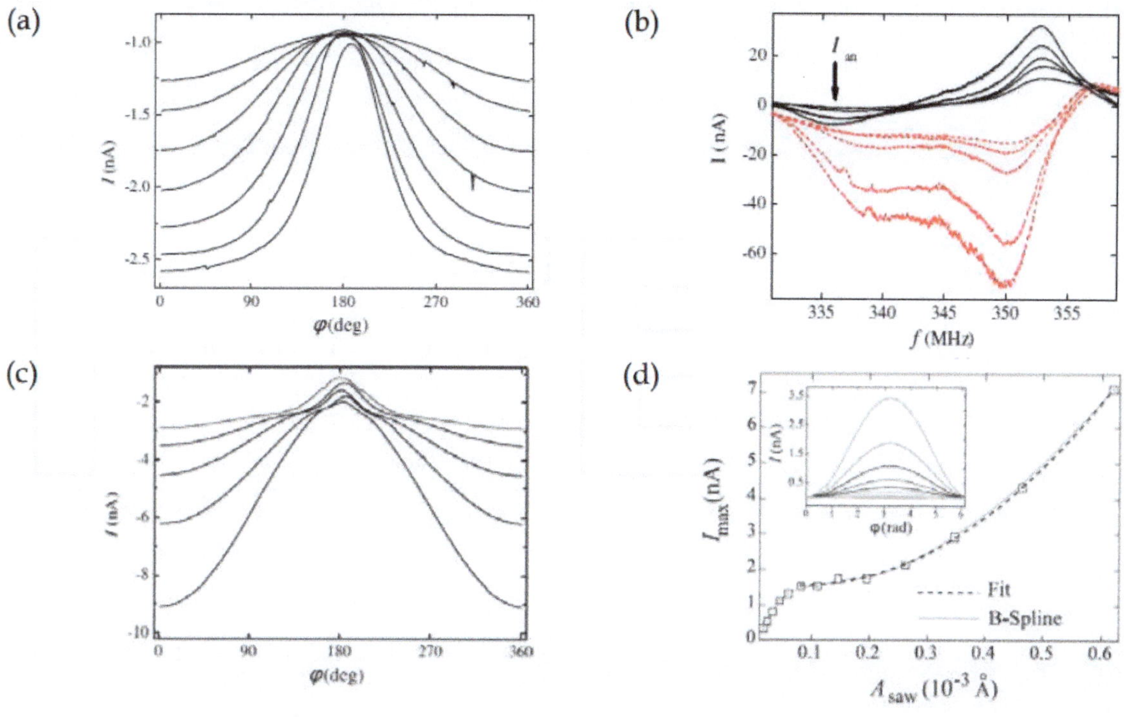

Figure 5: Results for a suspended beam that is 1.2 μm long and 300 nm wide. (a) Anomalous acoustoelectric current as a result of phase adjusting two counter propagating SAWs. (b) Acoustoelectric current vs. applied frequency of the IDTs. There is always a negative current value to the left and is indicative of the anomalous current. (c) Formation of a shock wave seen by its Sinc(x) shape. The device's response jumps to a higher order mode and returns to Sin(x). (d) Derived anomalous acoustoelectric current amplitude as a result of the calculated SAW amplitude. All images are taken from Beil et al., 2008, Copyright (2008) by The American Physical Society.

This zero current occurs because the two SAWs interfere destructively and create a nearly smooth surface with no deformation. At 0°, and also 360°, the current is at a maximum since the two SAWs interfere constructively allowing a maximum in the surface deformation. The anomalous acoustoelectric current can be used to probe the SAW amplitude of a suspended beam or nanostructure. The accompanying graphs show the acoustoelectric current from the device. It has been shown that the anomalous acoustoelectric current through the device in relation to the SAW amplitude can be described by Eqs. 12-14; where Eq. 12 is the general equation for an anomalous acoustoelectric current, 13 is the derived equation, and 14 is the strain equation of the beam.

$$j_{an}(L/2) = -\frac{\sigma_0 e_{z4}}{2\varepsilon_{GaAs}}(\Pi_{zzzy} S_{zy}(L/2)^2) \tag{13}$$

$$S_{zy}(z,t) = 2A_t\left(\left(\frac{6z^2}{L^3} - \frac{6z}{L^2}\right)\right)(Cos(\varphi/2) - Cos(k_{SAW}L + \varphi/2)) \tag{14}$$

Where σ_0 is the unperturbed conductivity, ε_{GaAs} is the permittivity of GaAs, Π is a tensor which describes the effect of strain S, and A_t is the transverse component of the SAW. An interesting phenomenon that occurs is the formation of shock waves in the suspended structure. From Figs. 5a and 5c it is seen that the device exhibits a Sin(x) type behavior in the anomalous acoustoelectric current at low RF powers. Once the RF power starts to increase it can be seen that the shape of the current starts to exhibit more of a Sinc(x) shape. This transition from a linear to a non-linear response indicates the formation of a shock wave in the suspended beam. As the power increases, the beam jumps into the next higher order mode and the current trace returns back to a linear response again. The shock wave formation is an indication that the beam will transition from one mode to another.

Surface Acoustic Waves in Magnetic Fields

We can use SAWs to probe the characteristics of a 2DEG under the presence of a magnetic field. Since a high mobility 2DEG is subject to Shubnikov-de Haas (SdH) oscillations, which creates changes in the conductivity, and SAWs are more sensitive to a conducting plane or surface, and the conductance of that plane, makes this combination an ideal candidate to investigate quantum effects of the 2DEG. The real interesting features are seen when integral Landau level filling factors are observed which causes a drop in conductivity of the 2DEG. At these quantized values the SAW responds strongly to the conductivity, σ. In Fig. 6 it can be seen that the SdH oscillations and the acoustoelectric current oscillations are nearly identical. The SdH measurement was measured using a standard four-point lockin technique, where as the acoustoelectric current was created by a SAW and was taken from two ohmic contacts on opposite sides of the sample. The peaks and valleys of the two measurements line up, for the most part, but there is a small offset. This offset is due to the fact that the SAW attenuation is not a linear with respect to the 2DEG conductivity, or magneto conductivity in this case, see Eq. 9.

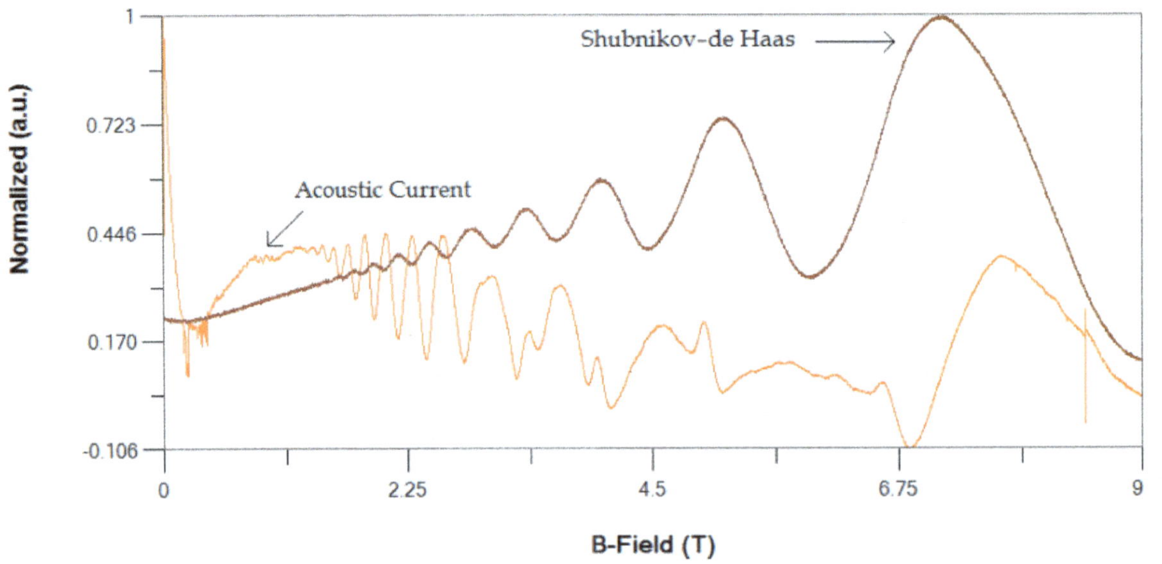

Figure 6: A magnetic field, of maximum value 9 T, was applied perpendicular to the surface of the device. The device is shown in figures 1d and 2a-2d and contains a 2DEG 40 nm below the surface. Upper trace is the Shubnikov-de Haas oscillations measured using standard lock-in techniques and normalized to 713.4 Ω. The lower trace is the acoustoelectric current generated from a SAW with -12 dBm applied to the right IDT at 1.488 GHz and is normalized to 10.47 nA.

Special attention has to be given to the acoustic current trace of Fig. 6 around 3 – 4 Tesla. There is a splitting of the peaks that occurs. This information does not appear in the SdH oscillation and is only isolated to a SAW effect. During this splitting the conductivity of the 2DEG is very low and σ_s becomes much smaller than σ_M, see Eq. 9. When this happens a maximum in SAW attenuation will occur which results in a reduced SAW amplitude. The center of a SAW current split is a minimum in the 2DEG conductivity; further details are discussed in Wixworth et al., 1989. From Fig. 6 there are two observations that need to be explained. One is the spike-like feature or discontinuity around the 8.5 T mark of the acoustic trace. This is caused from SAW reflections on the sample. This spike-like feature can be seen on other data plots as well with SAW currents. The final feature is the negative acoustic current that was measured. Since at this point the SAW was highly attenuated, causing little or no current to flow, caused the sample itself to heat. The negative current may be a combination of a thermal current and a small offset in the measuring equipment.

SURFACE ACOUSTIC WAVES IN QUANTUM ELECTRONICS

SAWs are used to produce a current in low dimensional electronic systems and NEMS. This can be used in various applications and for numerous different designs. The information presented in this section will deal with piezoelectric materials with an embedded two dimensional electron gas (2DEG); mainly GaAs/AlGaAs heterostructures unless otherwise stated. These measurements are carried out at liquid helium temperatures or lower, ≤ 4.2 K.

Quantum Point Contacts and Low Dimensional Channels

SAWs can be used to create a quantized current based upon Eq. 7. The advantage of the SAW inducing a quantized current is that this process can be used to populate and depopulate QDs and DQDs at higher frequencies then what can be used by applying an oscillating signal to the source-drain of the 2DEG. When using a SAW as the current source the acoustoelectric current can be pinched off in a Coulomb blockade just like a standard source-drain current can be. This pinch off is done via a quantum point contact (QPC) or a set of QPCs, which can be used to form QDs; see Fig. 7.

Quantum Point Contact Fabrication Fabricating

QPCs is done with the same methods as fabricating IDTs, see Sec. 2. Since QPCs are have small dimensions it is most common to use Electron Beam Lithography, or e-beam lithography, to create the structures. The exact dimensions of the QPC pair depends on what works best for the user and there is no set rules for design like there is for IDTs. When viewing Fig. 7 it can be seen that there are five sets of QPCs. A single QPC is seen as an electrode with another electrode opposite its position. All of the QPCs shown have a few common features; the tip of the QPC is small when compared to the rest of the electrode and the gap between the electrode tips is small as well. The tip is small so the electric field being emitted from the QPC is very localized, and the majority of the electrode is made wider so that it covers a wider portion of the 2DEG so the electrons are repelled. The gap between electrodes is small so pinch-off can be achieved with small voltages, more on this in Sec. 4.1.2.

Quantum Point Contact Operation

QPCs work by applying a negative voltage to them, this negative voltage produces an electric field that penetrates into the 2DEG causing electrons to be repelled at or around the QPC region; this can be seen if Fig 7(c). Here the black region represents the metallic electrode while the blue region represents the electric field and the green region represents the 2DEG or substrate. It can be taken as a good estimate that the electric field penetrates into the 2DEG from the QPC electrode at a 45° angle. As a larger negative voltage is applied to the QPC the effective screening electric field becomes larger causing the electron path within the 2DEG to become more constricted until single electron transport is achieved; this leads to a Coulomb Blockade, see Sec. 4.2. Then finally, the current is completely pinched off, see Figs 9 and 10.

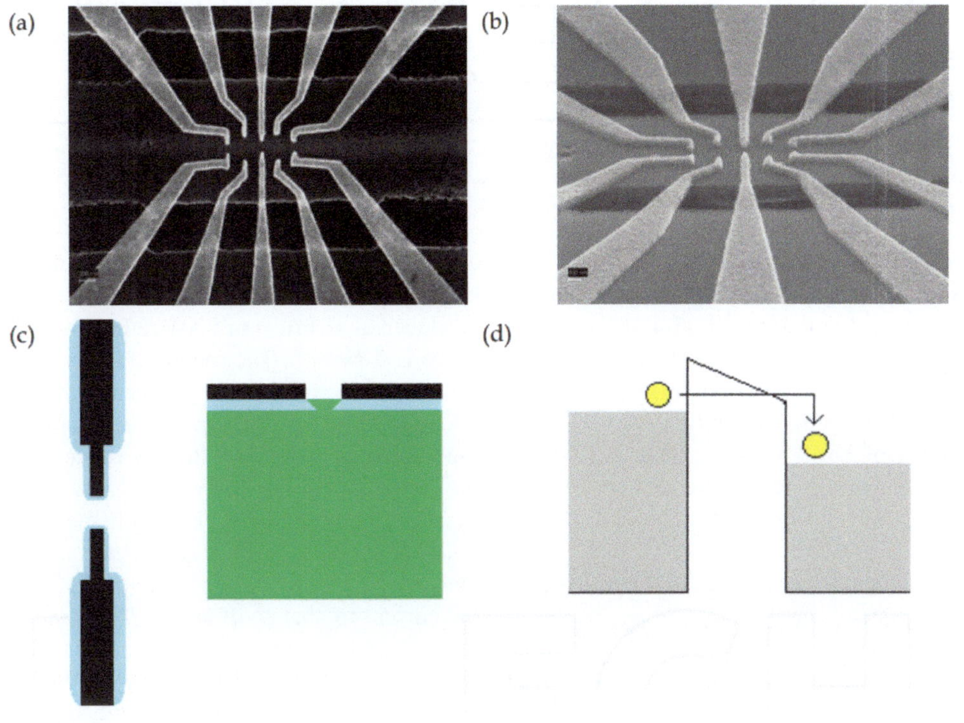

Figure 7: Two SEM images of the same DQD structure on top of a suspended tube section of an acoustic waveguide. The DQD is made up of five pairs of QPCs. The material is a GaAs/AlGaAs heterostructure with an embedded 2DEG about 40 nm below the surface. (a) Top view with a scale bar of 200 nm, (b) angled view with a scale bar of 100 nm. (c) schematic view of QPC with applied voltage, the blue area represents the electric field surrounding the QPC and penetrating into the 2DEG (green). (d) Tunneling of electrons across potential barrier created by applied voltage from the QPC.

The QPC creates a tunneling barrier which separates the source and drain regions of the sample. As the voltage magnitude becomes larger the strength, or height, or the tunneling barrier is increased and the width of the barrier becomes larger. When plotted a step like feature can be seen which corresponds to a single conductance step which has a value of $G = 2e^2/h$, where h is Plank's constant and e is the charge of an electron. Now the temperature must be low so the thermal energy, $E = k_b T$, of the background is smaller than the tunneling energy needed for the electrons to "jump" across the barrier, where k_b is the Boltzmann constant. As seen in Figs 9 and 10 this step like feature can be seen by doing an I-V measurement. By changing the temperature of the system the phonon energy is increased and causes scattering events to increase, or increase the electron-phonon interaction, and the steepening or smoothing of the step like feature is a direct measure of this.

Usage of Quantum Point Contacts and Surface Acoustic Waves

The use of QPCs offers a benefit of determining which SAW mode(s) are propagating in the sample. Different SAW modes, such as bulk, longitudinal, and transverse with propagate at different frequencies due to the fact that they have different sound velocities, see Eq. 1. Another factor, which affects the sound velocity, is the propagation direction of the SAW with respect to the crystal orientation of the material. In Fig. 8 a QPC had an applied voltage of -0.8 V, which puts the QPC into pinch-off mode. Since it is in pinch-off higher RF power is required to create a sufficiently strong SAW that will overcome the potential barrier. As the power is increased from -18 dBm to -10 dBm, three peaks emerge as transferring current through the tunnel barrier. From Eq. 7 we can calculate the electron count to be 6, 3, and 2 for RF powers of -10 dBm, -12 dBm, and -14 dBm, respectively (some rounding is taken into account, due to thermal errors in measurement).

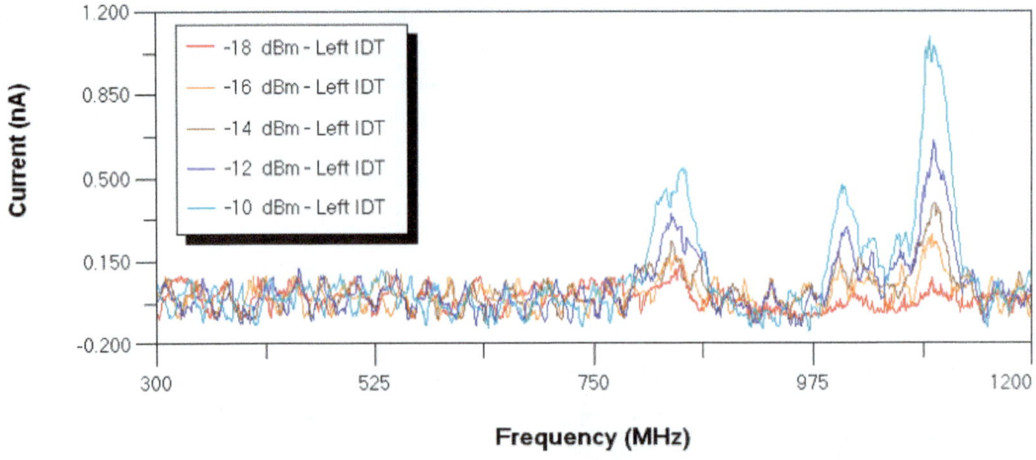

Figure 8: A frequency sweep of varying RF powers while the center QPC of the sample in Fig. 4 is held at -0.8 V. The first peak is at 840 MHz with a current of 540 pA and a velocity of 3,368 m/s, the second peak at 1.005 GHz with a current of 472 pA and a velocity of 4,020 m/s, the third peak is at 1.095 GHz with a current of 1.098 nA and a velocity of 4,380 m/s.

Now the three peaks represent different SAW modes. The highest frequency peak of Fig. 8 of 1.095 GHz represents a longitudinal wave with an acoustic velocity of 4,380 m/s and an angle of about 10° off from the (110) direction (Kuok et al., 2001). This small angle variation is due to a small misalignment during the lithography process. When viewing the lower peak of 840 MHz at a velocity of 3,368 m/s, this coincides with a fast transverse wave with, again, a 10° difference between the SAW direction and the (110) GaAs crystal orientation (Kuok et al., 2001).

Coulomb Blockade of Acoustoelectric Current When looking over Fig. 9, we see that single steps are observed just before the total acoustoelectric current becomes zero. Focusing on Fig. 9a, we see that the last step is at about 0.5 nA. When solving with Eq. 11 the first step yields an answer of n = 1, i.e. a single electron is being transferred. We can look at all of the remaining steps and solve for them as well which will reveal integer numbers for n and will increase by 1 for each step, as is expected. Fig. 9a also shows that an applied voltage across the source and drain contacts, or ohmics, has very little effect on the final quantized acoustoelectric current. We do see, however, that a larger, or smaller, gate voltage is required for final pinch-off in the system but this is due to the small offset in the Fermi energy because of the applied bias. The only real difference is the shift from negative current to a positive current which is easy explained by the fact that the source-drain bias is producing a current that is opposite in direction, when $V_{ds} < 0$ mV, to that of the acoustoelectric current. Another aspect is the RF power dependence on the quantized current. When viewing Fig. 9b there is a similar trend to that of Fig. 9a. The small change in RF power has only a small effect on the acoustoelectric current. This is because the RF power is directly proportional to the SAW amplitude and we do not really identify any significant difference in the number of electrons being transferred, or Coulomb steps, until a drop of about 2 dBm. So far an acoustoelectric current behaves in the same way as a standard source-drain bias current when a QPC is near pinch off. There is, however, another effect which can arise. As seen in Fig 9c a negative current arises and still exhibits the step like behavior. This negative current is the negative anomalous acoustoelectric current. This is said to be produced as an effect of SAW reflections, this is mostly seen in a two IDT system. The second IDT acts as a reflector much as the same way a reflection gradient is used for a unidirectional SAW, see Fig 1b. This can cause a standing wave to occur in the sample in such a way that the reflections effectively reduce the potential of the SAW and cause fewer electrons to be transported. With slight phase shifts added the standing wave with respect to the QPC a net negative potential in the energy landscape can exist on one side of the QPC which will cause the current to change direction. These steps are best observed for QPCs that are long when compared to the SAW wavelength. This makes it possible to observe the acoustoelectric step current without applying a magnetic field (Shilton et al, 1996). As mentioned earlier, the high mobility of the 2DEG will screen the acoustoelectric current but this effect is not as prevalent when inside of a long QPC channel. The current screening is reduced around the QPC region and thus electrons can be transported through the channel. A long channel can allow ballistic transport of the electron causing it to shuttle from one side of the QPC to another. Since the screening is minimal, the electron will be "dragged" through the channel and the current steps of Fig. 9 can be observed. Another phenomenon is oscillations in the acoustoelectric current once approaching the QPC pinch-off limit. As described in Shilton et al, 1996, and Maaø & Galperin, 1997, the acoustoelectric current oscillations occur at the same positions as the Coulomb steps. This oscillation is described as interference effects near the QPC at high frequencies which may be attributed to impurities in the 2DEG channel. The theory presented suggests that these oscillations are due to state transitions, both propagating and reflecting (Maaø & Galperin, 1997). The SAW may reflect from the QPC and create interference patterns; these patterns will affect the electron transmission probability through the QPC simply by changing the potential landscape. This mechanism is also sensitive to scattering events near the QPC, where energy quanta are emitted and absorbed between two waves. Others have contributed to the theory and more is being added.

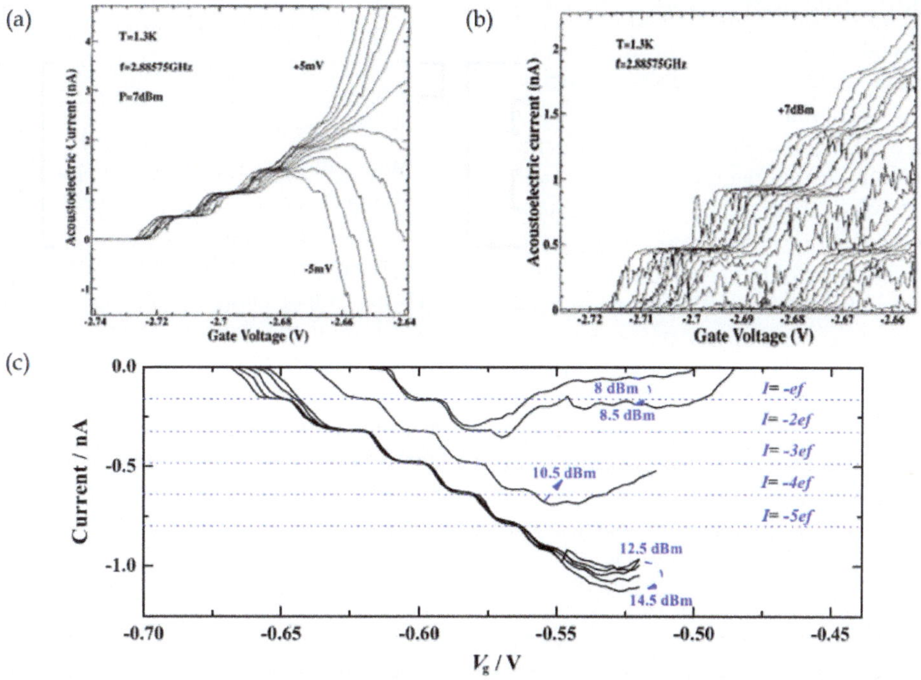

Figure 9: QPC pinch-off of acoustoelectric current: (a) pinch-off with varying source-drain voltages applied at the SAW, changes are in 1 mV increments (Talyanskii et al., 1997) Copyright (1997) by The American Physical Society. (b) Pinch-off with different SAW power levels, the left most trace is 7 dBm and decrements by 0.2 dBm (Talyanskii et al., 1997) Copyright (1997) by The American Physical Society. (c) A negative anomalous acoustoelectric current for different gate voltages and SAW powers; f = 1007.426 MHz, V_{ds} = 0 V, T = 1.7 K, (Song et al., 2010).

Fig. 10 shows the oscillations of the acoustoelectric current as it is being driven through a QPC (orange trace) as it approaches pinch-off. It can be seen that the step like features observed in Fig. 9 are have been replaced by oscillations. These oscillations are negative, or on the lower cycle, when the current from the source-drain bias is flat or non-changing. We then see that as the current starts to decrease the acoustoelectric current has a positive value, or is on the upper cycle of the oscillation. The inset graph of Fig. 10 shows the entire sweep of the QPC gate voltage. As one conduction channel pinches off at the V_g = -0.65 V we see that there is a large, nearly single, oscillation in the acoustoelectric current. Since SAWs are very sensitive to 2DEGs we get what seems to be an amplified signal when compared to a source-drain bias. This can be used to identify information that may otherwise have been too weak or nearly washed out from thermal effects.

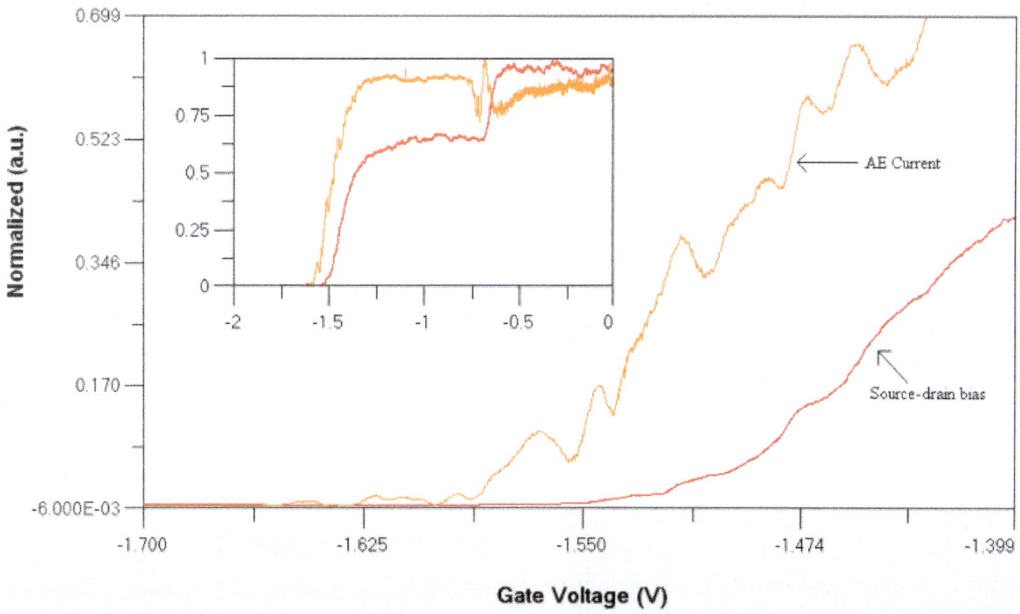

Figure 10: Acoustoelectric current oscillations as the QPC nears pinch-off. The data plot has been normalized to show the relationship between the acoustoelectric current oscillations (orange trace) and the pinch-off when using a voltage bias across the source-drain electrodes (red trace). The acoustoelectric current trace has been normalized to -8.832 nA and the source-drain bias trace has been normalized to 4.692 nA. Measurements were taken at 4.2 K with the sample shown in Fig. 7. Inset plot shows entire gate voltage sweep range

Population and Depopulation of Quantum Dots

SAWs have been used widely as a mechanism to control the population and depopulation of QDs and DQDs. The high speed, or frequency, of operation combined with quantized current production make SAWs suitable for use with quantum systems. In a traditional QD system the gate and source-drain voltages are changed to allow single electron population and depopulation. This has several drawbacks; one being it is hard to decouple the dot from the surrounding environment due to the precise control of the voltages. There is always some noise either in the system or from the electronics. Another problem is if one gate is changed then the others must be changed to ensure a constant electron count in each dot, if applicable. A SAW can be used as an alternative and has gotten more attention lately for the use in a QD or DQD system. The biggest reason for the attention is that SAWs can be used to interact with QD systems at much higher frequencies then what has been previously achieved by pulsing the gate electrodes. Since the acoustoelectric current can be well defined a QD system can use very high gate voltages to ensure no leakage current. We will take a look at the work done in Kataoka et al., 2007. This work describes very well the use of SAWs as a way to populate and depopulate a QD.

When the QD needs to be populated, the center plunger gate voltage value is lowered; causing the Fermi energy to lower below the Fermi energy of the surrounding 2DEG. The gate electrodes are still held at a high voltage value so the electrons cannot tunnel in or out of the QD, see Fig. 11e. The IDT is pulsed and as the SAW enters the QD region it changes the tunneling barrier created by the electrodes. The barrier is changed just enough that a single electron can tunnel into the dot, this is population; see Figs. 11a-b. The IDT is pulsed many times since the tunneling barrier is so high that the additional pulses are needed to increase the tunneling probability of the electron.

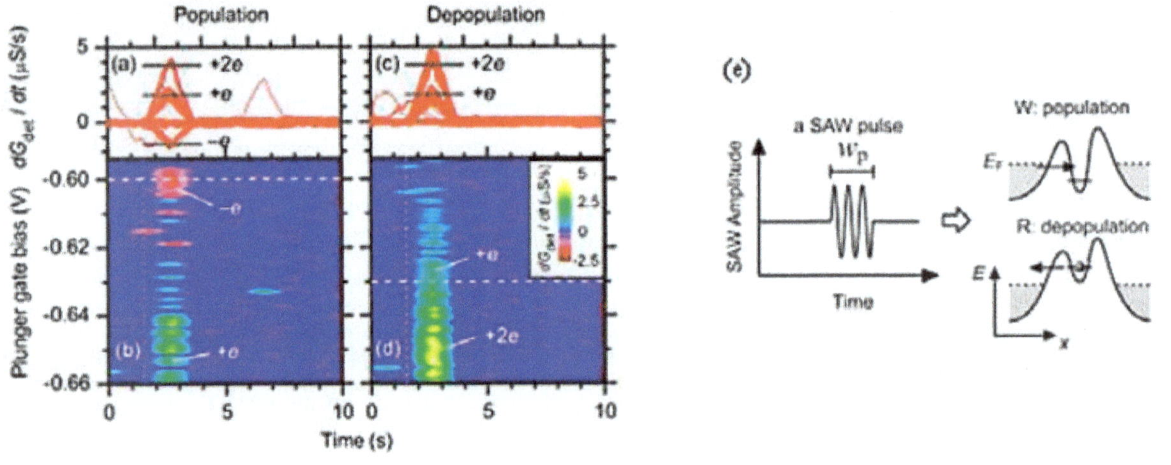

Figure 11: Results from the paper Kataoka et al., 2007, Copyright (2007) by The American Physical Society. (a) and (b) The population of the QD, the –e line is an electron entering the dot. (c) and (d) Depopulation of the QD. (e) Schematic of SAW for population and depopulation of the QD.

Likewise, as shown in Figs. 11c-d, the QD is depopulated by a similar mechanism. Here the plunger gate voltage is raised, made to be more negative, which causes the Fermi energy of the QD to be raised in comparison to the 2DEG Fermi energy outside of the dot, Fig. 11e. At this time the IDT is pulsed and the SAW enters the QD region. The SAW alters the tunneling barrier created by the gate electrodes when the potential of the SAW is superimposed onto the barrier, see Eq. 5. When the barrier is raised nothing happens since this will simply ensure the electron stays in the system, but when the negative cycle of the SAW superimposes with the barrier it decreases it which causes the electron to tunnel out. This method of operation has proven, quite nicely, that SAWs and high potential QPCs offer a more robust method of single electron population and depopulation. As shown in Figs. 11a-d the QD can maintain the electron confinement for a long period of time which can be difficult in a traditional setup due to noise reduction of the measurement electronics. The IDTs can be accessed from an outside system such as an antenna, for example, and interact with the QD allowing the system another degree of freedom from the traditional closed electronics. However, this may come at the cost of additional noise reduction equipment and filtering.

CONCLUSION

This chapter has shown all of the basic properties and uses of SAWs in nano-structures and nano-systems. The reader is shown the parameters required for fabrication, theory of operation, real results, and application. The use of SAWs is quite limitless in the area of NEMS. As people continue to refine the research through experiment and by continuing to add to the theory more applications will emerge.

ACKNOWLEDGMENT

We like to thank the National Science Foundation for support under a NIRT grant (No. ECCS-0708759) and the Air Force Office for Scientific Research for support under a MURI grant (No. FA9550-08-1-0337).

REFERENCES

1. Barnes, C. H. W., Shilton, J. M., & Robinson, A. M. (2000). Quantum computing using electrons trapped by surface acoustic waves. Physical Review B, Vol. 62, No. 12, (September 2000), pp. 8410-8419, ISSN 0163-1829
2. Beil, F. W., Wixforth, A., Wegscheider, W., Schuh, D., Bichler, M., & Blick, R. H. (2008). Shock Waves in Nanomechanical Resonators. Physical Review Letters, Vol. 100, No. 2, (January 2008), pp. 026801 1-4, ISSN 0031-9007
3. Campbell, C. (1998). Surface Acoustic Wave Devices for Mobile and Wireless Communications, Academic Press, Inc, ISBN 0-12-157340-0, San Diego, CA, USA
4. Kataoka, M., Schneble, R. J., Thorn, A. L., Barnes, C. H. W., Ford, C. J. B., Anderson, D., Jones, G. A. C., Farrer, I., Ritchie, D. A., & Pepper, M. (2007). Single-Electron Population and Depopulation of an Isolated Quantum Dot Using a SurfaceAcoustic-Wave Pulse. Physical Review Letters, Vol. 98, No. 4, (January 2007), pp. 46801-1 – 46801-4, ISSN 0031-9007
5. Kuok, M., Ng, S. C., & Zhang, V. L. (2001). Angular dispersion of surface acoustic waves on (001), (110), and (111) GaAs. Journal of Applied Physics, Vol. 89, No. 12, (June 2001), pp. 7899-7902, ISSN 0021-8979
6. Maaø, F. A. & Galperin, Y. (1997). Acoustoelectric effects in quantum constrictions. Physical Review B, Vol. 56, No. 7, (August 1997), pp. 4028-4036, ISSN 0163-1829
7. Salmon, V. (1946). A New Family of Horns. Jounal of the Acoustical Society of America, Vol. 17, No. 3, (January 1946), pp. 212-218, ISSN 0001-4966
8. Shilton, J. M., Mace, D. R., Talyanskii, V. I., Galperin, Y., Simmons, M. Y., Pepper, M., & Ritchie, D. A. (1996). On the acoustoelectric current in a one-dimensional channel. Journal of Condensed Matter Physics, Vol. 8, (March 1996), pp. L337-L343, ISSN 0953- 8984
9. Song, L., Chen, S. W., He, J. H., Zhang, C. Y., Lu, C., & Gao, J. (2010). The anomalous negative acoustoelectric current in single-electron transport. Solid State Communications, Vol. 150, (2010), pp. 292-296, ISSN 0038-1098
10. Talyanskii, V. I., Shilton, J. M., Pepper, M., Smith, C. G., Ford, C. J. B., Linfield, E. H., Ritchie, D. A., & Jones, G. A. C. (1997). Single-electron transport in a onedimensional channel by high-frequency surface acoustic waves. Physical Review B, Vol. 56, No. 23, (December 1997), pp. 15180-15184, ISSN 0163-1829
11. Wixforth, A., Scriba, J., Wassermeier, M., Kothaus, J. P., Weimann, G., & Schlapp, W. (1989). Surface acoustic waves on GaAs/AlxGa1-xAs heterostructures. Physical Review B, Vol. 40, No. 11, (October 1989), pp. 7874-7887

CHAPTER 15

WAVE ITERATIVE METHOD FOR ELECTROMAGNETIC SIMULATION

Somsak Akatimagool and Saran Choocadee

INTRODUCTION

Presently, the microwave circuit design is the most critical and necessary for modern communication systems. The problems with the design of circuits often include a lack of equipment and tools used to design and build them. Numerical methods related to electromagnetic waves have caused a revolution in microwave engineering; techniques such as FDTD (Finite Differential Time Domain) [1], TLM (Transmission Line Matrix) [2], and the Moment method [3] have been developed. Each method has respective disadvantages and limitations of usage foe both research and education. To perform the electromagnetic wave analysis of waveguide structures, several classic approaches can be found in the literature. A practical example of computer aid design is that the inductive iris in a rectangular waveguide has been analyzed with high accuracy through approximate modeling utilizing the full wave Mode Matching Method (MMM) [4] and Transmission Line Matrix (TLM) [5]. Therefore, the development and optimization of numerical methods are important for an efficient electromagnetic simulation tool [6-7].

In this chapter, we study and introduce an efficient electromagnetic simulation tool for analysis of inductive and capacitive obstacles and rectangular window in waveguides. The Wave Iterative Method (WIM) based on iterative procedure and wave propagation theory is proposed. This method has been combined with mode matching technique to characterize the obstacle in a rectangular waveguide. Also, the compact CAD tool and presenting the electromagnetic field distributions included in this topic.

Citation: Somsak Akatimagool and Saran Choocadee (2013). Wave Iterative Method for Electromagnetic Simulation, , Dr. Yi Zheng (Ed.), InTech, DOI: 10.5772/51450.

WAVE PROPAGATION AND WAVEGUIDE

Several years ago, the characteristics of waves in source-free, homogeneous regions of space have been discussed. Accordingly, we present this section by considering the reflection, transmission and incident experienced by electromagnetic waves on stratified dielectric surface.

There are many situations where a wave passes through several layers of different materials. One example is the passage of electromagnetic through a free space with transmission line mode or TEM. There are several ways use to analyze the correct results, but a very common method is an impedance transformation that integrates reflections into a single parameter to generate the numerical results.

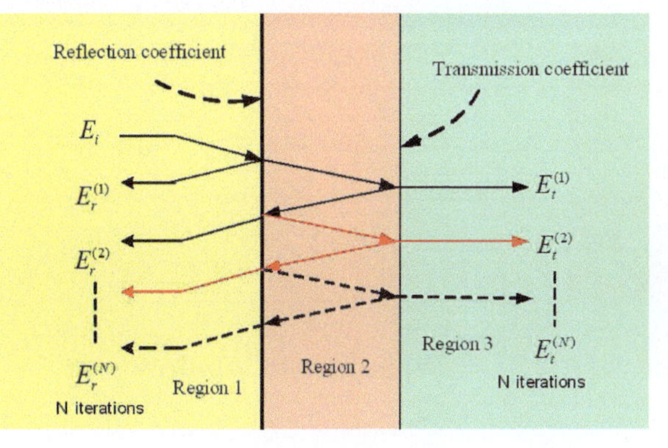

Figure 1. a) The wave propagation in stratified interface

Fig.1 shows a stratified dielectric region. When a wave is normally incident from the region 1 and transmitted into region 2 and region 3 respectively, an infinite number of transmitted, reflected waves in the same wavelength are produced at the boundary interface. At a time, the all incident waves (E_i), reflected waves(E_r) and transmitted waves (E_t) can be considered to be a single of each wave. We can develop the reflection coefficient expression by following the progression of the incident waves and its reflections. Similarly, the transmission coefficient expression can be presented by following the progression of the incident wave and its transmissions. Finally, we can use the relationship of expressions to calculate the properties of boundary interface.

Waveguides are used to transport electromagnetic energy along a fixed path that carries non-TEM modes, often called waveguide modes. Most important of waveguide properties is that they can support an infinite number mode of filed generated by diffraction form each interface within waveguide [8]. Waveguides are almost operated so that only a single propagation mode is present because the presence of more than one propagating mode causes dispersion. Typically, they must be operated over smaller bandwidths and smaller losses than transmission line, which makes them attractive for many applications. One of several examples is to place obstacle element in waveguides to make devices such as lumped element, filters, couplers or antenna. Fig.2(a) shows two thin metal fins placed on the top and bottom walls of a waveguide, called a capacitive window. Considering in the gap of obstacle, when the electric filed within the window increases, the energy stored between two conducting surfaces will increase. We observed that the equivalent circuit of obstacle is a shunt capacitor.

Similarly, Fig.2(b) shows an inductive window which the most magnetic fields can flow through in the obstacle width. The equivalent circuit of obstacle is a shunt inductor.

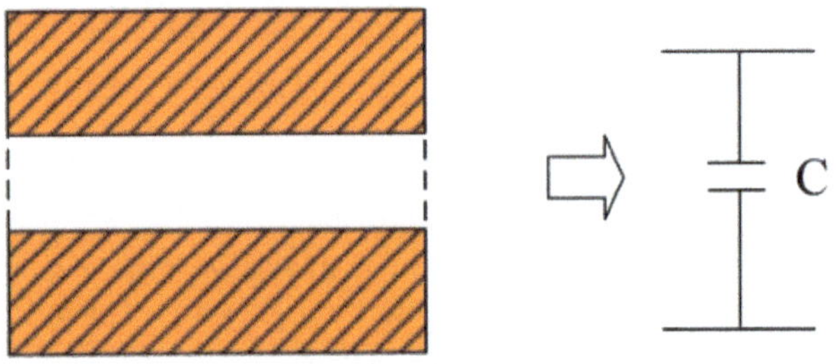

(a) capacitive window

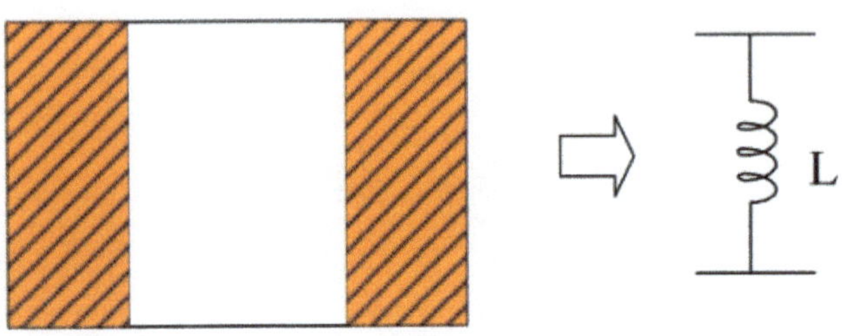

(b) inductive window

Figure 2. a) The capacitive and inductive windows

WAVE ITERATIVE METHOD (WIM)

An analysis of the electromagnetic wave properties within a waveguide consists of TE and TM field components. Most wave analyses are calculated in the spectral domain based on the series integration equation to present the electromagnetic field and to analyze the two ports network parameters such that the result of the frequency response reflects the characteristic of various planar circuits. In this section, we will present the cooperation of waves between the real domain and the spectrum domain.

The calculating concept for the electromagnetic wave propagation in a waveguide is based on the Wave Iterative Method (WIM) [9-11]. The operating process, as shown in Fig.3, present the amplitude and direction of the incident, reflected, and transmitted waves what propagate in the waveguide obstacle. On the obstacle, the waves are calculated in the real domain (pixels) and the waves in the free space are calculated in the spectrum domain (modes). To alternate between both domains, we use the Fast Fourier Transform (FFT) to reduce the computation time and show directly the electromagnetic field in the real domain.

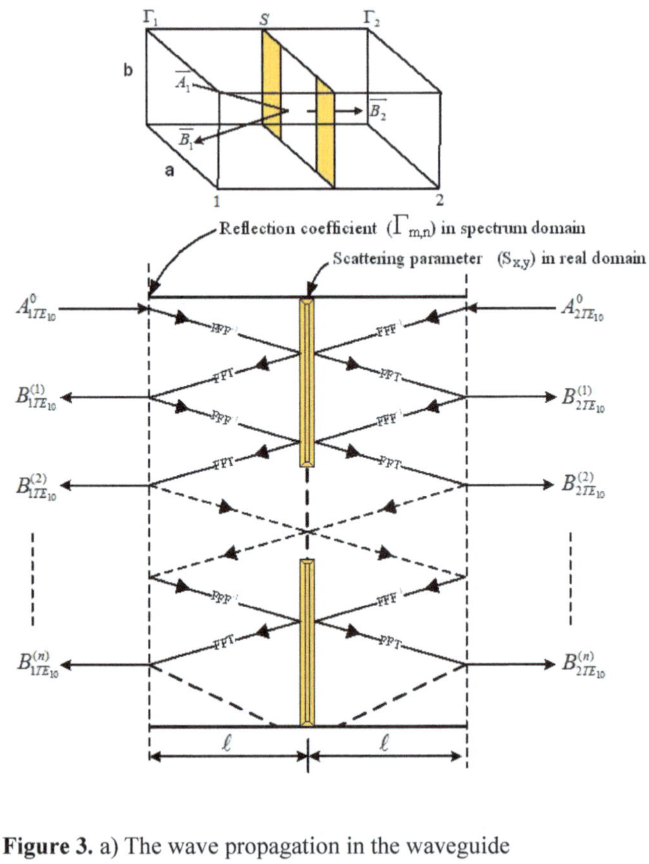

Figure 3. a) The wave propagation in the waveguide

In Fig.3, the first step, the TE10 field which is the dominant mode, propagate into the obstacle as iris circuit, and then the higher-mode waves caused by the scatter on the conductor surface will be incident and will reflect within the waveguide. An infinite numbers of reflected and transmitted waves are produced at the obstacle interface. Finally the calculation of waves will use the principle of mode matching at the input and the output ports of the waveguide.

The initial value of dominant mode, $TE_{1,0}$ is

$$A^{(0)}_{x,y} = -\sqrt{\frac{2}{ab}}\sin\left(\frac{\pi x}{a}\right), \quad (1)$$

where a and b are the dimension of the waveguide. The reflected waves $(B_{x,y})$ in n th iteration on the obstacle are

$$B^{(n)}_{x,y} = S_{x,y}A^{(n-1)}_{x,y}. \quad (2)$$

The scattering parameter ($S_{x,y}$) of waves in the real domain for a two-port network is defined as

$$\left[S_{x,y}\right] = \begin{bmatrix} -S_{C(x,y)} & S_{D(x,y)} \\ S_{D(x,y)} & -S_{C(x,y)} \end{bmatrix}, \quad (3)$$

where S_C is equal to 1 in the conductor area and S_D is equal to 1 for the free space. To transform the wave from the real domain to the spectrum domain, we use the Fast Fourier Transform (FFT) of the TE/TM modes as

$$B^{(n)}_{m,n} = Modal\_FFT\left(B^{(n)}_{x,y}\right). \quad (4)$$

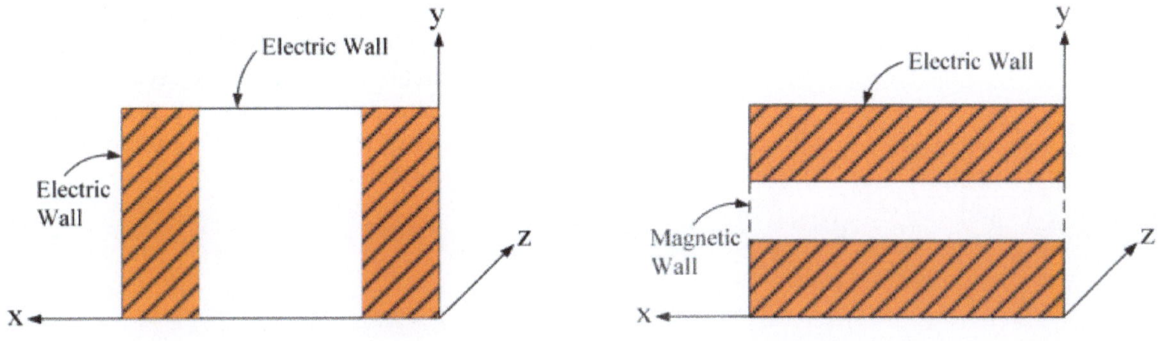

(a) Electric wall (b) Electric-magnetic wall

Figure 4. The electric and magnetic walls around the waveguide

The basic functions of TE field components in the electric wall [12], as shown in Fig.4(a) for an inductive obstacle, are

$$\vec{E}^{\alpha}_{x(x,y)} = K^{\alpha}_{x}\cos\left(\frac{m\pi x}{a}\right)\sin\left(\frac{n\pi y}{b}\right), \quad (5)$$

$$\vec{E}^{\alpha}_{y(x,y)} = -K^{\alpha}_{y}\sin\left(\frac{m\pi x}{a}\right)\cos\left(\frac{n\pi y}{b}\right), \quad (6)$$

where a refers to the TE mode and K is the constant value with respect to the x and y directions. The equations are defined as

$$K^{TE}_{x} = \frac{n\sqrt{\frac{2t}{ab}}}{b\sqrt{\frac{m^2}{a^2}+\frac{n^2}{b^2}}}, \quad K^{TM}_{x} = \frac{m\sqrt{\frac{2t}{ab}}}{a\sqrt{\frac{m^2}{a^2}+\frac{n^2}{b^2}}}$$

$$K^{TE}_{y} = \frac{-m\sqrt{\frac{2t}{ab}}}{a\sqrt{\frac{m^2}{a^2}+\frac{n^2}{b^2}}}, \quad K^{TM}_{y} = \frac{n\sqrt{\frac{2t}{ab}}}{b\sqrt{\frac{m^2}{a^2}+\frac{n^2}{b^2}}} \quad (7)$$

where t = 1 if m n, = 0 and t = 2 if m n, ¹ 0 .

The basic functions of TM field components in the electric-magnetic wall [12], as shown in Fig.4(b) for a capacitive obstacle, are

$$\vec{E}^{\alpha}_{x(x,y)} = -K^{\alpha}_{x}\sin\left(\frac{m\pi x}{a}\right)\sin\left(\frac{n\pi y}{b}\right), \quad (8)$$

$$\vec{E}^{\alpha}_{y(x,y)} = K^{\alpha}_{y}\cos\left(\frac{m\pi x}{a}\right)\cos\left(\frac{n\pi y}{b}\right), \quad (9)$$

where a refers to the TM mode and K is the constant value respect to the x and y directions. The equations are defined as

$$K^{TE}_{x} = \frac{n\sqrt{\frac{2t}{ab}}}{b\sqrt{\frac{m^2}{a^2}+\frac{n^2}{b^2}}}, \quad K^{TM}_{x} = \frac{m\sqrt{\frac{2t}{ab}}}{a\sqrt{\frac{m^2}{a^2}+\frac{n^2}{b^2}}}$$

$$K^{TE}_{y} = \frac{m\sqrt{\frac{2t}{ab}}}{a\sqrt{\frac{m^2}{a^2}+\frac{n^2}{b^2}}}, \quad K^{TM}_{y} = \frac{-n\sqrt{\frac{2t}{ab}}}{b\sqrt{\frac{m^2}{a^2}+\frac{n^2}{b^2}}} \quad (10)$$

where t = 1 if m n, = 0 and t = 2 if m n, ¹ 0 .

At the input port of the waveguide, the incident waves ($A_{x,y}$) of higher-order modes will feedback into the obstacle of the waveguide. The equation for incident waves ($A_{m,n}$) is

$$A_{m,n}^{(n)} = \Gamma_{m,n} B_{m,n}^{(n)} + A_{1,0}^{TE/TM}, \tag{11}$$

where $A_{1,0}$ is an initial exciton source. The reflection coefficient($\Gamma_{m,n}$) of waves in the spectrum domain at the input and the output ports of the waveguide can be written as

$$\Gamma_{m,n} = \Gamma_{m,n}^{TE/TM} = \frac{1 - Z_0 Y_{m,n}^{TE/TM}}{1 + Z_0 Y_{m,n}^{TE/TM}}, \tag{12}$$

where Z_0 is the intrinsic impedance of the dominant mode and $Y_{m,n}^{TE/TM}$ is the TE/TM modes admittance with the orders of m and n which can be expressed as

$$Y_{m,n}^{TE} = \frac{\left(\frac{m\pi}{a}\right)^2 + \left(\frac{n\pi}{b}\right)^2 - \omega^2 \mu\varepsilon}{j\omega\mu}, \tag{13}$$

$$Y_{m,n}^{TM} = \frac{j\omega\varepsilon}{\left(\frac{m\pi}{a}\right)^2 + \left(\frac{n\pi}{b}\right)^2 - \omega^2 \mu\varepsilon}. \tag{14}$$

Therefore, considering the waves on the obstacle, the waves ($A_{m,n}$) will be transformed by using the Inverse Fast Fourier Transform (IFFT) to analyze the basic functions of TE/TM field components by using Eq.(5)-(8), to come back to the real domain as

$$A_{x,y}^{(n)} = Pixel\_IFFT\left(B_{m,n}^{(n)}\right). \tag{15}$$

The implementation of the Wave Iterative method consists of a recurrence relationship of wave between the propagation in both sides of the waveguide and the propagation on the obstacle. From Eq.(2) and Eq.(11), the total waves at n th iteration are

$$A_{m,n}^{(n)} = \Gamma_{m,n} \hat{S} A_{m,n}^{(n-1)} + A_{1,0}^{TE/TM}, \tag{16}$$

where ˆ S is the spectrum operator of scattering coefficient.

Considering the wave propagation in the rectangular waveguide with zero thickness obstacles, the electromagnetic equivalent circuit of the obstacle section is presented to identify the impedance (Z) element and the input and the output sections of the waveguide are presented by dominant mode admittance, as shown in Fig.5.

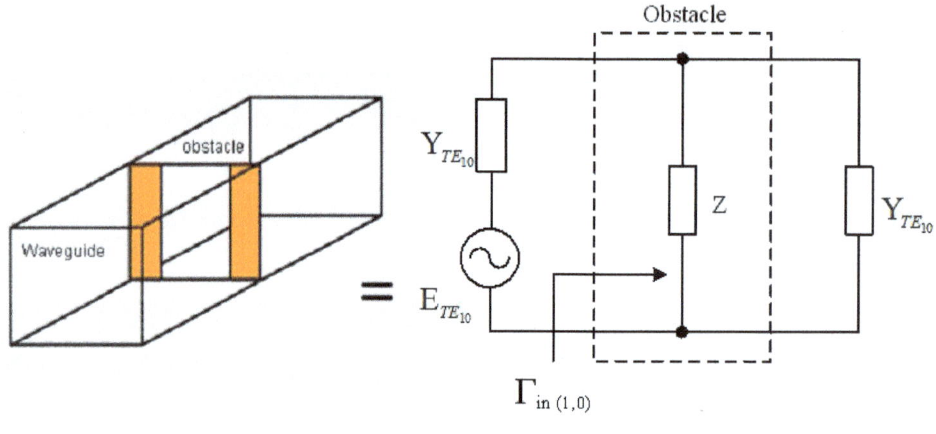

Figure 5. a) The electromagnetic equivalent circuit of an obstacle

At the convergence condition, the reflected wave on the obstacle at n th iteration, is tend toward zero, the total waves will be the steady state field. So, we can obtain the input reflection coefficient of obstacle circuit of dominant mode, $\Gamma_{in(1,0)}$ as,

$$\Gamma_{in(1,0)} = \frac{B_{1,0}^{TE/TM}}{A_{1,0}^{TE/TM}} = \frac{\sum_{n=1}^{N} B(n)_{1,0}^{TE/TM}}{A_{1,0}^{TE/TM}}. \tag{17}$$

The input impedance $Z_{in(1,0)}$ of zero thickness obstacles in the rectangular waveguide can be written as

$$Z_{in(1,0)} = Z_0 \left(\frac{\Gamma_{in(1,0)} - 1}{2\Gamma_{in(1,0)}}\right). \tag{18}$$

Then, the input reactance will be a positive imaginary value for an inductive element and a negative imaginary value for a capacitive element. If the input impedance is equal to zero or infinity, the obtained element will be a LC series or parallel resonant circuit respectively. The net 2.5D electric field distribution on the obstacle by summing the amplitude of the incident and the reflected waves of N iterations can be expressed as

$$E_{x,y} = \sqrt{Z_0} \sum_{n=1}^{N} \left(A_{x,y}^{(n)} + B_{x,y}^{(n)} \right). \tag{19}$$

Similarly, the net 2.5D magnetic fields can be expressed by subtracting the amplitude of the incident and the reflected waves of N iterations series as can be expressed as

$$H_{x,y} = \frac{1}{\sqrt{Z_0}} \sum_{n=1}^{N} \left(A_{x,y}^{(n)} - B_{x,y}^{(n)} \right). \tag{20}$$

WCD SIMULATION DESIGN

In this topic, we present a developed simulation program that conducts a numerical analysis by using the Wave Iterative Method (WIM), shown in the Fig.6. This developed simulation program, called WCD (Waveguide Circuit Design), consists of a main menu, parameter setup, and a design and display window, as shown in Fig.7. The WCD is constructed by using the GUI (Graphic User Interface) of MATLAB®. The user can setup the initial values that are used for calculating of the two ports network, calculate the obstacle characteristics and select the display windows of the simulated results. With the simulation program, it is possible to analyze and design the waveguide iris, waveguide filter and also visualize the 2.5D electromagnetic field distribution.

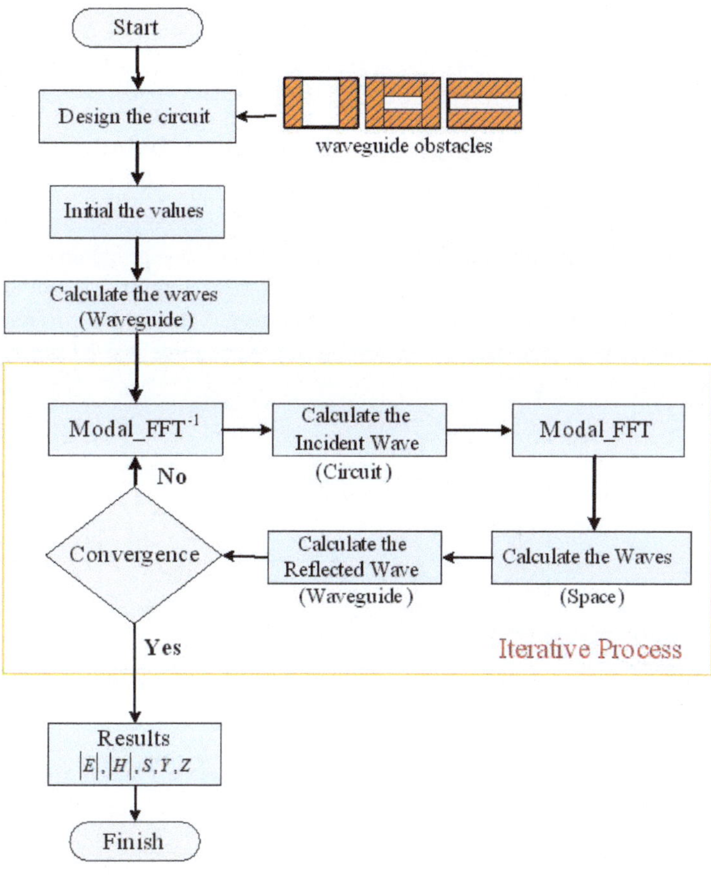

Figure 6. Flowchart of WCD simulated process

The simulated process of the WCD, as shown in Fig.6, is following:

1. Start the simulation program form the main menu which consists of an inductive, a capacitive and a window obstacle bottom, as shown in Fig.7 (a).
2. Design the waveguide obstacle structure what the user can determinate the desired obstacle dimensions, as shown in Fig.7(a).
3. Set the initial values; such as the size, length, frequency operating, dielectric constant, etc., as shown in Fig.7(a).
4. Calculate the incident wave of the dominant mode.
5. Calcula te the reflected and the transmitted waves on the obstacle surface in the real domain and in the free space in the spectrum domain. In this process, an infinite number of the transmitted and the reflected waves are produced on the obstacle interface. However the reasonable numbers of iteration are determined, the process will result in termination.
6. Present the simulated results; such as the electric and the magnetic field distribution, the impedance, the scattering and the admittance parameters, as shown in Fig.7(b).

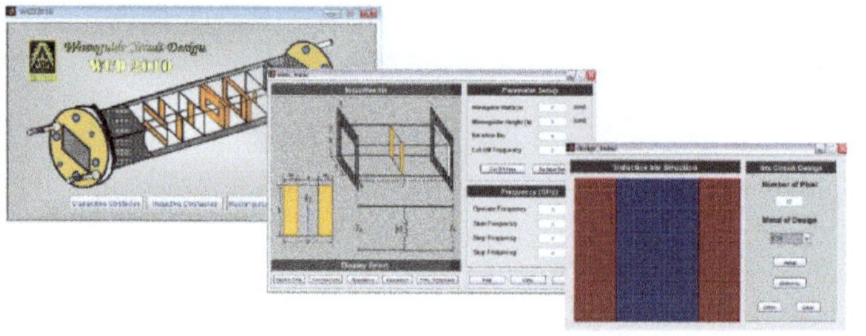

(a) Menu and parameter setup windows

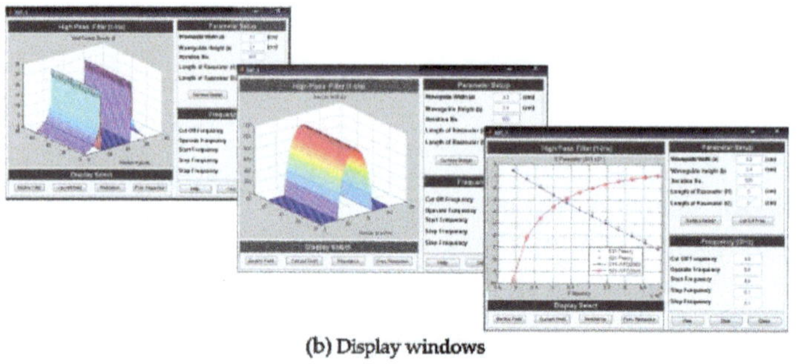

(b) Display windows

Figure 7. Waveguide Circuit Design (WCD) Tool

SIMULATION RESULTS

In this chapter, we present the analysis of the inductive, the capacitive obstacles and the rectangular window in the waveguide by using the developed WCD simulation and also introduce the electromagnetic distribution in the waveguide. The simulated WCD results will be compared with the CST simulation.

Inductive Obstacle Analysis

The vertical obstacle section transforms the inductive equivalent circuit, shown in Fig.8. The dimensions of rectangular waveguide consists of a width (a) equal to 6.4 cm., a height (b) equal to 3.2 cm. and the usable obstacle width (d) equal to 3.2 cm. and 4 cm. respectively. The cutoff frequency of the waveguide is 3.24 GHz.

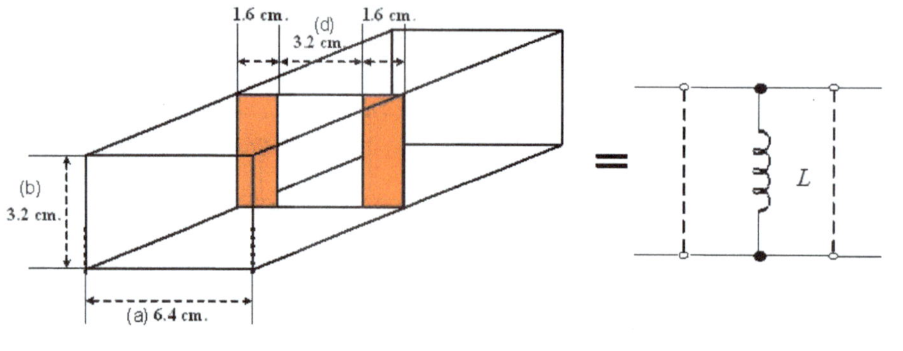

Figure 8. a) Inductive obstacle structure

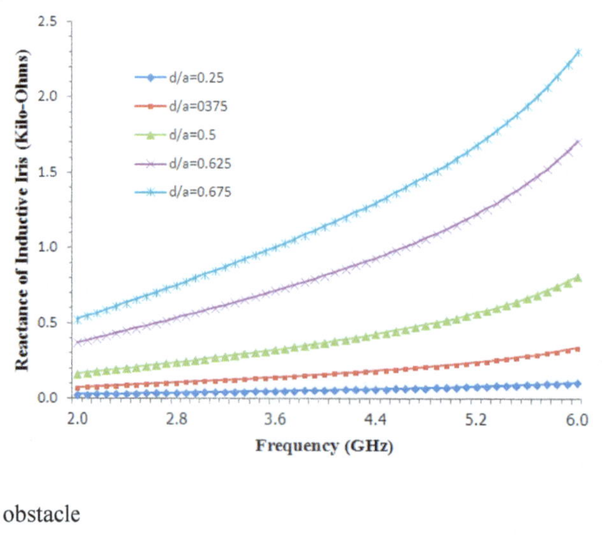

Figure 9. The inductance of inductive obstacle

Fig.9 presents the input reactance of the inductive obstacle in frequency range from 2-6 GHz following the variation the ratio of obstacle and the waveguide width. We observed that the obstacle width increased, the reactance of the inductive obstacle will increase that is consistent with the properties of the inductor.

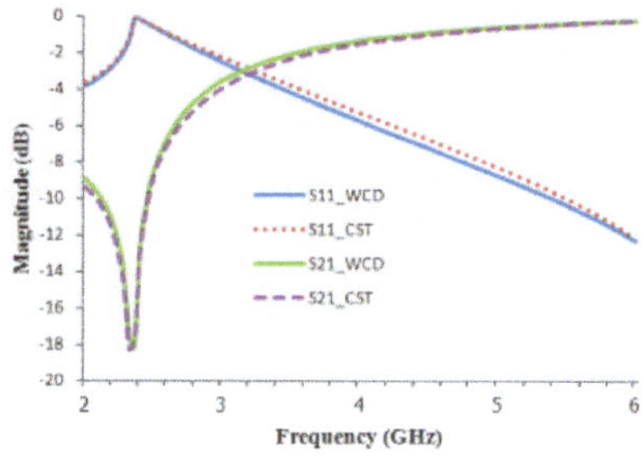

(a) Obstacle structure ratio of d/a = 0.5

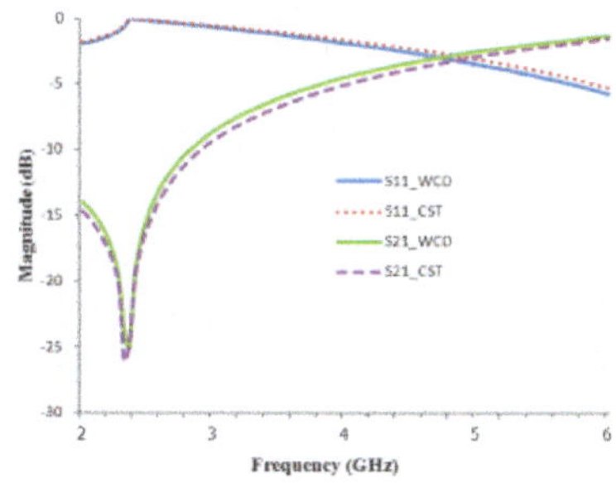

(b) Obstacle structure ratio of d/a = 0.375

Figure 10. Comparison of dB(S11), dB(S21) of the inductive obstacle between the WCD and the CST simulation

Fig.10(a) presents the comparison of dB(S11) and dB(S21) of the inductive obstacle structure between the WCD and the CST simulation In the case of a obstacle width equal to 3.2 cm., we can obtain the -3 dB cutoff frequency at 3.15 GHz. and Fig.10(b) presents the comparison of dB(S11) and dB(S21) of the inductive obstacle structure between the WCD and the CST simulation, at a obstacle width equal to 4.0 cm. We can obtain the -3 dB cutoff frequency at 4.78 GHz. This comparison revealed good agreement. We observed that the shunt inductive circuit is presented as a high pass filter.

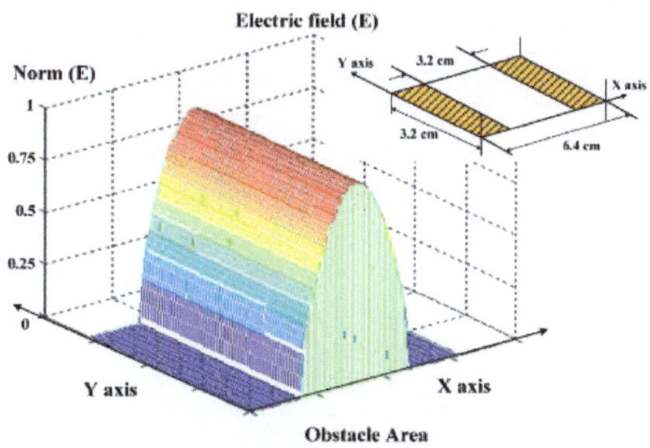

(a) Electric field

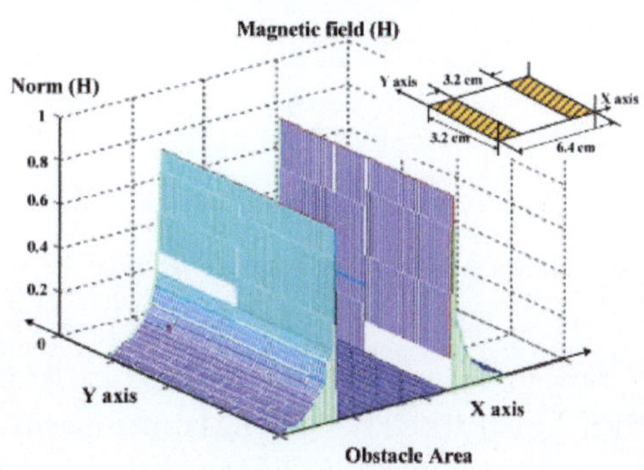

(b) Magnetic field

Figure 11. The 2.5D electromagnetic field illustration.

Fig.11 shows the 2.5D normalized electromagnetic field distributions on the inductive obstacle that are agreements with the electromagnetic field theory. The electric field density of TE10 fundamental mode, as shown in Fig.11(a) is maximum at the center obstacle and is minimum at the discontinuity of boundary obstacle. We observed that the electric field on the conducting surface is null. In Fig.11 (b),the amplitude of magnetic field on discontinuity of boundary obstacle is peak, and decreases until to zero following the distance on the both sides of the conductor.

Capacitive Obstacle Analysis

The horizontal obstacle section transforms the capacitive equivalent circuit, shown in Fig.12. The dimensions of rectangular waveguide consists of a width (a) equal to 6.4 cm., a height (b) equal to 3.2 cm., and the obstacle of iris circuit equal to 0.8 cm. and 0.4 cm. respectively.

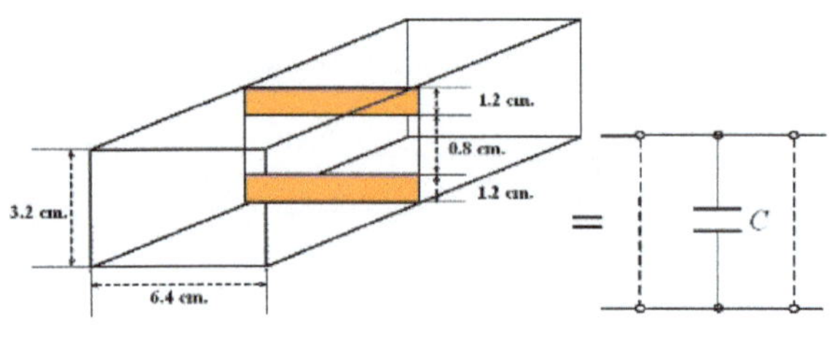

Figure 12. a) Capacitive obstacle structure

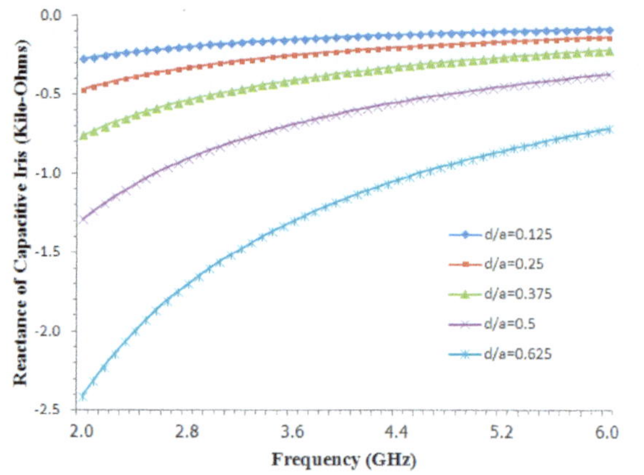

Figure 13. The capacitance of capacitive obstacle

Fig.13 presents the input reactance of the capacitive obstacle in frequency range from 2-6 GHz following the variation the ratio of obstacle and waveguide width. We observed that the obstacle width increased, the reactance of the capacitive obstacle will increase that is consistent with the properties of the capacitor.

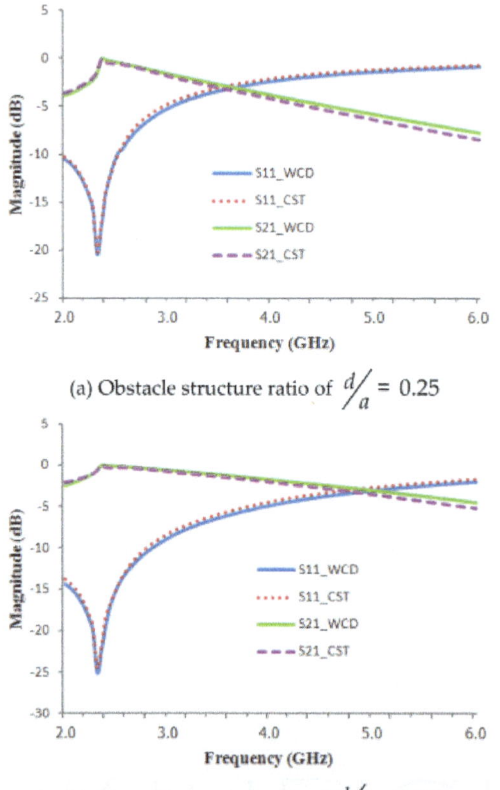

(a) Obstacle structure ratio of $d/a = 0.25$

(b) Obstacle structure ratio of $d/a = 0.125$

Figure 14. Comparison of dB(S11), dB(S21) of capacitive obstacle between the WCD and the CST simulation

Fig.14(a) presents the comparison of dB(S11) and dB(S21) of the capacitive obstacle structure between the WCD and the CST simulation. In the case of a width of obstacle structure is equal to 0.8 cm., we can obtain the -3 dB cutoff frequency at 3.43 GHz. Fig.14(b) presents the comparison of dB(S11) and dB(S21) of capacitive obstacle structure between The WCD and the CST simulation, of the obstacle structure width equal to 0.4 cm. We can obtain the -3 dB cutoff frequency at 4.82 GHz. This comparison revealed good agreement. We observed that the shunt capacitive circuit is presented as a low pass filter.

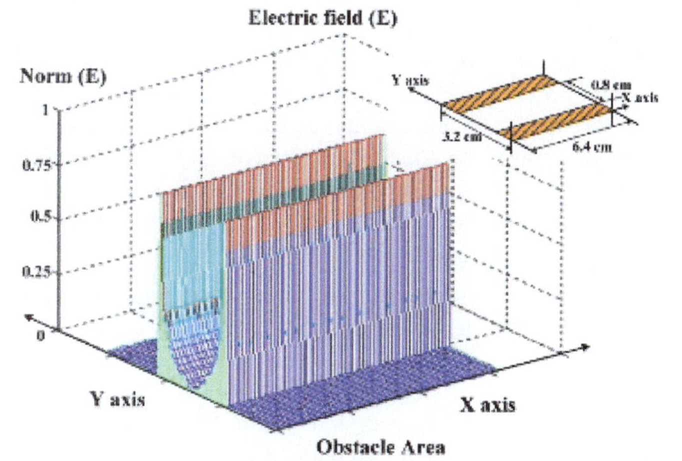

(a) Electric field

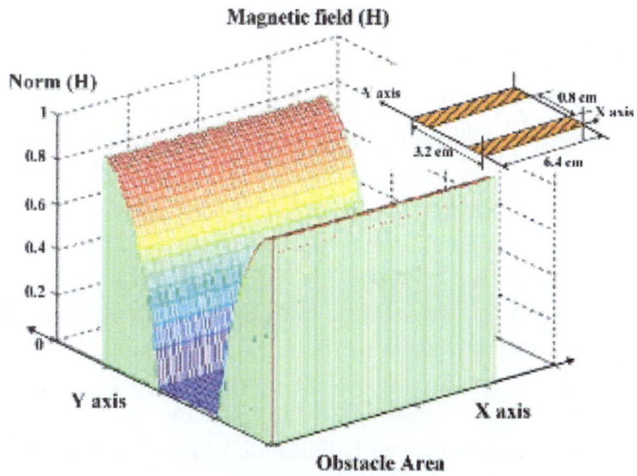

(b) Magnetic field

Figure 15. a) The 2.5D electromagnetic field illustration

Fig.15 shows the 2.5D normalized electromagnetic field distributions on the capacitive obstacle that are agreements with the electromagnetic field theory. The electric field density of TE_{01} fundamental mode, as shown in Fig.15(a), is minimum at the center obstacle and is maximum at the discontinuity of boundary obstacle, and we observed that the electric field on the conducting surface is null. InFig.15(b), the amplitude of magnetic field on the both sides of the conducting surface in y direction is maximal and decreases until to zero at the obstacle internal edges.

Rectangular Window Analysis

Finally, we study the rectangular window with a width of window equal to 0.9 cm. and a height equal to 0.675 cm., then we compare the analyzed results between the WCD and the CST simulation. The rectangular window section in the waveguide transforms the LC shunt parallel resonant circuit, as shown in Fig.16. We use the dimensions of rectangular waveguide consisting of a width (a) equal to 2.4 cm., a height (b) equal to 1.8 cm., the rectangular window dimensions are determine as in Fig.16. The cutoff frequency of the waveguide is 6.25 GHz. After simulating by the WCD tool, the input reactance of the rectangular window can be obtained at a resonant frequency of 15.5 GHz. An example of the reactance variation on the WCD display window is shown in Fig.17.

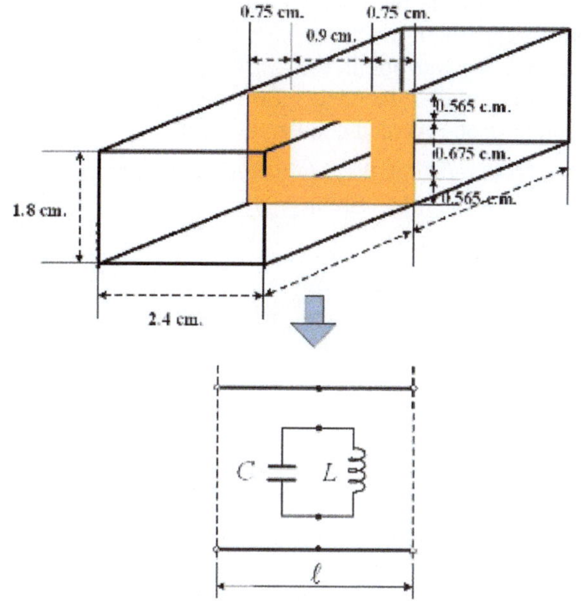

Figure 16. a) The rectangular window structure

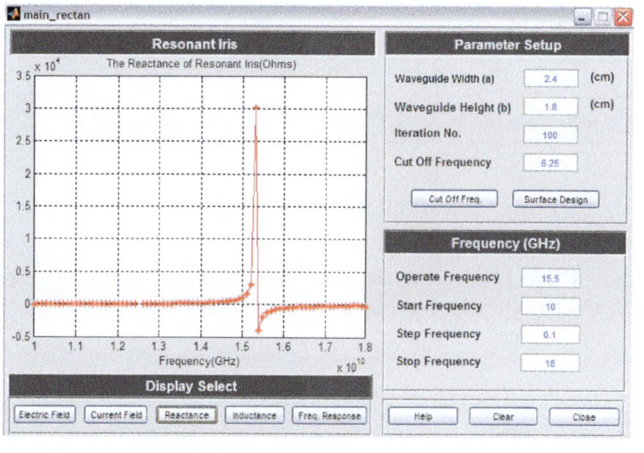

Figure 17. The reactance of rectangular window

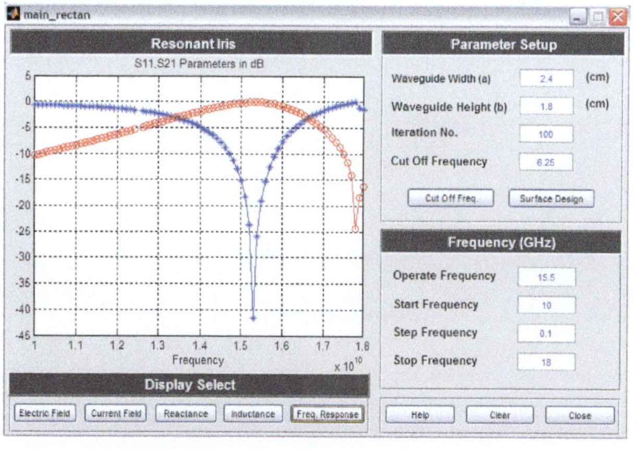

Figure 18. The dB(S11) and dB(S21) parameters of rectangular window

Fig.18 shows the dB(S11) and dB(S21) parameters of the operating frequency of 10-18 GHz, the rectangular windows is presented as a band pass filter, the -3 dB lower and upper cutoff frequency is at 15.25 GHz and at 16.12 GHz respectively. We observed that the center frequency is equal to 11 GHz and the bandwidth is equal to 2.98 GHz.

Fig.19 presents the comparison of dB(S11) and dB(S21) of the rectangular obstacle structure between the both simulations, we can obtain the -3 dB lower cutoff frequency at 13.25 GHz. and -3 dB upper cutoff frequency at 16.35 GHz. This comparison revealed good agreement.

We observed that the equivalent circuit at the resonant frequency is a shunt LC circuit.

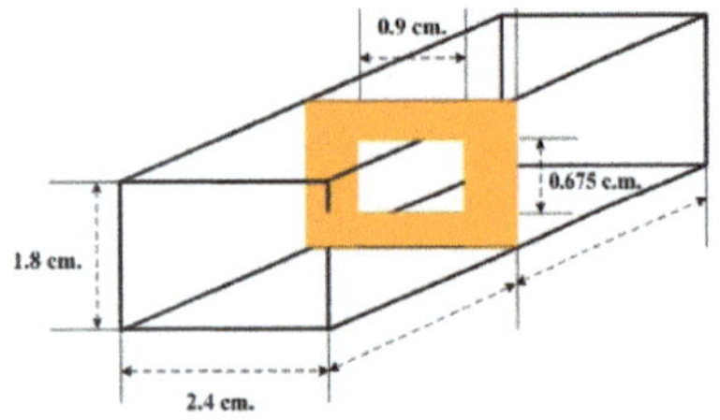

(a) Rectangular obstacle structure

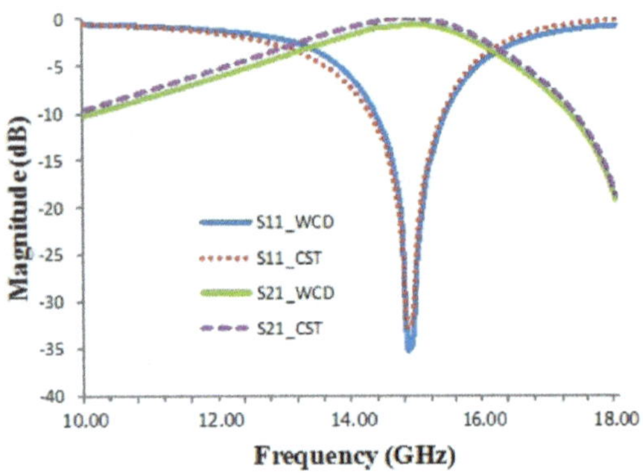

(b) Comparison of dB(S11), dB(S21) of rectangular obstacle

Figure 19. a) The rectangular obstacle structure and the comparison of dB(S11), dB(S21) of rectangular obstacle between the WCD and CST simulation

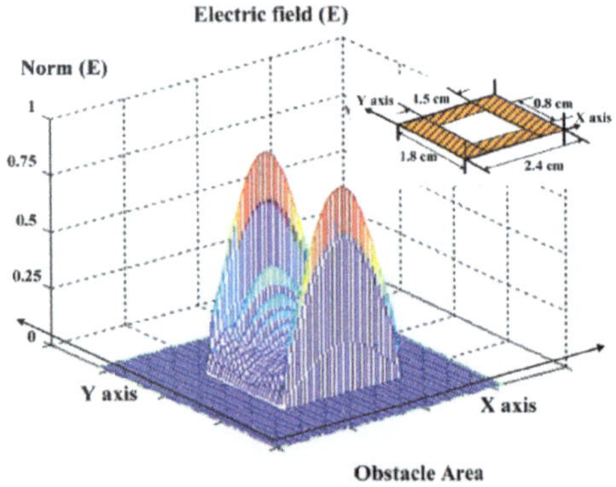

(a) Electric field

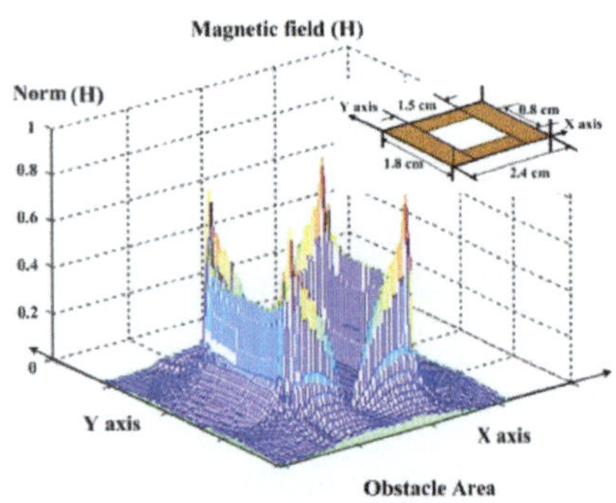

(b) Magnetic field

Figure 20. a) The 2.5D electromagnetic field illustration

Fig.20 shows the 2.5D electric and current distributions of TE01 fundamental mode on the rectangular window which are agreements with electromagnetic field theory. In Fig.20 (a), the normalized electric field peak is at the obstacle edge in x direction and minimum values are at the rectangular center window. In Fig.20(b), most current density distributions are presented at around conducting edge. The field density distribution can be used to lean qualitatively about the obstacle's operating behavior.

For efficiency of calculation, the computation time of the WCD simulation depends on the EM algorithm associating a pixel of obstacle structure to a class in an iterative process. In this simulation, the calculation at the 10 frequencies steps with 32*64 of cell numbers and 1 mm. of cell size will be equal to approximately 45 seconds of computing time, which is less than 2 times of the CST simulation (run on Pentium IV Processor, 1 GB of RAM at 3.0 GHz of speed).

CONCLUSION

The capability of developed WCD simulation based on the Wave Iterative Method (WIM) has been presented to the complex description of the 2.5D electromagnetic field and the characteristic analysis of the obstacle structures. The WCD simulation is properly conceived to apply for the inductive and the capacitive element in the rectangular waveguide without resorting to extensive computation times; it appears as an efficient alternative to the CST simulation. In the case of wave propagation, the WCD simulation can use shown the electromagnetic field distributions on the conducting obstacle. In conclusion, the WCD simulation can be used in the waveguide lumped circuit analysis and has proven to be with high accuracy and efficacy a useful tool for microwave circuit design and an educational aid.

ACKNOWLEDGEMENT

The authors would like to thank the Faculty of Technical Education, Rajamangala University of Technology Thanyaburi, Thailand, for providing the CST Microwave Studio® software for data analysis.

REFERENCES

1. T. Hiraoka, C. P. Chen, T. Anada, J. P. Hsu, Z. W. Ma, and C. C. Christopoulos, "Electric Field Distributions in Microwave Plana Circuits by Small Coaxial Probe and Comparison with FDTD Method," in Microwave Conference, European, pp.269-272, 2005.
2. J. W. Bandler, A. S. Mohamed, and M. H. Bakr, "TLM-based Modeling and Design Exploiting Space Mapping," Microwave Theory and Techniques, IEEE Transactions on, vol.53, pp. 2801-2811, 2005.
3. S. Seunghyun, K. Hyeong-Seok, J. Hyun-Kyo, and M. Un-Chul, "Frequency Domain Analysis of Microstrip Filters and Antennae using an Adaptive Frequency Sampling Moment Method," Magnetics, IEEE Transactions on, vol.42, pp.607-610, 2006. 352 Wave Propagation Theories and Applications
4. V. K. Chaudhary, P. Verma, and U. Balaji, "Field Theory Based CAD of Inductive Iris Waveguide Filter," in Asia-

Pacific Microwave Conference, APMC 2001, vol.1, pp.318- 321, 2001.

5. Sallier, J. Bornemann, and W. J. R. Hoefer, "Field-Based Waveguide Filter Synthesis in the Time Domain," AEU - International Journal of Electronics and Communications, vol.57, pp.119-127, 2003.
6. C. Hunter, L. Billonet, B. Jarry, P. Guillon, "Microwave Filters-Applications and Technology", Microwave Theory and Techniques, IEEE Transactions on, vol.50, pp. 794 – 805, 2002.
7. Y. P. Zhang, H. Guorui, Z. Wenmei and J. H. Sheng, "Electromagnetic Mode Theory of Periodically-Loaded Oversized Imperfect Waveguide and its Application to the Propagation of Radio Waves in Long Wall Coal Mining Face Tunnels", Antennas and Propagation, IEEE Transactions on, vol.58, no.5, pp.1816-1822, 2010.
8. N.Marcuvitz, "Waveguide Handbook", Short Run Press Ltd., London, UK, 1986.
9. S. Akatimagool, D. Bajon, and H. Baudrand, "Analysis of Multi-layer Integrated Inductors with Wave Concept Iterative Procedure (WCIP)," in Microwave Symposium Digest, 2001 IEEE MTT-S International, vol.3, pp.1941-1944, 2001.
10. S. Khamkleang and S. Akatimagool, "Microwave Planar Circuit Design Tool in the teaching of Microwave Engineering," in Electrical Engineering/Electronics, Computer, Telecommunications and Information Technology, ECTI-CON 2009, 6th International Conference on, pp.830-833, 2009.
11. S. Choocadee and S. Akatimagool, "Development of Efficiency EM Simulation Tool for Capacitive and Inductive Obstacle Analysis," in Electrical Engineering/ Electronics Computer Telecommunications and Information Technology (ECTI-CON), International Conference on, pp.1154-1158, 2010.
12. H.Baudrand,"Introduction of Microwave Circuit ", Toulouse, France, 1993.

CHAPTER 16

DIRECT OBSERVATION OF ISOLATED DAMON-ESHBACH AND BACKWARD VOLUME SPIN-WAVE PACKETS IN FERROMAGNETIC MICROSTRIPES

Philipp Wessels, Andreas Vogel, Jan-Niklas Tödt, Marek Wieland, Guido Meier & Markus Drescher

INTRODUCTION

Today's advanced electronic devices operate at high frequencies to complete complex tasks on reasonable time scales. These circuits solely use the charge of the electron to transport and process information. Future computer and memory concepts may use the electron spin instead of or in addition to its charge for data transport and storage as well as logical operations1·2·3·4·5. Such spintronic and magnonic devices benefit from the fact that frictional heating in magnetic transport is negligible compared to charge transport in conventional electronic devices where the resulting heat load imposes serious limitations in building ever smaller and faster circuits. Spin-waves appear as an ideal solution for fast magnetic information transport since their natural frequencies in the gigahertz regime are compatible with clock rates of modern computer processors and mobile communication technology. Extensive research has been carried out both theoretically and experimentally to characterize the dynamic parameters of spin-waves and first logic devices have been realized3·6·7·8·9·10·11·12·13 in which the spin-wave is excited locally via an Oersted field14·15·16 or through spin-torque17·18·19. Many studies consider the steady state where a spin-wave is continuously excited at microwave frequencies. Information transport and processing, however, relies on the dynamical modulation of a wave's amplitude or phase. For the realization of applicable spintronic devices, knowledge about the transient features is mandatory, i.e. spin-wave packets confined in space and in time are to be considered11·20·21·22·23.

In addition to fundamental insight into magnetic transport, spin-wave packets can serve as a simulator toolkit for electromagnetism where tunable dispersion relations in confined structures24·25·26 give rise to unusual wave packet dynamics. Backward volume (BV) spin-wave packets with small wave vectors$\mathbf{k}$, for example, exhibit a phase front motion with a phase velocity v_{ph} opposite to the propagation direction associated with the group velocityv_g. A negative scalar product in form of $v_{ph} \cdot v_g < 0$ is equivalent to a negative refractive index $n < 0$ in the optical regime27·28·29. This conceptual similarity shall enable model studies of waves30 also in the field of metamaterials where negative phase velocities occur31·32.

Parameters of spin excitations like spatial dimensions and damping mainly depend on the choice of materials and excitation mechanisms. Spin-waves in monocrystalline ferrimagnetic yttrium iron garnet $Y_3Fe_5O_{12}$ (YIG) films33 can be efficiently excited and feature a low damping constant; however, manufacturing of the monocrystalline layer is demanding and the wavelength is typically in the mm-range. Metallic ferromagnets like permalloy ($Ni_{80}Fe_{20}$) support micro-21·22·34·35 and nanoscale17·18·19 excitations meeting the size of functional units in computer and communication devices dictated by the high quantity of embedded logical units. The majority of experimental studies in ferromagnetic thin films focuses on Damon-Eshbach (DE) spin-waves36 where an external magnetic bias field $H_{ext}$is applied perpendicular to the propagation direction. Although backward volume (BV) spin-waves are most efficiently supported with a bias field parallel to the propagation path, they may even propagate without

Citation: Philipp Wessels, Andreas Vogel, Jan-Niklas Tödt, Marek Wieland, Guido Meier & Markus Drescher, Direct observation of isolated Damon-Eshbach and backward volume spin-wave packets in ferromagnetic microstripes, doi:10.1038/srep22117.

an external field[37] and are therefore desirable candidates for logic devices. BV spin-wave modes have been reported in YIG[38] including the observation of wave-packets[39,40] and solitons[41,42,43,44]. Downscaling of spin-wave dynamics using metallic ferromagnets is possible but the generation and control of BV wave packets is difficult[45,46,47,48] due to the lower excitation efficiency[37]and the small group velocity. Thus, a direct spatiotemporal observation of BV modes with counterpropagating phase- and group-velocity remains challenging and is often only indirectly achieved out of propagating DE spin-waves[49,50,51]. Recently, BV modes with counterpropagating phase- and group-velocities have been observed in ferrimagnetic insulators via direct excitation through circularly polarized femtosecond laser pulses[52]. Our work reports on similar excitations in metallic ferromagnets through a dedicated excitation scheme enabling the propagation of dispersive BV spin-wave packets with considerably smaller wavelength and damping. This is achieved through a novel optically driven high-current pulse source supporting excitations up to a bandwidth of 50 GHz via short magnetic field pulses of <70 ps FWHM duration, perfectly synchronized to the sub-ps optical probe pulses.

RESULTS

Experimental Setup

Combining magnetic microscopy with an excite & probe technique allows us to directly observe the spatiotemporal evolution of isolated wave packets. The experimental setup shown in Fig. 1 comprises a time-resolved scanning Kerr microscope[53] (TR-SKM) detecting the transient magnetization response normal to the surface (z-direction) through the magneto-optic Kerr effect (MOKE)[54,55], probed with femtosecond laser pulses in the visible spectral range (515 nm wavelength). In this work we excite spin-waves in an impulsive manner through a novel optically driven current source[53]. The green laser pulses are doubled in frequency to create synchronized, delayable ultraviolet (UV) pulses sufficiently energetic (4.8 eV photon energy) for releasing photoelectrons from a magnesium (Mg) surface exposed to a static electric field in a high-vacuum environment. The Mg photocathode is attached to a tapered coplanar waveguide (CPW), transporting the intense broadband current pulses to a magnetic sample in form of a rectangular permalloy layer of 35 × 10 μm²dimension and 30 nm thickness in 5 μm partial overlap with the waveguide conductor (see magnified inset in Fig. 1). These pulses are terminated in a serial connection of two radio-frequency resistors with 50 GHz bandwidth. Since current source, termination, magnetic sample and waveguide are combined on a single sapphire substrate, the choice of the waveguide impedance is arbitrary and enables the preparation and investigation of complex sample geometries. Here, an overall impedance of 100 Ω was chosen resulting in a CPW with a central 10 μm wide conductor surrounded by two ground planes separated by a gap of 48.5 μm, large enough to fit the permalloy sample layer without shorting conductor and ground striplines (seeSupplementary Fig. S1 for details of the central waveguide part). This excitation approach differs from established photoconductive switches such as an Auston switch commonly used for triggering spin precession dynamics[21,56]. The photocathode achieves superior peak currents[53] exceeding 0.5 A and avoids asymmetric pulse shapes as found in Auston switches due to long carrier lifetimes. Such pulses may influence the magnetization dynamics by a residual dynamic bias field in the trailing edge of the pulse[21,56] (compare alsoSupplementary Fig. S4). However, the photocathode can only be operated in an ultra-high vacuum environment and requires laser pulses in the UV spectral range.

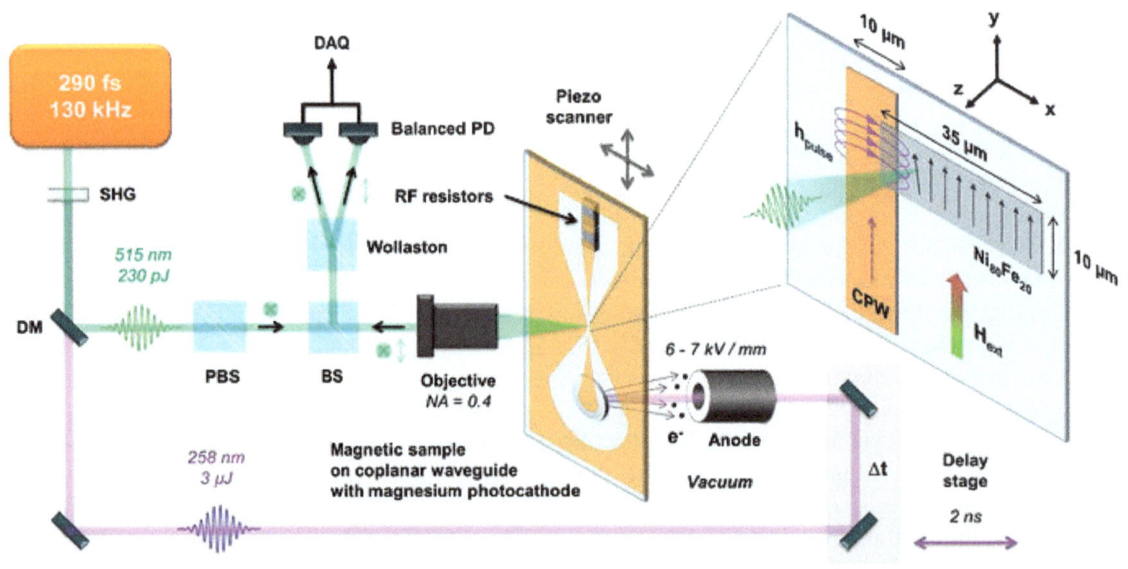

Figure 1: Schematic of the time-resolved scanning Kerr microscope (TR-SKM) and the sample. Visible pulses of a highly repetitive femtosecond laser amplifier are converted into the UV spectral range via second harmonic generation (SHG) to trigger a photocathode using the photo effect and an electrostatic extraction field. This allows the generation of intense and broadband current pulses for spin-wave excitation through a coplanar waveguide (CPW) terminated in two radio-frequency (RF) resistors. The residual visible light pulses are separated by a dichroic mirror (DM), polarization filtered by a polarizing beam splitter (PBS), and focused onto the magnetic sample layer. The reflected light is recollected and directed into a Wollaston prism through an intensity beam splitter (BS). The polarization change due to the magneto-optic Kerr effect (MOKE) is analyzed and collected by the data acquisition (DAQ) in a balanced photodetector (PD). The sample can be scanned through the probe laser focus by a two-axis piezo scanner.

The sample magnetization is prealigned by an external bias field $\mathbf{H}_{ext}$ generated by a set of permanent magnets and locally modified by the dynamic field of the current pulse $\mathbf{h}_{pulse}$ in the overlap region of waveguide conductor and permalloy layer. Once exposed to the excitation field, the magnetic moments start to precess; due to their mutual interaction this precessional motion is transferred to neighboring magnetic moments thus releasing a spin-wave packet. The response of the spin system is governed by the frequency f and the spatial field configuration of the source. While spatial dimensions dictate the excited wavelengths λ or wavenumbers $k = 2\pi/\lambda$, the angular frequency $\omega = 2\pi f$ drives the gyration motion of the magnetic moments. Only certain combinations of frequency and wave vector $\mathbf{k}$ allowed by the dispersion relation $\omega(\mathbf{k})$ plotted in Fig. 2a may lead to a propagating dispersive spin-wave packet. Here, the impulsive magnetic excitation emerges from femtosecond laser pulses and the pulse duration is limited by the bandwidth of waveguide and termination resistors as well as space charge effects in the photocathode and excitation pulses of 70 ps FWHM duration are achievable leading to a broadband excitation spectrum (see Methods section). Thus, the propagating spin-wave is dominated by the geometry of the excitation field given by its Fourier transform[21,22] shown in Fig. 2b. A spin-wave packet is then composed of the superposition of all excited wave vectors $\mathbf{k}$ convoluted by the excitation probability in this spectrum $P(\mathbf{k})$. The observable zcomponent $M_z(\mathbf{r}, t)$ reads

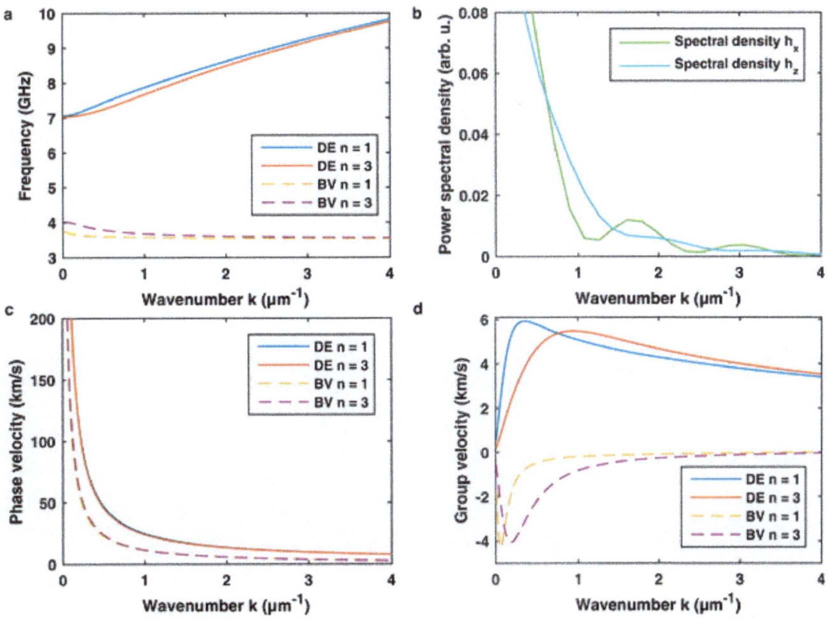

Figure 2: Spin-wave dispersion relation with excitation spectrum and derived phase- and group velocities. (**a**) Dispersion relation $f(k)$ for the first two quantized transversal Damon-Eshbach (DE) ($n = 1$ blue solid line, $n = 3$ orange solid line) and backward volume (BV) ($n = 1$ yellow dashed line, $n = 3$ purple dashed line) spin-wave modes of a permalloy ($\mu_0 M_s = 1.04$ T, $A = 13 \cdot 10^{-12}$ J/m) stripe with an effective width of $w_{eff} = 10.12$ µm, an effective length of $l_{eff} = 35.1$ µm and 30 nm thickness. The DE modes are calculated for an external field of $\mu_0 H_{ext} = 60$ mT amplitude reduced by the internal demagnetization field while the BV modes are calculated for an external field of $\mu_0 H_{ext} = 16$ mT amplitude. (**b**) Power spectral density $P(k)$ of the excited wavenumbers k by the magnetic field pulse determined by the Fourier transformation of the spatial magnetic field configuration for the h_x component (green solid line) and the h_z component (cyan solid line). (**c**) Phase velocities $v_{ph} = \omega(k)/k$ for all calculated spin-wave modes. (**d**) Group velocity $v_g = \partial\omega(k)/\partial k$ for all calculated spin-wave modes.

$$M_z(\mathbf{r}, t) \propto \int_0^{k_c} \underbrace{P(\mathbf{k}, \omega(\mathbf{k}))}_{\approx P(\mathbf{k})} \cdot \sin(\mathbf{k} \cdot \mathbf{r} - \omega(\mathbf{k})t) \cdot e^{-\alpha\omega(\mathbf{k})t} d\mathbf{k} \tag{1}$$

with a damping term[52] related to the dimensionless damping parameter $\alpha = 0.008$ for permalloy shown in equation (1). A cut-off wave vector $\mathbf{k}_c$ can be introduced for numerical integration because the excitation probability approaches zero for high wave vectors. Due to the broad frequency spectrum, the dispersion relation $\omega(\mathbf{k})$ defines the gyration frequency f connected to the specific wave vector k.

In addition to DE modes with co-propagating phase- and group velocities, high peak amplitudes enable us to also observe BV modes where the phase fronts do not emerge out of, but rather approach the excitation volume. This unusual fact emerges from the negative slope of the BV dispersion relation in Fig. 2a formally related to a negative group velocity $v_g = \partial\omega(k)/\partial k$ counteraligned to the phase front motion. Before focusing on the backward volume spin-wave packets, we demonstrate the capabilities of our setup by mapping the evolution of Damon-Eshbach modes.

Damon-Eshbach spin-wave packets

Figure 3 presents an illustration of the precessional motion (a) together with selected frames of a captured DE spin-wave packet (b) of Supplementary Video S5. Here, the external bias field is oriented iny direction, perpendicular to the propagation direction of the wave packet. The width of the permalloy layer extends from y=−5 to 5 μm with the center located at y=0 μm. The overlap with the waveguide conductor alongside the y axis ranges from x=0 to 5 μm. Upon arrival of the dynamic field pulse at the time t=0 ps, the M_z component of the magnetization starts to oscillate in a spatially confined source region where the permalloy structure partially covers the waveguide. As time elapses, a wave packet is released and propagates down the xaxis.

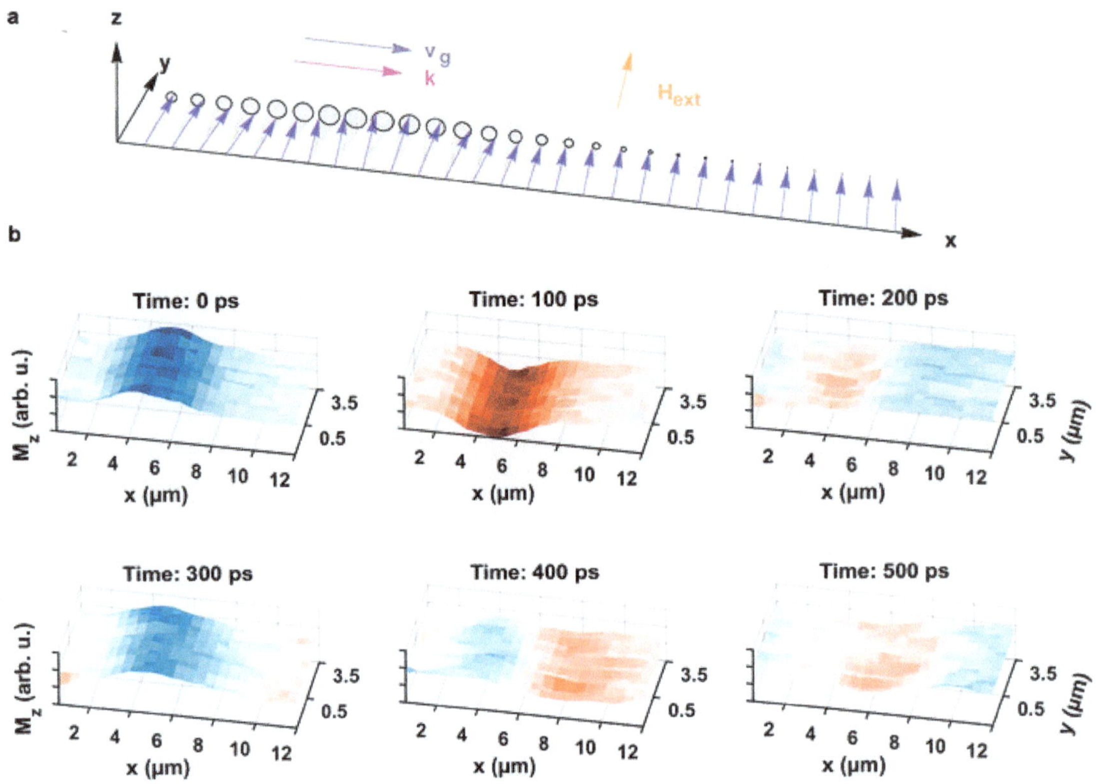

Figure 3: Propagation of a Damon-Eshbach spin-wave packet. (**a**) Precessional motion of the magnetic moments in a Damon-Eshbach spin-wave packet. The propagation direction of the phase fronts is determined by the wave vector $\mathbf{k}$ which is perpendicular to the external bias field $\mathbf{H}_{ext}$ and parallel to the group velocity $\mathbf{v}_g$ determining the propagation direction of the wave packet. (**b**) Selected frames of the Damon-Eshbach spin-wave movie illustrating the real space time evolution of the out-of-plane magnetization component M_z at six delay settings. In contrast to the frames in Fig. 6b the phase fronts emerge out of the excitation region toward higher x coordinates (see also Supplementary Video S5).

This process is depicted in more detail in Fig. 4. Here, the color coded detectable M_z component is plotted with respect to time and one spatial coordinate. Figure 4a shows the spin-wave propagation $M_z(x, t)$ in x direction along a fraction of the extended 35 μm long permalloy slab axis at a fixed transverse coordinate y=0 μm. With evolving time, new phase fronts emerge out of the excitation region (0 to 5 μm) toward larger x coordinates leading to a positive slope in the $x(t)$ diagram in Fig. 4a when tracing the oscillation maxima and minima. An evaluation of the slopes in terms of straight lines (black dashed lines in Fig. 4a) yields phase velocities ranging from $v_{ph} = \Delta x/\Delta t$ = 70 to 150 km/s (see Supplementary Fig. S2a). With elapsed time, the oscillation amplitude decreases in the source region around x=5 μm and almost vanishes there after 600 ps. But due to the interaction of the magnetic moments, a spin-wave packet starts to propagate toward increasing x positions. However, the oscillation there is not as confined in time as in the source region because the wave packet spreads. Additionally, the plot enables an extraction of the $\mathbf{k}$-vector magnitude by measuring the distance Δx between an oscillation maximum and minimum which is related to half the wavelength $\lambda/2$and thus to the wave vector magnitude via k= $2\pi/\lambda$. An evaluation yields k=0.3 to 0.6 μm^{-1} in accordance with the excitation spectrum due to the waveguide overlap with the permalloy layer.

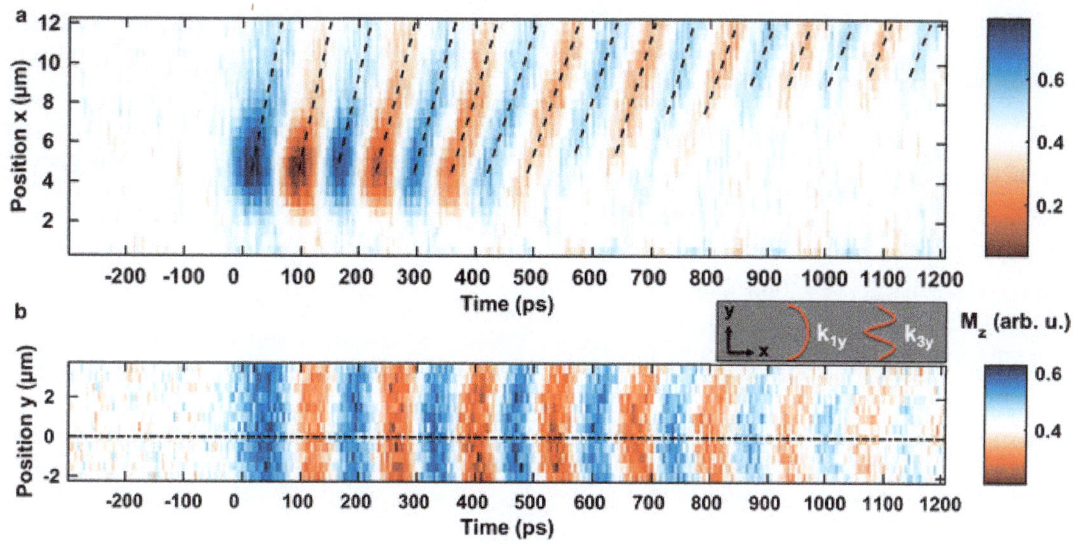

Figure 4: Details of the DE spin-wave packet propagation in a permalloy stripe. The color code identifies the detectable transient M_z component of the magnetization and the static bias field of $\mu_0 H_{ext} \approx 60$ mT amplitude is aligned in y direction. (**a**) Propagation in x direction at a fixed transverse position $y = 0$ µm in the center of the slab with respect to time. Black dashed lines trace the maxima and minima of the wave and thus relate to the phase velocity v_{ph}. (**b**) Temporal evolution of the spatial transverse mode distribution at a fixed longitudinal position ($x = 8$ µm). The amplitude of the first two modes k_{1y} and k_{3y} is depicted in the inset. The center of the slab marked by the black dash-dotted line is located at $y = 0$ µm slightly off from the center of the y scan range in this measurement.

Figure 4b traces the oscillation $M_z(y, t)$ at a fixed longitudinal position of x = 8 µm and shows the temporal evolution of the spatial modes along the 10 µm width of the permalloy structure parallel to the yaxis. We observe a transition from an initially constant phase over the slab into a curved phase front. This self-focusing of the spin-wave packet can be explained by an interference of the dominant first two quantized transverse modes[34,35,57] with a wavenumber of $k_{ny} = n\pi / w_{eff}\ \forall n \in \{1,3\}$ since only odd modes can be excited by a uniform radio-frequency field[58]. The effective stripe width w_{eff} is introduced to account for dipolar pinning effects at the sample boundaries reducing the magnetization because of the demagnetization field[25]and here computes to $w_{eff} = 10.12$ µm. Knowing that the amplitude of the mode scales with[58] $a_n \propto 1/n$, we can model the response of the magnetization oscillating with a frequency ω by[35,57]

$$M_z(x, y, t) \propto \left|\cos\left(\frac{\pi}{w_{eff}} \cdot y\right) \cdot \sin(kx - \omega t) + \frac{1}{3} \cdot \cos\left(\frac{3\pi}{w_{eff}} \cdot y\right) \cdot \sin(kx - \omega t + \varphi)\right|. \tag{2}$$

The first two transversal modes corresponding to the two cosine terms in equation (2) are depicted in the inset of Fig. 4b. In general, these modes will be phase-shifted by φ with respect to each other. In this model, a phase shift of $\varphi = 40°$ reproduces the observed curvature in the experiment.

Plotting the $M_z(t)$ component with respect to time at two different points $(x, y) = (8$ µm, 0 µm$)$ and $(12$ µm, 0 µm$)$ in the center of the slab along the propagation direction of the wave packet in Fig. 4a helps to further clarify the evolution of the wave packet. The corresponding TR-SKM data is shown in Fig. 5 including a fit according to

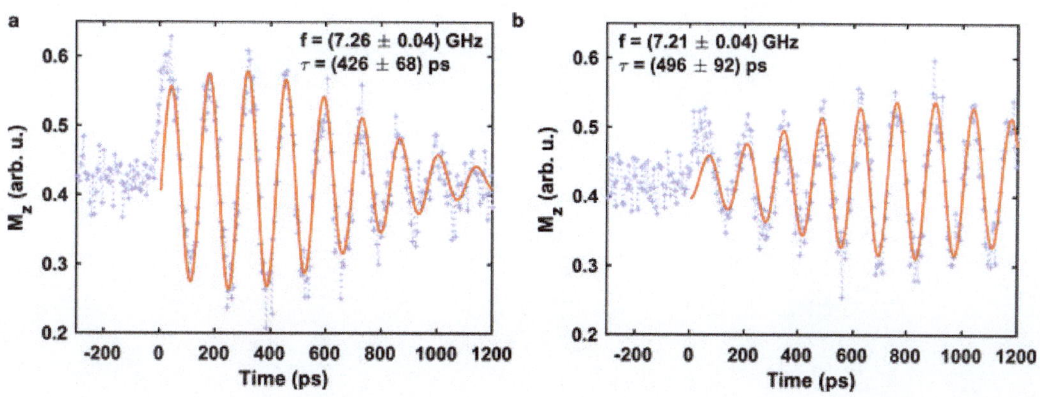

Figure 5: DE spin-wave packets. (**a**) At position $(x, y) = (8$ µm, 0 µm$)$ closer to the source region, and (**b**) at position (12 µm, 0 µm) further away from the source region in the center of the slab. The blue crosses represent the measured Kerr signal and the red solid line shows a fit to the data according to equation (3). The peak amplitude of the wave packet in (**b**) is reduced to 70% of the wave packet amplitude in (**a**).

$$M_z(t) = M_0 + M_A \cdot \exp\left(-\frac{(t-t_p)^2}{2 \cdot \tau^2}\right) \cdot \sin(\omega t + \varphi). \tag{3}$$

M_0 is an offset of the magnetization signal, M_A determines the peak amplitude of the wave packet, φ a phase offset, t_p the time of the wave packet envelope maximum, τ the root mean square (rms) wave packet duration and $\omega = 2\pi f$ the central angular oscillation frequency. Data points at negative time delays before the arrival of the excitation pulse have been excluded from the fit.

This method enables a global transient analysis of the above mentioned spin-wave parameters during formation and propagation of the wave packet. In particular, we are able to extract the carrier frequency f≈7.2 GHz which is in good agreement with the spin-wave dispersion relation in Fig. 2a and it is obvious that the wave packet duration τ increases while the peak amplitude M_A decreases along its way down the x axis. Furthermore, the center of the wave packet envelope t_p shifts to a later instance in time for an increased distancex from the source region meaning that it takes more time to travel to a distance further away from the excitation field. Hence, the group velocity vector points into the same direction as the phase velocity vector, namely into positive x direction. By relating the center of the wave packet t_p to the propagated distance x we can extract the conveyor speed of the whole wave packet which is described by the group velocity

$$v_g = \frac{\partial \omega}{\partial k} = v_{ph} + k\frac{\partial v_{ph}}{\partial k} = \frac{\Delta x}{\Delta t_p}. \tag{4}$$

Applying the fit of equation (3) to the measured wave packet data at different distances x provides a linear relation between x and t_p (seeSupplementary Fig. S2b) and yields a positive group velocity of $v_g = (6.0 \pm 0.8)$ km/s. This velocity is consistent with the group velocity calculated by computing the derivative of the spin-wave dispersion relation in Fig. 2d for the observed wavenumbers.

With knowledge of the spin-wave packet's oscillation frequency, the phase velocity can also be calculated via

$$v_{ph} = \frac{\omega}{k} = \frac{\Delta x}{\Delta t}. \tag{5}$$

By inserting the extracted wavenumbers from Fig. 4a ($k = 0.3$ to $0.6\,\mu m^{-1}$) and the carrier frequency (f≈7.2 GHz), phase velocities ranging from 75 to 151 km/s are obtained in agreement with the directly extracted velocities of the phase fronts in Supplementary Fig. S2a and with the phase velocities calculated using the dispersion relation inFig. 2c resulting in values between 70 and 150 km/s for wavenumbers between 0.2 and 0.4 μm^{-1}.

Backward volume spin-wave packets

For the acquisition of backward volume spin-wave packets, the external bias field is tilted by 90° and is now oriented in −x direction, antiparallel to the propagation direction of the wave packet. Note that the amplitude of the external field also changes due to the sample geometry (see Methods section). Figure 6 presents a space-time illustration of the precessional motion (a) together with selected frames of a captured BV spin-wave packet (b) of Supplementary Video S6. Upon arrival of the dynamic field pulse an oscillation of the $M_z$component is visible in the source region. As time elapses, a new phase front (in this case a blue color coded maximum) starts to approach the excitation region and propagates from larger to lower xcoordinates. When the maximum has arrived, a minimum (red color code) enters the scan window from the right border heading to the overlap region. From the six frames in Fig. 6b it is already obvious that the phase velocity describing the motion of the oscillation maxima and minima is directed along the negative x axis.

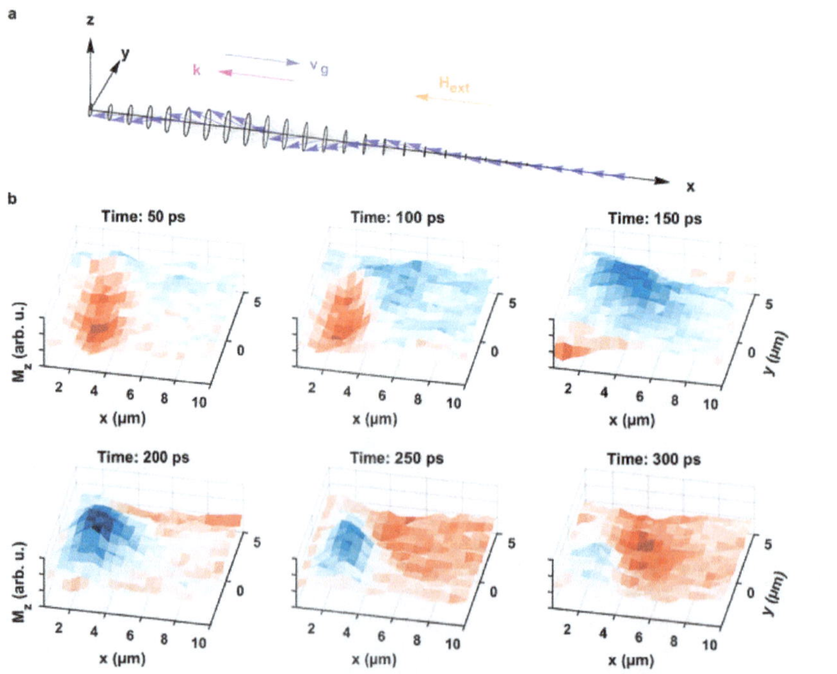

Figure 6: Propagation of a backward volume spin-wave packet. (**a**) Precessional motion of the magnetic moments in a backward volume spin-wave packet. The propagation direction of the phase fronts is determined by the wave vector k which is parallel to the external bias

field $\mathbf{H}_{ext}$ and antiparallel to the group velocity $\mathbf{v}_g$ determining the propagation direction of the wave packet. (**b**) Selected frames of the backward volume spin-wave movie illustrating the real space time evolution of the out-of-plane magnetization component M_z at six delay settings. It is evident that the phase fronts do not emerge out of the excitation region but rather approach the source from higher x coordinates indicating a negative phase velocity (see also Supplementary Video S6).

Figure 7 shows the BV spin-wave packet in the same representation used in Fig. 4 for the DE data. An oscillation of the magnetization can be observed in the overlap region of permalloy and waveguide conductor when the dynamic field pulse arrives at t = 0 ps and in this case the maxima and minima start to approach the excitation region from larger x coordinates with a negative phase velocity indicated by the slope of the black dashed lines. An evaluation of the slopes yields phase velocities ranging from $v_{ph} = \Delta x/\Delta t = -25$ to −60 km/s (seeSupplementary Fig. S3). The excited wave vector magnitude covered here is $k = 2\pi/\lambda = 0.2$ to 0.7 μm^{-1} corresponding to phase velocities ranging from $v_{ph} = \omega/k = -31$ to −110 km/s (f ≈ 3.5 GHz) in agreement with the directly extracted slopes and with the derived phase velocities from the spin-wave dispersion relation in Fig. 2c between −25 and −60 km/s for wavenumbers between 0.2 and 0.5 μm^{-1}.

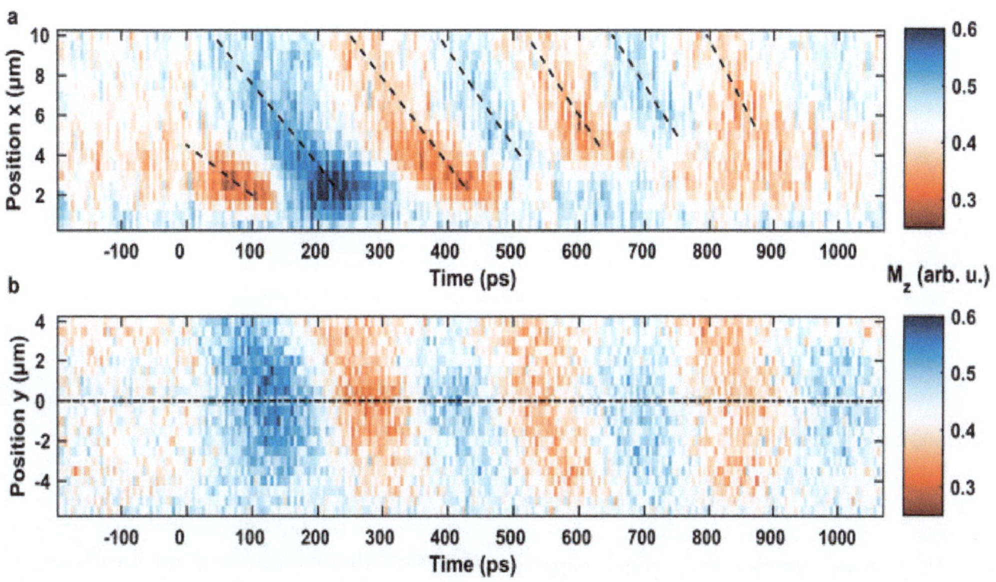

Figure 7: Details of the BV spin-wave packet propagation in a permalloy slab. The color code identifies the detectable transient M_z component of the magnetization and the static bias field of $\mu_0 H_{ext} \approx 16$ mT amplitude is aligned in $-x$ direction. (**a**) Propagation along the elongated x direction at a fixed transverse position $y = 0$ μm in the center of the element. Black dashed lines trace the maxima and minima of the wave and thus relate to the phase velocity v_{ph}. The negative slope of the phase fronts clearly substantiates a negative phase velocity. (**b**) Transverse spatial mode distribution at a fixed longitudinal position $x = 7$ μm with respect to time. The black dash-dotted line at $y = 0$ μm marks the center of the slab.

For the spatial mode distribution $M_z(y, t)$ in Fig. 7b we observe an almost uniform shape and a decay of the oscillation amplitude with time. It is conspicuous that the phase fronts in this image are slightly tilted and that also the phase fronts in Fig. 6b enter the scan window from the top right corner. This means that the propagation direction is not completely parallel to the x axis as would be expected for a perfectly collinear alignment of the external field. An explanation is given by the sensitivity of the group and phase velocity direction on the external field alignment due to the shape of the isofrequency line to which excitable modes are restricted24·59. The observed group velocity vector tilt of approximately 10° corresponds to a magnetic field tilt of 8°. Furthermore, a perfect collinear alignment would result in a vanishing torque on the dipole moments from the $h_x$component of the dynamic field. The excitation would then be mainly driven by the h_z component.

Similar to the Damon-Eshbach wave packets we can plot the $M_z(t)$component with respect to time at two different points $(x, y) = (7$ μm, 1 μm) and (10 μm, 1.5 μm) along the propagation direction of the spin-wave. Because of the slight phase front tilt discernible in Fig. 7b, the wave packet trajectory is tilted by 9.5° with respect to the x axis (compare also Fig. 6b). The result in Fig. 8 includes a fit according to equation (3) and emphasizes the evolution of the wave packet parameters. For the BV modes we obtain a carrier frequency of f ≈ 3.5 GHz in agreement with the spin-wave dispersion relation in Fig. 2a. Analogous to the DE mode the wave packet duration τ increases as the spin-wave emerges and propagates and the center of the wave packet envelope t_p shifts to later moments in time for an increased distance from the source region. The limited scan range showing only a fraction of the wave packet envelope with few oscillations, however, restrains a quantitative analysis of the group velocity as performed on the DE wave packets and only an upper limit of $v_g < 25$ km/s can be extracted. Altogether, we find a propagation of the wave envelope away from the source while the extracted phase fronts from Fig. 7aclearly approach the excitation region meaning that $\mathbf{k}$ is counteraligned to $\mathbf{v}_g$. This concludes that we have directly observed backward volume modes with negative $\mathbf{k}$-vectors that formally correspond to optical wave packets propagating in a medium with negative index of refraction. Note that the propagation of the wave packet envelope is given by the group velocity vector so that the flow of energy still points out of the source. A wave packet not emerging out of the excitation zone would violate causality and is also physically not possible in our geometry since the sample ends at x = 0 µm. Moreover, phase- and group velocity calculated by the dispersion relation in Fig. 2 formally exhibit a negative sign for the group velocity and a positive sign for the phase velocity. The physical meaning, however, is encoded in the oppositeness of the propagation direction and phase front motion.

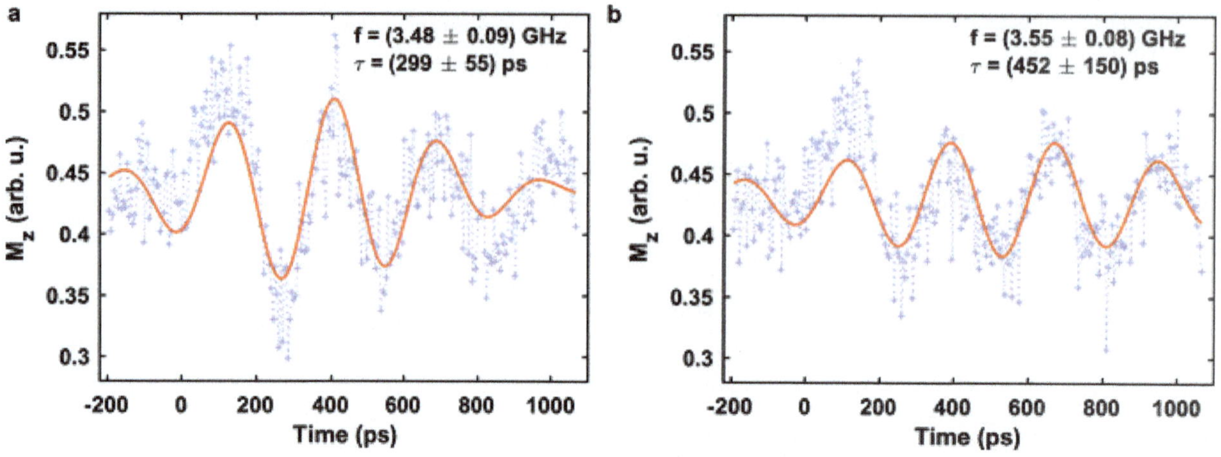

Figure 8: BV spin-wave packets. (**a**) At position (x, y) = (7 µm, 1 µm) closer to the source region, and (**b**) at position (10 µm, 1.5 µm) further away from the source region. The blue crosses represent the measured Kerr signal and the red solid line shows a fit to the data according to equation (3).

SUMMARY

In summary, we have directly observed micron-scale Damon-Eshbach and backward volume spin-wave packets with counterpropagating phase- and group-velocities in ferromagnetic materials through a novel pulsed excitation method. The optically driven high-current pulse source provides an efficient launch of BV as well as DE spin-wave packets using intense picosecond magnetic pulses created by strong current pulses out of a metallic photocathode. The spin-wave packets are traced by stroboscopic scanning Kerr microscopy with inherently synchronized excitation- and probe-pulses generated by a femtosecond laser system. The impulsive excitation with a direct observation enables following the spin-wave creation in time and observing the build-up of higher quantized modes in the permalloy sample leading to a curvature of the phase fronts. The BV mode with directly accessible phase fronts approaching the excitation region is a promising candidate for spintronic data transport and enables model studies of negative refraction analog to optical wave propagation inside negative index metamaterials.

METHODS

Fabrication

Rectangular permalloy ($Ni_{80}Fe_{20}$) samples of $35 \times 10\,\mu m^2$ dimension and 30 nm thickness are prepared onto a 400 µm thick sapphire (Al_2O_3) substrate ($\varepsilon_r = 9.4$) of $14 \times 7\,mm^2$ size using electron-beam lithography, physical vapor deposition (PVD), and lift-off processing. A coplanar waveguide (CPW) consisting of a 300 nm copper layer protected by a 5 nm gold cap layer is fabricated on top of the structure by ultraviolet (UV) photolithography, PVD, and lift-off processing. The 1.5 mm diameter magnesium (Mg) photocathode layer is deposited onto the source region of the waveguide by PVD through a perforated mask.

The waveguide design connects the macroscopic photocathode for pulse injection on one side with a resistive termination on the other side through a tapered section. In the latter, the waveguide conductor width is reduced down to 10 µm in order to increase the current density. The conductor is surrounded by two ground planes separated by a gap of 48.5 µm forming a CPW of 100 Ω impedance and 65 GHz cut-off frequency that provides enough space for the magnetic sample inside the gap. The overlap between permalloy layer and waveguide that determines the excitation spectrum of the k-vectors measures 5 µm. Careful impedance matching across the taper is vital for minimizing losses and preserving the pulse shape. Termination of the pulses is achieved in a serial connection of two 50 Ω radio-frequency resistors with a bandwidth of 50 GHz. The 400 µm long central layout of the sample is depicted in Supplementary Fig. S1.

Measurements

The Kerr microscope is based on a femtosecond chirped pulse amplification (CPA) ytterbium doped potassium-gadolinium tungstate (Yb:KGW) solid-state laser system that delivers pulses of 290 fs full-width-at-half-maximum (FWHM) duration at a repetition rate of 130 kHz and a wavelength of 1030 nm with a pulse energy of up to 46 µJ. By second- and fourth harmonic generation in two Beta-Barium-Borate (BBO) crystals, these pulses are converted into 515 nm green probe pulses and 258 nm UV pump pulses, respectively, that are inherently synchronized to each other since they emerge from the same source pulse. For stroboscopic measurements the pump pulse arrival time at the sample is adjustable with respect to the probe pulse using an optical delay stage that covers a time interval of 2 ns with an accuracy below one femtosecond53. To measure the Damon-Eshbach modes, a delay range of 1.5 ns is scanned in 5 ps steps; in case of the backward volume modes, a scan range of 1.27 ns is swept in 5 ps steps.

The probe pulses of 230 pJ pulse energy are polarization-filtered and focused onto the sample through the sapphire substrate by a microscope objective with a numerical aperture of 0.4 providing a spatial resolution of 550 nm53. The peak intensity in the focal spot of $1.6 \cdot 10^{11} W/cm^2$ is kept right below the damage threshold of the thin sample layer. The reflected light pulse is recollected by the same objective and its polarization is analyzed by a Wollaston prism and a balanced photodetector. This enables a measurement of the out-of-plane component of the magnetization M_z due to the polar Kerr effect. The sample can be scanned through the laser focus by an encoded two-axis piezo scanner with a travel range of $100 \times 100\ \mu m^2$ and a positioning accuracy of 25 nm. For the measurement of the Damon-Eshbach modes a window of $\Delta x = 12\ \mu m$ and $\Delta y = 6\ \mu m$ is scanned in 500 nm steps; in case of the backward volume modes a scan window of $\Delta x = 10\ \mu m$ and $\Delta y = 10\ \mu m$ is swept in 500 nm steps. The data acquisition time at each scanned data point is set to 200 ms so that one SKM pixel at a given pump-probe delay is averaged over 26,000 pump-probe cycles. An external bias field is applied by a set of two $5 \times 5 \times 1\ mm^3$ permanent magnets separated by a distance of 5 mm yielding $\mu_0 H_{ext} \approx 60$ mT for the DE wave packets and in a distance of 7 mm yielding $\mu_0 H_{ext} \approx 20$ mT for the BV modes. For BV wave packets the field is applied in −x direction and for DE measurements, the external field is aligned along the y axis. A directin-situ field measurement at the sample position is not possible due to space constraints. The field amplitude is estimated from reference measurements using the same geometry of the magnets and confirmed by simulations. A comparison with the backward volume spin-wave dispersion relation in Fig. 2a suggests a smaller external field of $\mu_0 H_{ext} = 16$ mT to fit the gyration frequency and the phase velocities in Fig. 2c. Due to the lack of a direct measurements of the local field this value seems to be more realistic.

The current pulses triggering the spin-waves are created by illuminating the Mg photocathode on the CPW conductor with UV pulses of 3 μJ pulse energy and applying an electrostatic extraction field of 6 to 7 kV/mm using a cylindrical anode in a vacuum environment with a base pressure of 10^{-7} mbar. Since the photon energy of the laser pulses of 4.8 eV exceeds the work function of the Mg layer (3.66 eV), electrons are released via the photoeffect leaving a positive charge hole. This vacancy is refilled with electrons through the CPW and leads to a propagating current pulse with a bunch charge of 15 pC corresponding to an average current of 2.2 μA. This results in a quantum efficiency of the photocathode in the 10^{-5} range in agreement with untreated Mg cathodes. An oxygen layer removal technique60 may further increase the quantum efficiency so that even stronger current pulses become available. In order to improve the detection sensitivity, the pump beam is chopped at a frequency of 1 kHz and the photodiode signal filtered by a lock-in amplifier referenced to the chopper frequency. The pulses are verified not to exceed 70 ps FWHM by a direct measurement with a 12 GHz sampling oscilloscope but simulations on the magnetic response of the permalloy layer (see Supplementary Fig. S4) suggest the actual pulse duration to be considerably shorter. The involved high frequencies are subject to the skin effect which confines the current to a thin surface layer of the conductor. The corresponding current density is estimated to exceed 10^{11} A/m² at the position of the excitation region of the magnetic sample, corresponding to a local magnetic field strength of $\mu_0 h_{pulse} \approx 15$ mT.

ACKNOWLEDGEMENTS

This work has been supported by the SFB 668 of the German Science Foundation, the FSP-301/302 of the German Ministry of Education and Research and the Cluster of Excellence 'The Hamburg Centre for Ultrafast Imaging (CUI)'. The authors are grateful to the Center for Optical Quantum Technologies (ZOQ) for providing the laboratory space and to the mechanical as well as the electrical workshops of the University of Hamburg.

REFERENCES

1. Chappert, C., Fert, A. & Van Dau, F. N. The emergence of spin electronics in data storage. Nature Mater. 6, 813–823, 10.1038/nmat2024 (2007).
2. Parkin, S. S. P., Hayashi, M. & Thomas, L. Magnetic domain-wall racetrack memory. Science 320, 190–194, 10.1126/science.1145799 (2008).
3. Hertel, R., Wulfhekel, W. & Kirschner, J. Domain-wall induced phase shifts in spin waves. Phys. Rev. Lett. 93, 257202, 10.1103/PhysRevLett.93.257202 (2004).
4. Cowburn, R. P. & Welland, M. E. Room temperature magnetic quantum cellular automata. Science 287, 1466–1468, 10.1126/science.287.5457.1466 (2000).
5. Cowburn, R. P. Magnetic nanodots for device applications. J. Magn. Magn. Mater. 242–245 Part 1, 505–511, 10.1016/S0304-8853(01)01086-1 (2002).
6. Chumak, A. V., Vasyuchka, V. I., Serga, A. A. & Hillebrands, B.Magnon spintronics. Nature Phys. 11, 453–461, 10.1038/nphys3347 (2015).
7. Chumak, A. V., Serga, A. A. & Hillebrands, B. Magnon transistor for all-magnon data processing. Nat. Commun. 5, 4700, 10.1038/ncomms5700 (2014).
8. Au, Y. et al. Resonant microwave-to-spin-wave transducer. Appl. Phys. Lett. 100, 182404, 10.1063/1.4711039 (2012).
9. Lenk, B., Ulrichs, H., Garbs, F. & Münzenberg, M. The building blocks of magnonics. Phys. Rep. 507, 107–136, 10.1016/j.physrep.2011.06.003 (2011).
10. Kostylev, M. P., Serga, A. A., Schneider, T., Leven, B. &Hillebrands, B. Spin-wave logical gates. Appl. Phys. Lett. 87, 153501, 10.1063/1.2089147 (2005).
11. Jamali, M., Kwon, J. H., Seo, S.-M., Lee, K.-J. & Yang, H. Spin wave nonreciprocity for logic device applications. Sci. Rep. 3, 3160, 10.1038/srep03160 (2013).
12. Vogt, K. et al. Realization of a spin-wave multiplexer. Nat. Commun. 5, 3727, 10.1038/ncomms4727 (2014).
13. Khitun, A., Bao, M. & Wang, K. Spin wave magnetic nanofabric: A new approach to spin-based logic circuitry. IEEE Trans. Magn. 44, 2141–2152, 10.1109/TMAG.2008.2000812 (2008).

14. Freeman, M. R., Hiebert, W. K. & Stankiewicz, A. Time-resolved scanning Kerr microscopy of ferromagnetic structures (invited). J. Appl. Phys. 83, 6217–6222, 10.1063/1.367716 (1998).
15. Hiebert, W. K., Stankiewicz, A. & Freeman, M. R. Direct observation of magnetic relaxation in a small permalloy disk by time-resolved scanning Kerr microscopy. Phys. Rev. Lett. 79, 1134–1137, 10.1103/PhysRevLett.79.1134 (1997).
16. Tamaru, S. et al. Measurement of magnetostatic mode excitation and relaxation in permalloy films using scanning Kerr imaging. Phys. Rev. B 70, 104416, 10.1103/PhysRevB.70.104416 (2004).
17. Demidov, V. E., Urazhdin, S. & Demokritov, S. O. Direct observation and mapping of spin waves emitted by spin-torque nano-oscillators. Nature Mater. 9, 984–988, 10.1038/nmat2882 (2010).
18. Madami, M. et al. Direct observation of a propagating spin wave induced by spin-transfer torque. Nature Nanotech. 6, 635–638, 10.1038/nnano.2011.140 (2011).
19. Bonetti, S. et al. Direct observation and imaging of a spin-wave soliton with p-like symmetry. Nat. Commun. 6, 8889, 10.1038/ncomms9889 (2015).
20. Nakashima, Y., Nagai, K., Tanaka, T. & Matsuyama, K. Optimum design consideration for interferometric spin wave logic operations. J. Appl. Phys. 109, 07D318, 10.1063/1.3549438 (2011).
21. Liu, Z., Giesen, F., Zhu, X., Sydora, R. D. & Freeman, M. R. Spin wave dynamics and the determination of intrinsic damping in locally excited permalloy thin films. Phys. Rev. Lett. 98, 087201, 10.1103/PhysRevLett.98.087201 (2007).
22. Perzlmaier, K., Woltersdorf, G. & Back, C. H. Observation of the propagation and interference of spin waves in ferromagnetic thin films. Phys. Rev. B 77, 054425, 10.1103/PhysRevB.77.054425 (2008).
23. Bauer, H. G., Chauleau, J.-Y., Woltersdorf, G. & Back, C. H.Coupling of spinwave modes in wire structures. Appl. Phys. Lett.104, 102404, 10.1063/1.4868250 (2014).
24. Kalinikos, B. A. & Slavin, A. N. Theory of dipole-exchange spin wave spectrum for ferromagnetic films with mixed exchange boundary conditions. J. Phys. C: Solid State Phys. 19, 7013, 10.1088/0022-3719/19/35/014 (1986).
25. Guslienko, K. Y., Demokritov, S. O., Hillebrands, B. & Slavin, A. N. Effective dipolar boundary conditions for dynamic magnetization in thin magnetic stripes. Phys. Rev. B 66, 132402, 10.1103/PhysRevB.66.132402 (2002).
26. Guslienko, K. Y., Chantrell, R. W. & Slavin, A. N. Dipolar localization of quantized spin-wave modes in thin rectangular magnetic elements. Phys. Rev. B 68, 024422, 10.1103/PhysRevB.68.024422 (2003).
27. Pendry, J. B. Time reversal and negative refraction. Science 322, 71–73, 10.1126/science.1162087 (2008).
28. Xu, T., Agrawal, A., Abashin, M., Chau, K. J. & Lezec, H. J. All-angle negative refraction and active flat lensing of ultraviolet light. Nature 497, 470–474, 10.1038/nature12158 (2013).
29. Shalaev, V. M. Optical negative-index metamaterials. Nature Photon. 1, 41–48, 10.1038/nphoton.2006.49 (2007).
30. Mansfeld, S. et al. Spin wave diffraction and perfect imaging of a grating. Phys. Rev. Lett. 108, 047204, 10.1103/PhysRevLett.108.047204 (2012).
31. Smith, D. R., Pendry, J. B. & Wiltshire, M. C. K. Metamaterials and negative refractive index. Science 305, 788–792, 10.1126/science.1096796 (2004).
32. Soukoulis, C. M. & Wegener, M. Past achievements and future challenges in the development of three-dimensional photonic metamaterials. Nature Photon. 5, 523–530, 10.1038/nphoton.2011.154 (2011).
33. Serga, A. A., Chumak, A. V. & Hillebrands, B. YIG magnonics. J. Phys. D: Appl. Phys. 43, 264002, 10.1088/0022-3727/43/26/264002 (2010).
34. Demidov, V. E., Demokritov, S. O., Rott, K., Krzysteczko, P. &Reiss, G. Mode interference and periodic self-focusing of spin waves in permalloy microstripes. Phys. Rev. B 77, 064406, 10.1103/PhysRevB.77.064406 (2008).
35. Krohn, A. et al. Observation of spin-wave propagation in permalloy microstripes (2009). arXiv:0905.2172.
36. Damon, R. & Eshbach, J. Magnetostatic modes of a ferromagnet slab. J. Phys. Chem. Solids 19, 308–320, 10.1016/0022-3697(61)90041-5 (1961).
37. Liu, H., Riley, G. & Buchanan, K. Directly excited backward volume spin waves in permalloy microstrips. IEEE Magn. Lett. 6, 4000304, 10.1109/LMAG.2015.2495162 (2015).
38. Stancil, D. D., Henty, B. E., Cepni, A. G. & Van't Hof, J. P.Observation of an inverse doppler shift from left-handed dipolar spin waves. Phys. Rev. B 74, 060404, 10.1103/PhysRevB.74.060404 (2006).
39. Serga, A. A., Schneider, T., Hillebrands, B., Demokritov, S. O. &Kostylev, M. P. Phase-sensitive brillouin light scattering spectroscopy from spin-wave packets. Appl. Phys. Lett. 89, 063506, 10.1063/1.2335627 (2006).
40. Schneider, T., Serga, A. A., Neumann, T., Hillebrands, B. &Kostylev, M. P. Phase reciprocity of spin-wave excitation by a microstrip antenna. Phys. Rev. B 77, 214411, 10.1103/PhysRevB.77.214411 (2008).
41. Scott, M. M., Kalinikos, B. A. & Patton, C. E. Self-generation of bright microwave magnetic envelope soliton trains in ferrite films through frequency filtering. Appl. Phys. Lett. 78, 970–972, 10.1063/1.1347401 (2001).
42. Kolodin, P. A. et al. Amplification of microwave magnetic envelope solitons in thin yttrium iron garnet films by parallel pumping. Phys. Rev. Lett. 80, 1976–1979, 10.1103/PhysRevLett.80.1976 (1998).
43. Kalinikos, B. A., Kovshikov, N. G. & Slavin, A. N. Experimental observation of magnetostatic wave envelope solitons in yttrium iron garnet films. Phys. Rev. B 42, 8658–8660, 10.1103/PhysRevB.42.8658 (1990).

44. Büttner, O. et al. Collisions of spin wave envelope solitons and self-focused spin wave packets in yttrium iron garnet films.Phys. Rev. Lett. 82, 4320–4323, 10.1103/PhysRevLett.82.4320 (1999).

45. Sato, N., Ishida, N., Kawakami, T. & Sekiguchi, K. Propagating spectroscopy of backward volume spin waves in a metallic FeNi film. Appl. Phys. Lett. 104, 032411, 10.1063/1.4863215 (2014).

46. Park, J. P., Eames, P., Engebretson, D. M., Berezovsky, J. &Crowell, P. A. Spatially resolved dynamics of localized spin-wave modes in ferromagnetic wires. Phys. Rev. Lett. 89, 277201, 10.1103/PhysRevLett.89.277201 (2002).

47. Brächer, T., Pirro, P., Serga, A. A. & Hillebrands, B. Localized parametric generation of spin waves in a longitudinally magnetized $Ni_{81}Fe_{19}$ waveguide. Appl. Phys. Lett. 103, 142415, 10.1063/1.4824304 (2013).

48. Demidov, V. & Demokritov, S. Magnonic waveguides studied by microfocus brillouin light scattering. IEEE Trans. Magn. 51, 0800215, 10.1109/TMAG.2014.2388196 (2015).

49. Brächer, T. et al. Generation of propagating backward volume spin waves by phase-sensitive mode conversion in two-dimensional microstructures. Appl. Phys. Lett. 102, 132411, 10.1063/1.4800005 (2013).

50. Demidov, V. E. et al. Radiation of spin waves from the open end of a microscopic magnetic-film waveguide. Phys. Rev. B 80, 014429, 10.1103/PhysRevB.80.014429 (2009).

51. Schneider, T. et al. Nondiffractive subwavelength wave beams in a medium with externally controlled anisotropy. Phys. Rev. Lett. 104, 197203, 10.1103/PhysRevLett.104.197203 (2010).

52. Satoh, T. et al. Directional control of spin-wave emission by spatially shaped light. Nature Photon. 6, 662–666, 10.1038/nphoton.2012.218 (2012).

53. Wessels, P. Time-resolved imaging of magnetic nanostructures in the visible and soft X-ray spectral range. PhD thesis, University of Hamburg (2014).

54. Weinberger, P. John Kerr and his effects found in 1877 and 1878.Philos. Mag. Lett. 88, 897–907, 10.1080/09500830802526604 (2008).

55. Qiu, Z. Q. & Bader, S. D. Surface magneto-optic Kerr effect. Rev. Sci. Instrum. 71, 1243–1255, 10.1063/1.1150496 (2000).

56. Lee, K.-D. et al. Femtosecond pump-probe MOKE microscopy for an ultrafast spin dynamics study. J. Korean Phys. Soc. 49, 2402–2407 (2006).

57. Demidov, V. E., Demokritov, S. O., Rott, K., Krzysteczko, P. &Reiss, G. Self-focusing of spin waves in permalloy microstripes.Appl. Phys. Lett. 91, 252504, 10.1063/1.2825421 (2007).

58. Kittel, C. Excitation of spin waves in a ferromagnet by a uniform rf field. Phys. Rev. 110, 1295–1297, 10.1103/PhysRev.110.1295 (1958).

59. Veerakumar, V. & Camley, R. E. Magnon focusing in thin ferromagnetic films. Phys. Rev. B 74, 214401, 10.1103/PhysRevB.74.214401 (2006).

60. Srinivasan-Rao, T., Schill, J., Zvi, I. B. & Woodle, M. Sputtered magnesium as a photocathode material for rf injectors. Rev. Sci. Instrum. 69, 2292–2296, 10.1063/1.1148934 (1998).

CHAPTER 17

CONTROLLED FORMATION AND REFLECTION OF A BRIGHT SOLITARY MATTER-WAVE

A L. Marchant, T. P. Billam, T. P. Wiles, M. M. H. Yu, S. A. Gardiner, and S. L. Cornish

Keyword: Subject terms, Physical sciences, Atomic and molecular physics

INTRODUCTION

Solitons are non-dispersive wave solutions that arise in a diverse range of nonlinear systems, stabilised by a focussing or defocussing nonlinearity. First observed in shallow water1, solitons have subsequently been studied in many other fields including nonlinear optics, biophysics, astrophysics, plasma and particle physics2. They are characterized by well localized wavepackets in one-dimension (1D) that maintain their initial shape and amplitude for all time, even following collisions with other solitons. Bose–Einstein condensates (BECs) formed from dilute atomic gases support bright soliton solutions in 1D for attractive interatomic interactions (focussing nonlinearity), manifesting themselves as localized humps in the field amplitude. In contrast, dark solitons appear as localized reductions in an otherwise uniform field amplitude, preserved by a defocussing nonlinearity (repulsive interactions). The control with which these systems can be manipulated, combined with the unique properties of matter-wave solitons, leads to a rich testing ground for theoretical descriptions of quantum many-body systems3·4·5.

BECs are commonly described by a mean-field treatment6·7 leading to the well-known Gross–Pitaevskii equation (GPE) in which the atomic interactions are described by a nonlinear term proportional to the s-wave scattering length a_s and the condensate density. In the 1D, homogeneous limit the GPE takes the form of a nonlinear Schrödinger equation that supports a spectrum of exact soliton solutions. Experiments approach this mathematically ideal scenario by confining the condensate in an elongated, prolate trap typically with tight radial confinement. However, this quasi-1D geometry is usually accompanied by the presence of weak axial harmonic trapping, which removes the integrability of the system and prevents the appearance of true solitons. Nevertheless, solitary wave solutions remain that retain many similarities to the classical soliton solutions8·9·10·11, such as propagation without dispersion. The formation of bright solitary matter-waves, the three-dimensional (3D) analogue to the bright soliton solutions of the 1D nonlinear Schrödinger equation, from BECs allows one to explore an array of potential applications including novel interferometric devices12using narrow Gaussian potentials as beam splitters13, the study of short-range atom-surface potentials14 and the realization of Schrödinger-cat states5·15.

Previously, bright solitary matter-waves have been realised in three separate experiments using ^{7}Li (refs 16, 17) and ^{85}Rb (ref. 18). In each case, a Feshbach resonance was used to switch the interactions from repulsive ($a_s>0$) to attractive ($a_s<0$) in order to form solitary waves out of the collapse instability19. In two of these experiments16·18, multiple wavepackets were created, allowing the study of the dynamics during collisions in the trap. The observation of solitary waves raises many interesting questions regarding the relationship between the mathematical ideal and the experimental reality. It is unclear how soliton-like the solitary waves created in experiments with finite radial trapping and harmonic axial confinement are. An answer to this question needs to be established before potential applications utilizing solitary waves can be realised.

Citation: A.L. Marchant, T. P. Billam, T. P. Wiles, M. M. H. Yu, S. A. Gardiner, and S. L. Cornish, Controlled Formation And Reflection Of A Bright Solitary Matter-Wave, http://arxiv.org/pdf/1301.5759v1.pdf.

At a more fundamental level it remains to be tested whether or not the GPE treatment fully describes the solitary waves created in experiments. Solitary waves realised experimentally typically contain $\lesssim 1,000$ atoms, placing them well outside of the thermodynamic limit and potentially outside the reach of the mean-field description. Several theoretical studies of bright solitary waves beyond the mean-field description have now been performed, either including effects of quantum noise using the truncated Wigner method3or using approximate analytic and numerical methods to simulate the full many-body problem4 5. These generate results potentially in conflict with the behaviour predicted by the GPE treatment.

In this work, we report the controlled formation of bright solitary matter-waves from a ^{85}Rb BEC. The experimental geometry is such that the velocity of the wavepackets can be precisely controlled, a key factor in facilitating the future exploration of solitary wave interactions and collisions. In addition, we observe and model the controlled reflection of solitary waves from a broad Gaussian potential barrier, demonstrating their particle-like nature. These results pave the way for new experimental studies of bright solitary matter-wave dynamics to elucidate the wealth of existing theoretical work.

RESULTS

Controlled expansion of a tunable BEC

^{85}Rb is a prime candidate for solitary wave experiments owing to the existence of a broad Feshbach resonance at ~155 G in collisions between atoms in the F=2, $m_F=-2$ state. We use this resonance to form a stable, repulsively interacting condensate in a crossed optical dipole trap, shown in Fig. 1a. The condensate is then loaded into a quasi-1D waveguide, better suited geometrically to the observation of solitary waves. At the point of release into the waveguide, the magnetic bias field controlling the atomic scattering length is jumped to a new value (see Fig. 1b). As the BEC propagates along the waveguide, the value of a_s determines the rate of expansion of the condensate in the axial direction. We probe this expansion by measuring the condensate size (using destructive absorption imaging) as a function of time for different values of a_s as shown in Fig. 1c. Fitting the experimental data we can extract an expansion rate for the BEC, dependent on a_s and N. This is shown in Fig. 1d, along with a 3D GPE simulation of the expansion (the solid line). At $a_s=-11\ a_0$ and N=2,000 we see the expansion rate of the BEC becomes consistent with zero. This lack of dispersion with time indicates the formation of a bright solitary matter-wave.

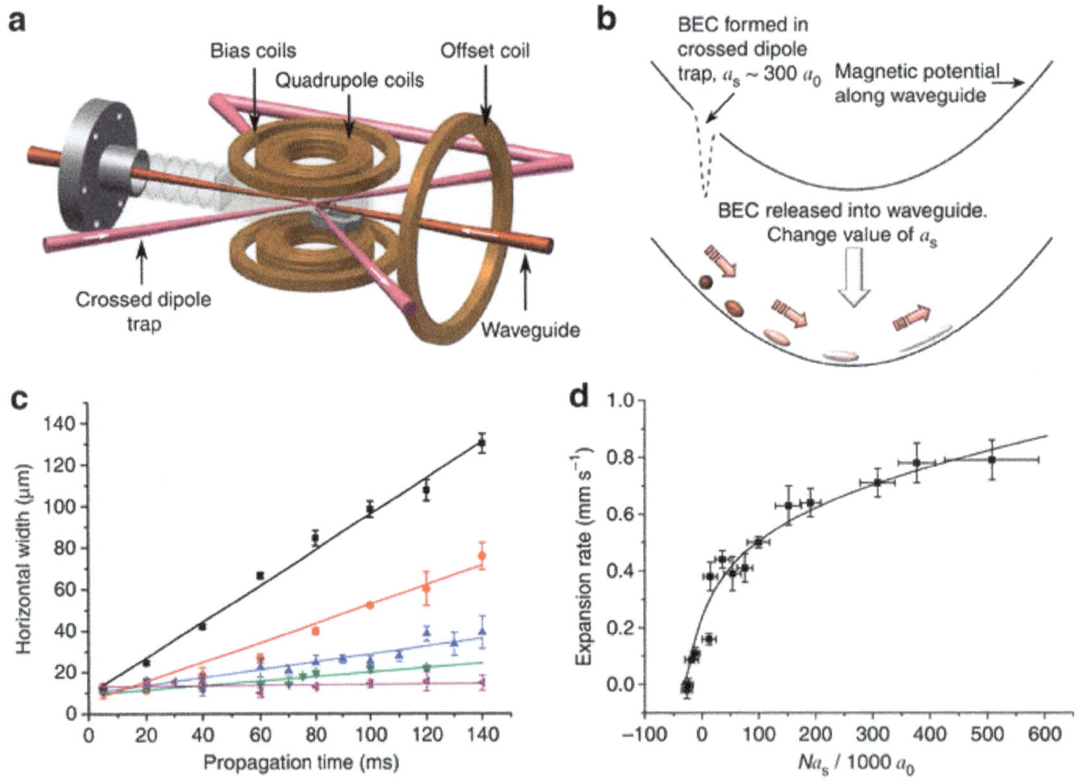

Figure 1: Expansion in the waveguide.

(**a**) Experimental setup showing the glass science cell, the crossed dipole trap used to create the BEC, the optical waveguide and the quadrupole, bias and offset coils. Also shown in the cell is a super-polished Dove prism (blue), mounted on a macor support, to be used for future experiments. (**b**) Schematic of the release of the condensate from the crossed dipole trap into the waveguide. (**c**) Condensate expansion in the waveguide for a_s=165 a_0 (black), 23 a_0 (red), 4 a_0 (blue), −7 a_0 (green) and −11 a_0 (purple). Solid lines are linear fits to the experimental data where the widths are rms values. (**d**) Condensate expansion rate in the waveguide as a function of atom number and scattering length. The solid line is the theoretical expansion rate calculated from a zero-temperature simulation of the experimental expansion using a cylindrically symmetric, 3D GPE. As in the data, the expansion rate is defined using the change in the width of the BEC between 10 and 100 ms after release into the waveguide potential, which is approximately linear over this time interval in all cases. Error bars in **c**,**d** are the s.e. calculated from repeated measurements.

Full size image

Figure 2 shows the propagation of this solitary wave, contrasted to that of a repulsively interacting BEC. As the repulsive wavepacket propagates the axial expansion causes a significant drop in optical depth not seen for the solitary wave. We observe the solitary wave propagating over a distance of 1.1 mm in a time of ~150 ms with very little distortion.

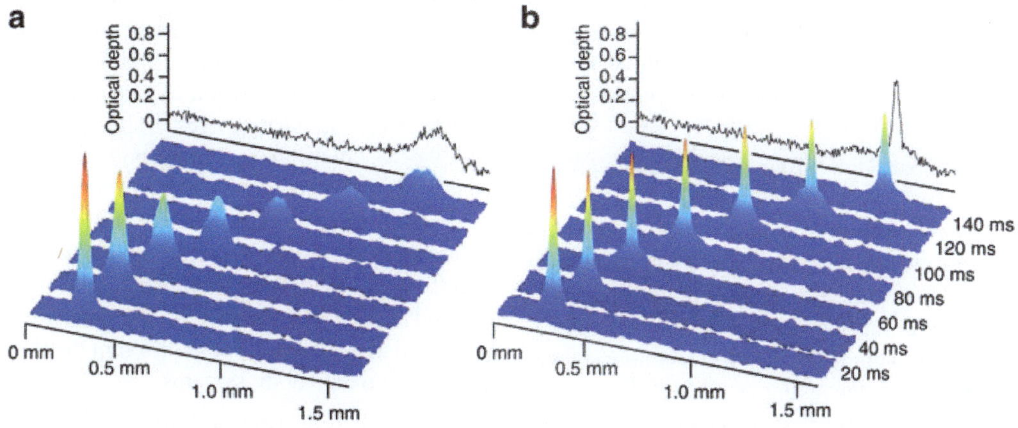

Figure 2: Propagation in the waveguide.

(**a**) As a repulsive BEC (a_s=58 a_0, N=3.5 × 10^3) propagates along the waveguide the atomic interactions cause the condensate to spread, leading to a drop in optical depth. (**b**) In contrast, the attractive interactions present in a bright solitary matter-wave (a_s=−11 a_0,N=2.0 × 10^3) hold the atomic wavepacket together as it propagates, maintaining its shape with time. Crosscuts shown are the horizontal optical depth profiles of the condensates after 140 ms propagation time along the waveguide.

Reflection from a broad and repulsive Gaussian barrier

To probe the stability of the solitary wave we investigate reflection of the wave packet from a repulsive Gaussian barrier with a 1/e^2 radius of 130 μm, shown in Fig. 3a. Figure 3b show the position of the solitary wave as a function of time in the presence of a 760 nK barrier potential. In this case, the barrier height is greater than the kinetic energy of the solitary wave and the wavepacket is cleanly reflected.

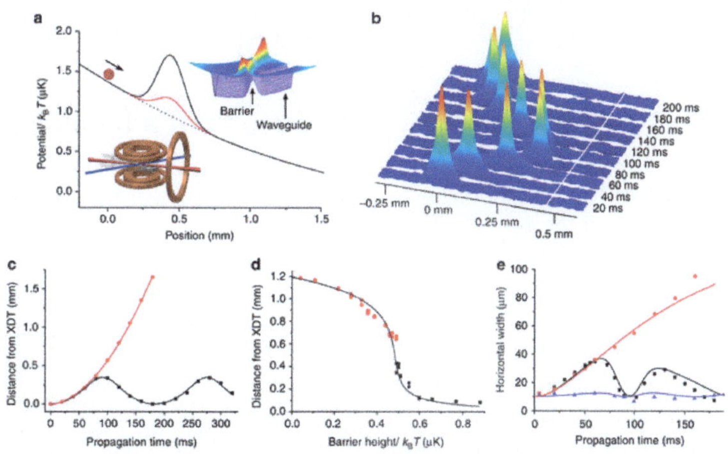

Figure 3: Reflection from a repulsive Gaussian barrier.

(**a**) Potential in the axial direction along the waveguide in the presence of the repulsive barrier. (Inset, upper: combined waveguide and Gaussian barrier potential. Lower: experimental setup.) (**b**) False colour images of a solitary wave reflecting from the barrier. The white line shows the location of the barrier centre. (**c**) Horizontal position, relative to the crossed dipole trap (XDT), of a solitary wave propagating in the waveguide in the absence (red) and presence (black) of the repulsive barrier. (**d**) The position of a solitary wave after 150 ms propagation time as a function of the barrier height. Red (black) points correspond to the solitary wave travelling over (being reflected from) the barrier. Solid lines in (**c**,**d**): theoretical trajectory calculated using a classical particle model with no free parameters. (**e**) Condensate width following reflection from the barrier. In the absence of a barrier, a repulsive BEC (a_s=58 a_0, N=3.5 × 10^3) will expand as it propagates (red). With the barrier in place, an oscillation in the condensate width is set up following the strong compression of the condensate at the barrier due to the shape of the potential (black). A solitary wave (a_s=−11 a_0, N=2.0 × 10^3) undergoing the same collision emerges unaltered (blue). Solid lines are the theoretical condensate widths calculated by solving the 3D (cylindrically symmetric) GPE.

Using a barrier much wider than the solitary wave size the atomic centre-of-mass coordinate behaves classically, with the solitary wave acting as a single particle 'rolling up a potential hill'. By varying the height of the potential barrier it is possible to select whether the solitary wave is reflected or allowed to travel over the barrier. The position of the solitary wave after 150 ms is shown in Fig. 3d as a function of barrier height. The solid line is a theoretical trajectory, calculated using Newtonian mechanics with no free parameters, and shows excellent agreement with the data.

In Fig. 3e we compare the effect of reflection from the barrier for a solitary wave and a repulsive BEC and contrast the change in width to the case of a repulsive BEC propagating along the waveguide in the absence of the barrier. The solid lines are the theoretical predictions for the condensate widths. We find that in the parameter regime of the experiment a 1D treatment is insufficient, and so we determine the theoretical widths using a 3D (cylindrically symmetric) GPE. This observation is consistent with other recent theoretical studies21. As expected, the solitary wave is robust against collisions with a repulsive Gaussian barrier and following the reflection maintains its shape, continuing to propagate without dispersion. In the absence of the barrier, the repulsive BEC expands steadily in time. (We attribute the disagreement between experiment and theory at longer times to a small thermal component making the measurement of the condensate width less accurate.) In the barrier reflection case, an oscillation in the condensate width is induced as a result of the larger spatial extent of the repulsive BEC causing it to be strongly compressed as it is reflected from the barrier. Such contrast in the behaviour of the repulsive BEC and the solitary wave reflection lends weight to previous theoretical prediction regarding the superior characteristics of solitary waves for observing quantum reflection from surfaces14.

DISCUSSION

There is currently much theoretical interest21·22·23·24·25 in the scattering of solitary waves from narrow potential barriers where, if the barrier width is on the order of the solitary wave width, quantum effects are observable. At high kinetic energy, soliton splitting is energetically allowed at narrow repulsive barriers. The effect of quantum tunnelling means the barrier can act as a beam splitter, dividing the soliton into two parts13·22. These multiple wavepackets can then be used to investigate the phase dependence of binary collisions23, the behaviour of collisions of two solitary waves on a barrier22·24 and would provide a solid first step towards the realization of a bright solitary wave interferometer. In the limit of low kinetic energy, a mean-field GPE treatment of the problem begins to break down26 and quantum behaviour, (described in the 1D limit by the Lieb-Liniger Hamiltonian27), becomes more significant. Here, splitting of the soliton is energetically forbidden and it becomes possible to create Schrödinger-cat states5·15.

The use of a narrow potential to controllably split a solitary wave presents an opportunity to investigate one of the key open questions arising from previous work; what governs the dynamics and stability of multiple solitary waves existing in the same trap? The long-lived nature of the solitary waves and their apparent stability during binary collisions has been the subject of a wealth of theoretical work3·28·29·30·31. Within the framework of the GPE, the observed stability of soliton collisions can only be explained by imposing a relative phase $\phi=\pi$between neighbouring solitary waves16·28·30 such that the collisions are effectively repulsive in character32. This stabilizes the solitons as it avoids the formation of a large density peak which may lead to collapse8·23·30·33. Several other studies address the apparent stability of solitary waves in binary collisions, offering different interpretations, which do not require the imposition of a relative phase $\phi=\pi$ between neighbouring solitary waves. The inclusion of quantum noise3 or accounting for many-body effects4 both result in effectively repulsive interactions between solitary waves, irrespective of initial phase. Interestingly, incoherent, fragmented objects are also predicted to form in the many-body formalism4. Further experimental studies are undoubtedly required to address the role of the relative phase in solitary wave collisions and to test the different theoretical descriptions of quantum many-body systems.

Although reflection and splitting experiments show the potential to settle the theoretical debate over the solitary wave formation and dynamics, the ability to probe such narrow and hence rapidly varying potentials using these wavepackets also lends itself to an obvious application in precision measurement. Atoms close to a surface are subject to the short-range Casimir–Polder and van der Waals potentials, which can be measured using the classical and quantum reflection of bright solitary matter-waves14. Our apparatus includes a super-polished Dove prism for such studies, see Fig. 1a. Further in the future, the ability to deliver and manipulate ultracold atoms near to a solid surface may open up new routes to probe short-range corrections to gravity34 due to exotic forces beyond the Standard Model.

METHODS

Production of a tunable BEC

We create a BEC with tunable atomic interactions using the method described in 35. A magnetic Feshbach resonance is used to tune both the elastic and inelastic scattering properties of the atomic sample to achieve efficient evaporation. Importantly, the resonance at 155 G in collisions between ^{85}Rb atoms in the F=2, $m_F=-2$ state gives control over the s-wave scattering length close to the zero crossing of $\sim 40\ a_0/G$.

The use of a magnetic Feshbach resonance means it is advantageous to work with a levitated crossed optical dipole trap. This is formed from a single 10.1 W, $\lambda=1{,}064$ nm laser beam (IPG: YLR-15-1064-LP-SF) used in a bow-tie configuration as shown in Fig. 1a. The term 'levitated' refers to the use of an additional magnetic quadrupole field whose vertical gradient is set to just less than that required to support atoms against gravity. This trap allows the magnetic field, and hence scattering length, to be changed independently of the trapping frequencies.

Loading the optical waveguide

To investigate the creation of solitary waves we begin by forming a BEC containing up to 10,000 atoms at a scattering length of $a_s \approx 300\ a_0$. The crossed beam trap in which the BEC is created has a roughly spherically symmetric geometry at the point of condensation, with final trap frequencies of $\omega_{x,y,z}=2\pi \times (31, 27, 25)$ Hz. This trap is ill-suited to the observation of bright solitary matter-waves and thus we transfer the condensate into a more quasi-1D waveguide created by an additional 1,064 nm laser beam, focused to a waist of 117 μm and intersecting the crossed trap at 45° to each beam. This enters the glass science cell through the back surface of an anti-reflection coated fused silica Dove prism (to be later used for the study of atom-surface interactions14).

To load the condensate into the waveguide the scattering length is ramped close to $a_s=0$ in 50 ms thus reducing the condensate size and creating a BEC approximately in the harmonic oscillator ground state of the crossed trap. The BEC is then held for 10 ms to allow the magnetic field to stabilize before simultaneously switching the waveguide beam on, the crossed beams off and jumping the quadrupole gradient in the vertical direction from $B'=21.5$ G cm^{-1} to 26 G cm^{-1}. Although it is advantageous in terms of the evaporation to be under levitated during the condensation phase, we must increase the gradient once we wish to transfer the atoms. This ensures a truer levitation of the atoms in the waveguide trap, thus maximizing the trap depth of the beam. In addition, the presence of the quadrupole gradient provides much of the, albeit weak, axial trapping along the beam, $\omega_{\text{axial}} = 1/2\sqrt{\mu B'^2/mB_0} = 2\pi \times 1$ Hz (ref. 37). Here, μ is the magnetic moment of atoms with mass m and B_0 is the magnetic bias field. The waveguide beam itself contributes <0.1 Hz to the axial trapping, hence the magnetic confinement dominates in this direction. At a beam power of 0.17 W, the waveguide and quadrupole potential produce a trap of $\omega_{x,y,z}=2\pi \times (1, 27, 27)$ Hz. Here, the radial trap frequency ($\omega_{y,z}$) approximately matches that of the crossed beam trap at the point of condensation.

Propagation in the waveguide

A small offset (2.6 mm) between the crossed dipole trap, that is, the waveguide loading position, and the quadrupole centre means that once loaded into the waveguide, the BEC propagates freely towards the magnetic field minimum along the direction of the waveguide, undergoing harmonic motion. As the BEC propagates its rate of expansion in the axial direction is determined by the scattering length. Although strictly speaking the expansion is nonlinear over the full-range of times measured, a linear approximation is valid over the range $10\text{ ms}<t<100\text{ ms}$ from which we can extract a 'rate'.

Control of the solitary wave velocity

The position of the magnetic field zero in the axial direction of the waveguide can be displaced by an amount determined by the magnetic field gradient in this direction, B′/2, and a moderate offset field, B_{offset}, according to $\Delta x=B_{\text{offset}}/(B'/2)$ (ref. 37). In this way, the amplitude, and hence velocity, of the solitary wave motion can be precisely controlled due to the dominance of the magnetic potential over the optical confinement of the waveguide along the axial direction. The maximum velocity is given by $v=A\omega_{\text{axial}}$, where A is the amplitude of the motion, set by the separation between the minimum of the magnetic potential along the axis of the waveguide and the release point from the crossed dipole trap. Using this technique the solitary wave can reach velocities of tens of mm s^{-1} when travelling through the centre of the harmonic potential or, alternatively, be brought to a near standstill, achieving velocities <0.5 mm s^{-1}.

Classical reflection from a Gaussian barrier

To produce the repulsive potential barrier we use a 532 nm Gaussian laser beam (derived from a Laser Quantum Finesse laser), focussed to a waist of 131 μm horizontally and 495 μm vertically, with a power of up to 2 W. The barrier is aligned to cross the waveguide in the horizontal plane at an angle of ~45° and is offset by 455 μm from where the BEC is released from the crossed dipole trap, see Fig. 3a. This angle is restricted by the available optical access close to the trap centre.

Theoretical modelling

The release of the BEC into the waveguide potential, and its subsequent expansion, was modelled at zero-temperature by solving the GPE in 3D using a cylindrically symmetric Fourier pseudospectral method. In all cases, the initial non-interacting ground state of a harmonic trap with axial (radial) frequency 30 (27) Hz (corresponding to the crossed dipole trap potential) was released instantaneously into another harmonic trap with axial (radial) frequency 1 (27) Hz and offset by 2.6 mm along the axial direction (corresponding to the waveguide potential). The scattering length was instantaneously changed to the appropriate value of a_s at the time of release.

In cases where the barrier was present this was modelled as a Gaussian 'light-sheet' potential centred on a plane perpendicular to the axial direction, offset from the initial harmonic trap by 2.145 mm, and with height 760 nK and width $131\sqrt{2}$ μm. Compared with the experimental barrier beam, this model neglects the vertical width of the beam, which is large compared with the radial extent of the BEC in the waveguide, and includes the geometric factor $\sqrt{2}$ to account for the 45° angle of the beam.

Expansion rates were calculated from the full-width at half maximum of the BEC axial density profile predicted by the GPE (obtained by integrating over the radial coordinate) after 10 and 100 ms of expansion. In all cases, the change in radius over this time interval was approximately linear. For the simulations in Fig. 3e, the width was calculated by convolving the BEC axial density profile predicted by the GPE with a 10 μm width Gaussian (to account for finite imaging resolution), and fitting a Gaussian distribution to the resulting profile using nonlinear least-squares.

REFERENCES

1. J. S. Russell, in Report of the fourteenth meeting of the British association for the advancement of Science, edited by J. Murray (1844), pp. 311–90.
2. T. Dauxois and M. Peyrard, Physics of Solitons (Cambridge University Press, 2006).
3. D. Cronin, J. Schmiedmayer, and D. E. Pritchard, Rev. Mod. Phys. 81, 1051 (2009).
4. S. Cornish, N. Parker, A. Martin, T. Judd, R. Scott, T. Fromhold, and C. Adams, Physica D: Nonlinear Phenomena 238, 1299 (2009), ISSN 0167-2789.
5. C. Weiss and Y. Castin, Phys. Rev. Lett. 102, 010403 (2009).

6. I. Streltsov, O. E. Alon, and L. S. Cederbaum, Phys. Rev. A 80, 043616 (2009).
7. J. Pethick and H. Smith, Bose-Einstein Condensation in Dilute Gases (Cambridge University Press, 2001).
8. L. Pitaevskii and S. Stringari, Bose-Einstein Condensation (Clarendon Press, Oxford, 2003).
9. T. P. Billam, A. L. Marchant, S. L. Cornish, S. A. Gardiner, and N. G. Parker, Bright solitary matter waves: formation, stability and interactions (arXiv:1209.0560, 2012).
10. K. E. Strecker, G. B. Partridge, A. G. Truscott, and R. G. Hulet, Nature 417, 150 (2002).
11. L. Khaykovich, F. Schreck, G. Ferrari, T. Bourdel, J. Cubizolles, L. D. Carr, Y. Castin, and C. Salomon, Science 296, 1290 (2002).
12. S. L. Cornish, S. T. Thompson, and C. E. Wieman, Phys. Rev. Lett. 96, 170401 (2006).
13. P. A. Ruprecht, M. J. Holland, K. Burnett, and M. Edwards, Phys. Rev. A 51, 4704 (1995).
14. B. J. Dabrowska-W¨uster, S. W¨uster, and M. J. Davis, New Journal of Physics 11, 053017 (2009).
15. I. Streltsov, O. E. Alon, and L. S. Cederbaum, Phys. Rev. Lett. 100, 130401 (2008).
16. J. L. Helm, T. P. Billam, and S. A. Gardiner, Phys. Rev. A 85, 053621 (2012).
17. T. P. Billam, S. L. Cornish, and S. A. Gardiner, Phys. Rev. A 83, 041602 (2011).
18. D. Martin and J. Ruostekoski, New Journal of Physics 14, 043040 (2012).
19. T. Ernst and J. Brand, Phys. Rev. A 81, 033614 (2010).
20. [20] J. Cuevas, P. G. Kevrekidis, B. A. Malomed, P. Dyke, and R. G. Hulet, arXiv:1301.3959 (2013).
21. P. Dyke, L. Sidong, S. Pollack, D. Dries, and R. Hulet, in 42nd Annual Meeting of the APS Division of Atomic, Molecular and Optical Physics (2011), vol. 56, No. 5.
22. B. Gertjerenken, T. P. Billam, L. Khaykovich, and C. Weiss, Phys. Rev. A 86, 033608 (2012).
23. E. H. Lieb and W. Liniger, Phys. Rev. 130, 1605 (1963).
24. U. Al Khawaja, H. T. C. Stoof, R. G. Hulet, K. E. Strecker, and G. B. Partridge, Phys. Rev. Lett. 89, 200404 (2002).
25. N. Parker, A. Martin, C. Adams, and S. Cornish, Physica D: Nonlinear Phenomena 238, 1456 (2009), ISSN 0167-2789.
26. N. G. Parker, A. M. Martin, S. L. Cornish, and C. S. Adams, Journal of Physics B: Atomic, Molecular and Optical Physics 41, 045303 (2008).
27. L. D. Carr and J. Brand, Phys. Rev. Lett. 92, 040401 (2004).
28. S. Dimopoulos and A. A. Geraci, Phys. Rev. D 68, 124021 (2003)
29. L. Marchant, S. H¨andel, S. A. Hopkins, T. P. Wiles, and S. L. Cornish, Phys. Rev. A 85, 053647 (2012).
30. Y.-J. Lin, A. R. Perry, R. L. Compton, I. B. Spielman, and J. V. Porto, Phys. Rev. A 79, 063631 (2009).
31. L. Marchant, S. H¨andel, T. P. Wiles, S. A. Hopkins, and S. L. Cornish, New Journal of Physics 13, 125003 (2011).

CHAPTER 18

FAST RETROSPECTIVELY TRIGGERED LOCAL PULSE-WAVE VELOCITY MEASUREMENTS IN MICE WITH CMR-MICROSCOPY USING A RADIAL TRAJECTORY

Patrick Winter, Thomas Kampf, Xavier Helluy, Fabian T Gutjahr, Cord B Meyer, Eberhard Rommel, Wolfgang R Bauer, Peter M Jakob and Volker Herold

Keyword: Pulse-wave velocity, Mouse Self-gating, Phase-contrast, CMR, Non-triggered, Retrospective, Radial, Aorta

BACKGROUND

Cardiovascular Disease (CVD) is still one of the main causes of death in industrial nations. One important indicator of cardiovascular risk is the aortic pulse-wave velocity (PWV) as a measure of arterial stiffness[1, 2].

In the past years genetically modified mice have turned out to be of great relevance as models for the research of this predictor for the formation of CVD. As a non-invasive method high-field CMR has gained great importance for the measurement of morphological and functional parameters of the cardiovascular system in small animals[3]. A triggered Cartesian PC-Cine sequence was proposed which uses the flow-area (QA) method to determine the local PWV[4]. In this method high spatial and temporal resolutions were achieved by the acquisition of multiple interleaved Cine experiments. With the development of more capable hardware and faster reconstruction algorithms non-Cartesian sequences were developed to accelerate the CMR experiment and improve CMR robustness against flow artifacts. Zhao et al. improved the temporal resolution and acquisition time for a measurement of the regional PWV by measuring flow in multiple slices with a triggered undersampled and asymmetric radial Phase Contrast MR sequence[5].

A drawback of prospectively triggered methods is the necessity of external probes, which monitor cardiac and respiratory motion. The most common method to trigger the CMR acquisition is the use of Electrocardiograms (ECG). However, rapid switching of the imaging gradients can cause interferences with the ECG probes, which can strongly corrupt the trigger signal[6]. In addition, in high field CMR, the small RF coil dimensions can make the attachment of triggering probes difficult. Therefore experience with animal handling is required to ensure a stable trigger signal. The use of the CMR signal itself offers the possibility of a wireless trigger signal, which does not require external probes and is therefore easier to handle. Since the peak value of the echo signal corresponds to the sum of the transverse magnetization across the whole field of view (FOV), moving and changing voxels cause fluctuations of the peak signal intensity corresponding to the motion[7]. The signal variations can be assigned to respiratory and cardiac motion within the excited slice. With the application of fast repetitive radial read-outs a navigator signal can be extracted which can be used for retrospective triggering. This technique was used for the generation of self-gated cardiac Cines in mice without additional hardware[8].

The goal of our work was to design a fast self-gated PC-Cine method for PWV measurements in the aortae of mice, which is independent of additional trigger hardware and is applicable to high field CMR, where susceptibility disturbances are significantly increased. Since a stable trigger signal is required for the determination of this parameter, the self-gated method promises to improve the robustness of the measurement. However, the challenge for flow CMR in the aorta without triggering probes is that the slice of interest is

Citation: Winter P, Kampf T, Helluy X, et al. Fast retrospectively triggered local pulse-wave velocity measurements in mice with CMR-microscopy using a radial trajectory.. 2013;15(1):88. doi:10.1186/1532-429X-15-88.

often located remotely from the beating heart and therefore might contain only a small portion of changing voxels, which are able to contribute to a self-gating signal. Additionally, high temporal and spatial resolutions as well as small echo times in order to reduce image artifacts are necessary, which imposes high requirements to the flow compensated spatial encoding gradients. To ensure high-resolution measurements and reduced sensitivity to susceptibility variations, a highly asymmetric radial pulse sequence was developed to enable the acquisition of a self-gating signal without prolonging the read-out.

METHODS

CMR

Pulse sequences

All measurements are carried out at a 17.6 T small animal system (Bruker Avance 750 WB, Bruker BioSpin MRI GmbH, Rheinstetten, Germany) with a 1 T/m gradient system and a home-built transverse electromagnetic (TEM) resonator with an inner diameter of 25 mm. The imaging protocol started with 2D-FLASH experiments to localize the descending aorta. The local PWV was measured in a slice perpendicular to the vessel (see Figure1). Based on the Cartesian Phase Contrast sequence proposed in[9] a motion-compensated radial PC-Cine-FLASH sequence was designed for through-plane velocity encoding (see Figure2). The scan protocol consists of three sequentially applied velocity-encoding steps with varying first gradient moments (M_1=0 s/m and M_1=±0.3 s/m), corresponding to a maximum encoding velocity (VENC) of 1.7 m/s. The radial sequence enables either a triggered flow-encoding mode based on the interleaved acquisition scheme of[4] or retrospective flow measurements without triggering.

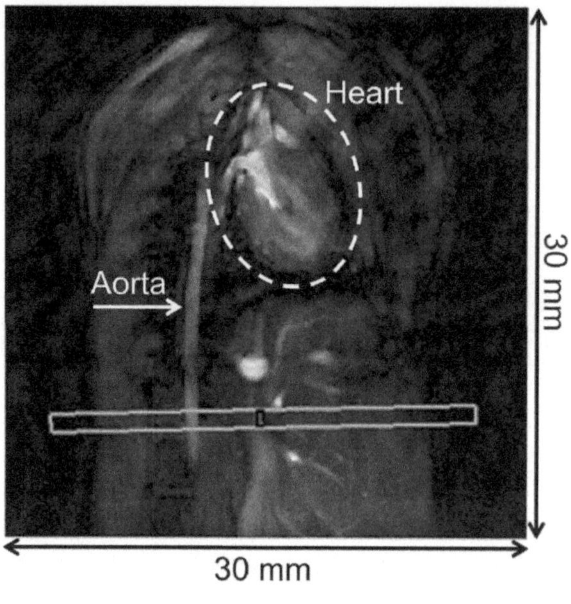

Figure 1 Slice positioning for the local PVW measurements in the abdominal aorta.

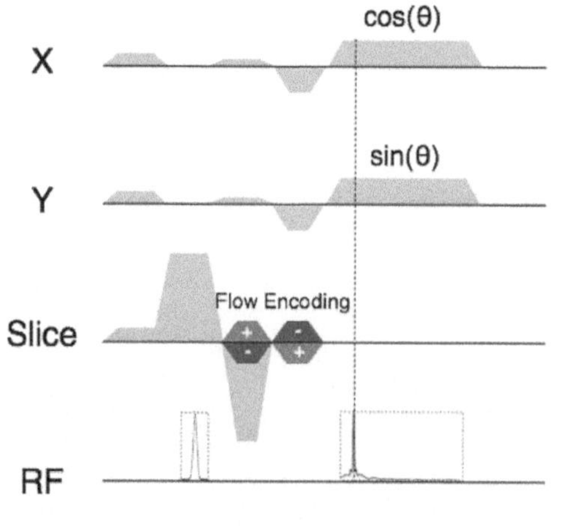

Figure 2 Pulse-sequence for the retrospective measurement of blood flow and the cross-sectional areas. The measurement is motion-compensated in all three encoding directions. Flow measurements are done with a bipolar gradient in slice direction, which is added to the slice-selection gradient. After gradient spoiling and slice selection tripolar gradients are applied in the logical X- and Y-direction with G_x=G·cos(θ) and G_y=G·sin(θ), where θ is the projection angle and G the maximum strength of the read-out gradient. The gradient echo is acquired at a position of 10% of the read-out duration.

To assess the quality of the retrospectively triggered PC-Cine measurement, prospectively triggered Cartesian and radial PC-Cine experiments were conducted for comparison. The Cartesian measurement was acquired using a 160×160 matrix and a field of view of 24×24 mm². The radial data was reconstructed on the same 160×160 grid with the same field of view and spatial resolution as with the Cartesian experiment. All images were zero-filled to a matrix size of 256×256 for the following evaluation. The slice thickness was 1 mm in all experiments. Further relevant parameters of all three measurement types are listed in Table1.

Table 1 Scan parameters for the three flow-encoding sequences

Parameters	Cartesian	Radial triggered	Retrospective
TR [ms]	5	3	3
TE [ms]	1.7	1.2	1.2
Sampling bandwidth [kHz]	75.8	75.8	75.8
Flip angle [°]	30-40	30-40	30-40
Echo position [% of the read-out]	25	10	10
Number of read-out points	160	90	90
Number of projections	160	160	16000*
Number of Cine segments	5	3	-
Frames per Cine Segment	8	14	-
Number of velocity-encoding steps	3	3	3
VENC [m/s]	1.7	1.7	1.7
Spatial resolution [μm^2]	150×150**	150×150‡	150×150‡
Field of view [mm^2]	24×24	24×24	24×24
Slice thickness [mm]	1	1	1
Measurement time [minutes]	6.7±1.1	4.3±0.6	2.9

*Including 500 dummy scans to establish the steady state. **Zero-filled to a 256 × 256 matrix. ‡ Reconstructed on a 160 × 160 grid and zero-filled to a 256 × 256 matrix.

Retrospectively triggered measurement

For the retrospective triggering mode a train of n_{Proj}=16000 radial projections per velocity-encoding step was acquired without cardiac and respiratory gating with a short repetition time (TR=3 ms).

To reduce the echo time and therefore the strong influence of B_0 inhomogeneities at 17.6T, which are especially significant in vessels close to the lung, the echo was set to 10% of the read-out duration by decreasing the areas of the read dephase gradients (see Figure2). A waiting period of 15 s was applied between each velocity-encoding step to reduce the effect of RF coil heating and SAR due to the fast application of RF pulses[10]. The total scan time including the waiting times was 3 minutes.

Prospectively triggered measurement

All prospectively triggered measurements were done with a pressure sensitive balloon to measure the thoracic motion during breathing and the heartbeats. The trigger signal was post-processed with an amplification unit (Rapid Biomedical, Rimpar, Germany). With the triggered Cartesian measurement an effective temporal resolution of 1 ms was achieved by the acquisition of 5 interleaved Cines with 8 movie frames each, covering 40 ms of the heart cycle around the systole. The total scan time depends on breathing and the heart rate and was 6-8 minutes in this study. For the prospectively triggered radial measurement three interleaved Cines with 14 movie frames each were acquired to obtain the same temporal resolution as for the Cartesian measurement, covering 42 ms of the heart cycle. For each image 160 equally distributed spokes were acquired, therefore the undersampling factor was af ≈3.12, respective to the Nyquist criterion. The total scan time was 3-5 minutes.

Gradient delay correction

As a non-Cartesian imaging method the quality of the radial measurement is compromised by trajectory errors caused by hardware imperfections such as delays of the spatial encoding gradients. To improve image quality, the gradient switching times were determined priorly with a homogenous $CuSO_4$ doped water phantom using two different methods as proposed in[11] and[12], respectively. Measurements using either technique yielded gradient delay values of about t_x=8 μ s for the X-, t_y=10 μ s for the Y- and t_z=8 μ s for the Z-gradient. The measured parameters were afterwards used for a correction of the gradient areas by adjusting the amplitudes of the read dephase gradients, as described in[11].

Trigger signal processing

For the retrospective reconstruction several steps were performed to obtain the trigger signal:

Modified golden angle based radial read-out

In mice anesthetized with isoflurane the heart typically contracts with a repetition rate of 300-600 beats per minute, which is usually too fast for golden-ratio based real time Cine applications[13]. Therefore, to achieve sufficient SNR and temporal resolution, the aortic motion is monitored over several seconds. For the reconstruction of a Cine frame, the projections acquired during many heart beats were afterwards sorted by their phase in the cardiac cycle and used to fill the k-space. Since a standard golden-ratio based angular distribution might result into an undesired clustering of the retrospectively selected projections, a modified golden angle based read-out was precalculated.

Using $\theta_0=0°$ as initial value, an angle list was calculated by consecutively adding the golden angle $\Delta\theta\approx 111.246°$[13]. Assuming that the variations from the average interval T_{RR} between two heart beats are small, the angles were afterwards sorted into $N \approx T_{RR}/TR$ interleaved sublists using the sorting:

$$\theta_n(k \cdot T_{RR}) = \theta_{n,0} + k \cdot \Delta\theta, \quad (1)$$

where n=1,…,N is the number index of the sublists,k=1,… ,K≈nProjN the index of the projections per sublist and $\theta_{n,0}=\theta_{n-1}(K \cdot T_{RR})$ the last angle value of the previous sublist. The idea of this sorting of projection angles is that in case of an almost constant heart rate a large amount of retrospectively selected projections for one time frame would belong to the same sublist and would therefore be golden-ratio distributed. An estimate of the average interval T_{RR} was assessed with the pressure sensitive balloon signal, which was observed with the trigger unit to monitor cardiac and respiratory motion of the animal. Depending on the heart rate, the measured projections consist of 33-38 sublists with 420-475 golden angle-distributed projections per sublist.

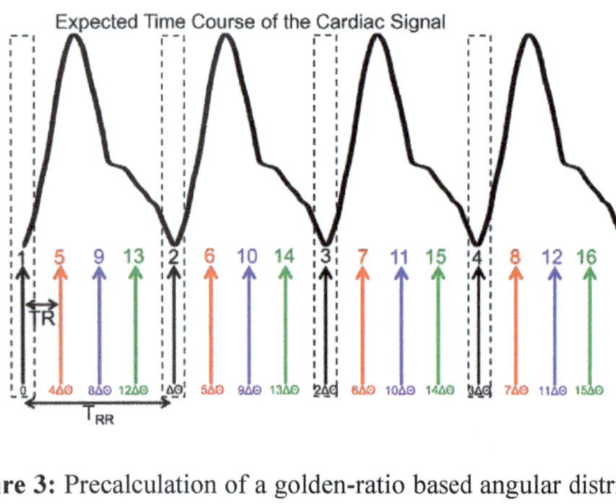

Figure 3: Precalculation of a golden-ratio based angular distribution to achieve a more uniform coverage of k-space with the projections selected for Cine reconstruction. Above: Expected time course of the cardiac signal, estimated from the average interval T_{RR}. Below: Using $\theta_0=0°$ as initial value, a list of projection angles was calculated by consecutively adding of the golden angle ($\Delta\theta\approx 111.246°$). Subsequently, the projections of the calculated angle list were sorted into $N\approx T_{RR}/TR$ interleaved sublists (demonstrated for N=4 sublists). Each sublist consists of $K\approx n_{Proj}/N$ golden-ratio-distributed projections (shown for K=4). In case of a constant signal each projection of one sublist belongs to the same heart cycle phase, as shown for the systolic phase and the black projections.

Signal processing

For retrospective triggering the k-space center of the radial projections was taken. The real part, the imaginary part and the magnitude values were investigated to find the trigger signal with the strongest modulation due to blood flow. The first 1.5 s (i.e., 500 projections) of the data were excluded from data processing since during this time the signal shows a strong transient to the steady state.

Since the original signal is often compromised by high harmonic disturbances (Figure4a), a Butterworth low-pass filter[14] was applied to suppress high-frequency interferences (F4b). Afterwards a Gaussian-shaped window function was used to reduce low-frequency modulations caused by respiratory motion (F4c). F4d and e show the magnitude responses of the used frequency filters. For the Butterworth filter a pass band frequency of 10-13 Hz and a stop band frequency of 15-20 Hz were found to provide the best filtering results. The high-pass filter has the shape $w(f)=1-\exp(-(f/\sigma)^2)$, with a window width of σ=4-8 Hz.

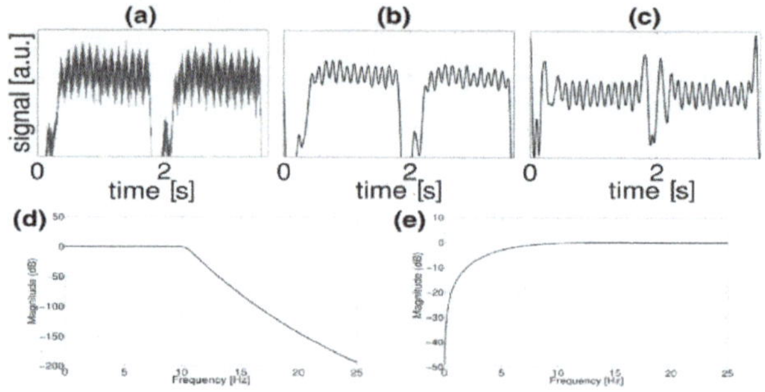

Figure 4 Filtering of the radial self-gating signal. (a): The raw signal is disturbed by higher harmonic oscillations. **(b)**: A Butterworth low-pass filter is applied to suppress high frequency interferences. **(c)**: Low frequency oscillations caused by respiratory motion are reduced with a gaussian-shaped window function. **(d)**: Magnitude response of the low-pass filter. Used pass band frequency: 10–13 Hz. Stop band frequency: 15-20 Hz. **(e)**: Magnitude response of the Gaussian window function ($w(f)=1-\exp(-(f/\sigma)^2)$). The used window width was σ=4-8 Hz.

K-space selection

For the assignment of the acquired projections to a corresponding heart cycle phase, the trigger signal extracted from the radial k-space data, which was acquired with a sampling rate of TR=3 ms, was interpolated to 1/10 of TR using a linear interpolation. To select a particular phase within the heart cycle, trigger points were determined with thresholds (threshold value: 90% of the local maximum of the interpolated self-gating signal). The first value above the threshold was chosen as the trigger point (Figure5a). After determination of the trigger positions each radial projection was assigned to a relative position in the heart cycle between 0 and 1 (see Figure5b). Data points during respiratory motion were gated out (relative position = -1). For the reconstruction of relative time frames, the relative heart cycle positions were divided into 100 equidistant selection intervals (Figure5b). Projections located within a temporal distance of ±0.5 ms around the corresponding heart cycle phase were assigned to each selection interval. Using this sliding window selection, an average number of about 120 projections per image were assigned to 100 time frames of the aortic motion covering the whole heart cycle.

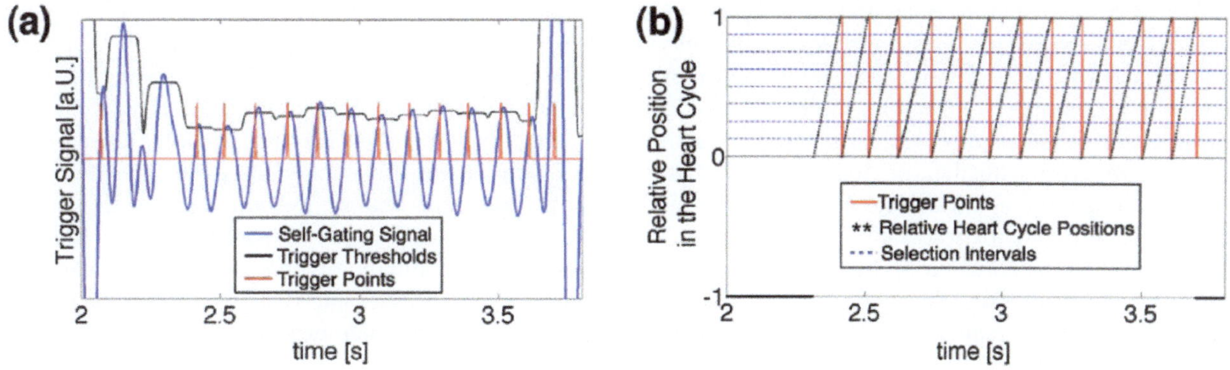

Figure 5 Determination of the trigger points. (a) Definition of trigger thresholds (black). As threshold value 90% of the local maxima of the cardiac signal (blue) is taken. As trigger points (red) always the first value above the threshold is chosen. **(b)** After definition of the trigger points (red) each read-out is assigned to a relative position in the heart cycle between 0 and 1 (black). Data during respiratory motion is gated out (relative position =-1). For projection selection the relative positions in the heart cycle were divided into 100 intervals (blue, displayed for 8 intervals).

Reconstruction

All radial reconstructions were done in MATLAB (The MathWorks, Inc., Natick, MA) with a Non Uniform Fast Fourier Transform (NUFFT,[15]) using the software toolbox provided by Fessler et al.[16]. The complex k-space data was re-gridded with a Kaiser-Bessel window[17]. The density compensation function was calculated using the iterative algorithm described by Pipe et al.[18].

Due to imperfect shimming a retrospective correction of B_0 offsets and linear inhomogeneities was applied for reconstruction. Therefore an off-resonance map was determined through the acquisition of two triggered gradient-echo images ϱ_1 and ϱ_2 with a relative echo time difference of Δ TE=0.5 ms, as described in[19]. With the phase of the conjugate complex product,ΔΦ(r⃗)=angle(ϱ∗1·ϱ2), the off-resonance frequencies can be calculated usingΔω0(r⃗)=ΔΦ(r⃗)ΔTE. A phase unwrapping was applied, when necessary[20]. Using this a priori information the off-resonance corrected imageϱ^(r⃗) can be estimated with the conjugated phase reconstruction proposed in ([21], Eq. 7):

$$\hat{\varrho}(\vec{r}) = \sum_j s(t_j)e^{i\Delta\omega_0(\vec{r})t_j}C(j,\vec{r}). \tag{2}$$

Hereby∑j denotes a summation over the measured k-space points, $s(t_j)$ the measured CMR signal and t_j the discrete sampling times of the measurement.C(j,r⃗)=eik⃗ (tj)r⃗ wj is a complex weighting matrix which contains the information about the k-space trajectoryk⃗ (tj) and the real-valued sampling densities $w_j$calculated with the Pipe algorithm[18]. Since an exact execution of the conjugated phase reconstruction has high computational requirements due to the large dimensions of the trajectory and the off-resonance term[21], Equation (2) is solved numerically using the NUFFT and an approximation of the exponential termeiΔω0(r⃗)tj, which is described in detail in[22, 23].

PWV calculation

For the calculation of the local pulse-wave velocity the flow-area (QA) method proposed by[24] was used. Assuming that a flow pulse through the aorta during the early systolic wave is reflection-free and unidirectional[24], the PWV can be determined using the approximation:

$$\mathrm{PWV} = \frac{dQ}{dA}, \tag{3}$$

where dQ is the change of the volume flow during the early systole and dA is the change of the cross-sectional area of the vessel. For a segmentation of the aorta all images were zero-filled to a matrix size of 256×256. The cross-sectional area A(t) and the volume flow Q(t) were determined using AMIRA (Mercury Computer Systems, Chelmsford, USA) for a segmentation of the vessel for each time frame. All segmentations were done 4 times by the same observer to reduce the standard error. The velocity information was calculated with a linear fit to the phase data of each pixel within the aorta.

For a PWV determination the early upstroke of the measured Q(A)-curve was fitted with a linear function. Since the Q and A data estimated from the reconstructed Cine images is superimposed with high frequency random noise, both curves were smoothed using a moving average filter[25] in the temporal domain, as mentioned in[4].

Animal handling

A group of 6 C57BL/6J mice (Charles River Laboratories) aged 6 months was imaged in vivo. Anesthetization was done with isoflurane (1.5-2% volume 0_2 (2L/min)). The temperature was kept constant at 35°C using the gradient cooling unit. All mice were measured vertically (head up). All experimental procedures were in accordance with institutional and internationally recognized guidelines and were approved by the Regierung von Unterfranken (Government of Lower Franconia, Würzburg, Germany) to comply with German animal protection law. The reference number of the permit of the animal experiments is 55.2-2531.01-14/10.

RESULTS

Figure6a displays a representative section of the self-gating signal of a motion compensated measurement (M_1=0 s/m). The signal modulations due to breathing and the blood flow through the aorta are clearly visible. Figure6b shows a histogram of the time intervals between each found trigger point, which was calculated from the navigator signals of all three velocity-encoding steps. The average interval between two trigger peaks was 107.2±4.1 ms (mean value and variation estimated with a Gaussian fit). The distribution of the trigger point distances is in range of natural variations of the heart rate. Through the reconstruction of 100 movie frames of the aortic motion an average frame rate of 933±36 frames/s was achieved. In Figure7representative flow compensated magnitude images of the aorta during the systolic heart phase are shown for the Cartesian, the triggered radial and the retrospective radial measurement (Figure7a to c upper row). The scan time was 7 minutes for the Cartesian measurement and 4 minutes for the triggered radial measurement. The retrospective measurement was done in only 3 minutes. For the non-triggered reconstruction an average number of 120 spokes could be used, which corresponds to an undersampling factor of more than 4 respective to the Nyquist criterion. The bottom images of Figure7a-c show phase maps corresponding to the same measurement. In the phase map of the retrospectively triggered measurement streaking artifacts are visible, however, the aortic flow is only slightly affected by these. The Figures8a and b display an overlay of the flow (Q) and the area (A) waveform for the triggered and retrospective measurement. The area curves of the three measurements are slightly shifted among each other due to uncertainties of the aorta segmentation. Also smaller values of the peak volume flow are noticeable with the prospectively triggered measurements, which might be the result of instabilities of the balloon signal. However, for a determination of the PWV only relative changes of the area- and flow curves are important[4]. All three flow-area plots show consistency with the hypothesized linear behavior during the early systole (Figure8c-e). For a determination of the pulse-wave velocity only the first data points after the beginning of the upstroke are relevant. With each the triggered Cartesian and the triggered radial measurement 8 data points (8 ms) of the early upstroke of the systolic-flow pulse could be used for a linear fit. The frame rate of the retrospective measurement was high enough to take 7 data points (≈7.5 ms) for the fit. Table2 shows the averaged results for the calculated parameters of all 6 monitored animals. All three methods have sufficient temporal resolution to sample the upstroke of the systolic volume flow with 5-8 data points, respectively. The average PWV determined with the Cartesian measurement was 2.4±0.2 m/s and for the prospectively triggered radial measurement 2.3±0.2 m/s. The PWV measured with the retrospectively triggered method was 2.3±0.2 m/s, which is in good agreement with the results of the prospectively triggered measurements.

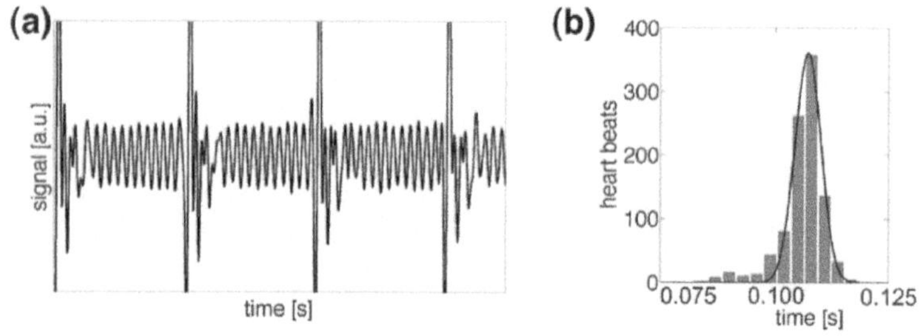

Figure 6 Measured self-gating signal. (a) Representative self-gating signal obtained in a C57BL/6J mouse (only motion compensated measurement shown). (b) Histogram of the trigger point distances. The navigator signal of all three velocity-encoding steps was taken (3×48 s measurement time). A large narrow peak is visible that can be assigned to the blood flow.

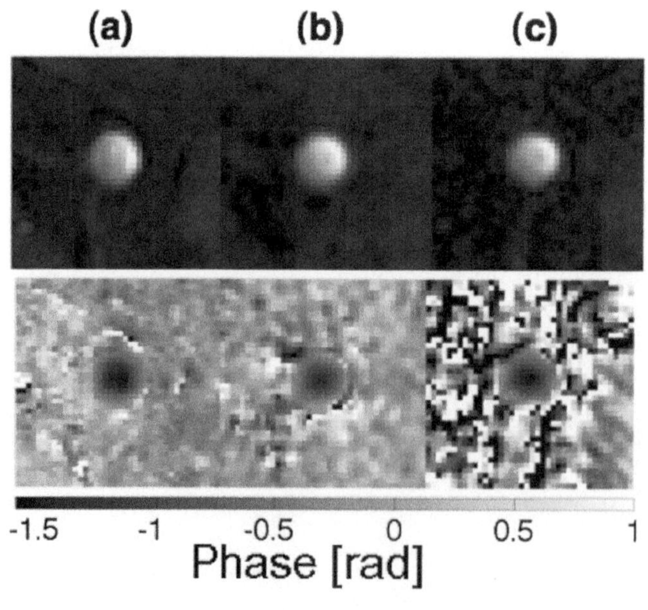

Figure 7 Images of the abdominal aorta. (a)-(c) above: Representative magnitude images of the motion compensated measurement for the time frame with the maximum velocity, acquired with a triggered Cartesian sampling **(a)**, a triggered radial sampling **(b)** and the retrospective radial measurement **(c)**. The spatial resolution of all images was 150×150 μ m², interpolated to 94×94 μ m² using zero-filling. **(a)-(c)** below: Velocity-encoded PC MR phase images belonging to the same slice and time frame.

DISCUSSION

PWV values

This work shows that a self-gated measurement of the local pulse-wave velocity in the aorta is feasible. The average PWV values of the retrospectively triggered measurement are in good agreement with the triggered Cartesian and radial measurements. In[4] PWV values of 2.6±0.2 m/s are published for 8-months-old WT-mice using the prospectively triggered Cartesian PC-Cine measurement, which is in the same range as the data presented here. Zhao et al.[5] published values of 3.54±0.97 m/s for the PWV of 9-months-old WT-mice. In ultrasound measurements PWV values of 2.86±0.14 m/s are found for 8-month old B6D2F1 mice[26].

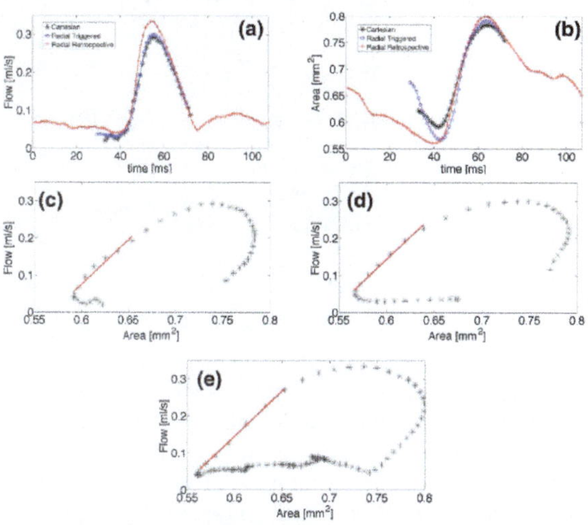

Figure 8 Results of the QA measurements. Representative results of a flow-area measurement in the abdominal aorta of a C57BL/6J mouse (age 6 months). Above: Volume flow Q(t) **(a)** and corresponding cross-sectional area A(t) **(b)**(overlay of the triggered Cartesian, the triggered radial and the retrospective radial measurement). Below: Q-Acurves to determine the local pulse-wave velocity. **(c)** With the triggered Cartesian measurement. **(d)** With the triggered radial measurement. **(e)** With the retrospective radial measurement.

Table 2 Average results of the local PWV measurements in the abdominal aorta, determined with 6 animals

Parameters	**Results**
Weight (g)	24.9±3.1
Peak CSA [mm^2]	
Cartesian Measurement	0.75±0.05
Radial Triggered Measurement	0.75±0.04
Retrospective Measurement	0.73±0.04
Peak volume flow [cm^3/s]	
Cartesian Measurement	0.27±0.04
Radial Triggered Measurement	0.28±0.03
Retrospective Measurement	0.28±0.04
PWV [m/s]	
Cartesian Measurement	2.4±0.2
Radial Triggered Measurement	2.3±0.2
Retrospective Measurement	2.3±0.2

Results are presented as mean ± SE. PWV: Pulse-wave velocity. CSA: Cross-sectional-area.

Trigger quality

In contrast to measurements close to the beating heart only view moving voxels can be found in a slice in the abdominal region. Therefore the waveform of the navigator signal is less distinctive then in a cardiac Cine measurement and depends on the scale of the pulsating vessel. Since the navigator signal depends on changes of blood signal within the slice, the trigger quality is sensitive to the applied flip angle. Robust trigger signals were only found with flip angles of 30° and higher. With flip angles less then 20°, the retrospective triggering often failed.

Temporal resolution

The monitoring of the pulse-wave propagation requires high temporal resolution. With the triggered interleaved PC-Cine measurements described in[4] a frame rate of 1000 frames/s can be achieved. Since always a fixed number of 100 time frames were reconstructed, the frame rate of the retrospective measurement depends on the heart rate. In the examined group the average cardiac interval was 101-115 ms, therefore the absolute frame rate was 870-990 frames/s, which was sufficient to take 5-8 data points during the upstroke of the systolic flow for the linear fit. For measurements of animals with lower heart rates (175 - 200 ms) one should increase the number of reconstructed Cine frames to ensure sufficient temporal resolution, which might result into longer measurement times.

The native temporal resolution of the retrospective measurement is restricted to the repetition time of the read-out, which was 3 ms in this experiment. However, since the heart rate shows small variations due to naturalfluctuations of the heart function, the sampled states in the cardiac cycle slightly vary over the time, allowing a denser sampling of the aortic motion. As a result the radial projections could be assigned to relative positions in the heart cycle using a sliding window width of only 1 ms. Furthermore one could implement small TR variations in the timing of the pulse sequence to randomize even more the sampling of the heart periods. In Figure9 the average positions in the heart cycle and the standard deviations for the heart cycle positions of the selected radial projections of one retrospective Cine measurement are plotted as a function of the frame index (between 1 und 100). The mean variation per frame was ±0.75% of the heart cycle, which corresponds to a deviation of ≈±0.8 ms. Since the temporal frames therefore had an overlap of ≈50%, the reconstructed Cine is slightly blurred and the measured area-and flow-curves appear to be smoothed. However, a low-pass filter is usually applied to the flow data of the triggered measurements as well to suppress high frequency random noise[4], which results into a similar smoothing effect.

Measurement time

A drawback of the triggered Cartesian PWV measurement is the long scan time. To achieve a temporal resolution of 1 ms, 5 segmented Cines were necessary. The average scan time of the Cartesian experiment was 6.7±1.1 minutes (mean ± SE). Since the early systolic time frame is very short, only few data points can be used for linear fitting. Therefore the heart rate has to be kept constant during the whole measurement. With the use of a highly asymmetric radial trajectory, the scan time of the triggered PWV measurement could be reduced to 4.3±0.6 minutes without losing temporal and spatial resolution. Using the retrospective measurement the whole PWV experiment can be done in only 2.9 minutes, which is more than two times faster than the triggered Cartesian protocol.

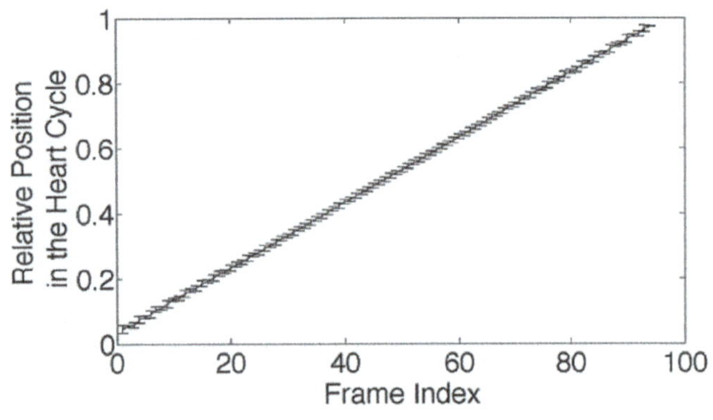

Figure 9 Temporal blurring of the reconstructed relative time frames. Average position in the heart cycle with standard deviation as a function of the reconstructed relative time frames. Since the mean variation is 0.75% of the heart cycle, the Cine frames have an overlap of 50%.

Contrast

Compared to the prospectively triggered methods, where the MRI signal does not reach the steady state due to cardiac and respiratory gating, the MRI signal of the retrospective flow measurement is in steady state all the time. Therefore each retrospectively reconstructed image contains a comparable contrast, while contrast changes are noticeable with the triggered and respiratory-gated measurements. In addition, since the surrounding static tissue appears to be darker than with the triggered sequences, the retrospectively triggered method enables an easier segmentation of the aorta.

Deviations of the radial trajectory

The retrospective technique does not require external ECG probes and is therefore less sensitive to animal handling. However, off-resonance and local field inhomogeneities can compromise the quality of the reconstructed images and lead to deviations of the radial trajectory. Using highly asymmetric projections, the echo time can be shortened from 1.7 ms to 1.2 ms, reducing the effect of inhomogeneities and flow on the measurement. With the aid of B_0 maps remaining susceptibility artifacts can be corrected retrospectively. To demonstrate the effect of the B_0 correction an exemplary map of the off-resonance frequencies$\Delta f(\vec{r})=\Delta\omega 0(\vec{r})2\pi$ is shown in Figure10 (top: complete map, bottom: section around the aorta). Around the aorta the frequency offset lies around 600 Hz, causing severe blurring artifacts in the uncorrected radial image which make an segmentation of the vessel impossible (see images in the middle). Using the measured B_0 map and the algorithm mentioned above the blurring can be significantly reduced, allowing a better distinction of the vessel from the surrounding tissue (see right images). However, the acquisition of the required triggered gradient-echo images for this optional B_0 correction results into an additional measurement time of 1-2 minutes. Even with correction, B_0 inhomogeneities remain the main cause for uncertainties of aorta segmentation and difficulties with PWV measurements in vessels close to lung tissue. However, issues with susceptibility artifacts in vessels nearby lung tissue were also reported for the triggered Cartesian measurement described in[4]. Another cause of trajectory deviations are timing errors caused by delays of the spatial encoding gradients[11]. By measuring the gradient switching times trajectory distortions can be reduced, however, residual trajectory errors and errors of the phase contrast maps caused by eddy currents[12, 27] can remain.

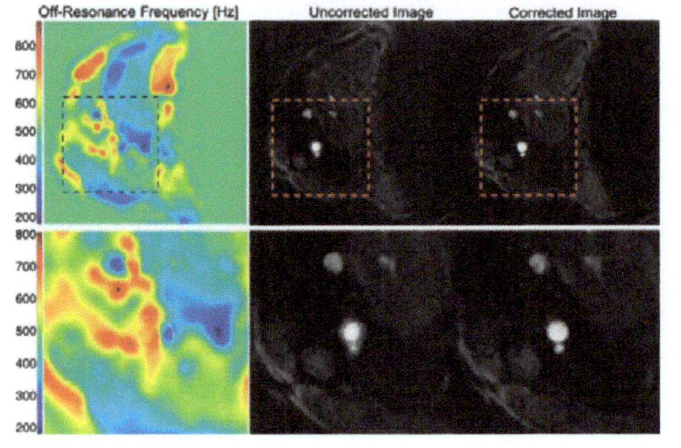

Figure 10 Off-resonance correction reconstruction, which was determined with two gradient echo images with echo time difference Δ TE=0.5 ms. Above: complete map. Below: section around the aorta. Middle: Uncorrected magnitude image of the radial measurement. Around the aorta severe blurring artifacts are visible. Right: Using the measured off-resonance map and the previously mentioned algorithm the radial images can be de-blurred.

Advantages of the radial acquisition

In[28] and[29] a Cartesian self-gating method was proposed which measures a navigator echo during refocusing of the slice-selection gradient without the need of non-uniform trajectories. In[8] the rectilinear self-gating signal could be demonstrated to be more robust than the radial signal since it does not suffer from trajectory off-center effects. However, in this method the acquisition of the navigator signal was achieved by a temporal separation of the slice-refocusing gradient and the read dephase and phase encoding gradients. With a flow-encoding sequence based on a tripolar slice-selection gradient and a bipolar flow-encoding gradient this separation would result in an undesirable prolonging of the read-out duration and in a longer echo time. With the radial self-gating sequence proposed in this work a self-gating signal can be acquired without extending the timings of the flow-encoding sequence. Through the application of high echo asymmetry and short read dephase gradients the echo time could be reduced, which helps to reduce misregistration effects due to blood flow[30]. In addition, since every radial projection samples the k-space center and therefore has equal information content, an easier reconstruction of time frames is possible compared to the retrospective Cartesian sampling. As a further advantage, the proposed self-gated flow-encoding sequence profits from the advantages of the radial trajectory, due to the higher incoherency of the flow and motion artifacts[31, 32] and a higher robustness against undersampling[33].

CONCLUSIONS

We present a robust self-gated PC-Cine method that enables a measurement of the local pulse-wave velocity in the murine aorta without the need of additional triggering hardware. This new technique allows the reconstruction of high-resolution Cines of the aortic motion with a frame rate comparable to triggered methods achieving similar results for the PWV values.

DECLARATIONS

Acknowledgements

This study was supported by the German Research Foundation (DFG) in the scope of Sonderforschungsbereich 688 (TP B4, B5, Z2), the Bundesministerium für Forschung und Bildung in the scope of BMBF01 E01004, the University of Würzburg in the funding program Open Access Publishing and the Comprehensive Heart Failure Center (Deutsches Zentrum für Herzinsuffizienz). We thank Carmen Bundschuh for the preparation of the animals, Norbert Elsaesser and Stephan Richter for developing a Matlab tool for phase unwrapping and David Alexander Kann and Katrin Winter for the help in preparing the manuscript.

Competing interests

The authors declare that they have no competing interests.

Authors' contributions

PW developed the radial flow-encoding sequences and conducted all experiments. ER, WRB, VH and PMJ enrolled and organized this study. TK and VH helped with the data evaluation. XH and VH helped with the pulse-sequence development, XH, CBM and FTG provided the software for trigger signal evaluation. All authors took part in critical review and drafting of the manuscript and have read and approved the final manuscript.

REFERENCES

1. Blacher J, Guerin AP, Pannnier B, Marchais SJ, Safar ME, London GM: Impact of aortic stiffness on survival in end-stage renal disease. Circulation. 1999, 99: 2434-9. 10.1161/01.CIR.99.18.2434.
2. Laurent S, Katsahian S, Fassot C, Tropeano AI, Gautier I, Laloux B, Boutouyrie P: Aortic Stiffness is an independent Predictor of fatal Stroke in essential hypertension. Stroke. 2003, 34: 1203-6. 10.1161/01.STR.0000065428.03209.64.
3. Greve JM, Les AS, Tang BT, Draney Blomme MT, Wilson NM, Dalman RL, Pelc NJ, Taylor CA: Allometric scaling of wall shear stress from mice to humans: quantification using cine phase-contrast MRI and computational fluid dynamics. Am J Physiol: Heart and Circ Physiol. 2006, 291: H1700-7. 10.1152/ajpheart.00274.2006.
4. Herold V, Parczyk M, Mörchel P, Ziener CH, Klug G, Bauer WR, Rommel E, Jakob PM: In vivo measurement of local aortic pulse-wave velocity in mice with MR microscopy at 17.6 tesla. Magn Reson Med. 2009, 61: 1293-9. 10.1002/mrm.21957.
5. Zhao X, Pratt R, Wansapura J: Quantification of aortic compliance in mice using radial phase contrast MRI. J Magn Reson Imaging. 2009, 30: 286-91. 10.1002/jmri.21846.
6. Polson MJR, Barker AT, Gardiner S: The effect of rapid rise-time magnetic fields on the ECG of the rat. Clin Phys Physiol Meas. 1982, 3 (3): 231-4. 10.1088/0143-0815/3/3/008.
7. Larson AC, White RD, Laub G, McVeigh ER, Li D, Simonetti OP: Self-gated cardiac cine MRI. Magn Reson Med. 2004, 51: 93-102. 10.1002/mrm.10664.
8. Hiba B, Richard N, Janier M, Croisille P: Cardiac and respiratory self-gated cine MRI in the mouse at 7T. Magn Reson Med. 2006, 55: 506-13. 10.1002/mrm.20815.
9. Herold V, Mörchel P, Faber C, Rommel E, Haase A, Jakob PM: In vivo quantitative three-dimensional motion mapping of the murine myocardium with PC-MRI at 17.6T. Magn Reson Med. 2006, 55: 1058-64. 10.1002/mrm.20866.
10. Swoap SJ, Overton M, Garber G: Effect of ambient temperature on cardiovascular parameters in rats and mice: a comparative approach. Am J Physiol: Regul, Integr Comp Physiol. 2004, 287: R391-6. 10.1152/ajpregu.00731.2003.
11. Peters DC, Derbyshire JA, McVeigh ER: Centering the projection reconstruction trajectory: reducing gradient delay errors. Magn Reson Med. 2003, 50: 1-6. 10.1002/mrm.10501.
12. Duyn JH, Yang Y, Frank JA, van der Veen JW: Simple correction method for k-space trajectory deviations in MRI. J Magn Reson. 1998, 132: 150-3. 10.1006/jmre.1998.1396.
13. Winkelmann S, Schaeffter T, Koehler T, Eggers H, Doessel O: An optimal radial profile order based on the golden ratio for time-resolved MRI. IEEE Trans Med Imaging. 2007, 26: 68-76.
14. Butterworth S: On the theory of filter amplifiers. Exp Wireless and the Wireless Eng. 1930, 7: 536-41.
15. Fessler JA, Sutton BP: Nonuniform fast fourier transform using min-max interpolation. IEEE Trans Signal Process. 2003, 51: 560-74. 10.1109/TSP.2002.807005.
16. Fessler JA: [http://web.eecs.umich.edu/~fessler/],
17. Jackson JI, Meyer CH, Nishimura DG, Macovski A: Selection of a convolution function for fourier inversion using gridding. IEEE Trans Med Imaging. 1991, 10: 473-8. 10.1109/42.97598.
18. Pipe JG, Menon P: Sampling density compensation in MRI: Rationale and an iterative numerical solution. Magn Reson Med. 1999, 41: 179-86. 10.1002/(SICI)1522-2594(199901)41:1<179::AID-MRM25>3.0.CO;2-V.
19. Schneider E, Glover G: Rapid in vivo proton shimming. Magn Reson Med. 1991, 18: 335-47. 10.1002/mrm.1910180208.
20. Jenkinson M: Fast, automated, N-dimensional phase-unwrapping algorithm. Magn Reson Med. 2003, 49: 193-7. 10.1002/mrm.10354.
21. Noll DC, Meyer CH, Pauly JM, Nishimura DG, Macovski A: A homogeneity correction method

for magnetic resonance imaging with time-varying gradients. IEEE Trans Med Imaging. 1991, 10: 629-37. 10.1109/42.108599.

22. Fessler JA, Sutton BP, Noll DC: Fast, iterative image reconstruction for MRI in the presence of field inhomogeneities. IEEE Trans Med Imaging. 2003, 22: 178-88. 10.1109/TMI.2002.808360.
23. Fessler JA, Lee S, Olafsson V, Shi HR, Noll DC: Toeplitz-based iterative image reconstruction for MRI with correction for magnetic field inhomogeneity. IEEE Trans Signal Process. 2005, 53 (9): 3393-402.
24. Vulliémoz S, Stergiopulos N, Meuli R: Estimation of local aortic elastic properties with MRI. Magn Reson Med. 2002, 47: 649-54. 10.1002/mrm.10100.
25. Smith SW: Moving average filters. The Scientist and Engineer's Guide to Digital Signal Processing. 1997, San Diego: California Technical Publishing, Chapter 15:277-284. [http://www.dspguide.com],
26. Reddy AK, Li YH, Pham TT, Ochoa LN, Trevino MT, Hartley CJ, Michael LH, Entman ML, Taffet GE: Measurement of aortic input impedance in mice: effects of age on aortic stiffness. Am J Physiol- Heart Circ Physiol. 2003, 285: H1464-70.
27. Bernstein MA, Zhou XJ, Polzin JA, King KF, Ganin A, Pelc NJ, Glover GH: Concomitant gradient terms in phase contrast MR: analysis and correction. Magn Reson Med. 1998, 39: 300-8. 10.1002/mrm.1910390218.
28. Esparaza-Coss E, Ramirez MS, Bankson JA: Wireless self-gated multiple-mouse cardiac cine MRI. Magn Reson Med. 2008, 59: 1203-6. 10.1002/mrm.21562.
29. Bovens SM, te Boekhorst BCM, den Ouden K, van de Kolk KWA, Nauert A, Nederhoff MGJ, Pasterkamp G, ten Hove M, van Echteld CJA: Evaluation of infarcted murine heart function: Comparison of prospectively triggered with self-gated MRI. NMR in Biomed. 2010, 24: 307-315.
30. Firmin DN, Nayler GL, Kilner PJ, Longmore DB: The application of phase shifts in NMR of flow measurement. Magn Reson Med. 1990, 14: 230-41. 10.1002/mrm.1910140209.
31. Nishimura DG, Jackson JI, Pauly JM: On the nature and reduction of the displacement artifact in flow images. Magn Reson Med. 1991, 22: 481-492. 10.1002/mrm.1910220255.
32. Glover GH, Pauly JM: Projection Reconstruction Techniques for Reduction of Motion Effects in MRI. Magn Reson Med. 1992, 28: 275-89. 10.1002/mrm.1910280209.
33. Peters DC, Korosec FR, Grist TM, Block WF, Holden JE, Vigen KK, Mistretta CA: Undersampled projection reconstruction applied to MR angiography. Magn Reson Med. 2000, 43: 91-101. 10.1002/(SICI)1522-2594(200001)43:1<91::AID-MRM11>3.0.CO;2-4.

CHAPTER 19

ENERGY FLUX IN THE COCHLEA: EVIDENCE AGAINST POWER AMPLIFICATION OF THE TRAVELING WAVE

Marcel Van Der Heijden and Corstiaen P. C. Versteegh

Keywords: Cochlear mechanics, Laser vibrometry, Basilar membrane, traveling wave, cochlear amplifier, group velocity

INTRODUCTION

The cochlea is both a transducer that converts sound to neural activity and a frequency analyzer that separates acoustic components. Its elongated fluid-filled cavities are separated by a thin elastic structure, the basilar membrane (BM), whose motion is coupled to sensory cells. The BM supports traveling waves that have two crucial, but poorly understood properties. First, their amplitude changes drastically during propagation, exhibiting a peak that shifts position with frequency. This frequency mapping underlies the spectral analysis. Second, soft sounds evoke sharper peaking than do intense sounds. This intensity dependency reflects the ear's dynamic range compression.

In the 1980s, the failure of classical fluid-mechanical models to account for these unusual wave properties led to the introduction of "active models," in which the peaking of the wave is associated with a region of negative damping (Kim et al. 1980b; Neely 1985). In this scenario, motile processes in outer hair cells (OHCs) amplify the waves. The injection of mechanical energy by this "cochlear amplifier" is assumed to improve the sensitivity to soft sounds, and its saturation is invoked to explain compression (Ashmore et al. 2010).

Citation: Van der Heijden, M. & Versteegh, C.P.C. JARO (2015), Energy Flux in the Cochlea: Evidence Against Power Amplification of the Traveling Wave, 16: 581. doi:10.1007/s10162-015-0529-5.

Cochlear amplification has slowly gained acceptance and is now the dominant view. There is a recent trend to present cochlear amplification as proven (e.g., Hudspeth 2013) even though the evidence quoted in favor of it comes from older studies: the loss of cochlear sensitivity following OHC damage (Evans and Harrison 1976) and the existence of spontaneous emissions (Kemp,1979). Until recently, that same evidence was more cautiously described as impressive, but inconclusive (Robles and Ruggero 2001; Ashmore 2008; Shera 2007), since both physiological vulnerability and spontaneous emissions leave room for alternative explanations not based on amplification. For instance, while it is undisputed that OHCs control BM motion, that is not to say that they also drive the vibrations in the sense of supplying the mechanical energy. In an alternative scenario, OHCs control BM motion by functioning as brakes that cause mechanical energy to be absorbed rather than injected (Allen 1979). A combination of the two roles, i.e., amplification at low intensities and variable attenuation at high intensities, is also conceivable. Spontaneous emissions certainly seem to reveal processes capable of producing mechanical energy (Talmadge et al. 1991), and there have been many modeling attempts to link them to some form of amplification (Shera 2003a; Duke and Jülicher 2008), but to conclude from their mere existence that the cochlea systematically amplifies its acoustic input at all frequencies is rather tentative. Spontaneous emissions may also be side effects of other forms of mechanical control exerted by OHCs, for instance, negative feedback of an automatic brake system. Particularly in the presence of delayed coupling and tuned circuits, feedback, even when negative, easily gives rise to ringing and other instabilities (Doyle et al. 1992). It is also noteworthy that spontaneous emissions are extremely rare in normal-hearing nonprimate laboratory animals (Martin et al. 1988) and that even in normal-hearing humans, their incidence is ~40 % (Wier et al. 1984).

Arguably, the most problematic aspect of the putative amplifier (and a good reason to keep an open mind toward alternatives) has been its physiological implementation (Ashmore et al.2010). Amplification requires phase-locked motile feedback at high frequencies (>150 kHz in some species). While somatic OHC motility by itself may be fast enough (Frank et al. 1999), it is difficult to see how the AC component of the OHC receptor potential evoked by near-threshold sounds can have sufficient amplitude to drive motility at such high frequencies, as it is shunted by the membrane capacitance (Cody and Russell 1987). It is unknown whether the alternative mechanism, hair bundle motility (Kennedy et al. 2006), can operate at these very high frequencies. For a parametric (rather than cycle-by-cycle) operation of OHCs, such as braking or adjusting the radial profile of BM motion (Ren and Gillespie 2007), high-frequency limitations are not a problem.

Rather than relying on circumstantial evidence, a number of studies have attempted to estimate the amount of amplification based on measurements in the auditory periphery. Combining auditory nerve recordings and otoacoustic emissions, Allen and Fahey (1992) found a negative result, although their conclusions have been disputed by others (Shera 2003b; de Boer et al.2005). Cochlear mechanical measurements are an obvious choice for tests of amplification. In sensitive cochleae, the range of sensitivity, assessed by normalizing BM vibrations to the middle ear response, exceeds 50 dB, and compressive growth persists to at least 100 dB sound pressure level (SPL) (Rhode 2007). Thus, if a simple "saturating amplifier" were the only explanation of cochlear compression, one would expect it to provide at least 50 dB of amplification at low intensities (as is indeed the case for simplified positive-feedback models like that of Cooper1998) and persist to amplify to very high intensities. Estimates of power gain at low intensities derived from BM and neural data are variable: 40 dB (Brass and Kemp 1993), 12 dB on average with confidence intervals spanning [−4, ∞] dB (Shera 2007), and 0.4–17.7 dB (de Boer and Nuttall 2001). The intensity range over which BM recordings and intracochlear pressure measurements appear to indicate amplification, is restricted to low and moderate intensities in some studies (e.g., Olson 2001), whereas in other studies, it appears to extend to high intensities (de Boer and Nuttall 2000; Dong and Olson 2009). The large variability among these studies is not well understood but suggests that further work is needed and that novel methods are welcome.

Active cochlear models implement amplification by introducing a negative real part of cochlear impedance (negative damping) over a limited, frequency-dependent cochlear region just basal to the peak of the wave. When traversing this region, the wave picks up energy. In this region, then, there should be a local power gain, i.e., a positive gradient in energy flux. Here, we tested this prediction by determining the energy flux of the traveling wave and by comparing this flux both to the power input to the middle ear (net gain) and across adjacent locations on the BM (local gain). We base our analyses on BM recordings at two adjacent locations, which allow a more direct analysis of the energy transport under scrutiny than previous analyses based on single-point BM recordings or neural data.

METHODS

BM Recordings

Details of the animal preparation, experimental setup, and stimuli are described in a recent publication (Versteegh and Van der Heijden 2013). BM motion was measured from pairs of locations spaced 145–252 μm in seven cochleae of Mongolian gerbil (Meriones unguiculatus; ~60 g) in the 12–21-kHz region, using a Doppler laser interferometer. All procedures were approved by the Erasmus MC laboratory animal committee. Animals were anesthetized, and the pinna was removed, followed by opening the bulla, which gave access to the round window. After tearing the round window membrane, reflective beads were inserted into the cochlea and allowed to settle on the BM. A glass cover slip placed over the round window stabilized the air-fluid interface. The use of reflective beads allowed the recording of phase-locked, sub-nanometer BM vibrations in response to sounds of very low intensities (down to 0 dB SPL) by improving the signal to noise ratio and preventing the interference of spurious reflections from neighboring structures. The specific mass of the beads (1.03 times that of water), their small size (20–25 μm, i.e., an order of magnitude smaller than the wavelength of the traveling wave), and their incompressibility minimize their interference with the traveling wave and BM motion; indeed, these beads were shown to have little effect on BM motion in sensitive cochleae (Cooper1999). For data to be included in the analysis, bead position had to be stable during data collection (variations <10 μm, monitored using the online camera built into the vibrometer and from the adjustment of the micrometers for horizontal beam positioning). The physiological condition of the cochlea was judged from the lower intensity limit at which the BM response showed compressive nonlinearity (Rhode 2007). Only data were accepted from sensitive cochleae that showed compressive nonlinearity down to 10-dB SPL per component or lower (Fig. 1).

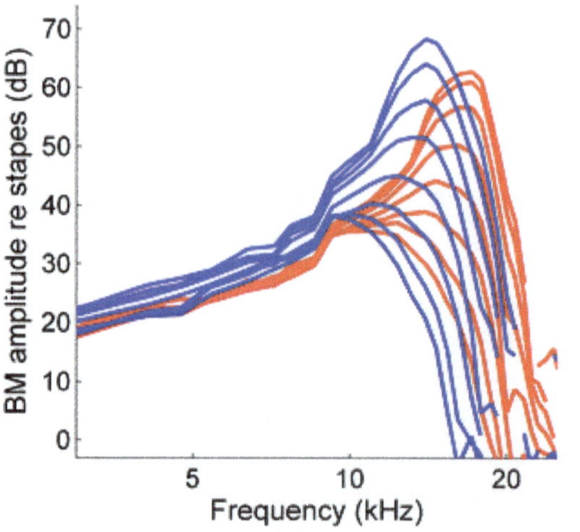

Figure. 1 BM responses showing high sensitivity and compressive nonlinearity at low intensities. BM responses from two adjacent locations, normalized to stapes response. Nonlinear compressive growth of the cochlear response is apparent from the systematic decrease of stapes-to-BM gain with sound intensity. Any cochlear trauma causes a linearization at lower levels, which would cause the lower-intensity curves to overlap. Conversely, the persistence of nonlinearity down to the lowest intensities is a stringent test of cochlear sensitivity. Our

criterion for data inclusion was the persistence of compression down to 10-dB SPL per component or better. Data shown were obtained from one of the five cochleae used for group velocity measurements (Fig. 5). Intensity was varied from 0- to 70-dB SPL per component in 10-dB steps. At both locations, the amplitude curves showed compression down to 0-dB SPL, the lowest intensity used (animal RG12433).

In order to be accepted for analysis, spectral components from the recordings must show Rayleigh significant ($p<0.001$) phase locking to the stimulus (Versteegh and Van der Heijden2012). These acceptance criteria exclude any data from insensitive and damaged cochleae, from which a significant phase locked responses to 10-dB-SPL stimuli—whether linear or not—cannot be recorded at all. All data were collected within 2.5 h from the tearing of the round window membrane. Best frequency (BF) was determined from the peak of the velocity-frequency curves normalized to stapes motion.

Custom MATLAB software computed stimuli that were sent to a TDT System 3 (24-bit D/A channel at 111.6 kHz; Tucker-Davis Technologies, Alachua, FL, USA). A probe sealed with Vaseline to the bony rim of the ear canal delivered sound stimuli. After correction for the acoustical transfer of the probe, the spectrum varied <4 dB in the 5–25-kHz range. A single-point laser vibrometer (OFV-534; Polytec, Waldbronn, Germany) connected to a velocity decoder (VD-06; Polytec) and TDT System 3 (24-bit A/D channel at 111.6 kHz) measured BM velocity of two locations in response to the same stimuli consecutively.

For the data underlying the group velocity measurements (Fig. 3), stimuli were tone complexes with an average frequency spacing of either 700 or 300 Hz and a bandwidth of ~2500 Hz, presented at a total intensity of 16–86 dB SPL in 10-dB steps. Irregular spacing of frequency components ensured that combination tones up to the third order did not coincide with any of the primary components (Meenderink and Van der Heijden 2011). A single stimulus lasted 60 s but could be repeated up to three times and the responses averaged. Group delays were determined using two methods. In the temporal method, the envelopes were extracted from the BM velocity response using a Hilbert transform and squared to lead to instantaneous power (up to an irrelevant scaling factor). Cross-correlation functions of the squared envelopes between the recording locations were computed, and the location of the maximum was determined. The fluctuations recorded at all three locations (stapes and the two beads) were highly similar (Fig. 3B; correlations between squared envelope >0.9). Thus, the power fluctuations of the stimulus were not deformed by excessive dispersion or nonlinearity, allowing the straightforward assessment of energy travel time from the response envelopes. In the spectral method, phase difference between the locations was plotted against frequency, and a third-order polynomial was fitted to these curves. The slope of these fits was then evaluated at BF. The two methods are compared in Figure 3D. As expected from the physical significance of group velocity as the speed of energy transport (Whitham 1974), the two methods were equivalent. The average data in Figure 3F were linearly extrapolated from 16–86 to 10–90 dB SPL in order to be applied to the energy flux analysis shown in Figure 4.

The local transfer functions (see "Results" section) shown in Figures 6 and 7 were obtained using the same type of irregularly spaced tone complexes, this time spanning a larger frequency range. In all recordings, the lowest-intensity stimuli were presented first, and the responses at both locations were recorded before moving to the next (higher) stimulus level. This order of stimulus presentation prevents any temporary reduction of cochlear sensitivity induced by prolonged high-intensity stimulation (Versteegh and Van der Heijden 2013) from affecting lower-intensity recordings.

Analysis of Fluid Motion

This section describes the details of the estimates of kinetic energy density leading to the findings presented in "Results" section (Fig. 4A–C). Estimating kinetic energy required analyzing the motion of the fluid surrounding the BM. Large parts of the BM are surrounded by free fluid or by supporting cells which, lacking any structurally stiff parts, behave like malleable bags of fluid (Steele and Taber 1979a). Supporting cells that have structural stiffness are organized in tunnels that do not impede longitudinal flow and/or in structures with abundant gaps, allowing the fluid to flow around them. Motivated by this anatomy (Lim 1986), the fluid was treated as freely moving. This reduces the analysis to solving the Laplace equation for irrotational fluid motion, using the two-dimensional pattern of BM reported by Ren et al. (2011a) as boundary conditions.

The analysis employs the analytical framework described by Steele and Taber (1979a) for computing three-dimensional fluid motion from the transverse motion profile of the BM and the geometry of the rigid boundaries. Fluid motion is described by a velocity potential obeying Laplace's equation $\nabla^2\phi=0$. Rectangular boxes represent the scalae; (x,y,z) denote longitudinal, radial, transverse directions (Fig. 2). Width and height, $L_1=L_2=500$ μm, were chosen to match the cross-sectional area in the gerbil's basal turn; the inner bony shelf width $w_S=250$ μm (Plassmann et al. 1987).

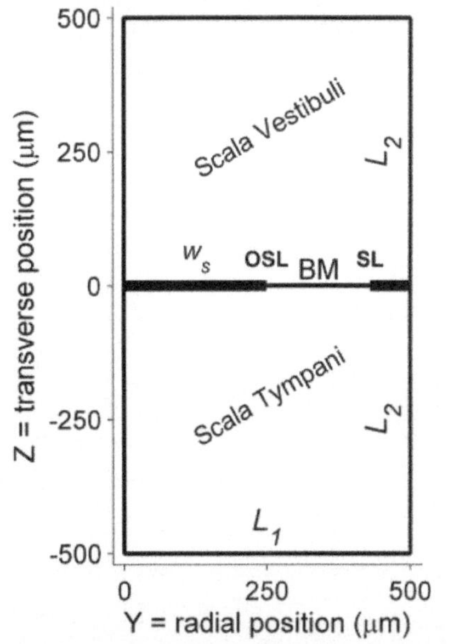

Figure. 2 Schematic cross section of cochlear ducts used for calculating the fluid motion near the BM. This cross section illustrates the boundary conditions for the Laplace equation describing the irrotational fluid motion. All boundaries are rigid except the basilar membrane (BM), whose motion is prescribed by the two-dimensional data of Ren et al. (2011a). The BM is suspended between the osseous spiral lamina (OSL) and the spiral ligament (SL). Dimensions L_1, L_2, and w_s are provided in the text.

The boundary conditions are

$$\frac{\partial\phi}{\partial y} = 0 \text{ for } y = 0, L_1;$$

$$\frac{\partial\phi}{\partial z} = v_{\mathrm{BM}}\eta(y)\cos(kx - \omega t) \text{ for } z = 0; \quad \frac{\partial\phi}{\partial z} = 0 \text{ for } z = \pm L_2. \tag{1}$$

Here, v_{BM} is the amplitude of transverse BM velocity at its radial maximum; η(y) combines the normalized radial profile of BM displacement with the vanishing displacement of the bony shelves. The cosine represents the longitudinal traveling wave having wave number k and angular frequency ω. Although this analysis is strictly valid only for a homogeneous duct with nonvarying properties, its application to the cochlea with its stiffness gradient and subsequent longitudinal wavelength variation is justified by the gradual nature of these gradients. As explained in Sec. III of Steele and Taber, these are the same conditions that justify the WKB approximation. Equation 7 of Steele and Taber provides the solution ϕ as a series of elementary functions, and we used this expression to compute the spatial distribution of kinetic energy shown in Figure 4A. Whereas Steele and Taber used standard modes of an elastic beam for the radial profile η(y), we inserted the radial profile of BM motion measured by Ren et al. (2011a) and values of wave number k derived from these same data. For the time average kinetic energy per unit length E_K, Eq. 16b of Steele and Taber provides the expression

$$E_{\mathrm{K}} = \tfrac{1}{2}\rho h_{eq} v_{\mathrm{BM}}^2 \int_0^{L_1} \eta^2(y)dy, \tag{2}$$

with ρ the fluid mass density and h_{eq} an equivalent thickness of fluid moving with the BM. h_{eq} depends on η(y) and on the wavelength λ=2π/k (Eq. 14b of Steele and Taber 1979a). The normalized radial profile η(y) was obtained from Figure 2C of Ren et al. (2011a), yielding $\int_0^{L_1} \eta^2(y)dy = 35$, which may be viewed as an effective width of the fluid motion. The wavelength λ was obtained by fitting parabolas to the phase curves in Figure 4B of Ren et al. (2011a) and evaluating their slopes at the 16-kHz location at 2500 μm from the stapes. With increasing intensity (10–90 dB SPL), λ increased from 254 to 403 μm, causing h_{eq} to vary from 31 to 43 μm. BM displacement amplitudes ξ_{BM} were taken from the raw data underlying Figure 4A of Ren et al. (2011a), evaluated over a 20-μm stretch around the 2500-μm location. Displacement was converted to velocity using $v_{BM} = \omega\, \xi_{BM}$. The confined character of the spatial distribution of E_K (Fig. 4A) and the fact that $kL_1 >> 1, kL_2 >> 1$ imply that the scalae do not constrain the fluid motion; the wave is a deep-water wave and genuinely three-dimensional (Steele and Taber 1979b). Indeed, varying L_1 and L_2 by ±50 % and w_S from 50 to 400 μm affected E_K by less than 0.2 dB. Thus, the exact geometry of the fixed boundaries (Fig. 2) is not critical to the estimate.

Relation between Wave Amplitude and Group Velocity

This section presents the mathematical underpinning of the analysis of group velocity gradient and local amplitude gain presented in "Results" section when discussing the low-intensity curves of Figure 6. We assume that the power flux P is constant (i.e., the wave is neither amplified nor attenuated) and analyze the consequences of this Ansatz. P is the product of group velocity and energy density; from the equality of time average kinetic and potential energy, given a local displacement amplitude A and local stiffness s,

$$P = \frac{1}{2}UsA^2. \tag{3}$$

When the power flux P is constant (neither amplified nor attenuated), the amplitude ratio G_{12} between adjacent locations 1 and 2 obeys

$$G_{12}^2 = (A_2/A_1)^2 = \frac{s_1 U_1}{s_2 U_2}. \tag{4}$$

Here, 1 and 2 are the basal and apical locations, respectively. In the interpretation of data as shown in Figures 6 and 7, it is important to realize that the amplitudes measured at two adjacent locations may differ by an additional unknown factor due to slight differences in radial position of the two reflective beads. This factor is independent of frequency and intensity (Cooper 2000). From the measurements, therefore, G_{12} can be determined up to an unknown constant factor.

In the familiar example of sea waves entering a shallow beach, gravity plays the role of restoring force ("stiffness"), and its constancy, $s_1 = s_2$, reduces the RHS of Eq. 4 to the ratio of group velocities. In that case, the spatial variation of group velocity is entirely caused by the dependence of effective fluid mass on water depth. In the cochlea, s_1/s_2 differs from unity owing to the longitudinal stiffness gradient. This is taken into account as follows. In the low-frequency limit, the waves are long and one-dimensional. There is no dispersion: at a fixed location, group velocity U_{LF} and phase velocity c_{LF} are equal and constant in the low-frequency limit:

$$U_{\mathrm{LF}} = c_{\mathrm{LF}} = \sqrt{s/m}, \tag{5}$$

where m is the cross-sectional fluid mass per length unit, and the subscript LF denotes the low-frequency limit. Using Eq. 4, the local amplitude gain $G_{12,LF}$ in the low-frequency limit becomes

$$G_{12,\mathrm{LF}}^2 = \left(\frac{s_1}{s_2}\right)^{3/2}. \tag{6}$$

This expression is independent of frequency and wavelength; it corresponds to the flat, linear, low-frequency portion of the local gain functions (Figs. 6 and 7; Figs. 2 and 3 of Ren et al.2011b). When normalizing G_{12} by $G_{12,LF}$ (which also eliminates the unknown geometric factor arising from possible differences in radial position of the two beads), one obtains

$$\begin{aligned}(G_{12}/G_{12,\mathrm{LF}})^2 &= \sqrt{s_2/s_1}\,\frac{U_1}{U_2}\\ &\equiv \gamma_{12}\frac{U_1}{U_2}.\end{aligned} \tag{7}$$

Apart from the factor γ_{12}, the normalization of the gain reduces the expression for the amplitude gain to the ratio of local group velocities, just like the case of sea waves approaching the beach. For nearby recording locations, γ_{12} is slightly smaller than one. Based on a frequency map created by longitudinal stiffness variations proportional to the square of BF (Emadi et al. 2004), γ_{12} equals the ratio of the BFs at the two locations

$$\begin{aligned}\frac{U_1}{U_2} &= \frac{f_{\mathrm{best},1}}{f_{\mathrm{best},2}}\frac{\tau_{\mathrm{HF}}}{\tau_{\mathrm{LF}}}\\ &= \frac{1}{\gamma_{12}}\frac{\tau_{\mathrm{HF}}}{\tau_{\mathrm{LF}}},\end{aligned} \tag{8}$$

which gives $\gamma_{12}=0.82$ and $\gamma_{12}=0.79$ for the data shown in Figures 6 and 7, respectively.

When relating the phase and gain curves of the local transfer functions in Figure 6, it is important to realize that the asymptotic slopes of the phase curves τ_{LF} and τ_{HF} do not directly correspond to U_1 and U_2: the latter are the group velocities at the two locations at a given frequency; the former correspond to group velocity in two frequency ranges at a given location. The two pairs are related by tonotopy ("scaling"). On account of the scaling invariance of U/ω(see Eq. 15 below), their ratios are related by

$$\begin{aligned}\frac{U_1}{U_2} &= \frac{f_{\mathrm{best},1}}{f_{\mathrm{best},2}}\frac{\tau_{\mathrm{HF}}}{\tau_{\mathrm{LF}}}\\ &= \frac{1}{\gamma_{12}}\frac{\tau_{\mathrm{HF}}}{\tau_{\mathrm{LF}}},\end{aligned} \tag{9}$$

leading to the simple relation between normalized peak gain and asymptotic slopes

$$(G_{12}/G_{12,\mathrm{LF}})^2 = \frac{\tau_{\mathrm{HF}}}{\tau_{\mathrm{LF}}}. \tag{10}$$

Equation 10 is the mathematical underpinning of the scaling argument used in the discussion of Figure 6 in the main text.

Degree of Damping of Waves

This section clarifies the derivation of the degree of damping presented in "Results" section when discussing the high-intensity curves in Figure 6. A vibrating system is called critically damped when its amplitude decreases by a factor $e^{2\pi}$ during each cycle of its (unforced) oscillation, so the rate of decay of the energy equals $20\log(e^{2\pi})=54.6$ dB per cycle. In an underdamped (overdamped) system, the losses are smaller (larger). The degree of damping is expressed by the dimensionless damping coefficient $\zeta=L/54.6$, where L is the temporal decay rate of energy decay of the system in decibel per cycle. Critical damping corresponds to $\zeta=1$. These definitions apply to traveling waves without modification. Now, the energy, in addition to being dissipated, is moving at the group velocity (Lighthill 1978). In a nondispersive wave, group velocity equals phase velocity, and the loss per spatial cycle equals the loss per temporal cycle. In a dispersive wave, the two are related through the ratio $\chi=c/U$ of phase velocity c to group velocity U, yielding a spatial decay rate of 54.6χ dB/cycle for a critically damped wave. Using the definitions of phase and group velocity in terms of phase, frequency, and distance (Whitham 1974), the ratio χ can be readily extracted from local phase transfer functions (φ_{12} versus f) using $\chi=U/c=\varphi_{12}/(f\,\partial\varphi_{12}/\partial f)$, where f is the frequency, φ_{12} is the phase difference between the locations, and $\partial\varphi_{12}/\partial f$ is the slope of the phase-frequency curve. For the 80-dB-SPL curve at 14 kHz (Fig. 6D), one obtains $\chi=3.0$. Thus, for these waves, critical damping would correspond to a spatial power decay rate of 164 dB/cycle. The observed rate of 36 dB/cycle (see "Results" section) shows that the wave is underdamped, having a dimensionless damping coefficient of $\zeta=36/164=0.22$.

Relation between Local Amplitude Gain and Local Phase Difference

This section describes the derivation of the predictions of the phase curves presented in "Results" section (Fig. 7B). Scaling invariance (Zweig 1976) states a trade-off between the frequency dependence and place dependence of the phase φ of BM displacement, which equals the phase of the excess pressure when damping is small (Lighthill 1978), of the form

$$\varphi(x,\omega) = -\Phi(\alpha x+\nu), \tag{11}$$

where α is the gradient of the frequency-place map, and

$$\nu = \log\omega/\omega_{\mathrm{ref}}, \tag{12}$$

with ω_{ref} an arbitrary reference frequency. The local wave number (spatial frequency) k then becomes

$$k = \partial\Phi/\partial x = \alpha\Phi'(\alpha x+\nu), \tag{13}$$

where the prime denotes the derivative. The group velocity U equals

$$U = \partial\omega/\partial k, \tag{14}$$

hence

$$U^{-1} = \partial k/\partial\omega = \alpha\Phi''(\alpha x + \nu)\,\partial\nu/\partial\omega = \frac{\alpha}{\omega}\Phi''(\alpha x + \nu)\,. \tag{15}$$

Combining Eqs. 7 and 15 and defining

$$\tilde{G}_{12} = \frac{G_{12}}{\sqrt{\gamma_{12}}G_{12,\mathrm{LF}}}, \tag{16}$$

it follows

$$\tilde{G}^2_{12}(\nu) = U(x_1,\nu)/U(x_2,\nu) = \Phi''(\alpha x_2 + \nu)/\Phi''(\alpha x_1 + \nu)\,. \tag{17}$$

Defining

$$h(\nu) = \tfrac{1}{2}\log\Phi''(\nu + \alpha x_1)\,, \tag{18}$$

it follows

$$\log\tilde{G}_{12}(\nu) = h(\nu + \alpha D) - h(\nu)\,, \tag{19}$$

where D is the distance x_2-x_1 between the recording locations. The phase difference Φ_{12} between locations x_1 and x_2 equals

$$\begin{aligned}\Phi_{12}(\nu) &= \Phi(\alpha x_2 + \nu) - \Phi(\alpha x_1 + \nu)\\ &= \alpha\int_{x_1}^{x_2} dx\Phi'(\alpha x + \nu)\\ &= \alpha\int_{\nu_0}^{\nu} d\mu\int_{x_1}^{x_2} dx\Phi''(\alpha x + \mu),\end{aligned} \tag{20}$$

where the lower bound $\nu_0=\log(\omega_0/\omega_{ref})$ of the first integral is indefinite, leading to an arbitrary integration constant. Combining Eqs. 18 and 20 yields

$$\Phi_{12}(\omega) = \alpha\int_{\log\omega_0}^{\log\omega} d\mu\int_0^D dx e^{2h(\alpha x+\mu)}. \tag{21}$$

The local transfer data (Figs. 6 and 7) consist of the phase difference Φ_{12} and amplitude gainG_{12}, both as a function of frequency. Equations 19 and 21 link these two measurable quantities in terms of the auxiliary function h(ν). In order to predict Φ_{12} from the G_{12} data, we first obtained h(ν) by numerically solving Eq. 19. This yielded h(ν) up to addition of an arbitrary periodic function with period αD. This ambiguity was resolved by selecting the most regularh(ν) obeying Eq. 19, i.e., by minimizing

$$\int d\nu(h''(\nu))^2 \tag{22}$$

over the range of ν values dictated by the data. This fixes h(ν) up to an additive constant (in the numerical computations, solving Eq. 19 and minimizing Eq. 22 were realized by interpolating the data on a fine grid and solving the discrete versions of these equations using linear algebra). The h(ν) thus obtained was inserted into Eq. 21 to yield the predicted local phase transfer$\Phi_{12}(\omega)$. The undetermined integration constant of Eq. 21 and the arbitrary additive constant toh(ν) result in an indefinite offset and scaling factor, respectively. Their values were chosen to best fit the phase data in a least square sense.

RESULTS

Net Power Gain

All traveling waves transport energy. The energy flux is the product of the energy density and the speed at which it travels. In dispersive media like the cochlea, the speed of the energy differs from the visible speed of wave crests. Instead, it equals the propagation speed of entire wave packets (Lighthill 1978), hence its name "group velocity."

We determined group velocity in sensitive cochleae by measuring vibrations of the stapes and two neighboring BM locations (Fig. 3A) in response to narrowband sound stimuli. These stimuli are an ongoing series of wave packets because of their magnitude fluctuations. The same fluctuations were found in the recorded waveforms. In order to quantify the power fluctuations, we extracted the Hilbert envelope from the recordings at all three locations. The squared envelopes were highly similar across the three locations (Fig. 3B; correlations >0.9). Thus, the power fluctuations of the stimulus were not deformed by excessive dispersion or nonlinearity, allowing the straightforward assessment of energy travel time from the response envelopes. The alternative method employs the slopes of phase-frequency curves (Fig. 3C). The two methods produced equivalent results (Fig. 3D), as expected from the generality of the group velocity concept and its physical significance as the velocity of energy transport in traveling waves under a wide range of conditions including damping and nonlinearity (Lighthill 1965; Whitham1974). Group delay depended strongly on sound intensity (Fig. 3E). Group velocity in five sensitive cochleae increased with intensity, varying from 0.9 m/s at 10 dB SPL to 2.1 m/s at 90 dB SPL (Fig. 3F). Thus, the acoustic energy, which approaches the ear at 340 m/s, is decelerated to a mere walking pace prior to sensory detection. Comparable low-intensity group velocities of ~1 m/s can be inferred from published data of sensitive cochleae (Ren et al. 2011a; Rhode and Recio 2000). Our high-intensity values are somewhat lower than the rough 3-m/s estimate by Lighthill (1981) based on 85-dB-SPL data from a damaged cochlea (Rhode 1978).

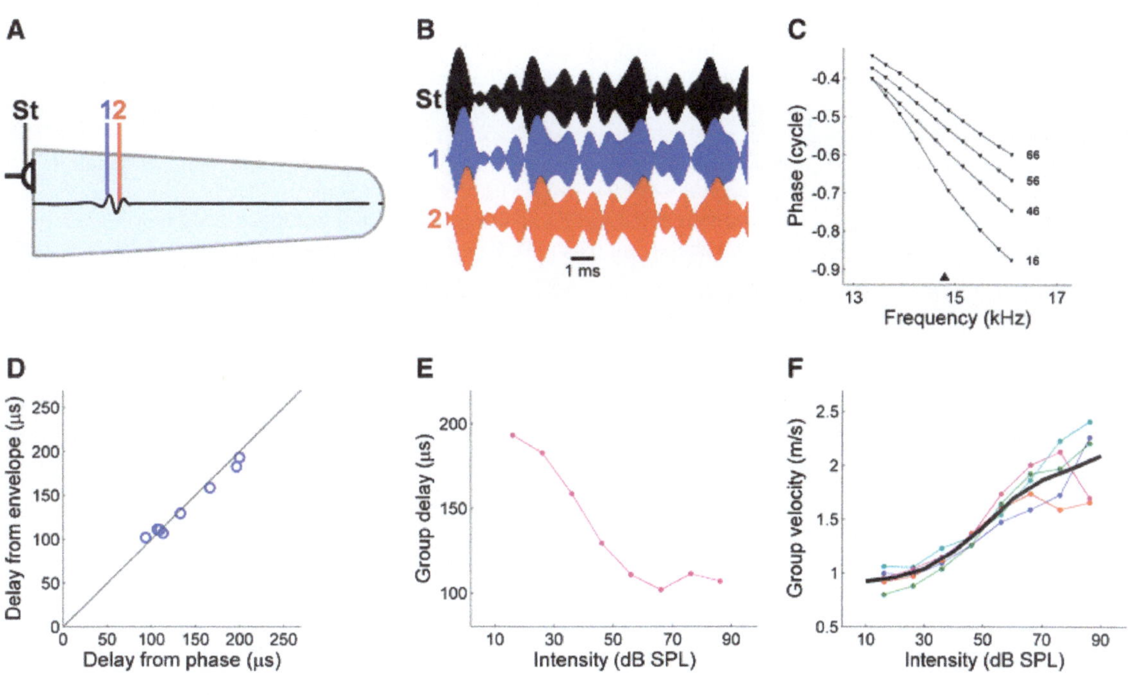

Figure. 3 Measuring the speed of cochlear energy transport. **A** Schematized cochlea with recording locations indicated: stapes (St) and two adjacent basilar membrane (BM) locations (1, 2).**B** Waveforms recorded at these locations evoked by narrowband sounds. Relative delays of magnitude fluctuations reflect energy travel times. **C** Phase difference between locations 1 and 2 versus frequency at various intensities (indicated in dB SPL). Trianglemarks best frequency (BF). **D** Comparison of group delays estimated from phase-frequency plots (abscissa) and from temporal comparison of magnitude fluctuations (ordinate). Unity line is shown for reference (black line). **E** Intensity-dependent group delay between locations 1 and 2. **F** Group velocity at BF from five cochleae, BFs 12.7–18.2 kHz. Thick black line: mean, linearly extrapolated to the 10–90 dB-SPL range (**B–E**: animal RG12448).

Next, we estimated both the kinetic and potential energy densities in the cochlea. The equality of their time averages (Lighthill 1978) provides an important cross-check. In order to estimate the kinetic energy, three-dimensional motion of the fluid surrounding the BM was computed from known spatial profiles of BM motion (Ren et al. 2011a) using a fluid-dynamic analysis (see "Analysis of Fluid Motion" section in "Methods"). From the three-dimensional fluid motion, the cross-sectional distributions of velocity magnitude were determined (Fig. 4A). Fluid motion was found to decay exponentially with distance from BM, characterized by an "equivalent thickness" (Steele and Taber 1979b). Wavelength varied strongly with intensity (Ren et al. 2011a); equivalent thickness varied in proportion (Fig. 4B). Thus, fluid motion was confined to smaller cross sections for softer sounds. Kinetic energy distributions were computed for each intensity and spatially integrated.

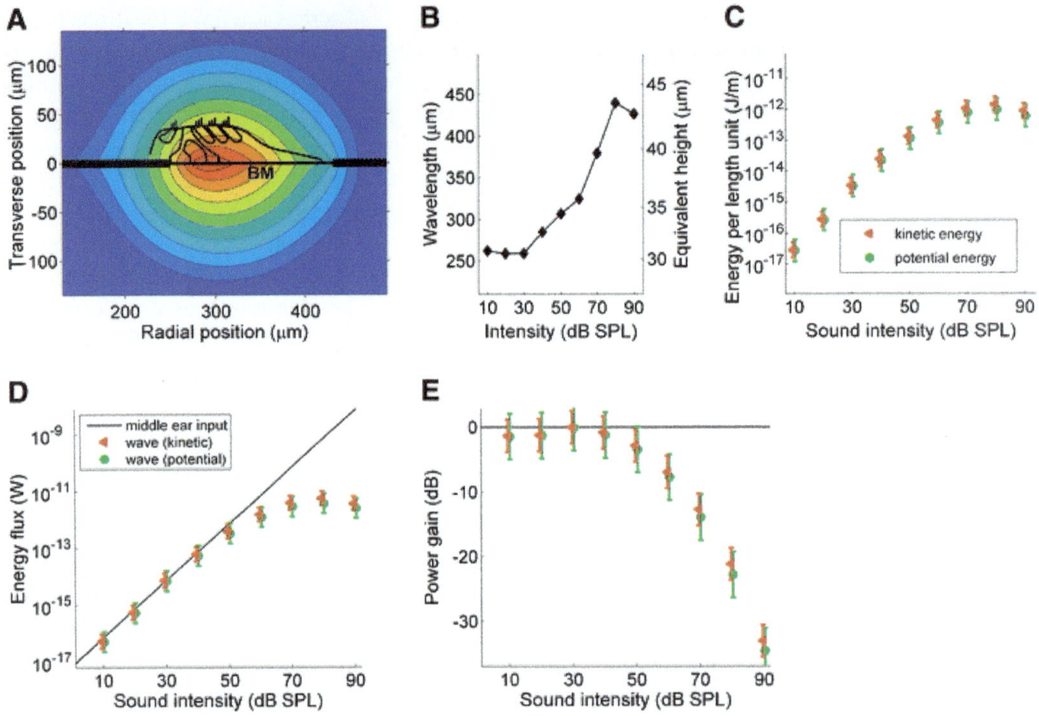

Figure. 4 Net power gain in the traveling wave. **A** Cochlear cross section showing the spatial distribution of fluid velocity magnitude (16 kHz, 50-dB-SPL tone). Contour spacing 3 dB (twofold reductions in kinetic energy). Organ of Corti sketched for reference. **B** Intensity dependence of wavelength at 16 kHz. Right abscissa: equivalent thickness of fluid motion (see text). **C** Energy density estimates derived from BM data of Ren et al. (2011a).Triangles: kinetic energy per unit length, summed over the cross section, of the 16-kHz wave at peak location, with error bars based on across-animal variation of BM displacement. Circles: potential energy, with error bars combining across-animal variation of BM displacement with spread in BM stiffness data. **D** Energy flux of the 16-kHz traveling wave at the 16-kHz place of the gerbil cochlea, derived from the data in the previous panel, using the same symbols and colors. Solid line: middle ear power input. **E**Net power gain from middle ear to traveling wave peak, derived from the data in previous panel. Error bars in **D**, **E** derive from across-animal standard errors in Ren et al. (2011a).

Potential energy was obtained by combining BM stiffness data (Emadi et al. 2004) with data on BM displacement (Ren et al. 2011a). A point stiffness of 0.79 N/m at the 16-kHz place of the gerbil cochlea measured by Emadi et al. (2004) with a probe tip diameter of 25 to 50 µm amounts to a stiffness per unit length $s=2.1\times10^4$ N/m^2. Alternative estimates from the literature are addressed in "Discussion" section. The time average potential energy per unit length equals $\frac{1}{4}\, s\xi_{BM}^2$, where ξ_{BM} is the displacement amplitude at the peak of the wave, extracted from Figure 4A of Ren et al. (2011a). The potential and kinetic energy estimates matched well (Fig. 4C). Their slight (1.7 dB) divergence toward high intensities may be attributed to the fact that in vitro stiffness data, measured using a glass fiber, do not incorporate the effect of stimulus intensity, whereas the wavelength data of Figure 4B suggest that in vivo stiffness increases somewhat with intensity.

The energy flux of the wave at its peak was obtained by multiplying group velocity and energy density (Fig. 4D). These estimates were compared to the power input to the middle ear $P_{ME}=\mathrm{Re}(p^2/Z_{ME})$, with p the RMS sound pressure near the eardrum and Z_{ME} the middle ear impedance. From Figure 11 of Ravicz et al. (1992), Z_{ME} at 16 kHz is real and equal to 6×10^7 Pa s/m^3. At 0 dB SPL, the RMS sound pressure is 20 µPa, yielding $P_{ME}=6.67\times10^{-18}$ W at 0 dB SPL. Alternative estimates from the literature are addressed in "Discussion" section.

Up to 40 dB SPL, traveling wave power was slightly less than the acoustic middle ear input; at higher intensities, it fell behind. The net power gain from middle ear to the peaking wave (Fig. 4E) was never positive. Thus, we found no indication of a net power gain. Averaged over intensities ≤40 dB SPL, the gain was -1.0 ± 0.6 dB. Thus, 20 ± 11 % of the power is lost. With increasing intensity, the gain dropped to −34 dB. Thus, at 90 dB SPL, only 0.04 % of the energy entering the ear actually reached its characteristic place in the cochlea.

Local Power Gain

The amplitude of the traveling wave is known to exhibit a local peak near its best place, particularly at low SPLs (e.g., Ren et al. 2011a). The rising flank of the peak necessarily exhibits a local amplitude gain. This, however, does not necessarily imply that there is also a local powergain. That the wave amplitude may well grow without injecting energy is illustrated by sea waves approaching the beach. Their growth is not caused by a coastal amplifier. The propagation speed depends on depth: when the wave enters shallower water, group velocity decreases (Lighthill 1978), causing an energy densification ("congestion") that boosts the amplitude. Although this geometry of shoaling does not apply to the cochlea, the inverse relation between local group velocity and local wave amplitude is a general property of traveling waves (Whitham 1974), and there may exist in the cochlea other factors that affect group velocity. Amplitude peaking by wave deceleration occurs in passive resonant models (Lighthill1981), but the idea is more general: it also applies to scenarios in which wave dispersion has nothing to do with resonance (Ranke 1950; Van der Heijden 2014).

Experimentally, wave deceleration in the cochlea is apparent from panoramic neural measurements, obtained by comparing the responses to the same stimulus across many single nerve fibers. These data show that the cochlear traveling wave slows down quite abruptly just prior to peaking (Kim et al. 1980a; Van der Heijden and Joris 2006; Palmer and Shackleton2009; Temchin et al. 2012). An example of this abrupt deceleration is shown in Figure 5, which reproduces spatial phase profiles in a single cochlea from Kim et al. (1980a). The key question is: can the local amplitude growth be accounted for by the local deceleration of energy transport? If so, there is no obvious role for power amplification. We addressed this question by studying local amplitude gain and group velocity in finer detail.

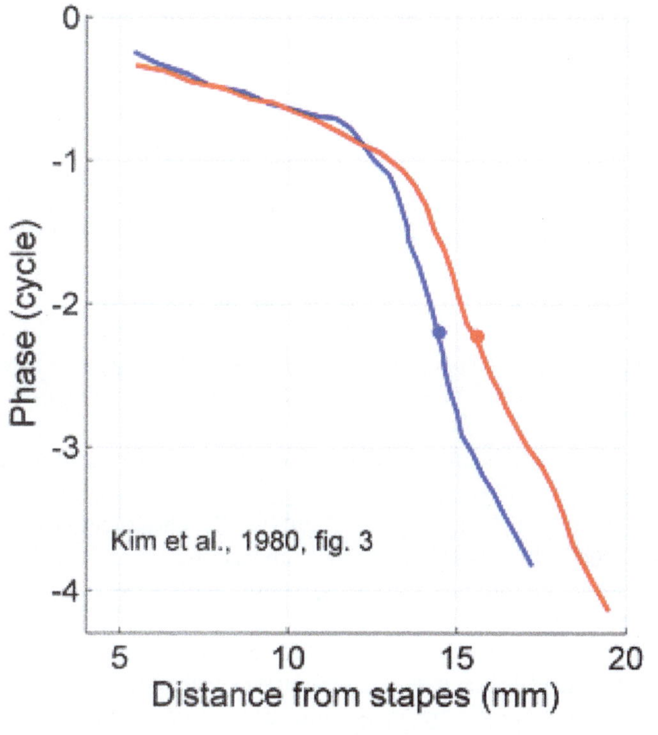

Figure. 5 Deceleration of cochlear traveling waves shown by panoramic neural data. These data were extracted from Figure 3 of Kim et al. (1980a). The curves show the phase of many AN fibers of a single ear of the cat in response to the same tone pair of 2100 and 2700 Hz. The abscissa is the cochlear position derived from the BFs of the fibers. The distinct bend of the curves reflects the deceleration of the waves, which occurs over a narrow region just basal to the characteristic place of the tone marked by the filled circles.

We measured in vivo BM vibrations at two adjacent locations, this time using a wideband multitone stimulus presented at various SPLs (see "BM Recordings" section in "Methods"). Figure 6A shows two sets of magnitude curves, normalized to stapes motion. Stapes-to-BM amplitude gain decreased with increasing intensity, reflecting the compressive nature of the BM response. Compression was evident at the lowest intensities used (0 dB SPL per tone), indicating the high sensitivity of this cochlea. Figure 6B shows the companion phase curves. From these two sets of single-point data, we constructed the local transfer functions as introduced in Ren et al. (2011b) by plotting the amplitude ratio and phase difference between the two locations against frequency.

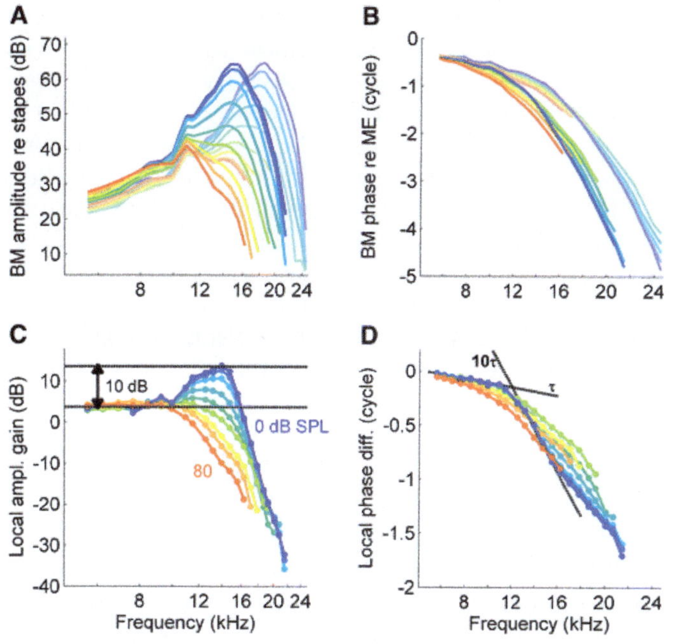

Figure. 6 Local amplitude gain and the deceleration of energy transport. **A** BM amplitude normalized to stapes motion in response to tone complexes presented at 0 to 80 dB SPL per component in 10-dB steps. Each curve represents a single recording. The two sets of curves were obtained from two adjacent locations in the same cochlea (BF 15.0, 18.3 kHz), the darker curves corresponding to the more basal location. **B** Companion phase curves. **C** Local transfer functions obtained by plotting the local gain (i.e., amplitude ratio across the two locations) against frequency. Lower black line marks the low-frequency limit to which all the curves converge, independent of intensity. This linear limit serves as the reference for the excess gain of the intensity-dependent portions of the curves. Upper black line marks the largest (10-dB) excess gain occurring in this dataset.**D** Companion phase curves, showing the phase difference between the two locations. The sharp bend at low intensities signals the abrupt deceleration of energy transport just prior to the wave peaking. The black lines match the slopes of the 0-dB-SPL curves (dark blue) at both sides of the transition. Their slopes differ by a factor of 10, marking the tenfold deceleration of the 0-dB-SPL curve (animal RG12446).

The local amplitude gain functions (Fig. 6C) quantify the wave peaking. The low-frequency portion was flat and invariant with intensity. In this linear, passive range, the amplitude ratio between adjacent locations is fixed by the BM stiffness gradient. We used this "passive amplitude ratio" of this linear range (lower horizontal black line in Fig. 6C) as the reference for the amplitude gain in the nonlinear, "active" range (see "Relation Between Wave Amplitude and Group Velocity" section in "Methods" for mathematical underpinning). At higher frequencies, local gain became strongly intensity dependent. This compressive range is commonly associated with amplification. Positive amplitude gains were observed only at low sound intensities (<40 dB SPL). The largest gain (~10 dB above the low-frequency reference, marked by the two-sided arrow in Fig. 6C) occurred for the lowest intensity (0 dB SPL) just below local best frequency (BF). This maximum local gain captures the steepest part of the growing flank of waves on their way to peaking beyond the recording site.

The companion phase transfer functions (Fig. 6D) curve downward, independent of sound intensity. As the slopes correspond to group delay, this means that low-frequency energy travels faster than high-frequency energy. Considering the frequency-place map, this implies that the energy slows down as it travels, in agreement with the neural data reproduced in Figure 5. The deceleration occurs just prior (basal) to the wave's peak region. Deceleration was abrupt at the low intensities (Fig. 6D, blue curves) and turned smoother with increasing intensity. For the softest sounds (0 dB SPL), a <20 % increment in frequency produced a tenfold reduction in group velocity (black lines in Fig. 6D). Since the best frequencies of the two recording locations differed by ~20 %, the group velocity of waves near the transition frequency fell tenfold between them. This tenfold deceleration should cause a 10-dB increase in energy density, a prediction matched by the observed ~10-dB amplitude gain (two-sided arrow in Fig. 6C). The quantitative match between phase bends and amplitude gain was a general finding in sensitive cochleae. The local amplitude gain was accounted for by the deceleration of the wave, and we found no indication of local power gain.

With increasing intensity, amplitude gain diminished and turned into loss (Fig. 6C). Notice that the transfer remained strongly compressive: the gain, even when negative, continued to decrease with intensity. Thus, at these high intensities, there was a local amplitude reduction at all frequencies in the nonlinear range. This shows at once that at these intensities, the active mechanism that creates the compression acts as a brake rather than an amplifier (see "Introduction" section). In order to estimate the local power loss at high intensities, one needs to combine the local amplitude ratio with the change in group velocity (just as was done for the 0-dB-SPL case above). The 80-dB-SPL local phase transfer at 14 kHz shows a sevenfold reduction in group velocity. In the absence of damping, this deceleration of energy transport would lead to an 8-dB local amplitude gain. The observed 14-dB amplitude reduction (a gain of minus 14 dB) therefore corresponds with a 22-dB power loss caused by local damping. Given the 0.6-cycle local phase difference, this corresponds to a power loss of 36 dB per wave cycle, indicating that the local wave propagation at 14 kHz, 80 dB SPL, was underdamped with a dimensionless damping coefficient $\zeta=0.22$ (see "Degree of Damping of Waves" section of "Methods"). Notice that 14 kHz is below the BF of the more apical of the two recording sites, so the 14-kHz wave has not reached its own characteristic place. Thus, a substantial amount of the energy of high-intensity sounds is absorbed in the region situated basal to their tonotopic place. The large size of the local losses and the place where they occur are consistent with our earlier observation (Fig. 4E) that only a minute fraction of the acoustic energy of intense sounds reaches the characteristic location. The bulk is dissipated just beforehand.

To further analyze the positive amplitude gain at low intensities, we performed recordings using finer frequency spacing. The local transfer functions (Fig. 7A, B) showed the same trends as observed in Figure 6: a positive local amplitude gain just below the lowest BF and a sharp phase bend. Notice the subtle effects of the 20-dB variation of stimulus intensity: a reduction of the amplitude gain and a slight smoothing and reduction of the phase bend. We derived the mathematical relation between local amplitude gain and phase transfer from the Ansatz that there is neither net energy injection nor net dissipation, causing the spatial gradient in group velocity to be the major determinant of amplitude variation. We used this mathematical relation to predict from the observed amplitude gain the associated phase transfer (see "Relation Between Local Amplitude Gain and Local Phase Difference" section in "Methods"). The predictions (Fig. 7B, lines) correctly predict the location and shape of the bend. For the lowest two intensities, calculated phase curves were slightly too steep. This near miss disappeared when allowing for a 1.8-dB underestimation of the gain for the 0- and 10-dB-SPL curves (Fig. 7C). The near miss suggests a small contribution of a secondary factor (in addition to reduction of group velocity), e.g., a simultaneous reduction in effective stiffness or a very slight (<2 dB) amplification after all (see "Discussion" section). Importantly, the calculations faithfully reproduced the observed correlation between amount of gain and sharpness of the phase bend, again emphasizing the intimate relation between amplitude growth and deceleration. Their quantitative match again suggests that the amplitude gain is not created by energy injection into the wave, but by the densification following the deceleration of energy transport by the wave.

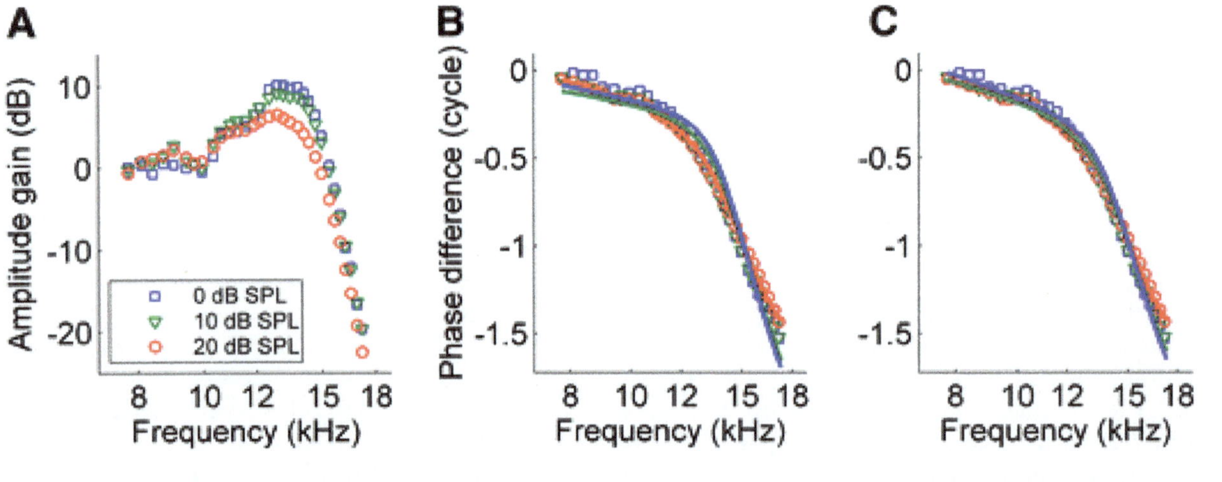

Figure. 7 Predicting local phase from local amplitude gain, assuming zero power gain. **A** Low-intensity local amplitude gain obtained with fine frequency spacing. Intensity per component is indicated in the graph. **B** Companion local phase curves. Symbols: data.Lines: prediction computed from the local amplitude data under the assumption that there is no power gain, i.e., that the only source of the local amplitude gain is the deceleration of the wave (see text). The predictions accounted for 96.9 % of the variance in the phase data. **C** As in **B**, but now allowing for a 1.8-dB near miss for the two lowest intensities (see text), which eliminates the small systematic deviations observed in **B**(animal RG12436).

DISCUSSION

Summary

We estimated the net power gain from middle ear to the peak of the traveling wave in two independent ways. The first method, based on kinetic energy, used the detailed spatial profile of BM motion (Ren et al. 2011a) to derive the motion of the surrounding fluid. In the second method, we estimated the potential energy from BM stiffness data (Emadi et al. 2004). These two independent energy estimates were combined with our own measurements of group velocity. The two methods yielded highly similar values for the energy flux, neither of which exceeded the middle ear power input at any intensity. At high intensities, there was a large (>30 dB) net loss. Thus, we did not find evidence for a net power gain at low intensities and clear evidence for a net power loss at high intensities.

In a second series of experiments, we performed two-point BM recordings and analyzed the local energy flux. We observed a steep deceleration of the wave that was sufficient to explain its peaking. We found no local power gain between adjacent points on the BM at any intensity. At high intensities, there was a strong local power loss (36 dB per traveled cycle just below BF).

Uncertainties in the Estimates of Net Power Gain

The determination of net power gain from middle ear to the peaking wave necessarily involved combining diverse sources of published data. This introduces some uncertainty, especially when the sources disagree. A fortunate circumstance was the availability of all necessary data for the 16-kHz range of the gerbil: no extrapolation was needed.

Uncertainties in BM stiffness estimation are due to methodological challenges including the need to use post mortem preparations, anisotropy of the anatomical structures, and sensitivity to ion concentrations of the bath (Emadi et al. 2004). The stiffness values of Naidu and Mountain (1998) were systematically higher than those of Emadi et al. (2004) employed here. Underneath the OHCs, 2.5 mm from the base, their stiffness values exceed that of Emadi et al. by a factor of 3.4. Had we used the data of Naidu and Mountain (despite the methodological issues concerning those data raised by Emadi et al.), the estimate of energy flux based on the potential energy (circles in Fig. 4D) would be elevated by 5.3 dB, yielding a 4.3-dB net power gain from middle ear to BM at low intensities. But this would also introduce a ~5-dB discrepancy between kinetic and potential energy estimates, because the former are independent of BM stiffness.

Middle ear losses present another source of uncertainty, equally affecting the estimates based on kinetic and potential energy. If, as suggested by de la Rochefoucauld et al. (2008), only 30 % of the power entering the middle ear is transferred to the cochlea, this would elevate the energy flux estimates by 5.2 dB, leading to a 4.2-dB value of the net power gain at low intensities. Figure 11 of de la Rochefoucauld et al. (2008) provides two alternative estimates of power input to the cochlea, based on previous work (Olson 1998; Dong and Olson 2006). At 16 kHz, 0 dB SPL, these cochlear power input estimates amount to 3.8×10^{-18} W and 1.2×10^{-18} W, respectively. Using these values (instead of the 6.7×10^{-18} W value used to compile Fig. 4E) would yield low-intensity, net power gain estimates of 1.4 and 6.5 dB, respectively. Notice that these alternative values were obtained using considerably more invasive techniques (e.g., drilling a hole in the bony wall near the stapes) than the data of Ravicz et al. (1992) that led to our minus 1-dB estimate of net power gain at low intensities. Perhaps more importantly, any uncertainties in acoustic power input are irrelevant in the measurement of local power gain between adjacent cochlear locations, which we found to be vanishing at low intensities and increasingly negative at higher intensities (Figs. 6 and 7). Obviously, there can be no net gain without local gain.

Considering the smallness of the potential corrections and the mutual consistency of our three independent estimates, it is difficult to reconcile our findings with even a modest (5 dB) amount of power amplification. In this respect, we fully confirm the conclusions of Allen and Fahey (1992). Our findings cast reasonable doubt on the existence of cochlear power amplification and invalidate the much larger (>20 dB) estimates of power gain found in previous studies as discussed below. In addition, our results show that the propagation of waves at high intensities is strongly dissipative. This is significant because the cochlea's dynamic compression persists to the high-intensity range. Apparently, the active mechanism works as a variable attenuator ("brake") rather than an amplifier at these intensities. Likewise, our findings invalidate claims of negative damping persisting to high intensities discussed below; if those claims were correct,positive local power gains should be observed.

Previous Estimates of Power Gain and Negative Partition Impedance

Brass and Kemp (1993) used a similar scheme to deduce energy flux from BM recordings and found significant power amplification. The crucial difference between their study and ours is the assessment of group velocity. Because they worked with single-point recordings from Robles et al. (1986), they were unable to assess group velocity directly and had to revert to extrapolations based on tonotopy. Moreover, the limited S/N ratio of the data (obtained with a Mössbauer source) necessitated considerable numerical smoothing of data in order to estimate group velocity. The resulting estimated dependence of group velocity on position (Fig. 2B of Brass and Kemp 1993) is very shallow. In contrast, we determined group velocity by a direct comparison of magnitude fluctuations between adjacent locations (Fig. 3), did not apply any smoothing, and found a steep transition in group velocity when varying frequency (Figs. 6 and 7). The steepness is a key observation of the current study, and the inferred abrupt deceleration of energy transport is a key step in reaching the conclusion that there is no local power amplification.

Several studies have used "inverse methods" to estimate BM impedance or the amount of power gain from BM data (Zweig 1991). While the estimates of power amplification thus derived are highly variable, some of these studies cannot be reconciled with our current findings: negative damping up to high intensities (e.g., 80–90 dB SPL in octave-wide bands, de Boer and Nuttall2000) and power gain estimates of up to 17.7 dB (de Boer and Nuttall 2001). Unlike Brass and Kemp

(1993) and the current study, the inverse method uses an explicit fluid dynamical model in which the cochlear partition is treated as point impedance $z_{BM}(x,\omega)$, a quantity that depends on frequency ω and place x, but not directly on wavelength. Estimates of $z_{BM}(x,\omega)$ are obtained by fitting the model to (extrapolated) single-point BM data. Claims of amplification follow from any negative real part of z_{BM} thus obtained ("negative damping").

The central assumption of the inverse method is the adequacy of point impedance to describe how the cochlear partition interacts with the fluid. This assumption is not self-evident (see also Brass and Kemp 1993), as it ignores the finite dimensions and internal structure of the organ of Corti. That simplification would be justified if the organ were small compared to the typical scale of the wave (the distance over which relative motion becomes comparable to absolute motion), but this is not the case for near-BF waves. In the transverse direction ("depth"), the motion of fluid participating in a wave falls off exponentially, having a penetration depth of $\sim\lambda/2\pi$ (Lighthill 1978; Steele and Taber 1979a). Near BF, the wavelength is ~250 μm (Fig. 4C), so the penetration depth equals ~40 μm. Because the height of the organ of Corti in the 16-kHz region well exceeds 50 μm (Edge et al. 1998), the amplitude of fluid motion varies at least threefold over its height. Thus, for these shorter wavelengths, the constituents of the partition need not move uniformly. Any nonuniform motion involves periodic internal deformations that are unlikely to be captured by a point impedance description. Only for much longer (lower-frequency) waves is the motion virtually uniform.

Intriguingly, the transition between these regimes, which takes place when the wavelength drops below a millimeter, is situated just basal to the peak—exactly in the region of alleged amplification. As long as one holds on to a point impedance description, it seems inevitable to invoke some degree of negative damping to be able to explain experimental data (de Boer 1995; Olson 2001). But if one relaxes the assumption of point impedance, negative damping may well become unnecessary. In terms of inverse methods, this amounts to allowing z_{BM} to depend not only on position and frequency, but also explicitly on wavelength: $z_{BM}=z_{BM}(x,\omega,\lambda)$. This generalized notion of partition impedance paves the way to explanations of amplitude growth that are more in line with the main experimental finding of this study: the abrupt deceleration of energy transport. An explicit implementation of this approach is the hydrodynamic waveguide model of Van der Heijden (2014). This passive, linear model has two coupled elastic beams ("membranes"). Its traveling waves exhibit mode shape swapping, a rapid change of the relative beam motion in the course of propagation. The transition reduces both the group velocity and the effective stiffness, and both reductions contribute to a local amplitude boost (the latter contribution, from the stiffness reduction, may explain the near miss mentioned in connection with Fig. 7). As discussed in that study, mode shape swapping mimics the behavior of active models: over a narrow spatial region, the wave amplitude is boosted. This boost, however, is not created by motile activity, but by the rapid transfer of power from a nondispersive and stiff vibration mode into a highly dispersive and compliant one. Yet, an observer watching one beam, but unaware of the other, would be tempted to attribute the sudden boost to a local power source. Being unaware of the other beam, the observer misses the spatially distributed (nonpoint-like) character of the impedance of the beam pair.

At first glance, the questioning of point impedance may appear to undermine our own analysis of net power gain (Fig. 2), because that does not take into account the finite size of the partition, either. Note, however, that the analysis deals with the vibration right at the peak of the traveling wave amplitude, whereas the region of alleged amplification is just basal to the peak. What makes the observer of the previous paragraph believe in amplification is not the magnitude of the peak per se, but the steep amplitude growth leading to the peak. Viewed from that perspective, the problem is not an unduly large peak magnitude, but the unexpected smallness of the wave magnitude just basal to the peak. The steep growth itself is more puzzling than its end product, the peak magnitude. While the acoustic power entering the cochlea is sufficient to create the peak magnitude, on its way there, the power appears to be somehow contained, i.e., prevented from generating the magnitude of local motion that it could afford. The sharp peak is then created by rapidly unleashing the power. The waveguide model of Van der Heijden (2014) demonstrates a possible physical mechanism of this unleashing, namely, a transition of vibration mode. At the peak (beyond the transition), the waves in that model are of the three-dimensional, "fanning" type (Steele and Taber 1979b), for which the cochlear partition is well approximated by a point impedance. It is in the transition region just basal to the peak that the point impedance description breaks down.

Cochlear Attenuation

Although it cannot be entirely excluded that the lack of power gain stems from a near-perfect balance between power amplification and ordinary dissipation, the most parsimonious interpretation of our findings is that there is no amplifier, that cochlear sensitivity is not realized by amplifying acoustic energy, but by spatially focusing it, and that dynamic compression is realized by locally adjusting the amount of dissipation to sound intensity. We end by briefly exploring the physiological and functional implications of such an interpretation. While not solving all known problems in cochlear mechanics, it offers alternative solutions to some and sheds new light on others.

The change of perspective from a saturating amplifier to a variable attenuator has remarkably little impact on the character of cochlear responses: both schemes predict compression, two-tone suppression, and distortion products (Van der Heijden 2005). But the underlying physiological mechanisms differ greatly. Mechanical amplification is physiologically demanding, as it involves positive feedback which is phase locked to high-frequency waveforms. Its problematic aspects include uncontrollable instabilities, severe shunting of the input to motile elements due to low-pass filtering, and lack of a clearly identified mechanism to couple high-frequency motile output to BM vibration (Ashmore et al. 2010). Cochlear attenuation is less demanding. It still requires an active process that regulates dissipation, but its feedback is negative, minimizing (though not necessarily eliminating) stability problems. Moreover, it involves a straightforward friction control, comparable to the brakes of a car. Just like operating a brake requires no synchronization to the wheel rotation, a cochlear attenuator can do without phase locking to high-frequency stimuli. This eliminates the remaining physiological problems listed above. OHCs, whose length can change with their membrane potential, play a crucial role in cochlear compression, and damaging them immediately reduces sensitivity. The exact mechanism by which OHCs control the vibrations is unknown, but if we assume that OHC shortening acts to increase local friction, this explains both the dynamic compression (through sound-induced depolarization) and reduced sensitivity with trauma (through loss of turgor). Neither phenomenon specifically favors amplification over attenuation.

The analysis of abrupt phase bends (Fig. 6) showed that the peaking of cochlear waves is quantitatively explained by the focusing of wave energy through selective deceleration. Sharp phase bends are also apparent in panoramic studies that pool phase data from large numbers of auditory nerve fibers in cat (Kim et al. 1980a; Van der Heijden and Joris 2006), guinea pig (Palmer and Shackleton 2009), and chinchilla (Temchin et al. 2012). Thus, abrupt wave deceleration is a general cochlear phenomenon. The smoothing of deceleration at high intensities (discussed in connection with Figs. 6 and 7) is also a general finding in mammals (Versteegh and Van der Heijden 2013). Smoothing has an interesting consequence: the basalward extension of the region over which the energy transport is slowed down. Because slowing down the energy transport enhances the spatial rate of dissipation, this suggests that the "premature deceleration" contributes to compressing the dynamic range, in addition to the direct control of damping. This additional mechanism would be especially useful at the highest intensities, when large local power losses are required to curb the spatial overlap of spectral components.

The insight that the cochlea exerts a form of mechanical sensitivity control predates models based on cochlear amplification (Rose et al. 1971; Kim et al. 1973). The functional necessity of a stage of dynamic range compression prior to transduction was recognized and analyzed by Allen (1979), who proposed a framework of "nonlinear damping [that] acts as a mechanical automatic gain control." The results of the present study support this view in which damping is an asset rather than a drawback. The detection of faint tones, however useful, is not the major task of most ears. In the daily life of many species, spectral analysis and dealing with noisy environments are much more common tasks, which impose nontrivial challenges of an entirely different nature. Because the cochlea is a waveguide, its function as a spectral analyzer requires that it absorb all acoustic power entering it. Any reflected component will interfere with the processing of higher frequencies; any component traveling beyond its proper region will interfere with lower frequencies. If the cochlea is to resolve individual components whose intensities differ by several orders of magnitude, the absorption must be both well placed and rigorous. Viewed in this light, dissipation is an indispensable tool rather than something that must be "overcome by amplification," and it is to be expected that the mammalian cochlea has developed a fine control over the amount of local dissipation.

REFERENCES

1. Allen JB (1979) Cochlear models—1978. In: Hoke B, de Boer E (eds) Models of the auditory system, vol 9, Scand Audiol Suppl., pp 1–16
2. Allen JB, Fahey PF (1992) Using acoustic distortion products to measure the cochlear amplifier gain on the basilar membrane. J Acoust Soc Am 92:178–188
3. Ashmore J (2008) Cochlear outer hair cell motility. Physiol Rev 88:173–210
4. Ashmore J, Avan P, Brownell WE et al (2010) The remarkable cochlear amplifier. Hear Res 266:1–17
5. Brass D, Kemp DT (1993) Analyses of Mössbauer mechanical measurements indicate that the cochlea is mechanically active. J Acoust Soc Am 93:1502–1515
6. Cody AR, Russell IJ (1987) The responses of hair cells in the basal turn of the guinea-pig cochlea to tones. J Physiol 383:551–569
7. Cooper NP (1998) Harmonic distortion on the basilar membrane in the basal turn of the guinea-pig cochlea. J Physiol 509:277–288
8. Cooper NP (1999) Vibration of beads placed on the basilar membrane in the basal turn of the cochlea. J Acoust Soc Am 106:L59–L64
9. Cooper NP (2000) Radial variation in the vibrations of the cochlear partition. Recent developments in auditory mechanics. World Scientific, Singapore, pp 109–115
10. De Boer E (1995) The "inverse problem" solved for a three-dimensional model of the cochlea. II Application to experimental data. J Acoust Soc Am 98:904–910
11. De Boer E, Nuttall AL (2000) The mechanical waveform of the basilar membrane. III. Intensity effects. J Acoust Soc Am 107:1497–1507
12. De Boer E, Nuttall AF (2001) Power gain of the cochlear amplifier. In: Breebaart D, Houtsma AJ, Kohlrausch A et al (eds) Physiological and psychological bases of auditory function. Shaker, Maastricht
13. De Boer E, Nuttall AL, Hu N et al (2005) The Allen-Fahey experiment extended. J Acoust Soc Am 117:1260–1266
14. De la Rochefoucauld O, Decraemer WF, Khanna SM, Olson ES (2008) Simultaneous measurements of ossicular velocity and intracochlear pressure leading to the cochlear input impedance in gerbil. J Assoc Res Otolaryngol 9:161–177
15. Dong W, Olson ES (2006) Middle ear forward and reverse transmission in gerbil. J Neurophysiol 95:2951–2961
16. Dong W, Olson ES (2009) In vivo impedance of the gerbil cochlear partition at auditory frequencies. Biophys J 97:1233–1243
17. Doyle JC, Francis BA, Tannenbaum AR (1992) Feedback control theory. MacMillan, New York
18. Duke TAJ, Jülicher F (2008) Critical oscillators as active elements in hearing. In: Manley G, Fay R, Popper A (eds) Active processes and otoacoustic emissions in hearing. Springer, New York

19. Edge RM, Evans BN, Pearce M et al (1998) Morphology of the unfixed cochlea. Hear Res 124:1–16
20. Emadi G, Richter CP, Dallos P (2004) Stiffness of the gerbil basilar membrane: radial and longitudinal variations. J Neurophysiol 91:474–488
21. Evans EF, Harrison RV (1976) Proceedings: correlation between cochlear outer hair cell damage and deterioration of cochlear nerve tuning properties in the guinea-pig. J Physiol 256:43P–44P
22. Frank G, Hemmert W, Gummer AW (1999) Limiting dynamics of high-frequency electromechanical transduction of outer hair cells. Proc Natl Acad Sci U S A 96:4420–4425
23. Hudspeth AJ (2013) The inner ear. Principles of neural science, 5th edn. McGraw-Hill, New York, pp 654–681
24. Kemp DT (1979) Evidence of mechanical nonlinearity and frequency selective wave amplification in the cochlea. Arch Otol Rhinol Laryn 224(1–2):37–45
25. Kennedy HJ, Evans MG, Crawford AC, Fettiplace R (2006) Depolarization of cochlear outer hair cells evokes active hair bundle motion by two mechanisms. J Neurosci 26:2757–2766
26. Kim DO, Molnar CE, Pfeiffer RR (1973) A system of nonlinear differential equations modeling basilar-membrane motion. J Acoust Soc Am 54:1517–1529
27. Kim DO, Molnar CE, Matthews JW (1980a) Cochlear mechanics: nonlinear behavior in two-tone responses as reflected in cochlear-nerve-fiber responses and in ear-canal sound pressure. J Acoust Soc Am 67:1704–1721
28. Kim DO, Neely ST, Molnar CE, Matthews JW (1980b) An active cochlear model with negative damping in the partition: comparison with Rhode's ante- and post-mortem observation. In: van den Brink G, Bilsen FA (eds) Psychological, physiological and behaviour studies in hearing. pp 7–14
29. Lighthill MJ (1965) Group velocity. IMA J Appl Math 1:1–28
30. Lighthill J (1978) Waves in fluids. Cambridge University Press, Cambridge
31. Lighthill J (1981) Energy flow in the cochlea. J Fluid Mech 106:149–213
32. Lim DJ (1986) Functional structure of the organ of Corti: a review. Hear Res 22:117–146
33. Martin GK, Lonsbyry-Martin BL, Probst R, Coats AC (1988) Spontaneous otoacoustic emissions in a nonhuman primate. I. Basic features and relations to other emissions. Hear Res 33:49–68
34. Meenderink SWF, van der Heijden M (2011) Distortion product otoacoustic emissions evoked by tone complexes. J Assoc Res Otolaryngol 12:29–44
35. Naidu RC, Mountain DC (1998) Measurements of the stiffness map challenge a basic tenet of cochlear theories. Hear Res 124:124–131
36. Neely ST (1985) Mathematical modeling of cochlear mechanics. J Acoust Soc Am 78:340–344
37. Olson ES (1998) Observing middle and inner ear mechanics with novel intracochlear pressure sensors. J Acoust Soc Am 103:3445–3463
38. Olson E (2001) Intracochlear pressure measurements related to cochlear tuning. J Acoust Soc Am 110:349–367
39. Palmer AR, Shackleton TM (2009) Variation in the phase of response to low-frequency pure tones in the guinea pig auditory nerve as functions of stimulus level and frequency. J Assoc Res Otolaryngol 10:233–250
40. Plassmann W, Peetz W, Schmidt M (1987) The cochlea in gerbilline rodents. Brain Behav Evol 30:82–101
41. Ranke OF (1950) Theory of the cochlea: a contribution to the hydrodynamics of the cochlea. J Acoust Soc Am 22:772–777
42. Ravicz ME, Rosowski JJ, Voigt HF (1992) Sound-power collection by the auditory periphery of the Mongolian gerbil Meriones unguiculatus. I: Middle-ear input impedance. J Acoust Soc Am 92:157–177
43. Ren T, Gillespie PG (2007) A mechanism for active hearing. Curr Opin Neurobiol 17:498–503
44. Ren T, He W, Gillespie PG (2011a) Measurement of cochlear power gain in the sensitive gerbil ear. Nat Commun 2:216
45. Ren T, He W, Porsov E (2011b) Localization of the cochlear amplifier in living sensitive ears. PLoS ONE 6, e20149
46. Rhode WS (1978) Some observations on cochlear mechanics. J Acoust Soc Am 64:158–176
47. Rhode WS (2007) Basilar membrane mechanics in the 6–9 kHz region of sensitive chinchilla cochleae. J Acoust Soc Am 121:2792–2804
48. Rhode WS, Recio A (2000) Study of mechanical motions in the basal region of the chinchilla cochlea. J Acoust Soc Am 107:3317–3332
49. Robles L, Ruggero MA (2001) Mechanics of the mammalian cochlea. Physiol Rev 81:1305–1352
50. Robles L, Ruggero MA, Rich NC (1986) Basilar-membrane mechanics at the base of the chinchilla cochlea. I. Input–output functions, tuning curves, and response phases. J Acoust Soc Am 80:1364–1374
51. Rose JE, Hind JE, Anderson DJ, Brugge JF (1971) Some effects of stimulus intensity on response of auditory nerve fibers in the squirrel monkey. J Neurophysiol 34:685–699
52. Shera CA (2003a) Mammalian spontaneous otoacoustic emissions are amplitude-stabilized cochlear standing waves. J Acoust Soc Am 114:244–262
53. Shera CA (2003b) Wave interference in the generation of reflection- and distortion-source emissions. In: Gummer AW, Dalhoff E, Nowotny M, Scherer MP (eds) Biophysics of the cochlea: from molecules to models. World Scientific, London, pp 439–453
54. Shera CA (2007) Laser amplification with a twist: traveling-wave propagation and gain functions from throughout the cochlea

55. Steele CR, Taber LA (1979a) Comparison of WKB and finite difference calculations for a two-dimensional cochlear model. J Acoust Soc Am 65:1001–1006
56. Steele CR, Taber LA (1979b) Comparison of WKB calculations and experimental results for three-dimensional cochlear models. J Acoust Soc Am 65:1007–1018
57. Talmadge CL, Tubis A, Wit HP, Long GR (1991) Are spontaneous otoacoustic emissions generated by self-sustained cochlear oscillators? J Acoust Soc Am 89:2391–2399
58. Temchin AN, Recio-Spinoso A, Cai H, Ruggero MA (2012) Traveling waves on the organ of corti of the chinchilla cochlea: spatial trajectories of inner hair cell depolarization inferred from responses of auditory-nerve fibers. J Neurosci 32:10522–10529
59. Van der Heijden M (2005) Cochlear gain control. J Acoust Soc Am 117:1223–1233
60. Van der Heijden M (2014) Frequency selectivity without resonance in a fluid waveguide. Proc Natl Acad Sci U S A 111:14548–14552
61. Van der Heijden M, Joris PX (2006) Panoramic measurements of the apex of the cochlea. J Neurosci 26:11462–11473
62. Versteegh CPC, van der Heijden M (2012) Basilar membrane responses to tones and tone complexes: nonlinear effects of stimulus intensity. J Assoc Res Otolaryngol 13:785–798
63. Versteegh CPC, van der Heijden M (2013) The spatial buildup of compression and suppression in the mammalian cochlea. J Assoc Res Otolaryngol 14:523–545
64. Whitham GB (1974) Linear and nonlinear waves. Wiley, New York
65. Wier CC, Norton SJ, Kincaid GE (1984) Spontaneous narrow-band otoacoustic signals emitted by human ears: a replication. J Acoust Soc Am 76:1248–1250
66. Zweig G (1976) Basilar membrane motion. Cold Spring Harb Symp Quant Biol 40:619–633
67. Zweig G (1991) Finding the impedance of the organ of Corti. J Acoust Soc Am 89:1229–1254

CHAPTER 20

BIFURCATION STUDY OF A NEURAL FIELD COMPETITION MODEL WITH AN APPLICATION TO PERCEPTUAL SWITCHING IN MOTION INTEGRATION

J. Rankin, A. I. Meso, G. S. Masson, O. Faugeras P. Kornprobst

Keywords: Multistability, Competition, Perception, Neural fields, Bifurcation, Motion

INTRODUCTION

Perception can evolve dynamically for fixed sensory inputs and so-called multistable stimuli have been the attention of much recent experimental and computational investigation. The focus of many modelling studies has been to reproduce the switching behaviour observed in psychophysical experiment and provide insight into the underlying mechanisms (Laing and Chow 2002; Freeman 2005; Kim et al. 2006; Shpiro et al. 2007; Moreno-Bote et al. 2007; Borisyuk et al. 2009; Ashwin and Lavric 2010). Bifurcation analysis (Strogatz 1994; Kuznetsov1998) and numerical continuation (Krauskopf et al. 2007) are powerful tools from the study of dynamical systems that have already proved effective in analysing rate models where the competing perceptual states are represented by discrete neural masses (Shpiro et al. 2007; Curtu et al. 2008; Theodoni et al. 2011b). Two commonly proposed mechanisms that drive the switching behaviour in these models are adaptation and noise and a strong argument is made in Shpiro et al. (2009) that a balance accounts best for experimental findings across different model architectures and different adaptation mechanisms. Existing studies using discrete neural masses have shown that switching in rivalry experiments can be described by a relatively simple dynamical system. However, the problem of multistable motion integration is different because the perceived direction of motion is represented on a continuous scale. We therefore asked the following questions. Can a minimal model with a continuous feature space describe switching behaviour in motion integration? Do qualitative changes in the dynamics predicted with bifurcation analysis correspond to changes in the mechanisms driving the switches?

Here, we will take advantage of the neural fields formalism (Amari 1971; Wilson and Cowan1972, 1973) in order to study neural competition in a model with a continuous feature space where adaptation and noise are implemented as mechanisms that can drive activity switches. The model describes the mean firing rate of a population of feature selective neurons. Deterministic versions of this feature-only model with spike frequency adaptation have been studied previously without input (Curtu and Ermentrout 2004) and with a unimodal input (Hansel and Sompolinsky 1998; Folias 2011). A key difference with existing rivalry models is that the competing percepts form tuned responses in a continuous feature space instead of being represented by discrete populations as in, for example, Shpiro et al. (2007), and Theodoni et al. (2011b). The more general model we use allows for perceptual transitions to occur in a smooth way as opposed to discrete switches between two isolated percepts. Starting from the results presented in Curtu and Ermentrout (2004), we will first introduce a simple (unimodal) input and investigate how the various types of solutions from the no-input case are modified. With the application of numerical bifurcation methods we find that although the boundaries between parameter regions featuring different types of responses are gradually distorted with increasing input strength, much of the global structure is

Citation: Rankin, J., Meso, A.I., Masson, G.S. et al. J Comput Neurosci (2014) Bifurcation Study Of A Neural Field Competition Model With An Application to Perceptual Switching In Motion Integration 36: 193, doi:10.1007/s10827-013-0465-5.

preserved. This allows for all possible types of behaviour, and parameter regions for which it can occur, to be comprehensively described across a wide range of model parameters controlling input gain, adaptation gain and the shape of the firing rate function. For a simple input we are able to match the models output to known response properties from the literature before considering the introduction of a complex (multimodal) input that gives rise to multistable behaviour.

In this paper we are interested in moving, ambiguous visual stimuli for which two or more distinct interpretations are possible, but where only one of these interpretations, or percepts, can be held at a time. Not only can the initial percept be different from one short presentation to the next, but for extended presentations, the percept can change, or switch, dynamically. This phenomenon of multistability has been observed and investigated with a number of different experimental paradigms, e.g. binocular rivalry experiments (Levelt 1968; Blake 1989, 2001), apparent motion (Ramachandran and Anstis 1983), motion plaids that are bistable (Hupé and Rubin 2003) or tristable (Hupé and Pressnitzer 2012) and the multistable barberpole illusion (Castet et al. 1999; Fisher and Zanker 2001; Meso et al. 2012b). During extended presentations of these stimuli, the dominant percept switches randomly and the dominance durations between switches have been shown to fit certain distributions dependent on the experimental paradigm (Levelt 1968; Leopold and Logothetis 1996; Logothetis et al. 1996; Lehky 1995; Zhou et al. 2004; Rubin and Hupé 2005).

Here, we will study the temporal dynamics of perception for the so-called multistable barberpole illusion, which has been investigated in complementary psychophysical experiments (Meso et al. 2012b). Some of these results will be presented alongside the modelling work. We will demonstrate how the general neural fields model can reproduce the main dynamical characteristics of the perceptual switches observed in the experiments. We use the mean switching rates reported in the experiments to constrain model parameters and propose specific mechanisms that can account for the behaviour in different contrast regimes. Importantly, we will show that the two contrast regimes identified experimentally, one in which the rate increases with contrast, the other in which the rate decreases with contrast, are linked to specific mechanisms with the model. Although a combination of noise and adaptation drive the switching, the dominant mechanism changes with contrast. Furthermore, we are able to quantify this in an experimentally testable way: the distribution of dominance durations fit different statistical distributions in each contrast regime.

In Section 2 section we give a mathematical description of the model before presenting general results that map the model's possible behaviours across parameter space in Section 3 and then applying the model to the study of multistable perception in Sections 4 and 5.

COMPETITION MODEL WITH CONTINUOUS FEATURE SPACE

The Neural Field Framework

The neural field equations provide an established framework for studying the dynamics of cortical activity, represented as an average membrane potential or mean firing rate, over a spatially continuous domain. Since the seminal work by Amari (1971), Wilson and Cowan (1972) and Wilson and Cowan (1973) a broad range of mathematical tools have been developed for their study, see reviews by Ermentrout (1998), Coombes (2005) and Bressloff (2012) along with Ermentrout and Terman (2010, Chapter 11) for a derivation of the equations. The equations describe the dynamical evolution of activity of one or more connected populations of neurons, each defined in terms of a spatial domain that can represent either physical space (on the cortex), an abstracted feature space (orientation, direction of motion, texture preference, etc.), or some combination of the two. This framework has proved especially useful in the study of neuro-biological phenomena characterised by complex spatio-temporal patterns of neuronal activity, such as, orientation tuning in the primary visual cortex V1 (Ben-Yishai et al. 1995; Somers et al. 1995; Hansel and Sompolinsky 1998; Veltz 2011), binocular rivalry (Kilpatrick and Bressloff 2010; Bressloff and Webber 2011), and motion integration (Giese 1998; Deco and Roland 2010).

Model Equations

In this section we describe a general neural competition model that switches between selected states for an ambiguous input. The model describes the time-evolution of a neuronal population defined across a continuous feature space in which a selected state corresponds to a tuning curve. A spatial connectivity is chosen that produces mutual inhibition between competing tuned responses so that a winner-takes-all mechanism leads to a definitive tuned response at any given time instant. Over time, shifts between tuned responses are driven by a combination of adaptation and noise.

We will consider a single neuronal population $p(v, t)$, defined across the continuous, periodic feature space $v \in [-\pi, \pi)$, whose evolution depends on time t. The variable $p(v, t)$ takes values in [0, 1] representing activity as a proportion of a maximal firing rate normalised to 1. We also define secondary adaptation $\alpha(v, t)$ and stochastic $X(v, t)$ variables. The time evolution of $p(v, t)$ is described by the following coupled system of integro-differential equations:

$$\begin{aligned} \tau_p \frac{\mathrm{d}}{\mathrm{d}t} p(v,t) &= -p(v,t) + S\big(\lambda[J(v) * p(v,t) - k_\alpha \alpha(v,t) \\ &+ k_X X(v,t) + k_I I(v) - T]\big), \end{aligned} \tag{1}$$

$$\tau_\alpha \frac{\mathrm{d}}{\mathrm{d}t} \alpha(v,t) = -\alpha(v,t) + p(v,t). \tag{2}$$

The principal Eq. (1) has time constant τ_p and has a standard decay term $-p$. A smooth, nonlinear sigmoidal firing rate function

$$S(x) = \frac{1}{1+\exp(-x)} \tag{3}$$

is used as plotted in Fig. 1a. The slope and threshold of the firing rate function are controlled by the parameters λ and T, respectively. The firing rate function processes lateral connections described by J and inputs from adaptation α, additive noise X and a time independent input I; the respective input gain parameters are k_α, k_X and k_I. The connectivity in the feature space v is represented by a convolutional operator J that approximates a Mexican hat connectivity (local excitation, lateral inhibition). As in Curtu and Ermentrout (2004), we use a 3-mode expansion and J takes the form

$$J(v) = J_0 + 2J_1 \cos(v) + 2J_2 \cos(2v), \tag{4}$$

see Fig. 1b. The adaptation dynamic in Eq. (2) describes linear spike frequency adaptation that evolves on a slow time scale τ_α. The additive noise $X(v, t)$ is an Ornstein-Uhlenbeck process that evolves on the same slow timescale τ_α as the adaptation, has mean $\langle X(v, t)\rangle = 0$, variance $\mathrm{Var}(X) = 1$ and no feature correlation; see Appendix A for further details. The input I depends only on the feature v and the so-called simple input studied in Section 3 is shown in Fig. 1c.

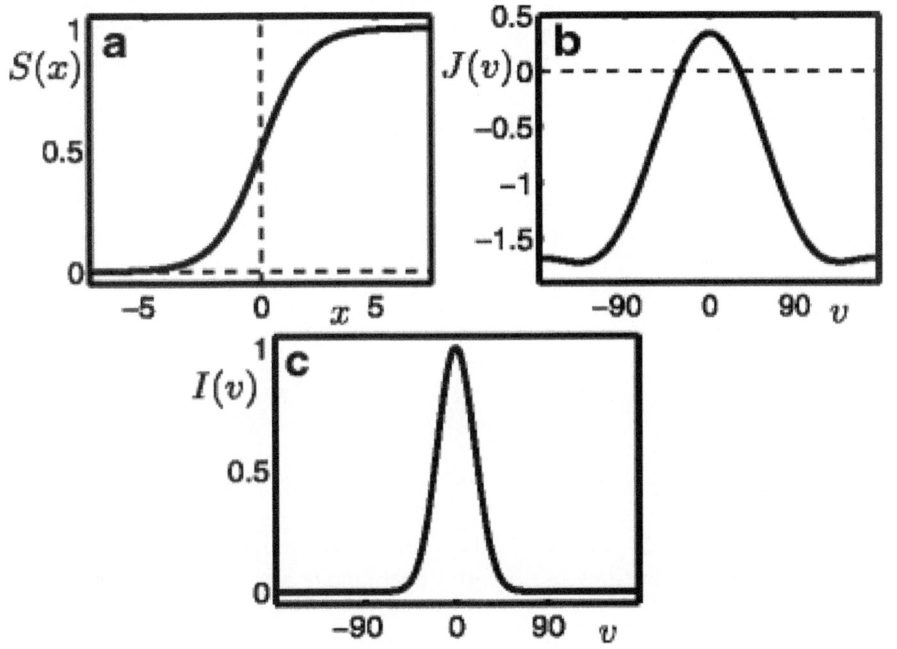

Figure. 1 Model features and stimulus. **a** The smooth (infinitely differentiable) sigmoidal firing rate function $S(x)$. **b** The convolutional kernel J is a three-mode approximation of a Mexican hat connectivity. **c** The simple stimulus is a Gaussian bump centred at $v = 0°$

Parameter Values, Initial Conditions and Numerical Computations

The parameter values used in the numerical computations in each section of the paper are given in Table 1. For the model simulations without noise ($k_X = 0$), as presented in Sections 3 and 4, we solve the system described by Eqs. (1–2) using the ODE23t solver in Matlab with default settings except the relative tolerance, which was set to 10^{-6}. A 200-point discretisation of v was used that satisfies error bounds for the computation of the integral term $J * p$ with a standard trapezoid method. For the simulations with noise ($k_X = 0.0025$), as presented in Section 5, the same discretisation is used with a standard Euler-Maruyama method and a fixed timestep of 0.5ms. In all simulations the initial conditions are set to a low level of activity $p_0(v) = 0.1$ (10% of the maximal firing rate) with a small randomised perturbation. Initial conditions are given by

$$p(v,0) = p_0(v), \quad \alpha(v,0) = 0, \quad X(v,0) = 0.$$

Table 1 Parameter values used in the numerical studies

Description	Parameter	Value	Section(s)
Zero-order coefficient of J	J_0	-1	Fixed
First-order coefficient of J	J_1	1/2	Fixed
Second-order coefficient of J	J_2	1/6	Fixed
Sigmoid threshold	T	-0.01	Fixed
Firing rate stiffness	λ	Free in [12,26]	Sections 3 and 4

...as function of contrast $c \in [0,1]$	$\lambda(c)$	[13,25]	Section 5
Adaptation strength	k_α	Free in [0,0.07]	Sections 3 and 4
	k_α	0.01	Section 5
Input gain	k_I	0 or 0.001	Section 3
	k_I	0.01	Sections 4 and 5
Noise strength	k_X	0	Sections 3 and 4
	k_X	0.0025	Section 5
Population time constant	τ_p	1ms	Fixed
Adaptation timescale	τ_α	100ms	Sections 3 and 4.3
Adaptation and noise timescale	τ_α	16.5s	Sections 4.4 and 5

In order to carry out a bifurcation analysis of the system (1)–(2) we use a numerical continuation package AUTO (Doedel et al. 1997) that allows us to compute branches of steady state and oscillatory solutions, and to detect and track bifurcations of these solutions. These computations are carried out in the absence of noise ($k_X = 0$) and in this case we can take advantage of the 3-mode approximation of J in Eq. (4) by expressing $p(v, t)$ and $\alpha(v, t)$ in terms of the same modes plus some orthogonal components $\hat{p}_\perp$ and $\widehat{\alpha}_\perp$:

$$\begin{aligned} p(v,t) &= \hat{p}_0(t) + \hat{p}_1(t)\cos(v) + \hat{p}_2(t)\sin(v) \\ &\quad + \hat{p}_3(t)\cos(2v) + \hat{p}_4(t)\sin(2v) + \hat{p}_\perp, \end{aligned} \tag{5}$$

$$\begin{aligned} \alpha(v,t) &= \widehat{\alpha}_0(t) + \widehat{\alpha}_1(t)\cos(v) + \widehat{\alpha}_2(t)\sin(v) \\ &\quad + \widehat{\alpha}_3(t)\cos(2v) + \widehat{\alpha}_4(t)\sin(2v) + \widehat{\alpha}_\perp. \end{aligned} \tag{6}$$

In Veltz and Faugeras (2010) it was proved that as $t \to \infty$ the orthogonal components decay to 0. Therefore, we can study steady-state and oscillatory solutions to (1)–(2) by solving a set of 10 ordinary differential equations in p^i and $\widehat{\alpha}_i, i = 0 \ldots 4$. The integral term $J * p$ was computed with the same 200-point trapezoid integration scheme used for in the ODE solver. Periodic orbits were typically computed with 150 mesh points (constants NTST = 50 and NCOL = 3) in AUTO. The reduced description is used only for the computation of the bifurcation diagrams.

GENERAL STUDY OF COMPETITION MODEL

Bifurcation Analysis and Numerical Continuation

Bifurcation analysis and its computational counterpart numerical continuation are crucial tools in the study of the neural field equations and dynamical systems in general. When varying a model parameter, a bifurcation is a special point at which there is a qualitative change to the types of response produced by the model. Numerical continuation allows one to locate, classify and track bifurcation points and, in this way to map out the exact boundaries between regions in parameter space with qualitatively different behaviour. This kind of information forms a basis for tuning a model's parameters; indeed, it is possible to ensure that parameter regions in which a desired behaviour is present are not isolated and ensure robustness with respect to small changes in the model set up. Numerical continuation has been used in general studies of the neural field equations (Veltz and Faugeras 2010; Veltz 2011), to investigate localised states (Laing and Troy 2003; Faye et al. 2013) and in a previous study of the short-term dynamics of the stimulus considered in this article (Rankin et al. 2013). One key advantage of using bifurcation and continuation techniques is that they allow for a model to be brought into an operating regime, close to bifurcation, where the model is most sensitive to changes in its input and where the combination of mechanisms involved in performing complex computations can be revealed. This general philosophy has been used to great effect in studies of orientation tuning in V1 (Veltz 2011, Chapter 9), simplified rate models of neural competition (Shpiro et al.2007; Theodoni et al. 2011b) and studies of decision making (Theodoni et al. 2011a).

Bifurcations relevant to this study

At a bifurcation point there is qualitative change to the types of response produced by the model. The types of response can differ in terms of 1) spatial properties, such as being tuned or untuned, 2) temporal properties, such as being steady or oscillatory and 3) stability, where stable implies responses that persist in time and unstable implies responses that are transient. Each type of response is associated with a solution of the underlying equations and the organisation of these solutions in state/feature space governs the dynamical behaviour. The main types of bifurcation that we encounter in this study are the pitchfork bifurcation and the Hopf bifurcation; these so-called codimension-one bifurcations occur at a given point as one parameter is varied. In this model the pitchfork is associated with a transition from a homogeneous state to a tuned state. Hopf bifurcations are associated with transitions from steady to oscillatory responses that can either travel in the feature space (travelling waves), or that remain static but oscillate in amplitude (standing waves). In the parameter plane (as two parameters are varied) these codimension-one bifurcations lie on curves. At points where these curves meet or intersect we encounter codimension-two bifurcations that act as organising centres close to which several solution types can be encountered. Due to the presence of translational and reflectional symmetry properties in the underlying equations when there is no input, the bifurcating solutions encountered have the same symmetry properties (Haragus and Iooss 2010). We will give an account of how these symmetry properties break down with the introduction of an input.

In Curtu and Ermentrout (2004) it was shown that in the absence of input ($k_I = 0$) the model possesses O(2)-symmetry and as a consequence several types of steady solutions and oscillatory patterns exist in different parameter regimes local to the codimension-two Bogdanov-Takens (BT) point (Dangelmayr and Knobloch 1987). As we will see, this BT point acts as an important organising centre in parameter space. First, in Section 3.3, we will give a summary of what is already known from Curtu and Ermentrout (2004) in terms of solution branches and bifurcation curves that are relevant to this study (the account will not be exhaustive, omitting several bifurcation curves involving exclusively unstable solutions). In Section 3.4, we will introduce an input and describe how the solutions existing in different parameter regions change and how the boundaries of these regions shift in parameter space. Although the system (1)–(2) has previously been studied with a unimodal input in (Hansel and Sompolinsky 1998) and more recently in (Folias 2011), the results we present allow us to build a more complete picture of the model's behaviour in terms of three parameters relevant to our study, the input gain k_I, the adaptation strength k_α and the sigmoidal slope λ.

No input ($k_I = 0$)

Figure 2 shows the different types of dynamical behaviour produced in different regions of the (λ, k_α)-parameter plane as demarcated by bifurcation curves. The model simulations shown in panels b–e were performed at the corresponding points B–E in panel a. In each case the simulation time was chosen such that the model reaches its stable behaviour during the simulation; the stable behaviour is either a steady state or an oscillatory state. In the white region the model produces an homogeneous (untuned) steady-state response at a low level of activity as shown in panel d. In the dark-grey region the model produces a steady-state response tuned to an arbitrary direction as shown in panel e. The boundary between the white and dark-grey regions is a pitchfork curve P; as λ is increased and the pitchfork bifurcation is encountered, the homogeneous steady-state becomes unstable and a ring of tuning curves forms the stable behaviour. In the light-grey region the stable behaviour is a travelling-wave solution with an arbitrary direction in v; the transient behaviour observed before reaching this stable state changes dependent on the chosen parameter values; see panels b and c. The boundary between the white and the light-grey regions is the coinciding Hopf-type curves H_{SW} an H_{TW1}. As λ is increased and the two coinciding bifurcation points are encountered the homogeneous steady states lose stability and two new branches bifurcate simultaneously: an unstable branch of standing wave solutions and a stable branch of travelling wave solutions; this is shown explicitly in Appendix B. In panel b, close to these curves, the unstable standing wave solution is seen as a transient behaviour before eventual convergence to the stable travelling wave solution. The boundary between the dark-grey and light-grey region is $H_{TW2}$and as $k_\alpha$is increased and the bifurcation is encountered the stable tuned response becomes spatially unstable and starts to travel in an arbitrary direction. In panel c the unstable tuned response is seen as a transient behaviour before starting to travel.

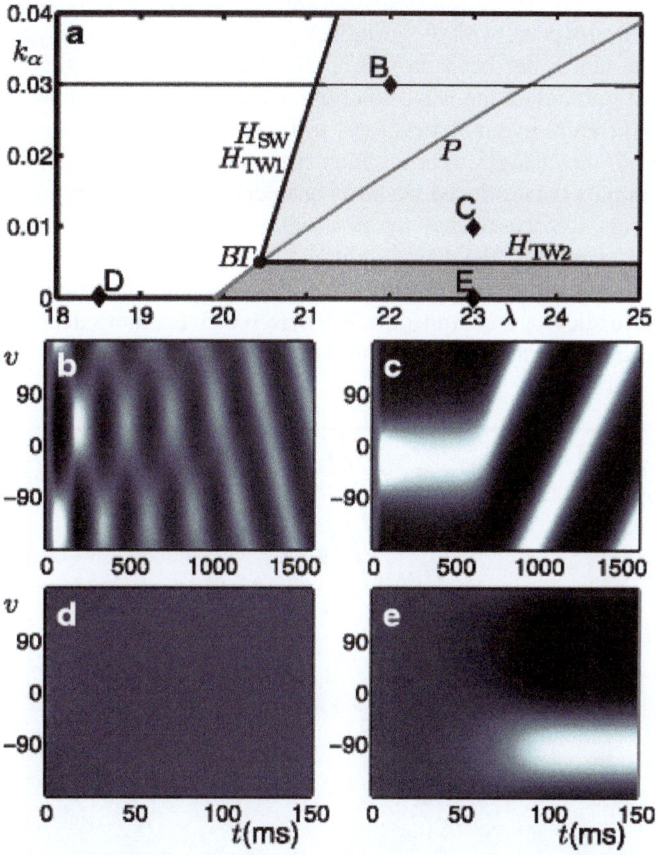

Figure. 2 Bifurcation diagram for the no-input case; summary of results from Curtu and Ermentrout (2004). **a** Bifurcation curves plotted in the (λ, k_α)-parameter plane demarcate regions with qualitatively different dynamics. The Hopf-type curves are H_{SW}, H_{TW1}(coinciding) and H_{TW2}, a pitchfork curve is P and these curves meet at the Bogdanov-Takens point BT. Panels **b–e** show the activity $p(v, t)$ indicated by intensity for model simulations at parameter values from the corresponding points B–E in panel **a**

Simple input with $k_I = 0.001$

Figure 3 shows a new bifurcation diagram after the introduction of the simple input I_{1D}shown in Fig. 1c with input gain k_I = 0.001. We are interested to see how the solutions identified in the previous section change and how their organisation in parameter space has been modified. The most notable result is that much of the structure from the no-input case has been preserved, albeit with subtle changes that are now discussed. In the white region (to the left of H_{SW1} and P^P^) there is now a low-activity response that is weakly tuned to the input centred at $v = 0$; see panel d. In the dark-grey region there is still a steady-state, tuned response, but now centred on the stimulus at $v = 0$. In the light-grey region the stable behaviour is still predominantly a travelling wave solution resembling those shown in Fig. 2b and c but with a slight modulation as the wave passes over the stimulus; the modulated solution will be shown later. Here we highlight a qualitatively different type of travelling wave solution that can be found close to the Hopf curve H_{TW2}, whereby the wave has been *pinned*to the stimulated direction, as a so-called*slosher* state (Folias 2011); see panel

c. Furthermore, an elongated region in parameter space has opened between H_{SW1} and the coinciding curves H_{sw2} and H_{TW1}, in which the stable behaviour is a standing wave, with one of its peaks aligned to the stimulus at $v = 0$. In the parameter regime studied here, a change in stability of the standing wave solution occurs with the introduction of an input, however, we note that it has been proved in Curtu and Ermentrout (2004) that stable standing wave solutions can also exist without input in different parameter regions. In order to describe in further detail the changes to bifurcation structure that occur when the stimulus is introduced we also consider several one-parameter slices in λ (indicated by horizontal lines) at fixed values of k_α taken from the diagrams already shown in Figs. 2a and 3a. The one-parameter bifurcation diagrams corresponding to these slices are in Appendix B where we give a more detailed account of way in which symmetries are broken when an input is introduced.

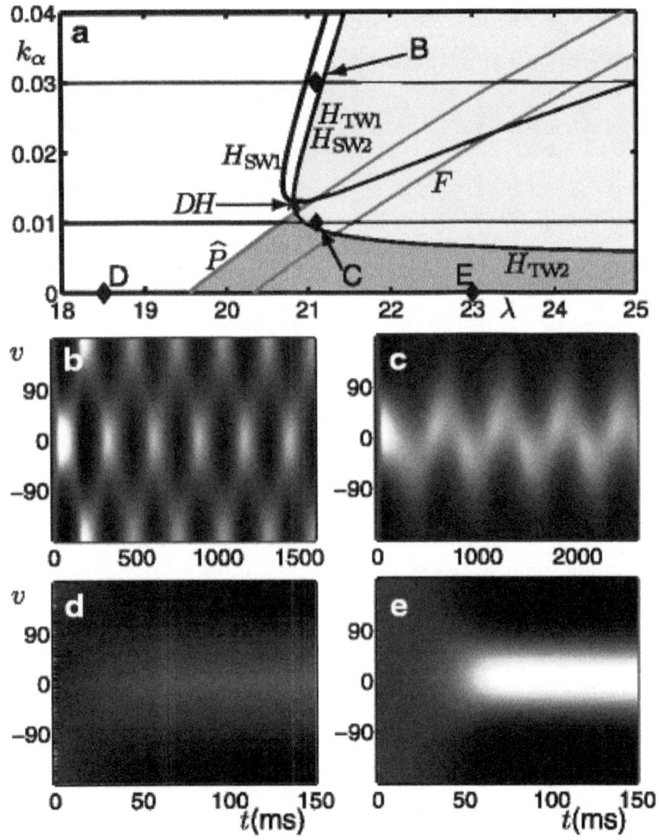

Figure. 3 Bifurcation diagram for the simple input case with $k_I = 0.001$. **a** Bifurcation curves plotted in the (λ, k_α)-parameter plane demarcate regions with qualitatively different dynamics. The Hopf curves are H_{SW1}, H_{TW2}, along with the coinciding H_{SW2} and H_{TW1}; a further Hopf in the light-grey region involves only unstable solutions and is not labelled; all the Hopf curves meet at a the double Hopf point *DH*. Two bifurcation curves resulting from the symmetry breaking of a pitchfork bifurcation are P^P^ and *F*; see Fig. 12 in Appendix B and accompanying text. Panels **b–e** show the activity $p(v, t)$ indicated by intensity for model simulations at parameter values from the corresponding points B–E in panel **a**

The bifurcation analysis with the same input studied throughout this section, as shown in Fig. 1c, continues in the next section. We present the case $k_I = 0.01$ but in the context of a motion stimulus.

COMPETITION MODEL APPLIED TO THE STUDY OF MULTISTABLE MOTION

Model of direction selectivity in MT

We extend the general study presented in Section 3 and demonstrate how the same model can be used to study a specific neuro-biological phenomenon for which perceptual shifts are observed. We now associate the model's periodic feature space $v \in [-\pi, \pi)$ with motion direction. We assume that the model's activity in terms of time-evolving of firing rates $p(v, t)$ are responses of direction-selective neurons in the middle temporal (MT) visual area. Indeed, MT is characterised by direction-selective neurons that are organised in a columnar fashion (Diogo et al. 2003). Here we only consider a feature space of motion direction and, thus, we assume the model responses to be averaged across physical (cortical) space. The chosen connectivity function (4) shown in Fig. 1b represents mutual inhibition between sub-populations of neurons associated with competing directions; this type of connectivity naturally gives rise to winner-takes-all responses tuned to one specific direction. Indeed, there is evidence that competing percepts have mutually inhibitory representations in MT (Logothetis et al. 1989; Leopold and Logothetis 1996). We use the models tuned response to dynamically simulate the mechanisms driving perception; cortical responses of MT have been linked specifically to the perception of motion (e.g. Britten (2003) and Serences and Boynton (2007)). We assume that over time any particular tuned response will slowly be inhibited as represented by the linear spike-frequency adaptation mechanism in the model. Furthermore, we assume there to be a fixed-amplitude stochastic fluctuation in the membrane potential that is modelled by additive noise (note that the noise is only introduced for the simulations presented in Section 5). We use as a model input pre-processed direction signals in the form expected from V1 (Britten 2003; Born and Bradley 2005). In Section 4.3 the model's response properties in terms of its contrast dependence and direction tuning properties will be matched to what is known about the direction selective behaviour of MT neurons from physiological studies (Albright 1984; Sclar et al. 1990; Diogo et al. 2003).

Definition of motion stimuli

We introduce two classical psychophysics stimuli where a luminance grating drifting diagonally (up and to the right in the example shown) is viewed through an aperture see Figs. 4a and b. In the first case, with a circular aperture, the grating is consistently perceived as moving in the diagonal direction D ($v = 0°$). In the second case, the aperture is rectangular and tilted relative to the grating orientation. The classical barberpole illusion (Hildreth 1983, Chapter 4) comes about as a result of the aperture problem (Wallach 1935; Wuerger et al. 1996): a diagonally drifting grating viewed through an elongated rectangular aperture is perceived as drifting in the direction of the long edge of the aperture. With a square aperture, the stimulus is multistable for short presentations on the order of 2–3s, where the dominant percepts are vertical V ($v = 45°$), horizontal H ($v = -45°$) and D ($v = 0°$) (Castet et al. 1999; Fisher and Zanker 2001). We denote this stimulus the multistable barberpole and it has been the subject of complementary psychophysics experiments (Meso et al. 2012b) from which some results will be presented in Section 5.2.

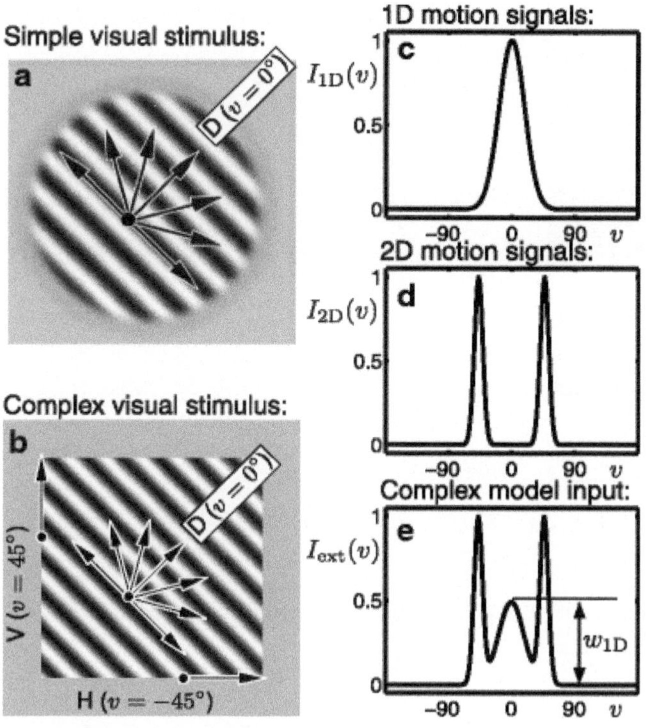

Figure. 4 Simple and complex motion stimuli. **a** A drifting luminance grating viewed through a circular soft aperture; the diagonal direction of motion D is consistently perceived. **b** A drifting luminance grating viewed through a square aperture; the dominant percepts are vertical V, horizontal H and diagonal D. **c** Representation of the 1D motion signals in direction space; the simple motion stimulus **a** is equated with I_{1D}. **d** Representation of the 2D motion signals in direction space. **e** Summation of the 1D and 2D motion signals with a weighting w_{1D} = 0.5; the complex motion stimulus **b** is equated with I_{ext}

It was shown in Barthélemy et al. (2008) that the motion signals from 1D cues stimulate a broad range of directions when compared with 2D cues that stimulate a more localised range of directions; see arrows in Figs. 4a and b. Based on these properties, it is proposed that the multistability for the square-aperture stimulus is primarily generated by competition between ambiguous 1D motion direction signals along grating contours on the interior of the aperture and more directionally specific 2D signals at the terminators along the aperture edges. We represent the 1D cues by a Gaussian bump $I_{2D}(v) = \exp(-v^2/2\sigma_{2D}^2)$ with σ_{1D} = 18° centred at v= 0° as shown in Fig. 4c (which is the same as Fig. 1c); on its own we call this a simple input that represents a drifting grating either filling the visual field (without aperture) or with an aperture that has no net effect on perceived direction such as the circular one shown in Fig. 4a. We represent the 2D cues by two Gaussian bumps $I_{2D}(v) = \exp(-v^2/2\sigma_{2D}^2)$ centred at v = 45° and$v$ = − 45° with width σ_{2D} = 6° as shown in Fig. 4d. Note that the functions I_{1D} and I_{2D} are normalised such that their maxima are 1 (not their areas). Figure 4(e) shows the complex inputI_{ext}represented as a summation of 1D and 2D motion signals with maximum normalised to 1 and a smaller weighting $w_{1D} \in [0, 1]$ given to 1D cues:

$$I_{ext}(v) = w_{1D} I_{1D}(v) + I_{2D}(v - 45) + I_{2D}(v + 45). \tag{7}$$

The weighting w_{1D}translates the fact that in motion integration experiments 2D cues play a more significant role that 1D cues in driving perceived direction of motion (Masson et al. 2000; Barthélemy et al. 2010). Here we represent this weighting in a simple linear relationship, but in future studies it may be relevant to consider the contrast response functions for 1D and 2D cues separately (Barthélemy et al. 2008).

Simple input case with k_I = 0.01: parameter tuning for motion study and contrast dependence

Figure 5a shows the two-parameter bifurcation diagram in λ and k_α for the same simple 1D input illustrated in Fig. 3a, but with the input gain increased by a factor of 10 to k_I = 0.01. The diagram shows the same organisation of bifurcation curves, but the oscillatory regions have shifted significantly towards the top-right corner. Once again, in the white region containing the point D1 there is a steady, low-activity, weakly tuned response (see lower curve in panel d). In the dark grey region containing the point D2 there is a steady, high-activity response with tuning width δ (see upper curve in panel d). We define δ as the width at half-height of the tuned response. Again, the boundary between these regions is demarcated by P^P^ in Fig. 3a. In the region containing the point B there is still a standing-wave-type solution, but it is modified by the input such that there are breather-type oscillations (Folias 2011) between a tuned and an untuned state over time; see Fig. 5b. In this context there is no physiological interpretation for this solution, and so we will ensure that the model is operated in a parameter region where it cannot be observed. In the region containing the point C there is still a periodic response with small-amplitude oscillations in v of a tuned response about v = 0; see Fig. 5c. This is a travelling-wave-type slosher solution as described in Section 3.4 that is pinned by the input at v= 0°; note the transient that makes one full excursion before being pinned. Closer to the bifurcation curve the solutions are immediately pinned and further away a phase slip can be encountered as shown in Fig. 13g–i in Appendix B. For a simple (unambiguous) input we again operate the model away from this oscillatory region of parameter space but find that when the complex input is introduced it is the slosher-type solutions that produce the desired switching behaviour.

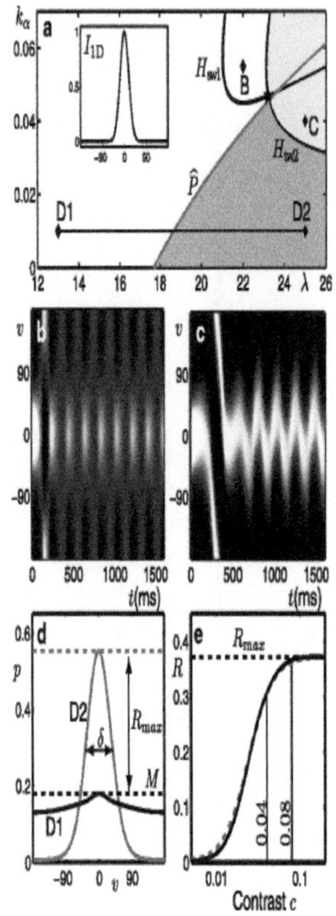

Figure. 5 Bifurcation study and contrast response for simple input I_{1D} (see inset of **a**) with input gain $k_I = 0.01$. **a** Two-parameter bifurcation diagram in terms of sigmoid slope λ and adaptation strength k_α shows qualitatively the same organisation of bifurcation curves as Fig. 3a. **b, c** Time-traces of the activity $p(v, t)$ indicated by intensity as computed at the corresponding points B and C labelled in **a**. **d** Steady-state responses in terms of the activity $p(v)$ at the corresponding points D1 and D2 labelled in **a**. **e** Contrast response in terms of normalised peak activity R for simple input (*solid curve*) fitted to a Naka-Rushton function (*dashed curve*) as described in Appendix C. The line between D1 and D2 in **a** is the operating range of the model. The response D1 shown in panel **d**corresponds to $c = 0$ and the response D2 in panel **d** corresponds to $c = 1$

In order to produce results and predictions that can be related directly to the experiments, where contrast was one of the main parameters investigated, contrast should also be represented as a parameter in the model. In Curtu et al. (2008) the input gain k_I is assumed to depend on contrast and for some critical value of $k_I$switching behaviour is observed. Here, we choose another option, arguing that motion signals arriving in MT, primarily from V1, are normalised by shifts in the sigmoidal nonlinearity (see Carandini and Heeger (2011) for a review) and therefore, the input gain k_I in Eq. (1) should remain fixed with respect to contrast. Indeed, by making the slope parameter λ depend on the contrast $c \in [0, 1]$, we are also able to reproduce the observed switching behaviour with increasing contrast.

We now fix $k_\alpha = 0.01$ such that we operate the model away from the oscillatory regions shown in Fig. 5 and describe how the model can be reparametrised in terms of contrast c. For some steady state $\bar{p}$, we define firing rate response R=max{$\bar{p}$}−M as the peak firing rate response above some baseline value M; max{$\bar{p}$}max is shown as a dashed line for solutions D1 and D2 in Fig. 5d and we set M=max{$\bar{p}$ D1}. As discussed in more detail in the Appendix C, the solution D1 at $\lambda = 13$ is consistent with an MT response to a very low contrast input ($c < 0.01$), whereas the solutions D2 at $\lambda = 25$ is consistent with a high contrast input ($c > 0.2$). By makingλ a specific function of c we are able to match the model's contrast response to known behaviour for MT neurons. As shown in Fig. 5e, we match the model's response to an appropriately parametrised Naka-Rushton function, which was used to fit contrast response data across several stages of the visual pathway including MT in (Sclar et al. 1990); again, refer to AppendixC for further details. The operating range for the model is indicated by a horizontal line at $k_\alpha = 0.01$ for λ □ [13, 25] in Fig. 5a. In Appendix C we also show that the tuning widths δ of the model responses are in agreement with the literature (Albright 1984; Diogo et al. 2003).

Complex input with $k_I = 0.01$

This section will focus on the bifurcation results for a complex input as shown in Figs. 6 and 7, but first we discuss the choice of timescale parameters. Up until now the results presented have been carried out with the main population time constant and adaptation timescale fixed at τ_p = 1ms and τ_α = 100ms, respectively. In the following sections the cortical time scale remains fixed at τ_p = 1ms and the adaptation timescale is tuned to a value of τ_α = 16.5s in order to match average switching rates in the experimental data presented in Section 5. Concerning the population time constant τ_p, minimal neuronal latencies in response to flashed or moving high contrast stimuli are 40–45ms in macaque visual area MT, see Kawano et al. (1994) and Raiguel et al. (1999). From our simulations, if we look at the typical early dynamics as shown in Fig. 6c1 and c2, the transition to the first direction tuned response occurs at $t_{□} \approx$ 20ms. The latency of this response depends both on τ_p and on the lateral spread of activity through the connectivity kernel, thus, the time constant we use is an appropriate order of magnitude. Concerning the adaptation timescale τ_α, slow adaptation of firing rates in the visual cortex occur on timescales that range from 1–10s for excitatory neurons and 2–19s for inhibitory neurons (Sanchez-Vives et al. 2000; Descalzo et al. 2005). Yang and Lisberger (2009) have shown that a 10s presentation of a moving stimulus reduces neuronal response strength in macaque MT and that this has an effect on pursuit eye movements. Such a time course is consistent with human psychophysics data on the relationship between adaptation duration and strength in visual motion processing, see (Mather et al. 1998). The chosen adaptation timescale is of an appropriate order of magnitude with respect to visual motion adaptation in cortical areas as well as for visual perception.

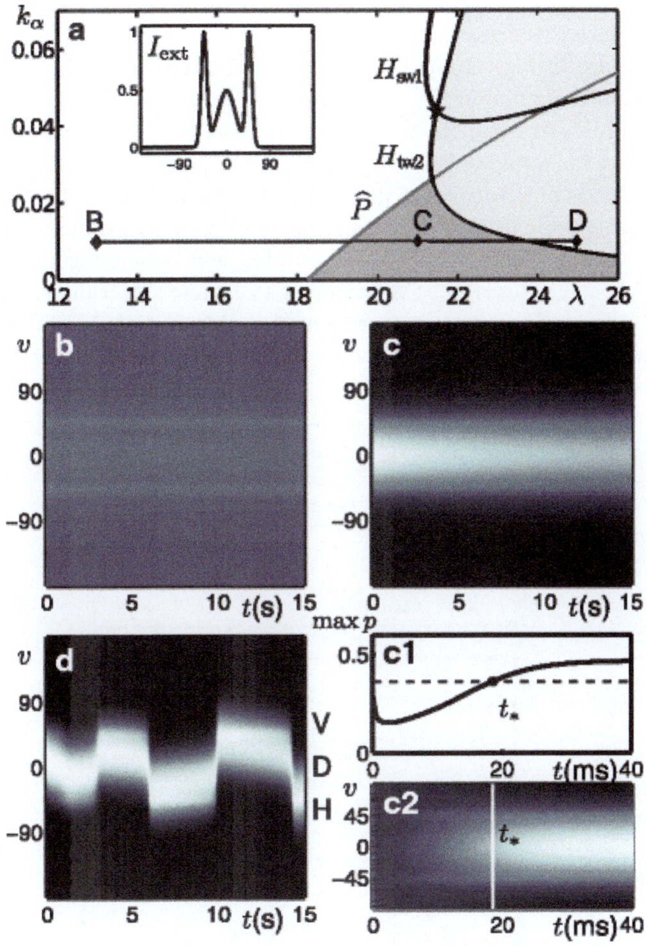

Figure. 6 Bifurcation study for complex input I_{ext} (see inset of **a**) with input gain $k_I = 0.01$. **a** Two-parameter bifurcation diagram in terms of sigmoid slope λ and adaptation strength $k_\alpha$shows qualitatively the same organisation of bifurcation curves as Figs. 3a and 5a. Line between points labelled B and D in **a** shows the operating region of the model as defined in Fig. 5a. **b–d** 15s time-traces of the activity $p(v, t)$ indicated by intensity as computed at the corresponding points B, C and D labelled in **a**. Panel c1 shows the maximum of the activity and panel **c2** shows detail from **c** for the first 40ms of the simulation. The first direction tuned response coincides with the max of the activity crossing through a threshold value of $\max\{p\} = M + \frac{R_{\max}}{2}$ as indicated by a horizontal dashed line in **c1**

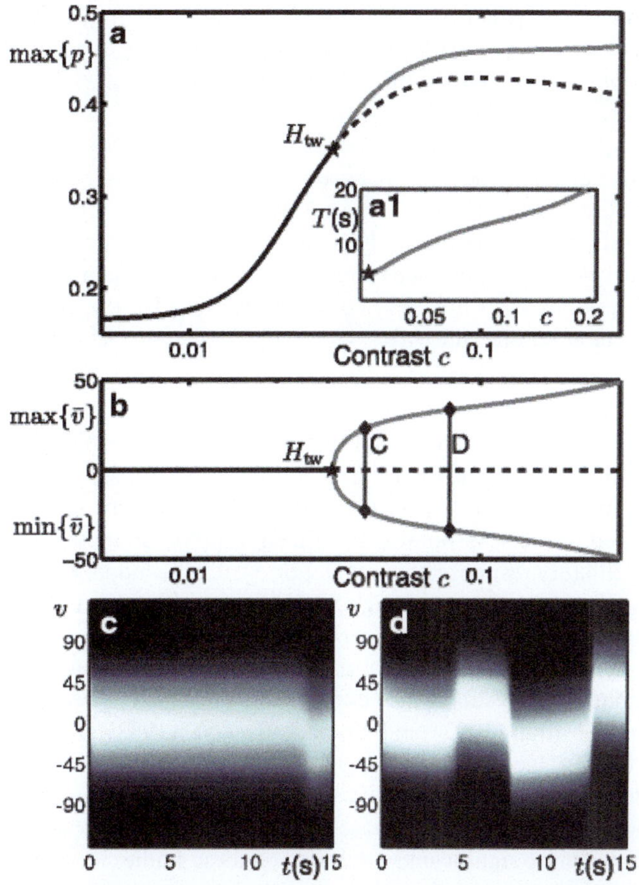

Figure. 7 Bifurcation diagram with complex input for model parametrised in terms of contrast. **a**, **b**One-parameter bifurcation diagrams show the same data plotted in terms of the maximum response and average direction, respectively; the steady-state response tuned to D is solid black when stable and dashed black when unstable. Stable branch of oscillations between H and V is grey. **a1** The period on the oscillatory branch. **c, d** Time-traces of the activity $p(v, t)$ indicated by intensity as computed at the corresponding points C and D labelled in **b**.

The change in the value of τ_α from 100ms to 16.5s does not qualitatively affect the bifurcation diagrams shown in earlier sections, but in order for these results to be relevant it is desirable that we work with a small-amplitude additive noise as governed by k_X. The single source of noise in the model evolves with its timescale τ_X set equal to τ_α. This choice was found to have a pronounced affect on the switching dynamics without the need for large values of k_X. Note that the time units displayed in figures up to this point are milliseconds, but will be seconds in the remainder of the paper.

Figure 6 shows the bifurcation diagram for a complex input and the different types of behaviour that are observed in the operating range of the model as defined in the previous section. In the presence of the complex input, we see the same four regions found for the simple input; compare Figs. 5a and 6a. The top-right-most region of the (λ, k_α)-plane in which oscillations about $v = 0°$ are observed has grown significantly. As for the simple input, there is a weakly tuned steady-state response in the region containing the point B and there is a tuned steady-state response in the region containing the point C; see panels b and c. However, the region containing the point D now shows an altered oscillatory behaviour. We see a model response that is initially centred at the direction D but after 2–3s shifts to H and proceeds to make regular switches between H and V, see panel d. The model's separation of timescales is now seen more clearly; the model spends prolonged periods at H or V during which the adaptation builds up and eventually induces a switch to the opposite state; with $\tau_\alpha = 16.5$s switches occur every ≈ 3s, but the transition itself takes only ≈ 50ms. We note that the time between switches is shorter than the timescale of adaptation, which implies that the switches occur whilst the adaptation is dynamic, i.e. still rising (falling) for the selected (suppressed) direction. Due to the dynamics being deterministic in the absence of noise ($k_X = 0$), the switches occur at regular intervals.

As described in the previous section we fix the operating region of the model with $k_\alpha = 0.01$ and for $\lambda \in [13, 25]$ as indicated by the horizontal line in Fig. 6a. As λ is increased from $\lambda = 13$ at b to$\lambda = 25$ at d (equivalently contrast increases from $c = 0$ to $c > 0.2$) there are transitions from a weakly tuned response at point B to a tuned response at point C to an oscillatory response at point D. For $c > 0.2$ the model response saturates as shown in Fig. 5e. We now incorporate another known aspect of contrast dependence in motion processing by varying the relative weighting between 1D and 2D cues in the input. The psychophysics experiments presented in Lorenceau and Shiffrar (1992) and Lorenceau et al. (1993) show that 1D cues (contour signals) play an important role in motion perception at low contrast that diminishes with increasing contrast. As contrast increases the 2D cues (terminator signals) play a more significant role. Based on these studies we propose that for the complex model input (7) the relative weighting of 1D cues should decrease linearly with contrast

$$w_{1D} = W_0 - W_1 c, \tag{8}$$

where $W_0 = 0.5$ and $W_1 = 1.1$. These specific values were chosen in order to match the experiments; see further comments in Section 5.3.

Figure 7 shows a one-parameter bifurcation diagram for the model working in the operating regime shown in Figs. 5a and 6a but now reparametrised in terms of contrast c as described above. At low contrast there is a stable steady-state response tuned to the direction D. The peak response $\max\{p\}$ increases with contrast. This steady-state response loses stability at a travelling-wave Hopf instability H_{tw}beyond which there is a stable oscillatory branch. The dependence of w_{1D} on the contrast affects the solutions in the two following ways. Firstly, the unstable branch associated with the D direction decreases in $\max\{p\}$ at large contrasts; see dashed curve in panel a. Secondly, the period and amplitude of the oscillations in $\bar{v}\,\bar{v}$ does not saturate but continues to increase with contrast as shown in the inset a1 and panel b. We also note that close to the bifurcation point H_{tw} there are long transients before the onset of oscillations, see panel c, and that further from the bifurcation point the onset of oscillations is faster, see panel d.

COMPARISON OF THE MODEL WITH EXPERIMENTAL RESULTS

Experimental results

Figure 8 shows a summary of experimental data obtained in psychophysics experiments with 15s presentations of the complex stimulus shown in Fig. 4b (Meso et al. 2012b). Four healthy volunteers who provided their informed consent were participants, of whom two were naive to the hypothesis being tested. All experiments were carried out with, and following CNRS ethical approval. An SR Eyelink 1000 video recorder was used for the eye movement recordings and psychophysics stimuli were presented on a CRT monitor through a Mac computer using Psychtoolbox version 3.0 running stimulus generation routines written in Matlab. Positions of the right eye were recorded and continuous smooth trajectories estimated after removing blinks, saccades (very fast abrupt movements) and applying a temporal low pass filter. The presented stimuli covered 10 degrees of visual angle (the size of the side of the square in Fig. 4b) and were presented at a distance of 57cm from the monitor. Each task was done over 8 blocks of up 15 minutes over which 36 trials spanning a range of six contrasts were randomly presented each time. In this paradigm, recorded forced choice decisions indicating shifts in perceived direction through the three directions H, D and V and the estimated eye directions were found to be coupled and both indicative of perceived direction. Further details of these experiments can be found in our previous presentation (Meso et al. 2012b) and a full description will appear in the experimental counterpart of this manuscript.

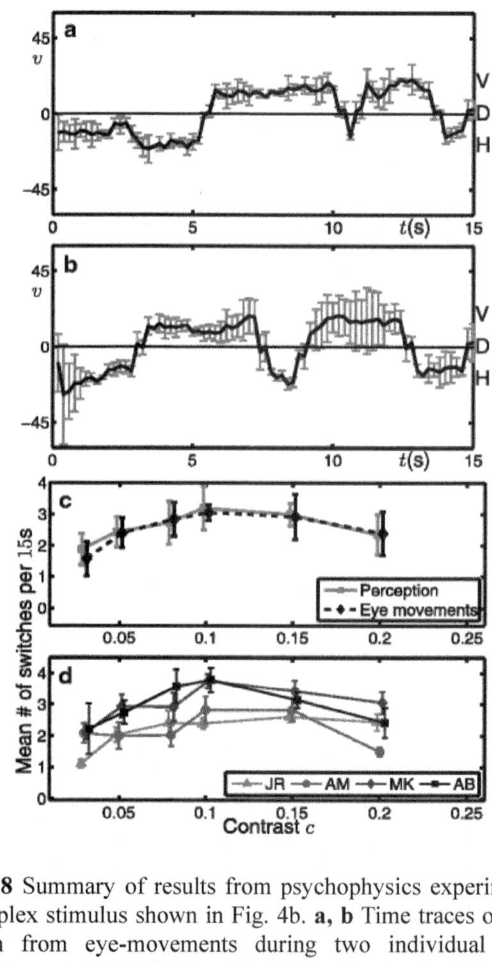

Figure. 8 Summary of results from psychophysics experiments for the complex stimulus shown in Fig. 4b. **a, b** Time traces of average direction from eye-movements during two individual stimulus presentations at $c = 0.08$. Error bars show the standard deviation of the computed direction of smooth components over 200 samples; the re-sampled value at 5Hz is the mean. **c** Relation between contrast and mean switching rate in terms of*perception* (reported by subjects) and as computed from *eye-movement* traces; grouped data is averaged across the four subjects with standard error shown. **d** Switching-rate data (from perception) separated out by subject with standard error for each subject shown

The temporal resolution of the eye traces is much higher than that of the reported transitions and allows for a relatively continuous representation of eye movement direction that can be compared with model simulations. Figure 8a and b show, for two different subjects, time traces of the time-integrated directional average of eye-movements from a single experimental trial at$c$ = 0.08. Switches in perception can be computed from these trajectories by imposing thresholds for the different percepts. Both trials show that the directions H and V are held for extended durations and regular switches occur between these two states. The switches involve sharp transitions through the diagonal direction D. The diagonal direction can be held for extended durations immediately after presentation onset. However, we note that the eye-movement direction during the first 1s of presentation has a more limited history in its temporal filtering. Short presentations of the same stimulus were investigated in a related set of experiments (Meso et al. 2012a) and modelling work (Rankin et al. 2013).

Figure 8 also shows the relationship between the averaged rate of switches between H and V over a range of contrast values $c \in \{0.03, 0.05, 0.08, 0.1, 0.15, 0.2\}$; in panel c the data is averaged across the four subjects and in panel d it is separated out by subject. The lowest contrast shown $c = 0.03$ corresponds to the smallest contrast value for which subjects were able to reliably report a direction of motion for the stimulus. For the grouped data, at low contrast ($c < 0.1$) the rate of switching increases with contrast with the rate being maximal at approximately $c \approx 0.1$. Beyond the peak, for contrasts $c >$ 0.1, the rate of switching decreases with contrast. For the data separated by subject shown in panel d, the subjects MK and AB have a peak rate around 3.5 switches per 15s presentation and the peak occurs at $c \approx 0.1$. For subjects JR and AM the peak rate is lower at around 2.5 switches per 15s presentation and there is a less prominent peak occurring at a higher contrast value $c > 0.1$. However, the common pattern reveals two qualitatively different regimes with respect to changing contrast. A low contrast regime for which the switching rate increases with contrast and a high contrast regime for which the switching rate decreases with contrast.

Model simulations with noise ($k_X = 0.0025$)

We now study the dynamics of the model in the presence of additive noise in the main neural field equation. Recall that the stochastic process in the model is operating on the same slow timescale τ_α as the adaptation and that the strength of the noise is $k_X = 0.0025$. Two cases will be studied, first the low contrast case at $c = 0.04$, close to the contrast threshold on the steep part of the model's contrast response; see Fig. 5e. Second, the high contrast case at $c = 0.08$, which is above the contrast threshold on the saturated part of the contrast response function. In the first case, noise is introduced in a parameter regime where the model is close to bifurcation and oscillations only occur after a long transient, see Fig. 7c. When operating in a nearby parameter regime close to bifurcation the noise causes random deviations away from the direction D and can drive the model into an oscillatory state more quickly. In the second case, noise is introduced in a parameter regime where the model produces an oscillatory response with a short transient behaviour, see Fig. 7d. In this regime the noise perturbs the regular oscillations either shortening or prolonging the time spent close to H and V.

Figure 9a–d shows 15s time traces of the population activity p for the cases $c = 0.04$ (first row) and $c = 0.08$ (second row). Note that each individual model simulation is quite different due to the noise, but we have selected representative examples that allow us to highlight key features in the model responses and compare the different contrast cases. In processing this simulated data we observe that the activity is initially centred around the direction D. After some transient period switching occurs primarily between H and V. In order to detect switches between the directions H and V a so-called perceptual threshold (PT) has been set at $v = \pm 10°$. The first switch from D to either H or V is detected the when the corresponding threshold is crossed for first time. Subsequent switches are only detected the next time the opposite threshold is crossed. Note that although other algorithms could be employed to detect these switches, we found that these do not have a great effect on the presented results.

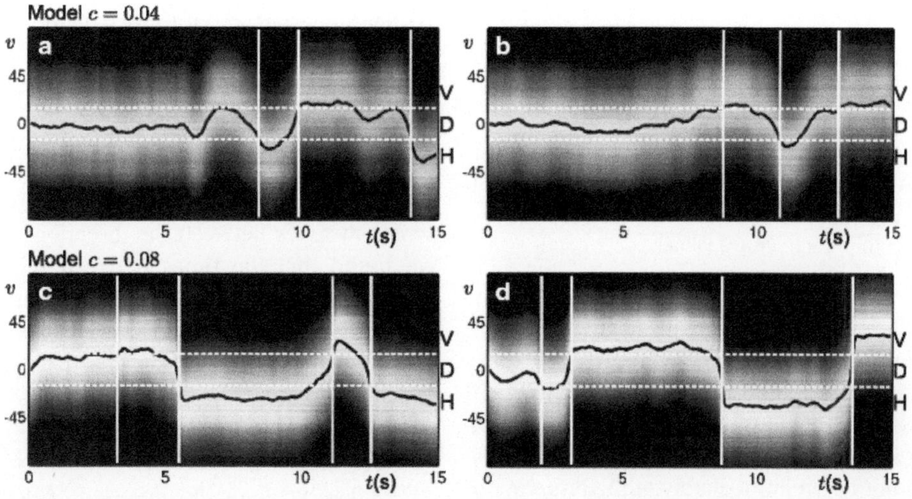

Fig. 9 Time traces from individual model simulations where intensity shows the population activity across direction space (vertical axis). The solid black line is the average of this activity (average direction $\bar{v}$) and the dashed lines indicate perception thresholds (PT) for detection of switches between the directions H and V; switches are indicated by vertical white lines. First and second rows shows examples from the low and high contrast cases, respectively

Across all the examples shown in Fig. 9, the average direction $\bar{v}$ oscillates in a random fashion and as time progresses the amplitude of these oscillations grows in v. For the case $c = 0.04$ there is a long transient and the first switch occurs for approximately $t \in [5s, 10s]$. For the case $c= 0.08$ the overall amplitude of the oscillations is larger and the first switch occurs for $t < 3s$. Note also that the level of activity shown as an intensity in Fig. 9 is higher in the $c = 0.08$ case. An important difference between the two contrast cases is that in the low contrast case, the transitions between H and V occur gradually when compared with the abrupt transitions in the high contrast case. This suggests that at low contrast the direction D could be seen during the transitions, where as in the high contrast case the switches occur directly from H to V.

With respect to the experimental data, the model consistently reproduces the characteristic behaviour of regular switches between the H and V. Furthermore, the sharp transitions through the diagonal direction D are also captured well by the model. Compare the second row of Fig. 9with the two examples shown in Fig. 8a and b.

Dependence of switching rate on contrast

Figure 10 shows the relationship between contrast and switching rate as computed with the model where the rate is expressed as the mean number of switches per 15s simulation. Panel a shows the relationship without noise ($k_X = 0$) and with noise ($k_X = 0.0025$). We show the average switching rate at discrete contrasts $c \in [0.02, 0.25]$ and at each contrast value we plot the switching rate averaged across 500 model simulations.

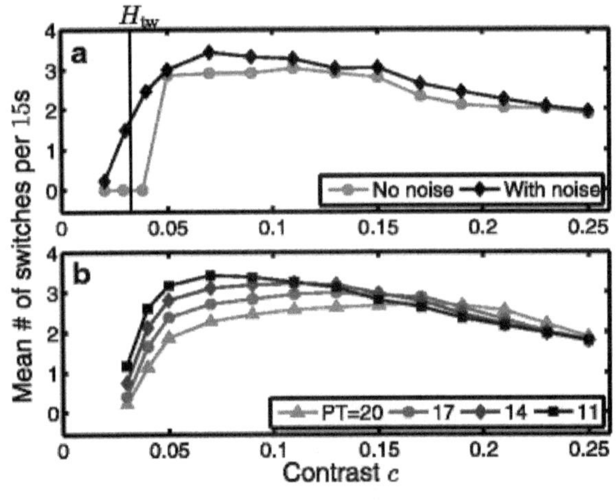

Figure. 10 Mean switching rates computed with the model and recorded from psychophysics experiments. **a** Switching rates computed with the model without noise and with noise. **b**Switching rate curves computed with the model for a range of *PT* values

The deterministic (no noise) case can be explained in terms of the bifurcation diagram shown in Fig. 7. At low contrast, no switching behaviour is observed as the model can only produce a steady-state response weakly tuned to the direction D. With increasing contrast, the onset of switching is abrupt, occurring just above $c = 0.04$ after the bifurcation H_{tw} at $c \approx 0.03$. Switching does not begin immediately at the bifurcation point, due to long transients for values of c nearby, see Fig. 7c. The switching rate remains constant at around 3 switches per 15s interval, and starts to drop off for contrasts $c > 0.12$. The reduction in switching rate for larger contrasts is due to the increasing period of the oscillations as shown in Fig. 7a1.

The introduction of noise with fixed intensity across all contrasts leads to an increase in average switching rate, see Fig. 10a. Here we only consider the average rate, but the distribution of the switching times will be discussed in Section 5.4. At larger contrasts, far from the bifurcation H_{tw}, the switches are primarily governed by an underlying adaptation driven oscillation and the increase in switching rate is minimal. At lower contrasts, the shortest period of the oscillations is predicted to be immediately after the bifurcation; see Fig. 7a1. However, close to bifurcation, the theory also predicts that a long transient behaviour will be observed before the onset of these fast switches. The transient behaviour is important here, due to the fact that we consider short 15s simulations. At very low contrasts, before H_{tw}, only the noise can drive a deviation from the diagonal direction leading to a switch. However, just after the bifurcation H_{tw}, the main contribution to the increased switching rate is noise shortening the transient period before the onset of adaptation-driven switching. Therefore, the bifurcation analysis provides an explanation for the fact that the peak in switching rate occurs shortly after the bifurcation where transients are curtailed by the noise and the period is shortest.

The model results with noise are able to accurately capture the two contrast regimes from the experimental data: an increase in switching rate at low contrasts and subsequent decrease in switching rate at higher contrasts, compare Fig. 10a black curve with Fig. 8a. The values of W_0,W_1 and τ_α were chosen in order to fit the experimental data, however, the two contrast regimes are robustly produced by the model independent of the specific values chosen. In Fig. 10(b) we show how, in the model, the relationship between switching rate and contrast changes with respect to *PT*. When *PT* is low the peak switching rate is highest and occurs at a low contrast value. As *PT* is increased, the peak rate decreases and also occurs at a higher contrast value; the relationship also appears to flatten out for larger *PT*. Figure 8b shows the reported switching rate curves from the experiments, separated out by individual subject. The data shows a range of peak switching rate between the subjects. For the two subjects with the highest switching rate (MK,AB), the prominent peak occurs at $c \approx 0.1$. For the other two subjects (JR,AM), the peak rate is lower, the response is flatter and the peak rate occurs at a larger value of c. We conclude that differences in perceptual threshold between subjects can account for inter-subject differences.

Distribution of switching times

In the previous section, we showed example model outputs for which switches between the directions H and V are detected. We found that the times between these switches vary and that, particularly in the low contrast case, the early transient behaviour can be very different from one simulation to the next. In order to investigate the distribution of the switching times we ran 1, 500 model simulations each of 15s and formed a data set by extracting the times between consecutive switches from each simulation.

Figure 11 shows histograms of the computed switching times t_{sw}. In the low contrast case approximately 1, 483 switches were recorded with mean time $\bar{t}_{sw} = 3.73s$ and SD = 2.89 (Coefficient of Variance COV = 0.56) and in the high contrast case 3, 154 switches were reported with mean time

$\bar{t}_{sw} = 4.07s$ and SD = 2.08 (COV = 0.51). Although the mean of t_{sw} is smaller in the high contrast case, more switches are detected because there is a shorter transient period before switching begins; the average time to the first switch in the low contrast case is 7.47s (SD = 2.87) compared with an average time of 3.02s (SD = 1.56) in the high contrast case. The aim now is to determine from which distribution the model data could have arisen. We follow the method presented in (Shpiro et al. 2009) and compare the model data with a Weibull probability distribution function (pdf), a gamma pdf and a log-normal pdf each with parameters chosen using a standard maximum likelihood estimate. By inspection, it appears that the data in the low contrast case are well fitted by a log-normal distribution and that the data in the high contrast case are well fitted by a gamma distribution. In order to confirm this we perform a Kolmogorov-Smirnov goodness-of-fit test. In the low contrast case the log-normal distribution provides best fit ($P = 0.087$) but the gamma and Weibull distributions can be rejected at the 5%significance level. In the high contrast case the gamma distribution provides best fit ($P = 0.635$) and both the log-normal and Weibull distributions can be rejected.

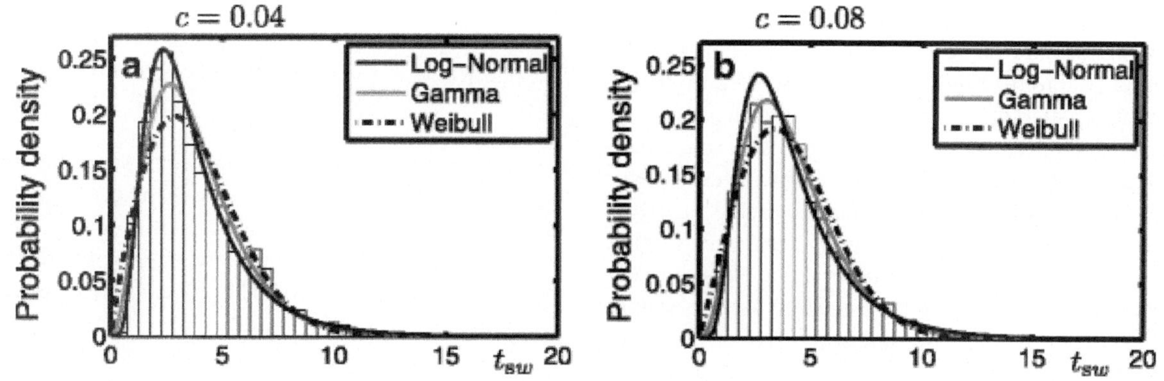

Figure. 11 Distribution of perceptual switching times and test distributions. Histograms show the distribution of switching times as computed from model simulations with $PT = 15°$ (see text for details). Candidate distributions are overlaid, where the shape and scale parameters have been chosen to best fit the model data. **a** Low contrast case in which the log-normal distribution provided a better fit. **b** High contrast case in which the gamma distribution provided a better fit

Clearly, studying only the mean and standard deviation for the two different contrast cases does not reveal a significant difference. However, we do find a change in the underlying distributions governing the switching times, which is indicative of a change in the dominant mechanism driving the switching. Typically switching behaviour that is driven by adaptation over noise will have a lower peak that occurs later and a smaller spread with shorter tail as characterised by the gamma distribution (Shpiro et al. 2009). However, when noise plays a more significant role, the peak is higher, earlier and the tail longer as characterised by the log-normal distribution. We also highlight the fact that the first switch occurs much earlier in the high-contrast case, this prediction could easily be tested experimentally.

DISCUSSION

Spatially extended neural fields models with a linear implementation of spike-frequency adaptation have been studied both in ring models (Hansel and Sompolinsky 1998; Curtu and Ermentrout 2004; Kilpatrick and Ermentrout 2012; Ermentrout et al. 2012) and infinite spatial domains (Pinto and Ermentrout 2001; Ermentrout et al. 2012). A simpler version of the model presented in this article, without an input or noise, was studied in Curtu and Ermentrout (2004). The existence of parameter regions with homogeneous, stationary tuned, travelling-wave and standing-wave responses was shown. In the absence of an input, these various different solution types are known to exist close to a so-called Bogdanov-Takens (BT) point in parameter space, which acts as a parametric organising centre for different types of dynamics. In the presence of simple inputs it has been shown that new solution types can be produced such as breathers and pinned travelling-wave solutions (Hansel and Sompolinsky 1998; Ermentrout et al. 2012). In this article we presented an in-depth numerical study of the complex organisation of these various solution types in three-dimensional parameter space. We were able to show that much of the structure local to the BT point is preserved with the introduction of a small, simple input, albeit in a subtly modified form. We gave an account of the changes that occur in terms of the complicated series of bifurcations that delineate regions of parameter space exhibiting qualitatively different dynamics. It was found that close to a travelling-wave-type Hopf bifurcation, solutions are pinned to the input. Furthermore, it was shown that for a standing-wave-type Hopf bifurcation giving rise to unstable solutions with no input, a region in parameter space with stable standing waves solutions was opened up when an input was introduced. Although great progress has been made analytically in the study of this class of model for simple inputs where a single location in feature space is stimulated (Hansel and Sompolinsky 1998; Ermentrout et al. 2012), the question of more complex inputs with stimulation of multiple locations provides a challenge. The advantage of the numerical approach used here is that we can directly extend earlier results when a complex input is introduced. In a recent study, perceptual multistability has been investigated in a model with synaptic depression and a two-location stimulus in a continuous feature space (Kilpatrick2012).

We subsequently investigated perceptual multistability for a stimulus that is multistable in terms of its perceived direction of motion and that has been the subject of recent psychophysical experiments, of which a summary was presented (see Meso et al. (2012b)). This specific application allows for the model's continuous feature space to be exploited; it allows for a truly dynamic consideration of perceived direction, which unlike binocular rivalry or ambiguous shapes, is known to be neurally represented on a continuous scale. We study the multistable barber pole, which consists of a diagonally drifting grating viewed through a square aperture. The stimulus is known to be multistable between the diagonal grating direction D and the horizontal H and vertical V aperture-edge directions. The characteristic perceptual response is dominated by D immediately after onset followed by regular switches between horizontal H and vertical V directions. In the model the complex multistable barberpole can be represented by three-bumps in the feature space of motion direction based on experimental insights about the different ways in which 1D and 2D motion cues are processed by the visual system. The simple input is used to tune model parameters and introduce a contrast parameter such that its behaviour matches the known contrast response properties from (Sclar et al. 1990). It is found that for a fixed adaptation strength, we are able to select a range of the nonlinear slope parameter such that the model's activity response can be matched to the qualitative and quantitative behaviour close to contrast threshold observed in physiological experiments. Once appropriately parametrised for the simple input, we find that for a complex input the model produces behaviour that is consistent with the characteristic perceptual responses described above.

We further investigated the relationship between contrast and the switching behaviour; in the experiments two different regimes were identified for the first time, at low contrast the switching rate increases with contrast and at higher contrasts the rate decreases with contrast. In the model we study a low contrast regime operating close to bifurcation and a high contrast regime above the contrast threshold. For both regimes we find common features in the switching behaviour produced by the model. Initially the percept D is dominant, but after some delay there is a shift to either H or V, typically within the first 1 − 8s (this behaviour is consistent with existing studies (Castet et al. 1999; Fisher and Zanker 2001; Rankin et al.2013)), after which regular switching occurs between H and V. In the high contrast case this regular switching starts earlier, which we would expect as the 2D cues associated with the aperture edges and H/V directions should be stronger with increased contrast. We also find that typically the transitions between H and V are relatively smooth, passing gradually through the direction D in the low contrast case when compared with the sharper transitions in the high contrast case. Furthermore, by studying the dynamics either with or without noise, we find that at high contrasts the mean rate of switching is governed by the adaptation-driven oscillations. Although the noise produces random deviations in these switching times, the mean rate is unaffected. However, at low contrast, where the model is operating close to bifurcation the noise has a larger effect on the dynamics. At low contrast the increasing regime is associated with noise curtailing transient behaviour driving the model into an oscillatory regime. We further quantified the difference between the two contrast regimes by showing that the switching times are best fitted by a log-normal distribution in the low contrast case and by a gamma distribution in the high contrast case.

CONCLUSIONS

In classical rivalry models competing states are modelled as discrete populations (Laing and Chow 2002; Shpiro et al. 2007; Moreno-Bote et al. 2007; Shpiro et al. 2009). The neural fields model at the core of this study has a continuous feature space, which allows multistability to be investigated in a motion integration problem where the different percepts are represented on a continuous scale. The minimal model incorporating spike frequency adaptation, additive noise and an input representing the multistable barberpole can capture characteristics of the switching observed in experiments: extended periods spent at the stimulated directions associated with different percepts and rapid switches between them. The bifurcation analysis allows for these dynamics to be related back to the travelling-wave-type solutions that are first pinned by a simple input and further modulated by the complex input in order to produce the desired behaviour. The bifurcation analysis predicts a change in the mechanisms driving switches between a low contrast and a high contrast regime characterised respectively by increasing and decreasing switching rates in the psychophysical experiments. The switches are driven primarily by noise at low contrast and adaptation at high contrast. The peak switching rate is predicted by the bifurcation analysis to occur just after bifurcation where the fastest adaptation-driven dynamics are reached after transients that are shortened by noise; in effect, when there is a balance between adaptation and noise.

The general approach applied to a neural field model in this paper — making use of bifurcation methods for tuning parameters such that the model operates close to bifurcation whilst simultaneously matching known response properties from physiological studies — will allow for much broader studies of multistable perception. In particular, extensions to models that consider physical space in conjunction with an abstracted feature space would allow for the particular spatial properties of multistable visual stimuli to be investigated. For example, more detailed neural fields models taking into account the spatial integration of motion stimuli such as (Tlapale et al. 2011) could be used to, within a single model architecture, investigate the underlying motion integration mechanisms that yield multistable perception across a broad range of stimuli, e.g., barberpoles, plaids, moving diamonds. Multistability has not been investigated in this kind of model to which the methods of numerical continuation and bifurcation analysis would be most applicable.

ACKNOWLEDGMENTS

The authors thank Karin Pilz for her helpful comments on the manuscript. JR, OF and PK acknowledge funding from ERC NERVI (grant no. 227747), from the EU FET-Proactive FP7 BrainScaleS (grant no. 269921) and from the EC IP project FP7 Mathemacs (grant no. 318723). GM acknowledge funding from the EU FET-Proactive FP7 BrainScaleS (grant no. 269921). AM was supported by the Aix-Marseille Université Foreign Post-doctoral Fellowship 2011 from 05/1104/12.

APPENDIX

A THE ORNSTEIN-UHLENBECK PROCESS

The noise $X(v, t)$ that appears in Eq. (1) is a classical Ornstein-Uhlenbeck process, see e.g. (Ermentrout and Terman 2010), that is described by the following stochastic differential equation

$$\tau_\alpha \mathrm{d}X(v,t) = -X(v,t)\mathrm{d}t + \sigma \mathrm{d}W(v,t), \tag{9}$$

where $W(v, t)$ is a feature uncorrelated Wiener or Brownian process. Note that the timescale τ_α is the same as the one for the adaptation α in Eq. (2).

The solution to Eq. (9) is readily found to be

$$X(v,t) = e^{-t/\tau_\alpha} X_0(v) + \sigma \int_0^t e^{-\frac{t-s}{\tau_\alpha}}\, dW(v,s), \quad t \geq 0$$

where $X_0(v)$ is the initial noise distribution, assumed to be independent of the Brownian. Its mean is therefore given by

$$\langle X(v,t)\rangle = \langle X_0(v)\rangle e^{-t/\tau_\alpha},$$

and its variance by

$$\mathrm{Var}(X(v,t)) = \mathrm{Var}(X_0(v))e^{-2t/\tau_\alpha} + \frac{\tau_\alpha \sigma^2}{2}(1 - e^{-2t/\tau_\alpha})$$

It is seen that as soon as t becomes larger than the timescale τ_α the mean becomes very close to 0 (it is even exactly equal to 0 for all times if the mean of the initial value is equal to 0), and the variance becomes very close to τασ22τασ22. By choosing $\sigma^2 = 2 / \tau_\alpha$ we ensure that $\mathrm{Var}(X(v, t))$ is close to 1 as soon as t becomes larger than the timescale τ_α.

B EXPLANATORY ONE-PARAMETER BIFURCATION DIAGRAMS

Figure 12 shows one-parameter bifurcation diagrams with zero adaptation gain $k_\alpha = 0$, first with no input in panel a and with a small simple input ($k_I = 0.001$) in panel c. In order to best represent the solution branches we plot them in terms of the even, first-order mode of the solutions p1^p1^ (the cos(v)-component). Figure 12a shows that for small λ there is a single, stable solution branch with p1^=0p1^=0; this corresponds to the flat (untuned) response shown in Fig. 2d. When λ is increased beyond the pitchfork P_0, this flat state loses its stability and a ring of tuned responses are created. Figure 12b shows the profiles of the different solutions that exist at $\lambda = 23$. The dashed curve B2 is the unstable flat state, the tuned state B1 centred at $v = 0$ corresponds to when p1^p1^ is largest and the tuned state B3 centred at $v = \pm$ 180 corresponds to when p1^p1^ is smallest. Due to the presence of translational symmetry, intermediate states centred at any value of v also exist; discrete examples of these are shown as grey curves, but note these exist on a continuous ring filling in the direction space v. The easiest way to see the effect of introducing the stimulus and the fact that this breaks the translational symmetry is by studying the states that exist in the small input case also at $\lambda = 23$ shown in Fig. 12d. Now the only stable solution is the tuned response D1 centred at $v = 0$, there is a counterpart unstable solution centred at $v = \pm$ 180 and all of the intermediate states have been destroyed. In the bifurcation diagram Fig. 12c the pitchfork bifurcation has been destroyed and there remain two disconnected solution branches. On the unstable branch the unstable solutions D2 and D3 are connected at a fold point F_0; this bifurcation is traced out as F in Fig. 3a. On the (upper) stable branch there is a smooth transition with increasing λ from a weakly- to a highly-tuned response centred at the stimulated direction. It is useful to detect where the increase in p1^p1^ is steepest as this signifies the transition to a tuned response. We denote this point P^0P^0 and as this is not strictly a bifurcation point we call it a pseudo pitchfork; it is still possible to trace out where this transition occurs in the (λ, k_α)-plane and this is plotted as P^P^ in Fig. 3a.

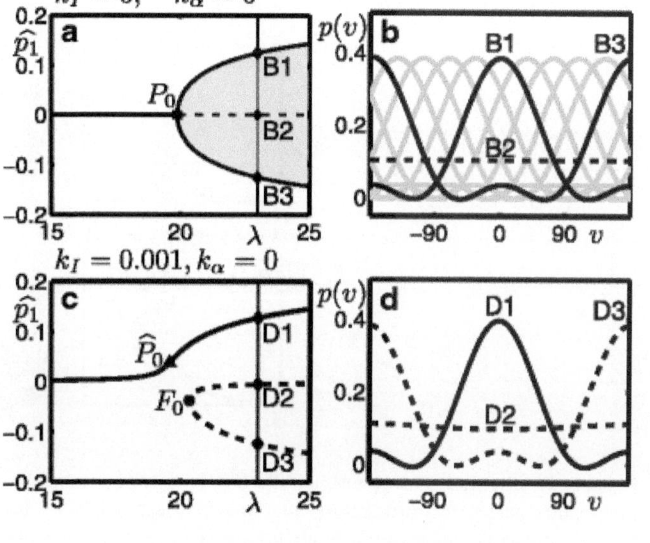

Figure. 12 Symmetry breaking of the pitchfork with introduction of a stimulus. **a** and **c** show bifurcation diagrams in λ for the no-input

and small input cases, respectively; stable states are solid curves and unstable states are dashed curves. **b** and **d** show the solution profiles in v-space at the labelled points for λ = 23; stable states are solid curves and unstable states are dashed curves. In panel **b** the solid black curves correspond to the solution B1, for which p^1p^1 takes its largest value and the solution B3, for which p^1p^1 takes its smallest value; several intermediate solutions are plotted as grey curves, see text

Figure 13 shows one-parameter bifurcation diagrams in λ for three different cases:

- No input cases with k_α = 0.03; see Fig. 13a–c; corresponds to the horizontal line through the point labelled B in Fig. 2a.
- First small, simple input case with k_α = 0.03; see Fig. 13d–f; corresponds to the horizontal line through the point labelled B in Fig. 3a.
- Second small, simple input case with k_α = 0.01; see Fig. 13g–i; corresponds to the horizontal line through the point labelled C slice in Fig. 3a.

In order to best represent the solution branches we plot them in terms of the maximum of the sum of the even and odd first-order mode of the solutions max{p1^+p2^} max{p1^+p2^} (the cos(v) and sin(v)-components). Figure 13a shows that two solution branches bifurcate simultaneously off the trivial branch at the twice-labelled point $H_{tw1}H_{sw}$. In panel b we show one period of a stable travelling wave solution from the branch corresponding to H_{tw1}; the wave can take either positive or negative (shown) direction in v. In panel c we show one period of an unstable standing wave solution from the branch corresponding to H_{sw1}; the standing wave oscillates such that 180°-out-of-phase (in v) maxima form alternatively. The phase in v of the entire waveform is arbitrary and, as for the pitchfork bifurcation, any translation of the whole waveform in v is also a solution. With the introduction of a stimulus, as for the pitchfork bifurcation discussed earlier, the translational symmetry of the solutions is broken resulting in changes to the solution structure. The bifurcation point H_{sw} shown in Fig. 13a splits into two bifurcation points H_{sw1} and H_{sw2} in panel d and solution profiles on the respective bifurcating branches are shown in panels e and f. On the first branch, which is initially stable close to H_{sw1}, one of the maxima is centred on the stimulated direction v = 0°. For the second unstable branch bifurcating from H_{sw2} the maxima are out of phase with the stimulated direction. The travelling wave branch is now a secondary bifurcation from the branch originating at H_{sw1} and the solutions now have a small modulation when the travelling wave passes over v = 0° (similar to the solution shown in panel i). In panel g, at a lower value of k_α, the steady-state solution branch is initially weakly tuned and forms a highly-tuned solution after P^0P^0. The tuned response becomes unstable at H_{tw2}; close to this bifurcation point the solution is pinned by the stimulated direction, as was shown in Fig. 3c, and as λ is increased further the amplitude in v of these oscillations about the stimulated direction increases, see panel Fig. 13h. When the point*PS* is reached in panel h the oscillations become large enough such that there is a phase slip and beyond this point we obtain a standard travelling wave solution once more, see panel i. Note that the travelling wave is still modulated as it passes over the stimulated direction. We reiterate the qualitative difference between the branches forming at H_{tw1} and H_{tw2}: for branches of travelling wave forming directly from an untuned or weakly tuned steady state as at H_{tw1} the solutions cannot be pinned to a stimulated direction, however, for branches of travelling wave forming directly from a tuned steady state as at H_{tw2} the solutions are pinned to a stimulated direction close to the bifurcation. Furthermore, we note that, with the introduction of a stimulus a region of stable standing wave solutions, in phase with the stimulated direction, are introduced between H_{sw1} and H_{tw1}; see branch segment through the point E in Fig. 13d.

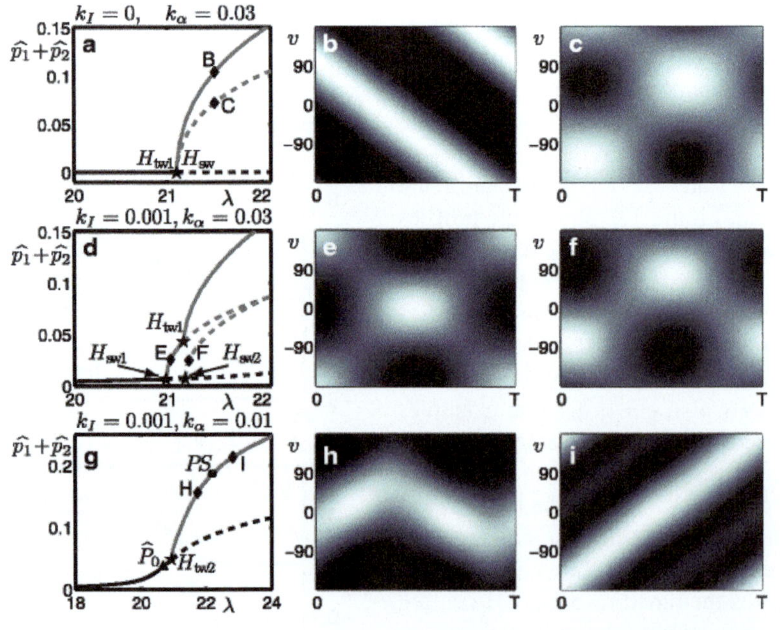

Figure. 13 Changes to standing- and travelling-wave branches born in Hopf bifurcations with introduction of the stimulus. The first column shows one-parameter bifurcation diagrams in λ where black curves are steady-state branches and grey periodic branches; stable solution branches are solid and unstable branches are dashed. Hopf bifurcations to travelling waves are H_{tw1} and H_{tw2}, and to standing waves are H_{sw}, H_{sw1}, and H_{sw2}. Second and third columns show one period T of the solutions at corresponding points on solution branches from one-parameter diagrams

C TUNING THE MODEL'S CONTRAST RESPONSE

Here we describe the way in which the model is reparametrised such that its response matches known physiological behaviour. In (Sclar et al. 1990) contrast response functions were computed across several stages of the visual pathway for a drifting sinusoidal grating. At cortical layers, individual cell recordings were made with the stimulus moving in the cell's preferred direction. The Naka-Rushton function was used to fit contrast response data and the response *R*as a function of contrast *c* is given by

$$R(c) = R_{\max}\frac{c^n}{c^n + c_{50}^n} + M, \tag{10}$$

where R_{max} is the maximal response cn50c50n is the contrast at which the response reaches half its maximum value and *n* is the exponent that determines the steepness of the curve. For visual area MT in the macaque these parameters were estimated to take the following values: R_{max} = 36, cn50=0.07c50n=0.07 and n = 3; the Naka-Rushton function is plotted for these values in Fig. 14f (dashed curve). Note that it is also necessary to adjust for the spontaneous firing rate *M*, which typically takes a value $M \in [0.05, 0.3]$.

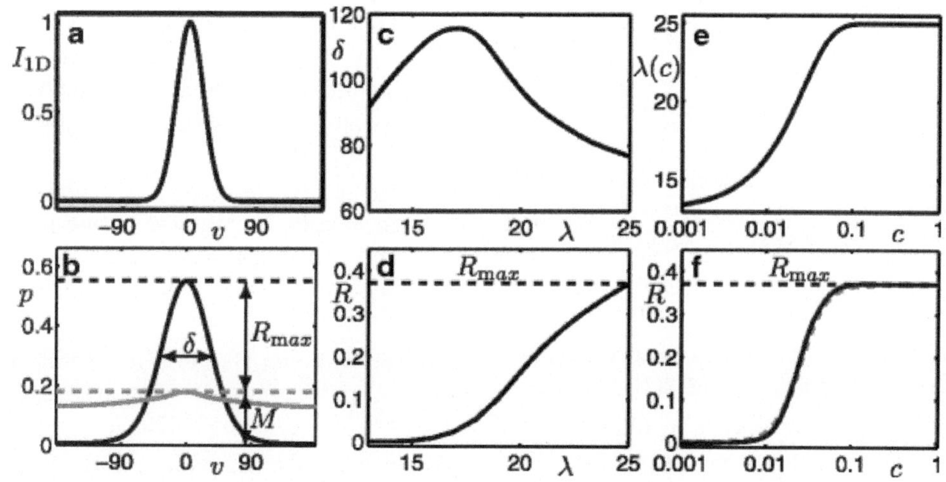

Figure. 14 Model response for drifting grating stimulus and tuning of λ-dependence on c. **a** Stimulus profile $I_{1D}(v)$ for a visual-field-covering drifting grating. **b** Steady state responses at λ =13 (grey curve) and at λ =25 (black curve); dashed horizontal lines indicate the maximum activity for these steady states. The tuning bandwidth, or tuning width at half-height, is $\delta$and the maximum level of activity above the spontaneous level (dashed grey line) is *R*. **c**Tuning width δ of response for $\lambda \in [\lambda_{min}, \lambda_{max}]$. **d** Normalised response *R* for $\lambda \in [\lambda_{min},\lambda_{max}]$. **e** Variation of λ with respect to c. **f** Contrast response curve with λ varying as in panel **e** (*black*) and the Naka-Rushton function with parameters as described in the text (*dashed grey*)

We wish to reproduce the known contrast response properties for a drifting grating stimulus represented by the input in Fig. 14a. For very low contrasts $c < 0.01$, the model should respond to a stimulus with a low level of activity that is weakly tuned or not tuned. Increasing c should result in a change in the model's response that is consistent with crossing the contrast threshold:

- an increasing level of activity consistent with the Naka-Rusthon function Eq. (10),
- a response that is tuned to the model's input with an appropriate tuning width (consistent with values reported in (Albright 1984; Diogo et al. 2003)).

We find that, for $k_\alpha = 0.01$ and $\lambda = 13$, the model's weakly-tuned, steady state response has an appropriate level activity that we take as the value for $M = 0.18$; see the lower curve in Fig. 14b. Furthermore, changing to $\lambda = 25$ we find that the model produces a tuned response that is consistent with a response above contrast threshold $c > 0.2$ with a maximum level of activity that agrees with the value of $R_{max} = 0.52 - M = 0.36$; see upper curve in Fig. 14b. We also find that varying λ in the range $\lambda \in [13, 25]$ that the tuning widths at half-height δ fall in the range $\delta \in [80°, 115°]$ (see Fig. 14c), which is consistent with the tuning widths obtained by individually probing a population neurons in the studies (Albright 1984) (mean 90°, standard deviation SD = 29°) and (Diogo et al. 2003) (mean 104°, SD = 35°). Figure 14d shows the model's response *R*plotted against $\lambda \in [13, 25]$ and we see that initially, close to $\lambda_{min} = 13$, the response has an appropriate shape, but for larger λ the response does not saturate as required. It is therefore necessary to introduce a function $\lambda(c)$ that saturates smoothly at the value $\lambda_{max} = 25$. We reparametrise the model so that λ depends on the contrast using a function that increases linearly from $\lambda = \lambda_{min}$ and saturates for large c at $\lambda = \lambda_{max}$. We effectively use the top half of a sigmoid function $S(x) = 1/(1 + \exp(-x))$. After fixing $\lambda_{min} = 13$ and $\lambda_{max} = 25$, it suffices to vary just the slope parameter μ in the following equation in order to find a good fit. The equation

$$\lambda(c) = \lambda_{\min} + (\lambda_{\max} - \lambda_{\min})\left(S(\mu c) - \frac{1}{2}\right), \tag{11}$$

is plotted in Fig. 14e with $\mu = 60$. By construction, we have set dependence of λ on c in order to obtain a function of the model response *R* in terms of *c* that matches a Naka-Rushton function appropriately parametrised for MT; see Fig. 14f (solid curve).

REFRENCES

1. Albright, T.D. (1984). Direction and orientation selectivity of neurons in visual area MT of the macaque. Journal of Neurophysiology, 52(6), 1106–1030.
2. Amari, S. (1971). Characteristics of randomly connected threshold element networks and neural systems. Proceedings of the IEEE, 59, 35–47.
3. Ashwin, P., & Lavric, A. (2010). A low-dimensional model of binocular rivalry using winnerless competition. Physica D: Nonlinear Phenomena, 239(9), 529–536.
4. Barthélemy, F., Perrinet, L., Castet, E., Masson, G. (2008). Dynamics of distributed 1D and 2D motion representations for short-latency ocular following. Vision Research, 48(4), 501–522.
5. Barthélemy, F., Fleuriet, J., Masson, G. (2010). Temporal dynamics of 2D motion integration for ocular following in macaque monkeys. Journal of Neurophysiology, 103(3), 1275–1282.
6. Ben-Yishai, R., Bar-Or, R., Sompolinsky, H. (1995). Theory of orientation tuning in visual cortex. Proceedings of the National Academy of Sciences, 92(9), 3844–3848.
7. Blake, R. (1989). A neural theory of binocular rivalry. Psychological Review, 96(1), 145.
8. Blake, R. (2001). A primer on binocular rivalry, including current controversies. Brain and Mind, 2(1), 5–38.
9. Borisyuk, R., Chik, D., Kazanovich, Y. (2009). Visual perception of ambiguous figures: synchronization based neural models. Biological Cybernetics, 100(6), 491–504.
10. Born, R., & Bradley, D. (2005). Structure and function of visual area MT. Annual Review of Neuroscience, 28, 157–189.
11. Bressloff, P. (2012). Spatiotemporal dynamics of continuum neural fields. Journal of Physics A: Mathematical and Theoretical, 45(3), 033001.
12. Bressloff, P., & Webber, M. (2011). Neural field model of binocular rivalry waves. Journal of Computational Neuroscience, 32(2), 233–252.
13. Britten, K.H. (2003). The middle temporal area: motion processing and the link to perception. The Visual Neurosciences, 2, 1203–1216.
14. Carandini, M., & Heeger, D. (2011). Normalization as a canonical neural computation. Nature Reviews Neuroscience, 13(1), 51–62.
15. Castet, E., Charton, V., Dufour, A. (1999). The extrinsic/intrinsic classification of two-dimensional motion signals with barber-pole stimuli. Vision Research, 39(5), 915–932.
16. Coombes, S. (2005). Waves, bumps, and patterns in neural fields theories. Biological Cybernetics, 93(2), 91–108.
17. Curtu, R., & Ermentrout, B. (2004). Pattern formation in a network of excitatory and inhibitory cells with adaptation. SIAM Journal on Applied Dynamical Systems, 3, 191.
18. Curtu, R., Shpiro, A., Rubin, N., Rinzel, J. (2008). Mechanisms for frequency control in neuronal competition models. SIAM Journal on Applied Dynamical Systems, 7(2), 609.
19. Dangelmayr, G., & Knobloch, E. (1987). The Takens–Bogdanov bifurcation with O(2)-Symmetry. Philosophical transactions for the royal society of london series A. Mathematical and Physical Sciences, 322(1565), 243–279.
20. Deco, G., & Roland, P. (2010). The role of multi-area interactions for the computation of apparent motion. NeuroImage, 51(3), 1018–1026.
21. Descalzo, V., Nowak, L., Brumberg, J., McCormick, D., Sanchez-Vives, M. (2005). Slow adaptation in fast-spiking neurons of visual cortex. Journal of Neurophysiology, 93(2), 1111–1118.
22. Diogo, A., Soares, J., Koulakov, A., Albright, T., Gattass, R. (2003). Electrophysiological imaging of functional architecture in the cortical middle temporal visual area of cebus apella monkey. The Journal of Neuroscience, 23(9), 3881.
23. Doedel, E., Champneys, A., Fairgrieve, T., Kuznetsov, Y., Sandstede, B., Wang, X. (1997). Auto 97: continuation and bifurcation software for ordinary differential equations (with homcont). http://cmvl.cs.concordia.ca/auto/.
24. Ermentrout, B. (1998). Neural networks as spatio-temporal pattern-forming systems. Reports on Progress in Physics, 61, 353–430.
25. Ermentrout, B., Folias, S.E., Kilpatrick, Z.P. (2012). Spatiotemporal pattern formation in neural fields with linear adaptation submitted.
26. Ermentrout, G.B., & Terman, D. (2010). Foundations of mathematical neuroscience. Interdisciplinary Applied Mathematics: Springer.
27. Faye, G., Rankin, J., Lloyd, D.J.B. (2013). Localized radial bumps of a neural field equation on the Euclidean plane and the Poincaré disk. Nonlinearity, 26, 437–478.
28. Fisher, N., & Zanker, J.M. (2001). The directional tuning of the barber-pole illusion. PERCEPTION-LONDON, 30(11), 1321–1336.
29. Folias, S. (2011). Nonlinear analysis of breathing pulses in a synaptically coupled neural network. SIAM Journal on Applied Dynamical Systems, 10(2), 744–787.
30. Freeman, A. (2005). Multistage model for binocular rivalry. Journal of Neurophysiology, 94(6), 4412–4420.
31. Giese, M. (1998). Dynamic neural field theory for motion perception. Springer.
32. Hansel, D., & Sompolinsky, H. (1998). Modeling feature selectivity in local cortical circuits. In Methods in neuronal modeling: from ions to networks (Chap 13, pp. 499–567). Cambridge: MIT Press.
33. Haragus, M., & Iooss, G. (2010). Local bifurcations, center manifolds, and normal forms in infinite dimensional systems. EDP Sci. Springer Verlag UTX series.
34. Hildreth, E. (1983). The detection of intensity changes by computer and biological vision systems (Vol. 22, pp. 1–27).

35. Hupé, J., & Rubin, N. (2003). The dynamics of bi-stable alternation in ambiguous motion displays: a fresh look at plaids. Vision Research, 43(5), 531–548.

36. Hupé, J., & Pressnitzer, D. (2012). The initial phase of auditory and visual scene analysis. Philosophical Transactions of the Royal Society B: Biological Sciences, 367(1591), 942–953.

37. Kawano, K., Shidara, M., Watanabe, Y., Yamane, S. (1994). Neural activity in cortical area mst of alert monkey during ocular following responses. Journal of Neurophysiology, 71(6), 2305–2324.

38. Kilpatrick, Z. (2012). Short term synaptic depression improves information processing in perceptual multistability. arXiv:12120076 submitted.

39. Kilpatrick, Z., & Bressloff, P. (2010). Binocular rivalry in a competitive neural network with synaptic depression. SIAM Applied Dynamical Systems, 9(4), 1303–1347.

40. Kilpatrick, Z., & Ermentrout, B. (2012). Hallucinogen persisting perception disorder in neuronal networks with adaptation. Journal of Computational Neuroscience, 32(1), 25–53.

41. Kim, Y., Grabowecky, M., Suzuki, S. (2006). Stochastic resonance in binocular rivalry. Vision Research, 46(3), 392–406.

42. Krauskopf, B., Osinga, H.M., Galán-Vioque, J. (2007). Numerical continuation methods for dynamical systems. Springer.

43. Kuznetsov, Y.A. (1998). Elements of applied bifurcation theory, 2nd edn. Applied mathematical sciences. Springer.

44. Laing, C., & Chow, C. (2002). A spiking neuron model for binocular rivalry. Journal of Computational Neuroscience, 12(1), 39–53.

45. Laing, C.R., & Troy, W.C. (2003). PDE methods for nonlocal models. SIAM Journal on Applied Dynamical Systems, 2(3), 487–516.

46. Lehky, S. (1995). Binocular rivalry is not chaotic. Proceedings of the Royal Society of London Series B: Biological Sciences, 259(1354), 71–76.

47. Leopold, D., & Logothetis, N. (1996). Activity changes in early visual cortex reflect monkeys' percepts during binocular rivalry. Nature, 379(6565), 549–553.

48. Levelt, W. (1968). On binocular rivalry (p. 107). The Hague-Paris: Mouton.

49. Logothetis, N., Schall, J., et al. (1989). Neuronal correlates of subjective visual perception. Science, 245(4919), 761–763.

50. Logothetis, N., Leopold, D., Sheinberg, D. (1996). What is rivalling during binocular rivalry? Nature, 380(6575), 621–624.

51. Lorenceau, J., & Shiffrar, M. (1992). The influence of terminators on motion integration across space. Vision Research, 32(2), 263–273.

52. Lorenceau, J., Shiffrar, M., Wells, N., Castet, E. (1993). Different motion sensitive units are involved in recovering the direction of moving lines. Vision Research, 33, 1207–1207.

53. Masson, G., Rybarczyk, Y., Castet, E., Mestre, D. (2000). Temporal dynamics of motion integration for the initiation of tracking eye movements at ultra-short latencies. Visual Neuroscience, 17(5), 753–767.

54. Mather, G., Verstraten, F., Anstis, S. (1998). The motion aftereffect: a modern perspective. MIT Press.

55. Meso, A., Rankin, J., Faugeras, O., Kornprobst, P., Masson, G.S. (2012a). Motion direction integration following the onset of multistable stimuli (I): dynamic shifts in both perception and eye movements depend on signal strength. European conference on visual perception. Perception, 41

56. Meso, A., Rankin, J., Kornprobst, P., Faugeras, O., Masson, G.S. (2012b). Perceptual transition dynamics of a multi-stable visual motion stimulus I: experiments. Visual sciences society 12th annual meeting. Journal of vision, 12(4), 748.

57. Moreno-Bote, R., Rinzel, J., Rubin, N. (2007). Noise-induced alternations in an attractor network model of perceptual bistability. Journal of Neurophysiology, 98(3), 1125–1139.

58. Pinto, D., & Ermentrout, G. (2001). Spatially structured activity in synaptically coupled neuronal networks: 1. traveling fronts and pulses. SIAM Journal on Applied Mathematics, 62(1), 206–225.

59. Raiguel, S., Xiao, D.K., Marcar, V., Orban, G. (1999). Response latency of macaque area mt/v5 neurons and its relationship to stimulus parameters. Journal of Neurophysiology, 82(4), 1944–1956.

60. Ramachandran, V., & Anstis, S. (1983). Perceptual organization in moving patterns. Nature, 304, 529–531. doi:10.1038/304529a0.

61. Rankin, J., Tlapale, É., Veltz, R., Faugeras, O., Kornprobst, P. (2013). Bifurcation analysis applied to a model of motion integration with a multistable stimulus. Journal of Computational Neuroscience, 34(1), 103–124. doi:10.1007/s10827-012-0409-5.

62. Rubin, N., & Hupé, J. et al. (2005). Dynamics of perceptual bistability: Plaids and binocular rivalry compared In Alais, D., & Blake, R. (Eds.), Binocular rivalry, (pp. 137–154). Cambridge: MIT Press.

63. Sanchez-Vives, M., Nowak, L., McCormick, D. (2000). Cellular mechanisms of long-lasting adaptation in visual cortical neurons in vitro. The Journal of Neuroscience, 20(11), 4286–4299.

64. Sclar, G., Maunsell, J., Lennie, P. (1990). Coding of image contrast in central visual pathways of the macaque monkey. Vision Research, 30(1), 1–10.

65. Serences, J., & Boynton, G. (2007). The representation of behavioral choice for motion in human visual cortex. Journal of Neuroscience, 27(47), 12,893–12,899.

66. Shpiro, A., Curtu, R., Rinzel, J., Rubin, N. (2007). Dynamical characteristics common to neuronal competition models. Journal of Neurophysiology, 97(1), 462–473.

67. Shpiro, A., Moreno-Bote, R., Rubin, N., Rinzel, J. (2009). Balance between noise and adaptation in

competition models of perceptual bistability. Journal of Computational Neuroscience, 27(1), 37–54.

68. Somers, D., Nelson, S., Sur, M. (1995). An emergent model of orientation selectivity in cat visual cortical simple cells. Journal of Neuroscience, 15(8), 5448.

69. Strogatz, S. (1994). Nonlinear dynamics and chaos. MA: Addison-Wesley Reading.

70. Theodoni, P., Kovács, G., Greenlee, M., Deco, G. (2011a). Neuronal adaptation effects in decision making. Journal of Neuroscience, 31(1), 234–246.

71. Theodoni, P., Panagiotaropoulos, T., Kapoor, V., Logothetis, N., Deco, G. (2011b). Cortical microcircuit dynamics mediating binocular rivalry: the role of adaptation in inhibition. Frontiers in human neuroscience, 4(145).

72. Tlapale, E., Kornprobst, P., Masson, G.S., Faugeras, O. (2011). A neural field model for motion estimation. In M. Bergounioux (Ed.), Mathematical image processing, springer proceedings in mathematics (Vol. 5, pp. 159–180).

73. Veltz, R. (2011). Nonlinear analysis methods in neural field models. PhD thesis, Univ Paris Est ED MSTIC.

74. Veltz, R., & Faugeras, O. (2010). Local/global analysis of the stationary solutions of some neural field equations. SIAM Journal on Applied Dynamical Systems, 9(3), 954–998.

75. Wallach, H. (1935). Über visuell wahrgenommene Bewegungsrichtung. Psychological Research, 20(1), 325–380.

76. Wilson, H., & Cowan, J. (1972). Excitatory and inhibitory interactions in localized populations of model neurons. Biophys J, 12, 1–24.

77. Wilson, H., & Cowan, J. (1973). A mathematical theory of the functional dynamics of cortical and thalamic nervous tissue. Biological Cybernetics, 13(2), 55–80.

78. Wuerger, S., Shapley, R., Rubin, N. (1996). On the visually perceived direction of motion by Hans Wallach: 60 years later. Perception, 25, 1317–1367.

79. Yang, J., & Lisberger, S.G. (2009). Relationship between adapted neural population responses in mt and motion adaptation in speed and direction of smooth-pursuit eye movements. Journal of Neurophysiology, 101(5), 2693–2707.

80. Zhou, Y., Gao, J., White, K., Merk, I., Yao, K. (2004). Perceptual dominance time distributions in multistable visual perception. Biological Cybernetics, 90(4), 256–263.

CHAPTER 21

OSCILLATION OF THE VELVET WORM SLIME JET BY PASSIVE HYDRODYNAMIC INSTABILITY

Andre´s Concha, Paula Mellado, Bernal Morera-Brenes, Cristiano Sampaio Costa, L. Mahadevan & Julia´n Monge-Na´jera

Keyword: Biological sciences, Zoology, Fluids and plasma physics

INTRODUCTION

Rapid motions in nature are seen in a variety of situations associated with escape and predation. Extreme examples include the chameleon tongue that uses an unusual muscle-spring configuration to capture prey[1], and the venus flytrap that stores elastic energy and uses an instability for the rapid closure of its leaf[2]. The velvet worm is an unusual example of how an organism projects itself by squirting a jet of slime in an oscillatory fashion, not only for capturing prey[3], but also for defence[4,5,6,7,8]. Animals use squirting mechanisms for elimination of waste products (for example, urination in vertebrates); reproduction (gamete expulsion); communication (for example, pheromone excretion in mammals); locomotion (for example, squids); defence (for example, horned lizards); and hunting (for example, archer fish)[9,10,11,12,13]. These jets are normally directed in a straight line. We are aware of only three exceptions: velvet worms (Phylum Onychophora), spitting spiders (Scytodes spp.) and spitting cobras (Naja spp.). Spitting spiders oscillate their fangs[14], and spitting cobras actively oscillate their heads[15]. Despite having been a subject of study for over a century, the mechanism underlying the rapid oscillatory squirting of slime by the velvet worm remains a mystery[4,16,17]. Indeed, Darwin even hypothesized the creation of the disordered web as a potential origin for the evolution of spider webs[18].

In this article, we demonstrate that the fast oscillations of the jet slime and papillae in Onychophora are the result of a syringe-like system that, by means of a geometric amplifier, allows for fast squirt using slow muscular contraction. This fast flow activates an elastohydrodynamical instability that explains the fast papilla oscillations during defence and attack. We present a physical simulacrum that reproduces the instability in the same range of parameters of the natural system opening new venues for self-supported microfluidic devices and applications[19,20,21,22].

RESULTS AND DISCUSSION

Worm attack kinematics

To capture the dynamics of the squirting process, we filmed several worm attacks. In Fig. 1a–d, we show a series of snapshots of an attack recorded using high-speed imaging (480 frames per second a.k.a. f.p.s.), with the average duration of a squirt for all specimens being $\Delta t_{ave}=0.064\pm0.005$. Furthermore, we see that the squirt does not remain oriented but instead sprays an entire region as shown in Fig. 1e. These measurements raise the natural question of the spatio-temporal evolution of the liquid jet and its control by the worm.

Citation: Concha, A. et al. Oscillation of the velvet worm slime jet by passive hydrodynamic instability. Nat. Commun. 6:6292 doi: 10.1038/ncomms7292 (2015).

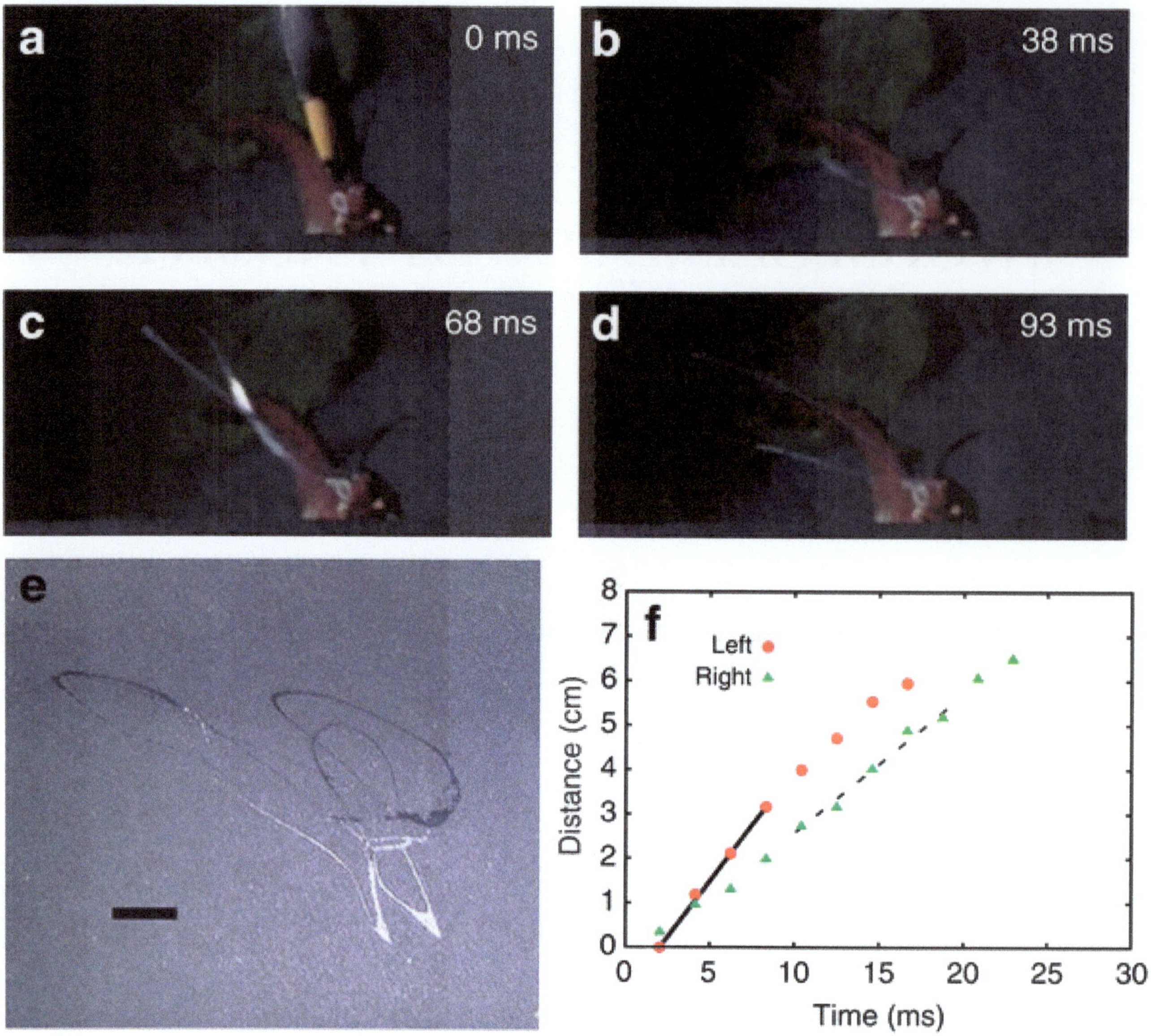

Figure 1: Slime jet oscillations and jet speed during worm attack.

Giant red velvet worm Peripatus solorzanoi used to record the squirting process. Full-body length is ~17.5 cm. In **a**, a soft paintbrush used to activate its attack is shown, and it was digitally removed in the other snapshots for clarity. (**a**–**d**) Different stages of the attack recorded at 480 f.p.s. The active part of the attack is completed in Δt_{squirt}~65 ms. (**e**) Slime pattern generated on a wall of the foam tunnel used to keep worms at focal distance (scale bar, 1 cm). Three or more oscillations of the slime jet are the typical outcome. (**f**) Liquid jet tip position as a function of time for the squirt. For this data, we used two cameras: one at 30 f.p.s. and the other at 480 f.p.s. The two cameras configuration allowed us to compute the jet velocity. The solid line corresponds to a squirt speed of V=5.0 m s^{-1} and the dotted line to a squirt speed of V=3.2 m s^{-1}. Solid red dots and green triangles correspond to left and right papilla, respectively.

Anatomy of the ejecting system

Actively controlled muscular action has long been invoked as the natural explanation for the spectacular way in which these worms quickly squirt slime and continues to be the favoured mechanism16. However, papillar oscillations are fast in comparison with any other motion of the worm ($f_{papilla}$/ $f_{walking}$~30–60) and with known time scales (~0.5 s) for the fastest muscles in the worm23, suggesting a conceptual difficulty with this hypothesis. Therefore, we examined the anatomy of the whole-squirt system and surrounding tissues. During squirting, the oral papilla extends from its folded shape to its full length (Fig. 2) of up to L~6 mm. In Fig. 2a, we see a large reservoir region (re) where slime is stored and a narrow duct that ends at the oral papilla, a syringe-like geometry that facilitates the acceleration of the slime for the fast squirt. In Fig. 2a,b, we show that muscle fibres in the oral papilla are similar to those found in the legs, but fewer in number. The relaxed papilla has an accordion shape (Fig. 2c,d) that is unfolded just before the squirting process, and can thus be easily packed while also having an inhomogeneous bending rigidity, with soft spots that make papillae more pliable and susceptible to bending as slime is squirted.

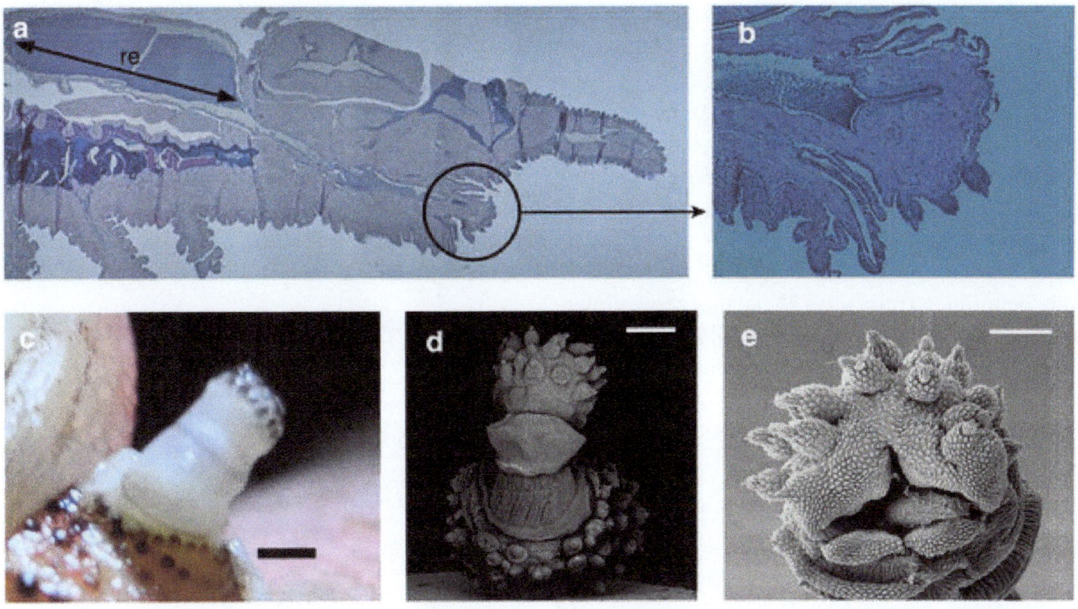

Figure 2: Squirt system and papilla structure.

(**a**) A haematoxylin and eosin-stained section of a Peripatus solorzanoi. The squirt system is composed of a large slime reservoir, re, whose length is depicted by a black arrow. The reservoir has a large diameter compared with the diameter of the duct that connects it with the oral papilla (black circle). (**b**) Longitudinal cut of an oral papilla of a Peripatus solorzanoi. Its wrinkled surface and sphincter-like tissues surrounding the inner duct are shown (scale 4mm). The black substance inside the duct corresponds to remnants of the slime. (**c**) Structure of a fresh oral papilla (red Peripatus solorzanoi) when deployed. Its semi-transparent and wavy structure is apparent (scale bar, 1 mm). (**d**) Oral papilla accordion structure of aEpiperipatus acacioi (scale bar, 200 μm). In this case the hinged accordion structure is resolved by scanning electron microscopy (Leica Leo 440) taken at 15 kV after coating with gold (~20 nm) in a Denton Desk IV gold sputter model. This specimen was used to confirm that the accordion-like structure present in Peripatus solorzanoi is generic to Onychophora regardless of its size. (**e**) Opening of the oral papillaD~350 μm and d~200 μm (scale bar, 100 μm).

Our anatomical findings are consistent with earlier evidence that slime papilla are modified limbs24,25 with a nervous system similar to that in their legs26. Detailed descriptions of the Onychophora muscular system23 show that the fastest muscles are located in the jaw with typical twitch time scales ~0.5 s, which, while fast for this primitive worm, are nearly 25 times slower relative to the papillary oscillation time scale ~20 ms. Given that the legs and the papillary muscles are even slower, consistent with the primitive nature of these worms5,27,28, we are left with an obvious question—how are rapid changes in direction that occur over a time scale of a few milliseconds possible without the existence of any specialized rapid muscular actuation or neural control?

Physical mechanism

A way around this conundrum is to realize that a physical mechanism can drive the rapid and nearly chaotic oscillations of the papilla just as a garden hose pipe oscillates when water squirts out of it rapidly. Indeed, the inertial effects associated with the exiting fluid jet drive the elastic hose pipe to flutter, a subject that has been well studied experimentally and theoretically at the macroscale29. In the current microscopic setting, the interplay between fluid forces and the papilla elasticity produce the characteristic oscillatory waving motion used to capture prey, and obviate the need for any fast moving controlled muscles. However, this mechanism requires fluid inertia to play a critical destabilizing role, that is, the ratio of inertial to viscous forces characterized by the Reynolds number $Re = VR/\nu > 1$ (where V is the characteristic fluid speed, ν the fluid kinematic viscosity and Rthe tube radius). In microfluidic geometries where typical sizes are small, unless the velocities are sufficiently large, inertial effects are unimportant.

Our microscopy studies (see Fig. 2a) show that the squirting system has a reservoir that contracts slowly driving the slime through a small duct that runs close to the centre of the oral papilla. This geometric amplifier can lead to an increase in the speed during squirting. This is functionally similar to structures found in the chelicerae of spitting spiders14, and suggests that this geometric amplifier mechanism could be relevant for a variety of other squirting organisms. Our observations are in contrast with previously reported studies30 that have persisted into the modern literature31, where no mention of the cross-section reduction has been made. Our micrographs show muscular structures around the slime reservoirs, which are functionally consistent with the contraction of this organ. These structures resemble the design of radial tires where a fibre network is used to reinforce the wall32, consistent with detailed study of muscular fibres at reservoir level33. Measurements of the squirted volume and reservoir geometry show that the contraction ratio$\delta R_{re}/R_{re} < 0.03$ (with $R_{re} \sim 2$ mm) changes in about 0.1 s, enough to produce speeds of $V \sim 5\ \mathrm{m\,s^{-1}}$, so that the Reynolds number Re~2,700. Since we do not see perfect synchronization between liquid jets coming from different papilla, whole-body contraction as the main driving force in squirting33is unlikely.

This leads to the conclusion that the instability arises due to a competition between fluid inertia and elastic resistance. When a liquid moves steadily through a flexible pipe at small V, flow-induced damping prevents any oscillations from growing. For large enough V, centrifugal and Coriolis forces make the pipe unstable for fluid speeds $V>V_c$. In the limit when the effects of gravity can be neglected, a simple scaling argument allows us to estimate the frequency of oscillations f by balancing the stabilizing elastic bending resistance with the destabilizing inertial forces, that is,

$$EI/\lambda^4 \sim MVf/\lambda \sim MV^2/\lambda^2, \qquad (1)$$

where EI is the bending stiffness, $\lambda \sim 2L$ is the approximate oscillation wavelength for the cantilevered papilla and M is the mass density per unit length of the fluid in the pipe. This yields

$$f \sim \left(\frac{EI}{M}\right)^{1/2} \frac{1}{4L^2} \qquad (2)$$

Similarly, a typical speed scale can be estimated as $u_0 \sim f\lambda = (EI/M)^{1/2}(1/L)$. Fig. 2, the oscillation frequency $f \sim 58$ Hz, papilla length L=6.0 mm, papilla outer diameter D=1.0 mm and papilla inner diameter d=0.5 mm, from where $\beta \sim 0.25$. The measured frequency allows to estimate the effective Young's modulus of the papilla to be $E \sim 20$ kPa consistent with measurements using small magnets to pull the papilla that yields $E \sim 40$ kPa, and thence the typical speed $u_0 \sim 0.5\,m\,s^{-1}$. The precise critical speed V_c depends on boundary conditions, which for the cantilever case gives $V_c \sim 2\pi u_0 = 3\,m\,s^{-1}$. At this critical speed, stability is lost via a Hopf type bifurcation[29]

For unsteady flows, such as when the jet is being accelerated inside the flexible papilla, the fluid acceleration dv/dt can destabilize the system at even lower jet speeds. Our measurements show that $V \sim 3.2–5.0\,m\,s^{-1}$. Therefore, even without muscular action the papilla will become unstable due to a simple physical instability[29].

Synthetic simulacrum

To show that it is indeed possible to drive these oscillatory instabilities on small scales, we made a synthetic papilla out of a soft elastomer (see Methods) in the form of a flexible micropipe with a rectangular cross-section that defines an oscillation axis for its softest bending mode. Our pipe was moulded out of polydimethylsiloxane (PDMS) with a Young's modulus E=288 kPa, with thickness h=1.42 mm, width w=1.60 mm, inner diameterd=0.81 mm and length L=9.5 mm. Our model system is simpler than the natural one in at least two aspects: there is no roughness along the inner part of the duct or an external accordion-like geometry. Our experimental results show that synthetic papilla becomes unstable and oscillates when the liquid reaches a speed of $V_c=8.6\,m\,s^{-1}$. This occurs in the same range of dimensionless parameters that for the natural organ in agreement with theoretical predictions[29], and shows that fast muscular action at the papillary level is unnecessary for oscillations. From measurements of external diameters (Fig. 2d), we found that B=EI can locally change up to 1/10 of the stiffness corresponding to a uniform soft papilla of diameter D_0, showing that the accordion-like microstructure will lower V_c. Microtubes with accordion-like shapes made out of the same polymer were made varying the degree of external roughness. Our data shows that V_c lowers as we increase the amplitude of external roughness. The flow of liquids in pipes with inner roughness has been extensively studied[34,35]. The effect of inner pipe imperfections is to decrease the effective cross-section of the tube, as well as to anticipate the transition to turbulence[35]. All these effects together lower V_c, being the most important for the fluid regime described, the accordion-like modulations of the tube. Naturally, this experiment also suggests a prototype of a fluid driven micromechanical actuator[36,37,38].

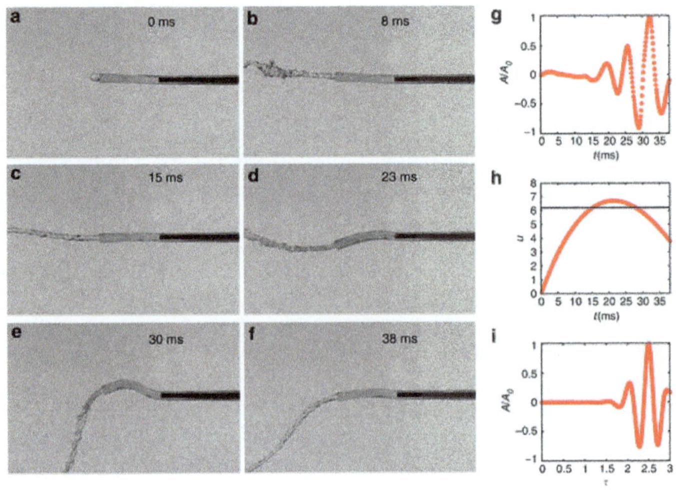

Figure 3: Bioinspired physical simulacrum of the oral papilla.

(a–f) A soft elastic cantilevered tube made out of polydimethylsiloxane (PDMS) becomes unstable as the fluid flowing through it increases its speed. Its height is h=1.42 mm, width w=1.60 mm, length L=9.5 mm and hole diameter is 0.81 mm. The fluid used in this experiment was water. **(g)** The vertical motion of a point at the centre line of the hose close to the tip. As fluid speed is increased, oscillations develop and grow. The vertical axis is the dimensionless ratio between oscillation amplitude A and its maximum A_0. **(h)** During an emptying cycle, the fluid speed varies as a function of time, with the dimensionless speed$u \varepsilon [0.0,6.9]$, with $u=V/u_0$. Horizontal black line depicts the theoretical threshold for instability in the case of steady flow. **(i)** Numerical simulations of the squirting dynamics associated with the dynamics of emptying shown in **h** using the governing equations. As in **g**, the vertical axis is the ratio between oscillation amplitude A and its maximum A_0. The horizontal axis is the dimensionless time $\tau=t/\tau_0$, whereτ_0 is the bending time scale.

Our observations, minimal theory and physical mimic show how velvet worms can spray a web rapidly to entangle their prey without resorting to a complex neuromuscular control strategy. Instead, they can simply harness the instability associated with rapid flow through a long soft nozzle that causes it and the exiting jet to oscillate. As this jet solidifies rapidly, it entraps the object dynamically. While there is a superficial similarity to spider webs, we see that here dynamics is of essence. Our findings and synthetic model might also pave the way to self-supported flexible microfluidic devices that can take advantage of similar instabilities to produce a variety of products such as microdrops[22,39,40,41], and non-woven fibre structures[19].

METHODS

Sample collection

Specimens of Peripatus solorzanoi (n=3) were collected in Guayacan, Siquirres, Costa Rica (10°3′21.38″ N, 83°32′44.04″ W, 500 m.a.s.l.). The specimen of Epiperipatus acacioi (n=1) was collected in Minas Gerais, Brazil (20°22′44″ S, 43°32′55″ W, 1220, m.a.s.l.). The worms were housed in thermoplastic polystyrene (PS) terraria (370 × 240 × 280 mm) with a window covered with a metallic net as air vent. The temperature varied between 18 and 23 °C. The terraria received artificial illumination with standard daylight fluorescent lamps and had a 3–4-cm layer of leaf litter extracted from the original habitat of the specimens. This litter contained small invertebrates that are the natural food of the worms. The humidity in terraria was kept close to saturation all the time.

Optical microscopy

Tissues were fixed in 10% neutral-buffered formalin, processed with paraffin wax, sectioned at 6 μm and stained with haematoxylin and eosin for histological examination42·43. We used optical microscopy (Olympus BX41, Olympus camera E330-ADU1.2x, software Olympus Studio 2) to obtain the geometrical parameters involved in the reservoir and papilla duct, and to analyse the tissue structure at papilla and reservoir level.

Electron microscopy

Tissue samples kept in 70% alcohol were collected for scanning electron microscopy (SEM). Selected samples were re-hydrated and cleaned with a 3% solution of sodium hypochlorite (NaClO) and distilled water. They were dehydrated with an increasing alcoholic series: 30, 50, 70, 80, 90 and 100%, respectively. Later they were critical point dried in a Baltec CPD 030 and then coated with gold (~20 nm thick) in a Denton Desk IV gold sputter system. Specimens were imaged in a SEM (Leica, LEO 440). The morphological terminology for structures follows the one in ref. 4444.

Image analysis

The measurement of squirting times was performed using high-speed imaging of P. solorzanoi during squirt. We elicited the attack from the specimen using a gentle paint brush on its back. We have used a Casio EX-ZR200 camera capable of capturing colour movies up to 1,000 f.p.s. For speed measurement we have used two cameras; one at 30 f.p.s. (lateral view) and a second high-speed camera (Casio EX-ZR200) at 480 f.p.s. The first camera allowed us to measure the distance between the worm and the high-speed camera, and with this data and the pixel location from the high-sped camera we computed the Euclidean distance between the papilla and the jet tip. The time evolution of the slime jet tip is shown in Videos 1 and 2. To measure oscillations in the synthetic system, we have used a Phantom Miro 310 high-speed camera. We used up to 8,000 f.p.s., as in this case we are free to use high-intensity illumination to capture detailed information about the oscillation amplitude and frequency.

Micropipe fabrication and characterization

The same method was used to fabricate rounded and rectangular cross-section channels. We began by fabricating a template of the channel of dimensions w~1 mm (width), h~1 mm (height) and L~30 mm (length). This was done using a standard micro-mill (φ=1.0 mm), on Plexiglas at extremely low feeding rates (5 mm per min) to avoid defects on the structure.

To make it easier to peel off the PDMS and the needle, we applied a soapy solution wash and let it dry; we repeated this procedure three times. The dried layer of soap42 does not allow PDMS to stick either to the template or to the needle. We have been able to produce micropipes of inner diameters (d) from 800 μm down to 300 μm in a reproducible way. We obtained these microchannels by using commercial needles of different sizes, which we mounted on a scotch tape that is covering the end of the channels. We punctured this film with the help of a microscope (for proper centring) and later we introduced the required size needle through that centring hole. When pulling the needle out of the PDMS channel, the thin soap layer worked as a lubricant allowing easy slip without damaging the inner tube surface. Later we cleaned the tube using isopropyl alcohol.

We mixed Sylgard 184 in a ratio 25:1 to get a soft polymer for the pipes. We hand mixed for 3 min and then used a centrifuge for mixing and bubble removal. We repeated the previous steps three times. Once completing this step, we carefully filled the templates using a syringe and then degassed the sample for 40–60 min. We checked for bubbles and possible leaking in the tape wall and then placed it in an oven at 70° C for 15 h. The PDMS removal was manually done. Some samples showed small lateral films attached to the main pipe. These leftovers when existent were removed by using a scalpel under the microscope. We double checked the integrity of the final micropipes after isopropyl alcohol and nitrogen cleaning by visual inspection under the microscope.

Model for fluid–papilla interactions

The motion of a pipe conveying fluid whose motions are restricted to be planar can be described by the following equation29:

$$\mathrm{EI}z^{(4)} + \left\{ MV^2 + \left[M\frac{dV}{dt} - (M+m)g \right](L-x) \right\} z^{(2)}$$
$$+ 2MV\frac{\partial^2 z}{\partial x \partial t} + (M+m)g\frac{\partial z}{\partial x} + v\frac{\partial z}{\partial t} + (M+m)\frac{\partial^2 z}{\partial t^2} = 0 \quad (3)$$

where z(x,t) is the lateral motion of the beam centre line, M the fluid mass per unit length, EI the flexural rigidity, m is the tube mass per unit length, L is the tube length, V is the fluid speed, gacceleration of gravity and ν is a phenomenological damping factor. We must emphasize that regardless of the cross-section shape the relevant quantity is the bending stiffness B=EI.

The boundary conditions in this problem are $z(0,t)=0$, $z^{(1)}(0,t)=0$, $z^{(2)}(L,t)=0$ and $z^{(3)}(L,t)=0$. The relevant physical parameters needed to specify the parameter space for this problem can be found in equation (3) in a dimensionless form. This is easily achieved by using the dimensionless time and coordinates $\eta=z/L$, $\chi=x/L$, $u=V/u_0$ and $\tau=t/\tau_0$. Where τ_0 is the bending time scale, and u_0 is a characteristic speed. The dimensionless equation reads:

$$\frac{\partial \eta^4}{\partial \chi^4}+\left\{u^2+\left(\beta^{1/2}\dot{u}(\tau)-\gamma\right)(1-\chi)\right\}\frac{\partial \eta^2}{\partial \chi^2}$$
$$+2\beta^{1/2}u(\tau)\frac{\partial^2 \eta}{\partial \tau \partial \chi}+\gamma\frac{\partial \eta}{\partial \chi}+\sigma\frac{\partial \eta}{\partial \tau}+\frac{\partial^2 \eta}{\partial \tau^2}=0 \quad (4)$$

Numerical simulation

We have solved equation (4) using different numerical approaches. We did compare different methods as the coefficients of the PDE are not constant and may abruptly change as a function of time. The results we show were obtained using the Garlekin method. We use as basis the set of functions provided by the problem:

$$\mathscr{L}\psi = 0 \quad (5)$$

with the boundary conditions (BC) $\psi(0)=0$, $\psi'(0)=0$, $\psi^{(2)}(1)=0$ and $\psi^{(3)}(1)=0$. Where:

$$\mathscr{L}=\frac{\partial^4}{\partial \chi^4}+\frac{\partial^2}{\partial \tau^2} \quad (6)$$

This problem endows us with a good expansion basis that fulfils the BC. After using the anzats,

$$\eta_{\text{nmax}}(\chi,\tau)=\sum_{j=1}^{\text{nmax}}\psi_j(\chi)q_j(\tau) \quad (7)$$

and using the inner product properties of $\{\psi_j(\chi)\}_{j=1}^{\text{nmax}}$ over the domain [0,1], we reduced equation (4) into a linear system using the first nmax basis functions. We emphasize that equation (4) captures well the onset of the instability. However, the oscillation amplitude does not saturate without including nonlinear terms in curvature29. To keep the model simple, we augmented equation (4) by a confining term $\alpha\eta^2$, and included the experimentally found damping σ.

REFERENCES

1. Mu¨ller, U. & Kranenbarg, S. Power at the tip of the tongue. Science 304, 217–219 (2004).
2. Forterre, Y., Skotheim, J., Dumais, J. & Mahadevan, L. How the venus flytrap snaps. Nature 433, 421–425 (2005).
3. Read, V. & Hughes, R. Feeding behaviour and prey choice in macroperipatus torquatus (onychophora). Proc. R. Soc. Lond. B 230, 483–506 (1987).
4. Belt, T. The Naturalist in Nicaragua: a Narrative of a Residence at the Gold Mines of Chontales; Journeys in the Savannahs and Forests; with Observations on Animals and Plants in Reference to the Theory of Evolution of Living Forms Vol. 561 (ICON Group International, 1888).
5. Bouvier, E. Monographie Des Onychophores Vol. I (Masson et Cie., 1905).
6. Claude-Joseph Observations Sur un peripate du chili. Ann. Sci. Nat. Zool. Biol. Anim. Ser 10, 285–299 (1928).
7. Alexander, A. Notes on onychophoran behavior. Ann. Natal Mus. 14, 35–43 (1957).
8. Alexander, A. Peripatus: fierce little giant. Ann. Natal Mus. 14, 35–43 (1958).
9. Demont, M. E. & Gosline, J. M. Mechanics of jet propulsion in the hydromedusan jellyfish, polyorchis pexicillatus: III. A natural resonating bell; the presence and importance of a resonant phenomenon in the locomotor structure. J. Exp. Biol. 134, 347–361 (1988).
10. Anderson, E. J. & Grosenbaugh, M. A. Jet flow in steadily swimming adult squid. J. Exp. Biol. 208, 1125–1146 (2005).
11. Dabiri, J. O. & Gharib, M. The role of optimal vortex formation in biological fluid transport. Proc. R. Soc. B 272, 1557–1560 (2005).
12. Sherbrooke, W. C., Middendorf, III G. A. & Guyer, C. Blood-squirting variability in horned lizards (phrynosoma). Copeia 2001, 1114–1122 (2001).
13. Vailati, A., Zinnato, L. & Cerbino, R. How archer fish achieve a powerful impact: hydrodynamic instability of a pulsed jet in toxotes jaculatrix. PLoS ONE 7, e47867 (2012).
14. Suter, R. B. & Stratton, G. E. Spitting performance parameters and their biomechanical implications in the spitting spider, scytodes thoracica. J. Insect Sci. 9, 1–15 (2009).
15. Young, B., Boetig, M. & Westhoff, G. Functional bases of the spatial dispersal of venom during cobra spitting. Physiol. Biochem. Zool. 82, 80 (2009).
16. Morera-Brenes, B. & Monge-Na´jera, J. A new giant species of placented worm and the mechanism by which onychophorans weave their nets (onychophora: Peripatidae). Rev. Biol. Trop. 58, 1127–1142 (2010).
17. Haritos, V. et al. Harnessing disorder: onychophorans use highly unstructured proteins, not silks, for prey capture. Proc. R. Soc. B 277, 3255–3263 (2010).
18. Moseley, H. On the structure and development of peripatus capensis. Philos. Trans. R. Soc. Lond. 164, 757–782 (1874).
19. Mellado, P. et al. A simple model for nanofiber formation by rotary jetspinning. Appl. Phys. Lett. 99, 203107–203107 (2011).
20. Kim, H.-Y., Lee, M., Park, K. J., Kim, S. & Mahadevan, L. Nanopottery: coiling of electrospun polymer nanofibers. Nano Lett. 10, 2138–2140 (2010).
21. Kang, E. et al. Digitally tunable physicochemical coding of material composition and topography in continuous microfibres. Nat. Mater. 10, 877–883 (2011).
22. Eggers, J. & Villermaux, E. Physics of liquid jets. Rep. Prog. Phys. 71, 036601 (2008).
23. Hoyle, G. & Del Castillo, J. Neuromuscular transmission in peripatus. J. Exp. Biol. 83, 13–29 (1979).
24. Budd, G. E. A palaeontological solution to the arthropod head problem. Nature 417, 271–275 (2002).
25. Scholtz, G. & Edgecombe, G. D. The evolution of arthropod heads: reconciling morphological, developmental and palaeontological evidence. Dev. Genes Evol. 216, 395–415 (2006).
26. Mayer, G. & Whitington, P. M. Neural development in onychophora (velvet worms) suggests a step-wise evolution of segmentation in the nervous system of panarthropoda. Dev. Biol. 335, 263–275 (2009).

27. Liu, J. et al. An armoured cambrian lobopodian from china with arthropod-like appendages. Nature 470, 526–530 (2011).
28. Fortey, R. & Thomas, R. The case of the velvet worm. Nature 361, 205–206 (1993).
29. Paidoussis, M. Fluid-Structure Interactions: Slender Structures and Axial Flow Vol. 1 (Academic press, 1998).
30. Manton, S. M. & Heatley, N. G. Studies on the onychophora. ii. the feeding, digestion, excretion, and food storage of peripatopsis with biochemical estimations and analyses. Philos. Trans. R. Soc. Lond. B 227, 411–464 (1937).
31. Barnes, R. S. K., Calow, P. P., Olive, P., Golding, D. W. & Spicer, J. I. The Invertebrates: a Synthesis (Wiley-Blackwell, 2009).
32. Clark, S. K. Mechanics of Pneumatic Tires (US Department of Transportation, National Highway Traffic Safety Administration, 1981).
33. Baer, A. & Mayer, G. Comparative anatomy of slime glands in onychophora (velvet worms). J. Morphol. 273, 1079–1088 (2012).
34. Nikuradse, J. Laws of Flow in Rough Pipes (National Advisory Committee for Aeronautics, Washington, 1950).
35. Kandlikar, S. G., Schmitt, D., Carrano, A. L. & Taylor, J. B. Characterization of surface roughness effects on pressure drop in single-phase flow in minichannels. Phys. Fluids 17, 100606 (2005).
36. Whitesides, G. The origins and the future of microfluidics. Nature 442, 368–373 (2006).
37. Leslie, D. C. et al. Frequency-specific flow control in microfluidic circuits with passive elastomeric features. Nat. Phys. 5, 231–235 (2009).
38. Ho, C.-M. & Tai, Y.-C. Micro-electro-mechanical-systems (mems) and fluid flows. Ann. Rev. Fluid Mech. 30, 579–612 (1998).
39. Kaufman, J. et al. Structured spheres generated by an in-fibre fluid instability. Nature 487, 463–467 (2012).
40. Kong, T. et al. Droplet based microfluidic fabrication of designer microparticles for encapsulation applications. Biomicrofluidics 6, 34104 (2012).
41. Yang, W., Duan, H., Li, C. & Deng, W. Crossover of varicose and whipping instabilities in electrified microjets. Phys. Rev. Lett. 112, 054501 (2014).
42. Bonhomme, O., Liot, O., Biance, A.-L. & Bocquet, L. Soft nanofluidic transport in a soap film. Phys. Rev. Lett. 110, 054502 (2013).
43. Santoro, M., Morales, J. A., Stacy, B. A. & Greiner, E. C. Rameshwarotrema uterocrescens trematode parasitism of the oesophageal glands in green sea turtles (chelonia mydas). Veterinary Record. 160, 59–60 (2007).
44. de Sena Oliveira, I. et al. Unexplored character diversity in onychophora (velvet worms): a comparative study of three peripatid species. PloS ONE 7, e51220 (2012).

CHAPTER 22

SOCIAL WAVES IN GIANT HONEYBEES (APIS DORSATA) ELICIT NEST VIBRATIONS

Kastberger G, Weihmann F, Hoetzl T

Keywords: Giant honeybee, Apis dorsata, Bee curtain, Shimmering Defence waves, Comb vibrations, Elastic plate model, Mechanic pendulum model, Frequency spectrum, Colony-intrinsic communication

INTRODUCTION

Giant honeybees (Apis dorsata) are one of the most ancient honeybee species (Ruttner 1988; Oldroyd and Wongsiri 2006; Kastberger et al. 2011a). They nest in the open (Seeley et al. 1982) and are, therefore, especially exposed to predators, such as mammals (Kastberger 1999), birds (Seeley et al. 1982; Kastberger and Sharma 2000) and wasps (Seeley et al. 1982; Kastberger et al. 2008, 2010). Their most prominent defensive behaviour against such threats is shimmering, which generates repetitive social waves with anti-predatory impact (Kastberger et al. 2008,2010, 2012, 2013; Weihmann et al. 2012). In shimmering, bees at the nest surface, predominantly younger cohorts (Lerchbacher et al., submitted) in the quiescent regions peripheral to the mouth zone (Kastberger et al. 2011b), show simultaneous and cascaded actions (Schmelzer and Kastberger 2009) whereby their abdomens are flipped upwards at an angle between 20° and 120° (Kastberger et al. 2011a, b). These individual mechanical actions of surface bees emerge to a visual display of waves which have the potential to repel or at least scare external predators such as wasps (Kastberger et al. 2008) or vertebrates (Kastberger et al.2013).

The mechanical expression of shimmering, represented in particular by its motion components in the z-direction (i.e. towards and away from the comb; Kastberger et al. 2011b) is still unclear in its ultimate goals. However, two mechanistic viewpoints can here be proposed:

- First, mechanical cues provoked by shimmering may be important for the rapidness of the wave propagation across the nest. Alternative strategies eventually processed by visual cues among adjacent neighbours, utilising stigmergic (Grasse 1959; Kastberger et al. 2013) principles, such as bucket-bridging (Kastberger et al. 2012) or eavesdropping (Peake 2005; Jones et al. 2011), would be much slower by one or two orders of magnitude than those which are actually observed in shimmering (Kastberger et al. 2012, 2013). The saltatoric principle of wave propagation (Kastberger et al. 2013), which may speed up the wave from a basically bucket-bridging process by a factor of 3, is supposed to be associated to the visual input of threatening cues and to the mechanical effects of the shimmering waves in the bee curtain.
- Second, shimmering could benefit the entirety of the colony by allowing colony-intrinsic propagation of information across the nest via its mechanical expression of wave components (Kastberger et al. 2011b). This proposition is summarized by the colony-intrinsic information hypothesis which predicts that the shimmering waves may also disseminate information about the momentary defence state of the colony to those nest members which do not participate in this collective action. Shimmering-passive cohorts usually make up more than 90 % of the colony and comprise curtain bees in the surface and subsurface layers on both sides of the nest.

Citation: Kastberger, G., Weihmann, F. & Hoetzl, T. Naturwissenschaften (2013), Social waves in giant honeybees (Apis dorsata) elicit nest vibrations 100: 595. doi:10.1007/s00114-013-1056-z.

In accordance with this proposition of a potential colony-intrinsic spread of information across the nest, the mechanical components of the shimmering process should affect the centrally positioned comb of the nest. The comb is the only architectural structure of an A. dorsata nest with the capacity to rapidly bridge information from the threatened to the non-threatened nest side. Therefore, we examined whether shimmering does produce vibrations at the comb. To qualify as signals which could potentially be utilised for intra-colonial communication, mechanical vibrations of the comb should exceed at least the perception threshold of Western honeybees (Sandeman et al. 1996) which can be estimated at dislocation amplitudes of >9 µm in the low-frequency range (<10 Hz).

MATERIAL AND METHODS

Experimental site

The experimental nests of giant honeybees (A. dorsata) were located at a hotel site in Sauraha, Chitwan, Nepal (at the border of the Chitwan National Park). Preliminary experiments had been performed in February 2009, but in November 2010, a single nest had been selected for a much broader in-depth investigation. This nest was attached to blocks of concrete making up a balcony of the hotel. The nest had a hemicyclic form, measured 83 × 60 cm (width × height), was approximately 2 weeks old and could have had a weight of 30 kg (see Seeley et al. 1982 and the 'Results' section). In the first days of observation, the comb was covered only by a single layer of honeybees, but a few days later, the colony had achieved a multilayer cover due to the progressive hatching of young bees (Fig. 1).

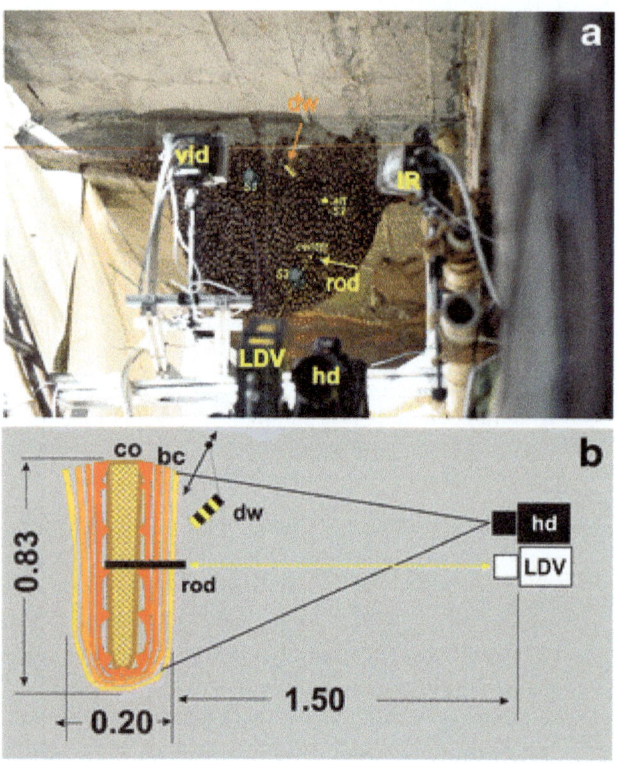

Figure. 1 Measurement set-up of the experimental A. dorsata nest. **a**hd HD video camera, vidblack and white video camera, IR infrared camera, LDV laser Doppler vibrometer, with the yellow straight line as the ray of the laser; the dummy wasp (dw, see orange arrow) was moved under computer control from the left to the right side of the nest, the orange line at the upper rim of the nest gives the wire of the cable car device by which the dummy had been moved. The mouth zone was positioned to the left of the violet curve. **b**Schematics of the set-up from the side view, with a perpendicular cross-section through the nest; distances in metres; a wooden rod (diameter, 8 mm; length, 10 cm) was stuck through the bee curtain (bc) into the comb (co) at position centre or alternatively at positions S1–3 with S2 = att, and a white plate was mounted on the front plane of the rod to reflect the laser beam

Video recording

The experimental nest was filmed with a high-definition (HD) video camera (Panasonic HVX 200) from a distance of 1.5 m, whereby the camera angle covered the whole nest (Fig. 1). This distance provided an undistorted view and was also distant enough to keep the colony undisturbed, as the bees accepted the camera as just a landmark. The recordings were made at 50 frames per second (FPS) with a resolution of 1,280 × 720 pixels.

Laser vibrometry

A wooden rod (8 mm thick and 10 cm long with a weight of 1.28 g) was stuck into the comb, with one end protruding slightly out the surface of the bee curtain. A piece of white paper was glued to its plane end to serve as a reflector for the laser beam of the laser Doppler vibrometer (abbreviated further on as 'LDV'; Polytec PDV 100; Fig. 1). In this way, the directional z-component of the movement of the comb (towards and away from the nest surface) was detected by the velocity values assessed by the Doppler phase shift between the emission and reflection of the laser signal. These data were assessed with the wooden rod positioned at selected points:

For the spectrum analysis, we used the positions S1–2 (14 November 2010) and included mass flight, pure quiescence and presentation of the dummy wasp. For further evaluations of the correlation between shimmering and comb vibrations, we used the position att, 10 cm below the attachment zone (experiments: 9 November 2010), and the position centre, central between the right-hand border of the mouth zone and the right-hand rim of the nest (experiments: 10 November 2010), at least 30 cm below the attachment zone (Fig. 1).

The LDV had a resolution of <0.05 µm/s. The raw data were produced by the LDV in sample intervals of 20 µs (resulting in a sampling frequency of $f_s = 1/(20 \times 10^{-6})$ [Hz]), processed by MATLAB through a low-pass Butterworth IIR filter with a cut-off frequency of $f_{LP} = 250$ Hz (with $f_{LP} = 0.01 \times f_s/2$) and displayed in data streams, frequency spectrograms (abscissa: experimental time; ordinate: frequency components; z-axis: power spectral density [PSD]) and single-sided Fourier spectra (abscissa: frequency components; ordinate: LDV signal [in millimetres per second]). The raw Butterworth filtered LDV data streams were offset-corrected and integrated for the calculation of dislocation values scaled in millimetres.

Dummy wasp stimulation

We stimulated the experimental colony by presenting a dummy wasp which was a cuboid Styrofoam block measuring 8 × 2 × 2 cm covered with yellow and black stripes (Fig. 1; Kastberger et al. 2011a, b, 2012, 2013). It was suspended beneath a horizontal wire fixed above the nest attached by a flexible thread, allowing the dummy to swing. It was moved by a miniature computer-driven cable car device at a constant velocity (0.1–0.5 m/s) along the horizontal wire, 20 cm in front of the nest near its upper attachment zone. This stimulation method was chosen to mimic a free-flying wasp scanning in front of the nest and has regularly provoked shimmering (Kastberger et al. 2011a, b). The dummy wasp (dw) was presented to the colony by drawing it from the parking site at the upper left side of the nest to the right side and back again (Fig. 1). Automated pinpointing of the dummy's location was achieved by assessment of its motion (see the next section) using predefined templates. This allowed the identification of the dummy regardless of the viewing angle and the associated light conditions (Fig. 2) and the assessment of its gravity position (dummy: x_{dw}, y_{dw}) during the whole presentation cycle.

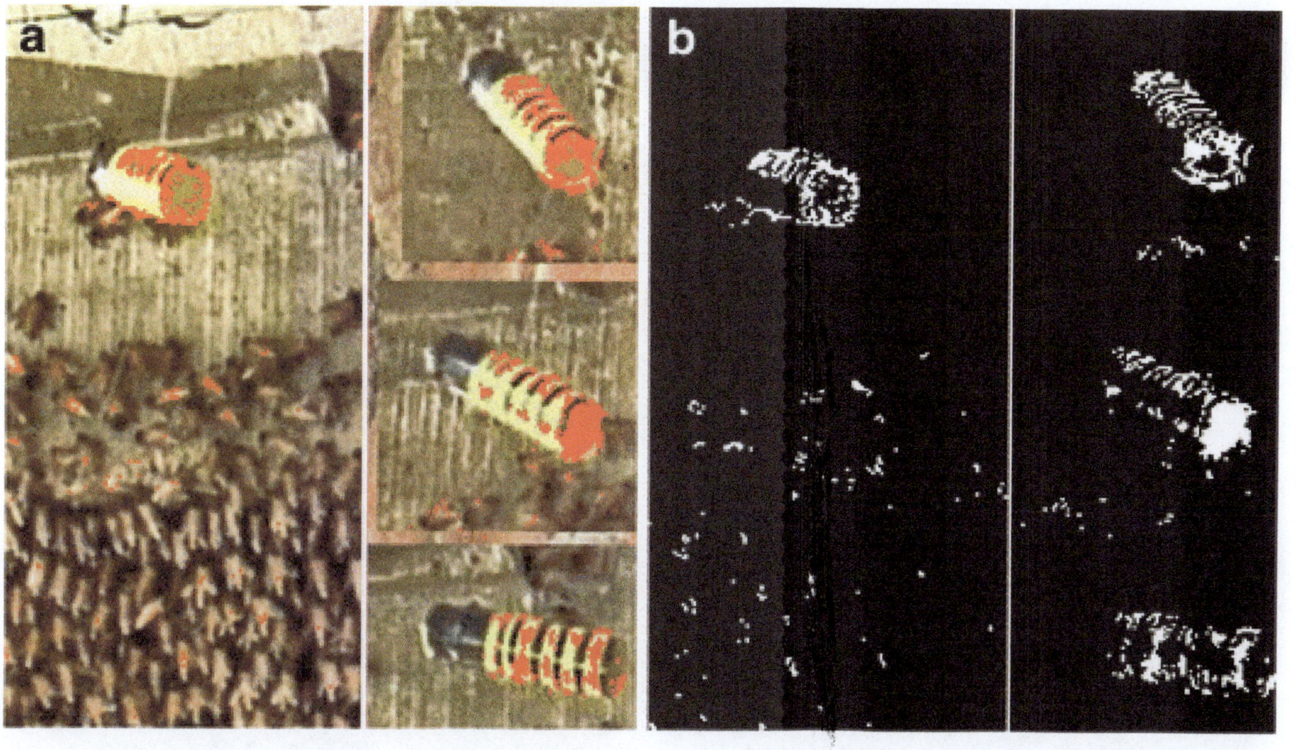

Figure. 2 Automated detection of the dummy wasp position. Four scenes under various lighting conditions and angles taken from real video images of the dummy wasp. The colour of the dummy was partially matched in the video film (marked as red areas in **a**) and segmented (marked as white areas in **b**); for comparison, see HD images of the dummy above the attachment zone of the nest in Figs. 1 and 3

Experimental Sessions and Assessment of Shimmering Motions

During shimmering, cohorts of bees in the surface layer of the nest flip their abdomens upwards in an angle which may exceed 90°. Synchronized and cascading processes emerge to a Mexican wave-like pattern. A light-emitting diode (LED) placed next to the comb was used to signal the start of the recording of the oscillations of the central comb. A video-recorded LED flash allowed synchronisation between video recording, image analysis and vibrometry. Each of the experimental sessions (n_s = 23) lasted for 6,292 frames (≡ 125.84 s) and included two phases: (a) an arousal phase, in which the colony had been stimulated by the presentation of the moving dummy wasp (Fig. 1), and (b) a subsequent quiescent phase, in which the dummy had been halted at its parking site at the upper left side of the nest (Fig. 1). The video films were re-formatted as sequences of jpg images using the Avid Media Composer editing software (Avid Technology, Inc.) in order to enable the pixel-based luminance analysis (ImagePro Plus, Media Cybernetics). Motion patterns of dummy (dw) and shimmering (sm) were quantified in terms of differences in pixel luminance (Δlum) between two sequential frames. Minor to no change in luminance values (Δlum ≤ 5) represented the 'motionless' state and were displayed as 'black' in the differential image. Change in luminance values of Δlum > 5 signalled 'movement'; they were segmented (Kastberger et al. 2011a, b,2012, 2013) as 'white' spots and charted regarding the coordinates of their gravity points (shimmering: $\mathbf{x}_{sm}$, $\mathbf{y}_{sm}$) and their pixel areas per frame. The sum of white areas in a difference image of the nest surface was taken separately for dummy and shimmering (A_{sm}) as the value for motion activity at the time the second sequential frame was captured (Fig. 2).

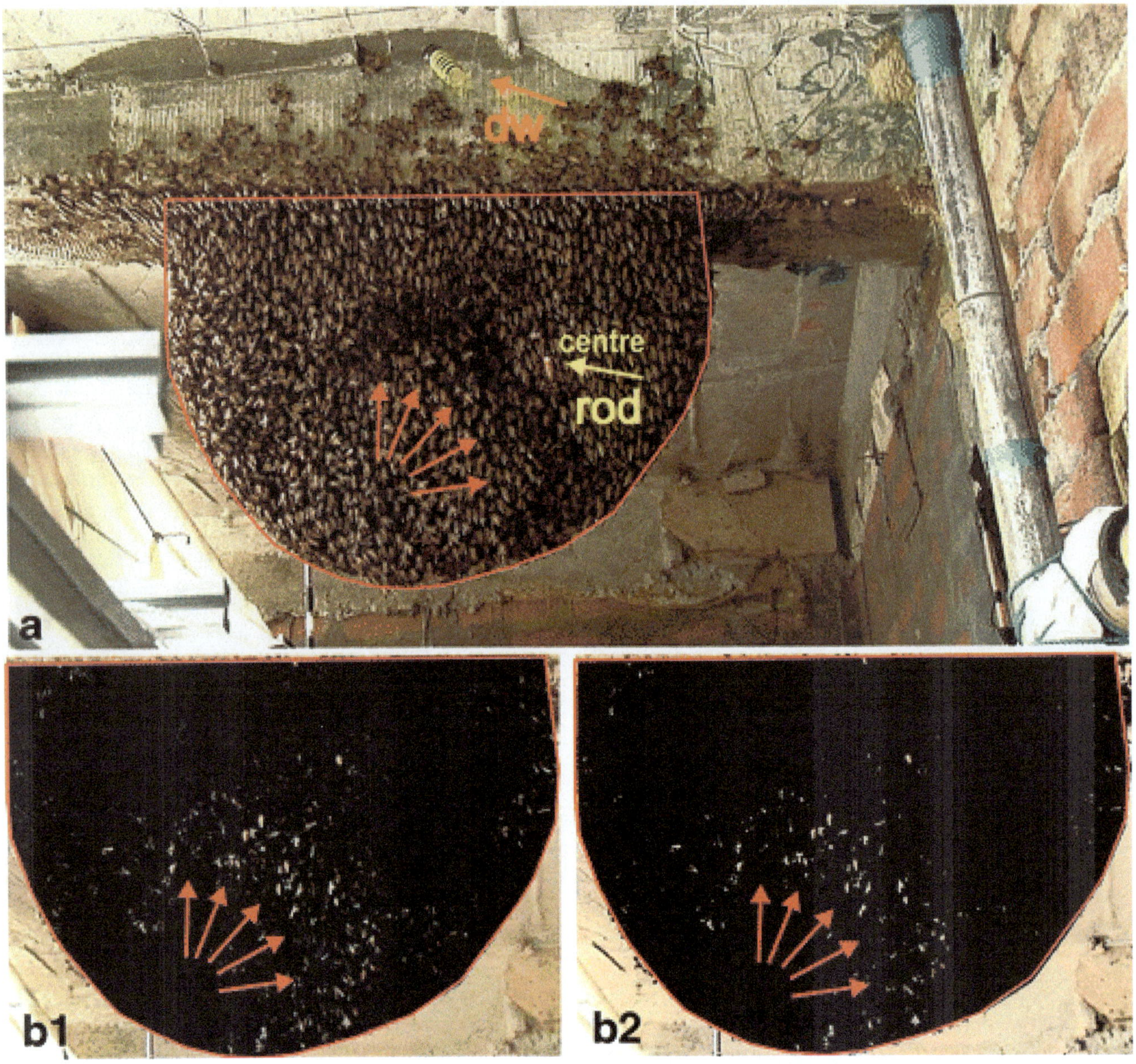

Figure. 3 Detection of motion activity at the nest surface. **a** Original HD image of the nest with the dummy wasp (dw, see arrow) from 1 of 50 experimental sessions. The wooden rod for the LDV measurement (see small yellow circle marked by the yellow arrow) was fixed in the comb (by pushing it through), here at position centre (see Fig. 1). On the front plane of the rod, the original bright reflection spot of the laser beam is visible. The darker shadow-like areas in the plane of the nest (in front of the red arrows) indicate the presence of a shimmering wave which spread in the image hemi-cyclically from bottom left to top right. The red line around the nest gives the area of interest (AOI) in which motion detection was processed by image analysis technique. **b** Differential luminance images of the nest (AOI only), achieved by pixel-wise subtraction of frames ($lum[f_i]-lum[f_{i-1}]$), revealing the motion-active surface bees at time t_i as bright pixel areas on the black background; $\mathbf{b_1}$ gives the differences in pixel luminance as grey shades (black: $\Delta lum=0$; white: $\Delta lum=255$), $\mathbf{b_2}$ shows the supra-threshold changes in luminance (black: $\Delta lum \leq 5$; white: $\Delta lum > 5$) quantifying motion by segmented pixel areas (Kastberger et al. 2011a, b; compare Fig. 2b)

COMPARISON OF SHIMMERING ACTIVITY AND COMB VIBRATION

For each experimental session, the motion patterns of the dummy (dw) and of shimmering (sm) at the nest surface were monitored together with the displacements of the central comb (D_c) during the arousal (ar) and quiescent (qu) phases. The two 'oscillatory' signals, the shimmering activity and the comb vibration, exhibited different base ('natural') periods (T_{sm}, T_c) and base ('natural') frequencies (f_{sm}, f_c), respectively. Therefore, the momentary amplitudes of the signals (A_{sm}, A_c) could not be directly compared. We established a demodulation procedure by maxima enveloping by searching for peaks in the course of the amplitudes of both signals (${}^{max}A_{sm}$, ${}^{max}A_c$) within running intervals ($[f_{i-2}, f_{i+2}]$, with f_i as the frame being examined) and by connecting the detected peaks by straight lines (Fig. 4c, e). The enveloped data of both signals (${}^{env}A_{sm}$, ${}^{env}A_c$) were then correlated for every time point t_i (at the frames f_i) throughout both experimental phases. For the basic statistics (mean ± SE), the data were stepped according to the intervals of the dummy presentation cycles in the arousal phase and in 5-s intervals for the subsequent quiescent phase (see vertical intercept lines in Fig. 4).

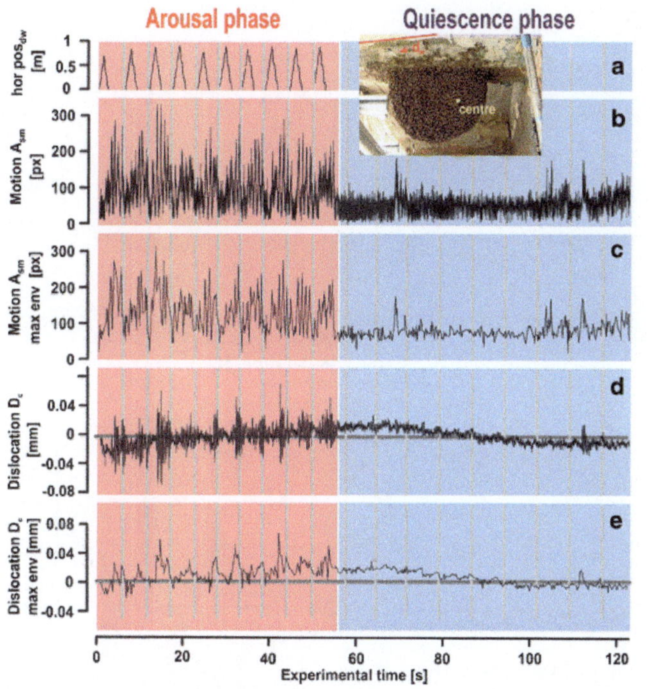

Figure. 4 Motion at the nest surface and comb vibrations in the arousal (pink background) andquiescence (blue background) phases of a single experimental session. **a** The horizontal position of the dummy wasp (hor pos$_{dw}$) over time during ten passes across and back in the front of the nest (which defined the arousal phase in the experiment) from position 0 m at the left side to position 1 m at the right side as depicted in the images of Figs. 1and 3. **b** Time course of motion activity due to shimmering (A_{sm}) at the surface of the experimental nest scaled in pixel area (see Fig. 3b$_2$). The peak performance was caused by the movement of the dummy wasp (pictured in Fig. 2). **c** Envelope function of the maximal values of the motion activity A_{sm} shown in **b** (max env A_{sm}, see command 1). **d**Vibrations scaled as dislocation (D_c) over time of the central comb in millimetres, as assessed by LDV with the reflective end of the rod as target at the centre position of the nest (see inset). **e** Envelope function of the maximal values of the vibrations shown in **d**(max env D_c, see command 1). The grey vertical lines in **a–e** define the interval limits in which correlations between shimmering and vibration activity (see Fig. 5).

RESULTS

Shimmering activity dislocates mass in the bee curtain

The mass of a giant honeybee (A. dorsata) nest is determined by two assembly parts (Ruttner1988; Kastberger 1999; Kastberger et al. 2011a, b), comb and bee curtain. The comb is built of wax cells with watery contents of larvae, honey and pollen. A single cell of the comb with a diameter of 6 mm and a length of 25 mm (values based on own observations) has a volume of 7.065×10^{-7} m^3 which sums up for the comb with a double layer of cells (assuming a mean mass density of 0.80 kg/L due to the part of the wax) to 40 kg/m^2 comb area.

The bee curtain consists of multiple layers of bees. In addition to the honey stores in specified comb areas in the genus Apis (Ruttner 1988), giant honeybees possess individual honey provisioning (own observations) as a social property which makes them particularly robust against short-term food shortages. Most of the shimmering-active curtain bees contain a mass of 100 mg or more of watery honey in their stomach (Oldroyd and Wongsiri 2006; Kastberger et al. 2011b), which accounts for the majority of the abdomen mass. Assuming that a single bee has a body weight of 150 mg, the bee curtain of a nest of 50,000 bees would weight approximately 7.5 kg. Based on these estimates, large giant honeybee nests could achieve a mass of more than $M_{nest} = 50$ kg (Ruttner 1988).

Colony members as honey storers become much heavier in their abdomens compared to their counterbalancing heads. This aspect is particularly important for the shimmering action, whereby surface bees flip their abdomens (abd) up at angles ranging from 20° to 120°, depending on their arousal level (Kastberger et al. 2011a, b). From a physics viewpoint, each bee acts here as a torsion pendulum (Kastberger et al. 2011b) with an acentric axis and two disparate masses on both sides of the rotation centre, the upward head and the downward abdomen. Abdomen flips cause a single oscillation or multiple oscillations of the mass of the abdomen in a curved motion. Mass inertia produces a reaction force, which initially presses the individual bee with the thorax towards the nest, but consecutively, the thrust of the pendulum mass initiates a centrifugal force which is directed away from the comb (see a movie of an abdomen-flipping model bee in Kastberger et al. 2011b). Both motion components affect the z-direction, which can be defined regarding the inertial system of the nest as towards and away from the comb (which defines the z-direction). In this way, a single bee would provoke a centrifugal force of $F_z = 1.266$ mN (Eq. 1), assuming an abdomen mass [M_{abd}] of 100 mg of honey and the respective flipping properties (angle [Θ_{abd}], 90°; duration [Δt_{abd}], 80 ms; radius of abdomen length [l_{abd}], 10 mm):

$$F_z = M_{\text{abd}} \times \omega_{\text{abd}}^2 \times l_{\text{abd}}, \text{ with } \omega_{\text{abd}} = \Theta_{\text{abd}} / \Delta t_{\text{abd}} \text{ as angular velocity} \qquad `(1)$$

In the collective action of shimmering (Seeley et al. 1982; Kastberger et al. 2011a, b) in which hundreds of surface bees are synchronised and cascading within a fraction of a second, the centrifugal forces of the shimmering-active surface bees may total up to more than 1 N and pull the subsurface layers of the bee curtain away from the comb (Kastberger et al. 2011a). For an A. dorsata nest, this magnitude of force is powerful enough to shift a significant mass of the bee nest for a fraction of a second to an observable extent. Under this mechanical condition, a giant honeybee nest might be considered as a driven physical pendulum (see Online Resource 1). This is even enhanced by the fact that the mechanical perturbation caused by the concerted wave-like patterns of abdominal flips occurs asymmetrically at only this side of the bee curtain at which the surface bees had been threatened by the visual cue.

Shimmering activity provokes comb vibrations

In the experiments presented in this paper, two phases of the nest are distinguished (Fig. 4; see Online Resource 2/Movie 1):

the arousal phase, in which the dummy wasp was presented in subsequent cycles provoking shimmering waves (Fig. 4a), and the quiescent phase, in which the dummy stayed positioned at the parking site (see the 'Material and methods' section). Motion activity at the nest surface was low in quiescent phases (Figs. 4, 5 and 6), but rose significantly during arousal in general, such as during mass flight activity (Kastberger et al. 1996) with diffuse, unsynchronized locomotion, or in shimmering activity, in which the abdomens predominantly of surface bees display coordinated, repetitive and, therefore, portioned motions of patterns over time (Kastberger et al. 2011a, b; Fig. 4b). All these motions of the curtain bees provoke mass shifts which give rise to associated vibrations at the comb (Fig. 4d).

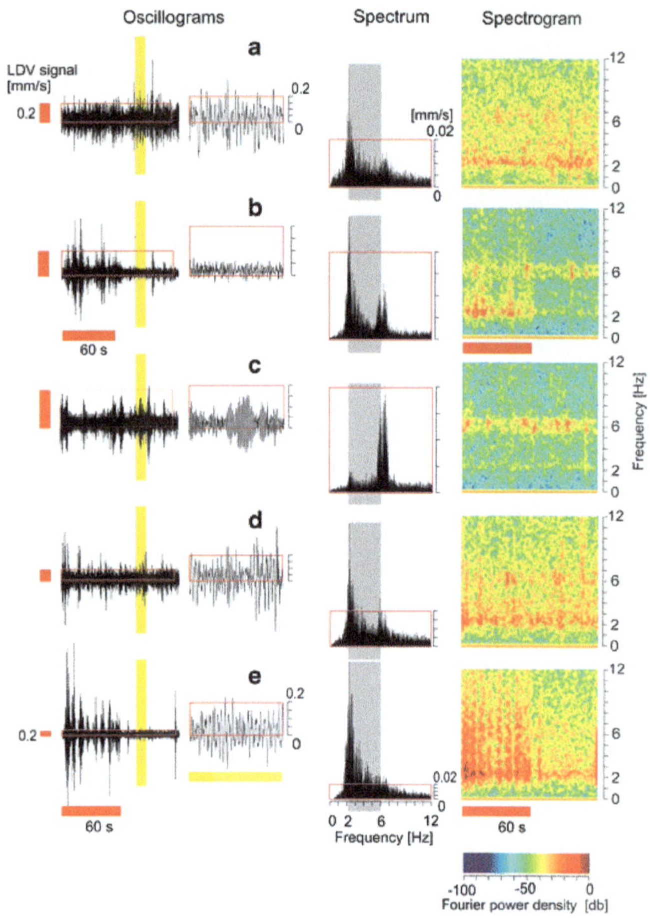

Figure. 5 LDV signals of the comb under different experimental conditions. **a** Quiescent conditions after the mass flight activity, LDV signals recorded at position S1 (Fig. 1); the dummy wasp was at its parking site. **b** The dummy wasp was presented for 60 s (see time scale asred horizontal bar), LDV measurement at position S1. **c** Mass flight activity without the presentation of the dummy wasp, LDV measurement at position S1. **d** Quiescent conditions without the presentation of the dummy wasp, LDV signals recorded at position S2 (Fig. 1). **e** The dummy wasp was presented for 60 s (see time scale as red horizontal bar), LDV signals recorded at position S2. 'Oscillograms' panels, left-side panels refer to LDV oscillograms (ordinate [A]: velocity in millimetres per second) of the whole experimental session (abscissa, 125 s), right-side panels refer to 10 s (from 80 to 90 s of the original session, as documented by the position and width of the vertical yellow bar). 'Spectrum' panels refer to frequency spectra of those sessions documented in the 125-s oscillograms; abscissa: frequency range in hertz; ordinate: amplitude parameter (velocity in millimetres per second). 'Spectrogram' panels: abscissa: experimental time; ordinate, frequency; z-axis, PSD from −100 to 0 db; PSD=10 log10 ABS $(A_i/A_0)^2$), with A[velocity in millimetres per second] as the amplitude of the Fourier coefficients; see scale of colour codes

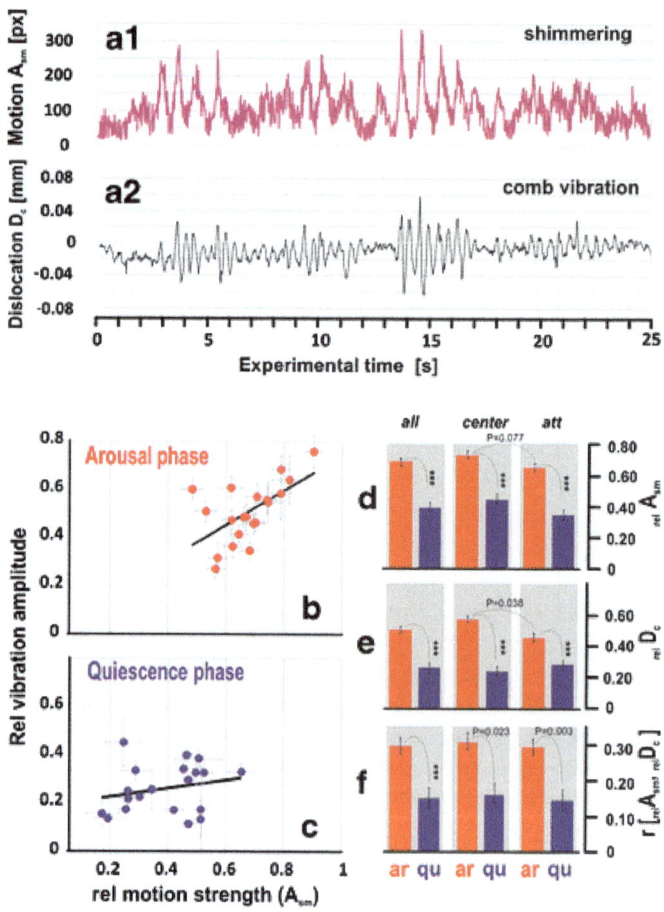

Figure. 6 Correlation of the comb vibrations and the shimmering activity. For measurement at the position centre, samples of motion activity at the nest surface are shown during shimmering (**a**$_1$) and the corresponding vibrations at the comb (**a**$_2$). **b**, **c** Abscissa, motion strength $(_{rel}s_m)$; ordinate, dislocation per 0.02 s: $_{rel}D_c$ of vibrometer signals in thearousal (ar, red) and quiescence (qu, blue) phases; for every experimental session (n_s=21), the envelope data of comb vibrations and shimmering motions per frame were correlated. Full circles arithmetical means, horizontal and vertical bars SEMs; regression functions of the mean values: arousal phase ($_{rel}D_c$=0.080+0.676s_m; R^2=0.413); quiescence phase ($_{rel}D_c$=0.192+0.176s_m; R^2=0.0549). **d–f** Comparison between arousal and quiescent phases: motion strength at the nest surface $_{rel}A_{sm}$ (**d**), comb vibration as relative dislocation $_{rel}D_c$ (**e**) and correlation coefficients r [$_{rel}A_{sm}$, $_{rel}D_c$] (**f**) of the means (n_s=21) at all positions and at the positions centre (n_s=9) and att (n_s=12); vertical black bars indicate SEMs; statistics: ***P<<0.01 (t test)

Generally, shimmering waves are produced by collective decision-making of the surface bees, whereby the natural repetition rate and the motion strength of the shimmering waves (Kastberger et al. 2008, 2011a, b) depend on the arousal level of the participating surface bees. We observed waves at a repetition rate of wps_{sm}=0.957±0.030 Hz (N_{wav}=1,287 waves in 1,361 s recording time, with wps as waves per second). The sequences of comb vibrations showed, particularly in the arousal phases, a rhythm in their modulation similar to that of the associated shimmering waves (Fig. 4b, d). This coincidence convincingly indicates that the comb oscillations were incited through the driving force of the shimmering process. The sinusoid oscillations at the comb had base frequencies of pps_c=2.156±0.042 Hz (N_p=2,887 periods in 1,341 s; with pps as periods per second) which are more than double the average repetition rate of shimmering waves.

In Fig. 5, five experimental sessions exemplify the comb vibrations displayed by oscillograms, spectrograms and frequency spectra of original LDV velocity data (see the 'Material and methods' section). The first session (Fig. 5a) refers to the basically quiescent condition at position S1 (Fig. 1) near the attachment zone. In this session, small shimmering waves happened which had been released by colony-intrinsic reasons. Two frequency bands are documented in the comb vibration, one at the basic frequency of slightly more than 2 Hz, as calculated previously, and a second band at roughly 6 Hz.

The second session (Fig. 5b) refers to the main experimental condition investigated in this paper in which the presentation of the dummy wasp gave rise to shimmering waves in thearousal phase, while in the post-presentation phase, the colony returned to quiescence. In this session, comb oscillation is displayed on both frequency bands, whereas it becomes clear that the shimmering waves provoked oscillations at the lower band, while in the post-presentation phase, these lower band activities terminated while the small oscillations at the 6-Hz band still continued. There are two reasons why these comb oscillations provoked by shimmering waves (Fig. 5b) are much smaller than those under Fig. 5e. First, the shimmering waves happened mostly at nest regions other than at position S1 (which was not in the centre of the shimmering patterns), and second, the position S1 was near the attachment zone of the nest.

The third session (Fig. 5c) refers to a period of mass flight activity in which the colony turns into a general state of arousal (Kastberger et al. 1996). Here, locomotor behaviour at the nest surface is combined with heavy flight activity, whereby bees start from the outer layers of the bee curtain and return to the nest some minutes later. In this state, the colony is in turmoil, revealing a characteristic diffuse mechanical activity of locomotor actions and small shimmering waves provoked by returning bees (Kastberger et al. 2011a). The striking difference here to all other samples of Fig. 5 is that the comb oscillations are dominantly represented at the frequency band at 6 Hz. This preference could be characteristic for the driving force as recruited under mass flight activity. However, the velocities of the comb oscillations are here at a low level compared to those at position S3 (Fig. 5d, e).

The fourth sample session (Fig. 5d) refers again to quiescent conditions without the presentation of the dummy wasp as documented under Fig. 5a. Also, here, small 'spontaneous' shimmering waves happened and the comb oscillations are reflected at both frequency bands. However, the vibrations reveal a much higher intensity (as displayed in the oscillograms and in the spectrum), which is obviously caused by the greater distance of position S3 (Fig. 1) from the attachment zone of the nest.

Lastly, session 5 (Fig. 5e) refers to position S2 and to the sequence of two phases, the arousal phase, in which shimmering waves were elicited by the presentation of the dummy wasp, and the subsequent quiescence phase, without the presence of the dummy. The shimmering-provoked comb oscillations reveal a broad frequency band with a clear peak at slightly above 2 Hz, while in the quiescent phase, both frequency bands were better contrasted (compare Fig. 5d). The shimmering-induced oscillations were quite intense, which is due to the fact that the measurement was taken from position S2 which was located near the main path of the shimmering waves and near the lower rim of the nest.

These vibration spectra of the compound physical system of the nest reveal two strikingly constant frequency bands: the low band of slightly above 2 Hz can be plausibly interpreted as the natural frequency of this compound system consisting of the comb and bee curtain because it is factually independent regarding the frequency performance from the driving forces, which may be provoked by shimmering waves (Figs. 4 and 5b, e) or by the mass flight activity (Fig. 5c). Both arousal states chosen differ in their time profiles of the driving forces: shimmering waves are rhythmic and strong and the mass flight activity provides more stochastic and much weaker influence. Nevertheless, in the frequency spectra, two prominent bands are visible, at >2 and 6 Hz, which is slightly more than double of the base frequency (Fig. 5a–e).

The samples in Fig. 5 let us assume that the driving forces of shimmering waves and mass flight activity feed their energy into different frequency bands: shimmering waves regard the low band of slightly above 2 Hz and mass flight activity regards the higher band at 6 Hz. The fact that both frequency bands can be detected under both, quiescent and arousal, conditions, though at different intensities and at different locations, supports the view that the compound system of the comb and bee curtain functions as an elastic plate oscillator.

Correlation of comb oscillations with shimmering waves

Both periodic processes, the comb oscillations and the shimmering waves, differed in their basic frequencies (Fig. 4). For correlation, time-dependent power levels of both signals were compared throughout the experimental sessions frame by frame, estimated by peak enveloping (see Fig. 4), by collecting the arithmetical means ± SE of the strength levels of the shimmering signals (A_{sm}) and of the vibration (A_c) in stepped intervals ($\Delta t = 0.02$, FPS = 50 Hz). The amplitude of the vibration between the two time points is further on termed as dislocation (D_c).

The two enveloped data sets ($n_{exp} = 21$ experiments; $n_{int} = 132,321$ time intervals of 0.02 s) correlated positively (for regression functions, see Fig. 6). The comb oscillations showed underarousal conditions due to shimmering waves significantly ($P < 0.001$, Student test) larger dislocations than under quiescent conditions (arousal: $D_c = 7.230 \pm 0.573$ µm per 0.02 s frame interval, which equals to $D_c = 0.3099 \pm 0.0286$ mm/s, $n_s = 388$ dummy wasp sessions; Fig. 6a;quiescence: $D_c = 3.489 \pm 0.283$ µm per 0.02 s, which equals to $D_c = 0.1745 \pm 0.142$ mm/s, $n_s = 150$ post-arousal sessions; Fig. 6b).

Spatial characteristics of the physical excitability of the comb

The two comb positions (centre and att; Figs. 1 and 6) represent, under the preposition of a physical pendulum (see Online Resource 1 and the 'Discussion' section), two sample distances taken from the pivotal axis. The basic frequencies in the arousal phase were slightly above 2 Hz (position centre, 2.112 ± 0.063 Hz; $n_p = 1,779$ oscillation cycles in 842 s; position att, 2.225 ± 0.034 Hz; $n_p = 1,108$ periods in 499 s) and did not significantly differ (P=0.131, Student test; n_{exp}=21) between both positions. The enveloped amplitudes of oscillations ($^{env}A_{sm}$, $^{env}D_c$; see the 'Material and methods' section) were related per session to the maximal value to compensate for differences in the recording conditions in the experimental sessions. The dislocations of comb vibration ($_{rel}D_c$; see Eqs. 2a, 2b, and 2c) under arousal were markedly higher at positioncentre (n_{exp}=9) than at position att (n_{exp}=11; P=0.059, Student test; Fig. 6d):

$$D_c[t_{i-1}, t_i] = \Delta A_c/(0.02)\ [\mathrm{mm/s}] \text{ with } \Delta A_c = (A_c[t_i] - A_c[t_{i-1}]) \text{ for every int} \quad (2a)$$

$$_{rel}D_c = D_c[t_{i-1}, t_i] / \max D_c[t_{i-1}, t_i] \quad (2b)$$

$$A_c[t_{i-1}, t_i] = \Delta A_c / (0.02^2)\ [\mathrm{mm/s^2}] \quad (2c)$$

The shimmering activities in the two series of experiments concerning the measurement positions centre and att had similar strength levels (P=0.077, Student test; Fig. 5c), which allows comparison of the associated dislocations. This consistency is also expressed by thearousal factor F_{ar} (Eq. 3; centre: F_{ar}=1.64; att: F_{ar}=2.05; P=0.83, chi-square test) which denotes a similar proportion between the motion levels in the subsequent experimental phases of arousal (ar) and quiescence (qu):

$$F_{ar} = {}_{rel}A_{sm}[ar] / {}_{rel}A_{sm}[qu] \quad (3)$$

As expected, the correlation coefficients between $_{rel}D_c$ values of comb vibration and the $_{rel}A_m$ values of motion strength at the nest surface yielded higher (P<0.05, Student test) magnitudes under arousal than under quiescence and had also similar proportions regarding both measurement positions (Fig. 6e).

Assessment of cycle-based dislocation and acceleration spectra of comb vibrations

To compensate for obvious noise effects (see Online Resource 1) and to prove perspectives for potential communication goals, the dislocation and acceleration spectra have to be considered regarding the basic ('natural') period of the vibration cycles of the comb ($T_c = 1/f_c$) under shimmering (arousal) conditions which was assessed as pps_c=2.156±0.042 Hz (see previous section). For that, we determined the peaks and sinks of the time integral of the LDV signal with an automated method according to the algorithm of command 1:

Command 1

- IF $(A_i - A_{i-3}) > A_{threshold}$ AND $(A_i - A_{i+3}) > A_{threshold}$ THEN PEAK $[f_i]$=TRUE
- IF $(A_{i-3} - A_i) > A_{threshold}$ AND $(A_{i+3} - A_i) > A_{threshold}$ THEN SINK $[f_i]$=TRUE

checking every frame f_i for exceeding the respective threshold conditions with $A_{threshold}$=5 μm.

This procedure factually introduced a digital low-pass filter (with $T_c > 20 * \Delta t_{ff}$) and documented the dynamics at the basic frequency of the comb. It allowed the assessment of dislocation (or of acceleration) values per oscillation cycle of the comb and utilized the $|\Delta A_{p,s}|$ values between a peak and the subsequent sink (or between a sink and the subsequent peak) for calculating the respective dislocations D_{vc} within vibration cycles (Eq. 4):

$$D_{vc} = |\Delta A_{p,s}| / \Delta t_c \text{ and } a_c = |\Delta A_{p,s}| / \Delta t_c^2 \text{ with } |\Delta A_{p,s}| = |A_{PEAK} - A_{SINK}| \text{ ar} \quad (4)$$

The resulting spectra of displacement and acceleration reveal (Fig. 7) that, under arousalconditions, 51.08 % of vibration cycles of comb oscillations exceeds the sensory threshold of honeybees (calculated after Sandeman et al. 1996). The spectra of Fig. 7 strongly differ (P≪0.01, Student test) for both positions under arousal (centre: n_{vc}=3,160; att: n_{vc}=3,416) and between arousal and quiescence (centre: n_{vc}=3,215; att: n_{vc}=2,895), but not underquiescence. The data suggest that the spectra are primarily shaped by differences in amplitudes of the vibrations and not by different noise levels.

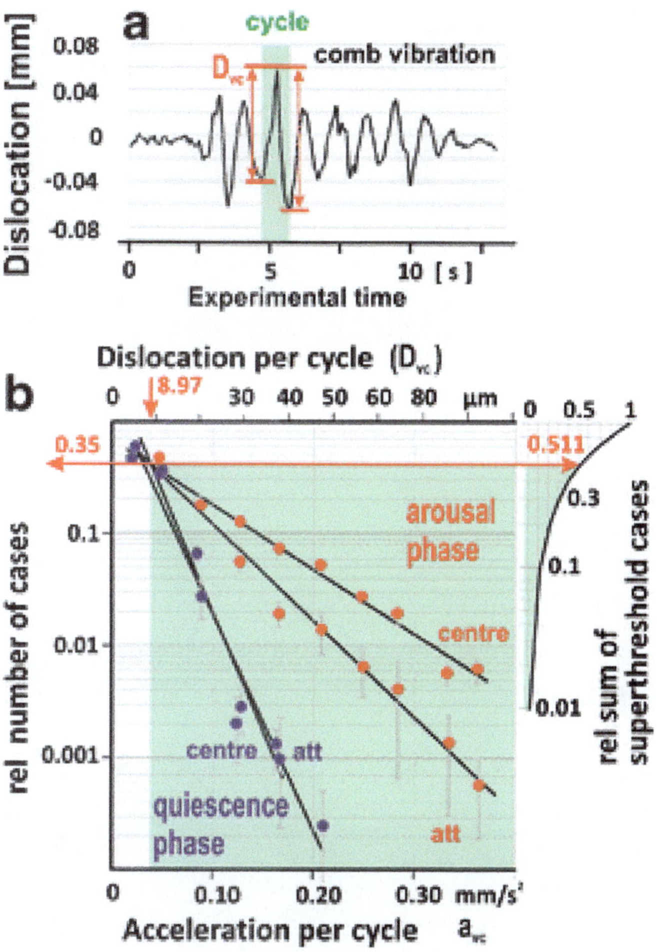

Figure. 7 Cycle-based dislocation spectra at the comb in the experimental A. dorsata nest as provoked by shimmering waves. **a** Definition of dislocation (D_{vc}) per cycle of comb vibration (f_{vc}=2.114 Hz; T_{vc}=0.473 s) as the distance [in micrometres] between a sink and the successive peak of the time-integrated DLV signal or between a peak and the successive sink. **b** Spectra of accelerations (a_{vc}) and dislocations (D_{vc}) of the comb of the experimental A. dorsata nest in the arousal and quiescence phases of experiments: ordinate, relative numbers $_{rel}N_{vc}$ of vibration cycle (vc) intervals of the respective class value (D_{vc}); abscissa, ten classes of dislocations D_{vc} per cycle of comb vibration. Full circles and vertical bars give the means±SEMs. Black lines regression functions of the means: $y=ce^{xd}$; with $y={}_{rel}N_{vc}$; $x=D_{vc}$; $R^2>0.94$; N_{vc}=12,686. The green area refers to supra-threshold (D_{vc}>8.97 µm per T_{vc}, after Sandeman et al. 1996) dislocation amplitudes which made up 51.1 % of cases (accessory diagram on the right)

DISCUSSION

Communicative impact of substrate-borne vibrations

The shimmering process in giant honeybees (Kastberger et al. 2008, 2011a, 2012, 2013) exemplifies the principle that fast and accurate spreading, gathering and sharing of information entails the success of group-living species (Evans et al. 2007). In an A. dorsata colony, the collective of curtain bees form for the purpose of repelling enemies (Kastberger et al. 2008,2011a, b) adaptive patterns in the visual, pheromone and mechanical domains. Visual cues trigger and flexibly adjust the Mexican wave-like shimmering process (Kastberger et al. 2010,2011a, 2013; Schmelzer and Kastberger 2009). Nasonov scenting (Kastberger et al. 1996) modulates the coherence and strength of the shimmering process. Mechanical traits, synchronised and cascading as displayed by the waves in the visual domain, are produced by the abdomen-lifting actions (Kastberger et al. 2011a, b) of bees primarily in the surface layer. These mechanical activities cause mass shifts in the nest which take place within fractions of a second (Kastberger et al. 2011a, 2012, 2013) and are asymmetrical regarding the nest geometry, as shimmering primarily concerns the threatened nest side. However, the forces associated with these mass shifts affect the whole nest and thus also the centrally positioned comb. The present paper questions here whether the shimmering-active surface bees do have the capacity to establish comb vibrations which are supra-threshold (Sandeman et al. 1996) cues to provide intra-colony information for the entirety of the nest mates (Figs. 1, 4, 5, and 6).

Use of substrate-borne vibrations for communicative goals is well known in arthropods. It has been described in sap-sucking bugs when signalling presence, attraction, alarm or defence between group members (Cocroft 1996; Cocroft and Rodriguez 2005; Hartbauer 2010) and also in termites (Evans et al. 2007), sawflies (Carne 1962) and caterpillars (Claridge 1985; DeVries et al. 1993; Cocroft and Rodriguez 2005; Yack et al. 2001; Fletcher et al. 2006). Social hymenoptera use vibration of substrates as signals, particularly for recruiting foragers (ants: Roces et al. 1993; Roces and Tautz 2001; dancing honey bees: Michelsen et al. 1986; Kirchner1993; Tautz et al. 1996; Nieh and Tautz 2000; Hrncir et al. 2006), whereas the individual insects' pounding, beating or knocking on a substrate are utilized as the sources of mechanical energy.

Honeybees contrast herein with other vibration-producing arthropods insofar as they utilise the comb as a substrate, which is built by the community itself, to which they directly cling for transmitting their communicative signals (Nieh and Tautz 2000). However, the shimmering process of giant honeybees and its relationship to comb vibration is different to the mechanical activities in the Western honeybees (Nieh and Tautz 2000). In giant honeybees, the shimmering-active individuals are positioned on the surface of the bee curtain. They have several layers of colony members underneath, which separate them from direct access to the comb and dampen the direct energy transfer from surface to comb. Nevertheless, the data presented in this paper provide evidence that the shimmering action yields sufficient power to drive the comb into measurable magnitude of vibration (Figs. 4, 5, 6, and 7).

Testing the colony-intrinsic information hypothesis in giant honeybees

In this paper, we provide first evidence that shimmering behaviour (Kastberger et al. 2011a, b,2012, 2013) of giant honeybees may essentially contribute to intra-colonial communication. The findings may support the colony-intrinsic information hypothesis (Kastberger et al. 2008) which assumes that shimmering has the potential to signal the momentary state of colony defence to the entire community, directing information to all curtain bees, even to those which have not actively participated in the wave. The concerted power of hundreds of shimmering-active surface bees conveys information through the multiple curtain layers. The centrally located comb in a giant honeybee nest is here the only element which does have the capacity of transmitting mechanical energy across the nest, in particular from the threatened to the contra-lateral, non-threatened side. This colony-intrinsic information hypothesis also addresses that both behaviours, shimmering and comb-building, may have customized themselves in relation to each other in the course of evolution, resulting in a cross-linked fine-tuning of their physical properties. Giant honeybees could herein serve as a prominent example for the performance of information transfer among colony members (summarised for arthropods by Evans et al. 2007and more generally by Alcock 2005).

However, there is still an essential question which begs for an answer in future work: Would this sort of mechanoreceptive stimulation across the comb as provoked in shimmering waves actually bias or even benefit curtain bees on the non-threatened side? Although the data of the present paper leave this aspect untouched and cannot clarify this, we nevertheless back up the possibility and importance of comb-borne vibrations for colony-intrinsic communication by enlisting relevant physical properties of giant honeybee nests which are specifically affected by shimmering.

Mechanical requirements enabling colony-intrinsic communication

Vibrations of the comb during shimmering exhibited a basic frequency of slightly over 2 Hz. They were excited by the natural repetition rate of the shimmering waves at slightly <1 Hz. This finding demonstrates independence between both frequencies which is an attribute for an obvious property of resonance of the comb. It is the nature of resonant systems (Nave 2001) to respond to external periodic influences, whereas the driven system tends to respond to frequency components of the external forces, which are close to and preferably slightly below the respective frequency bands (Nave 2001), and also to ultimately settle down to a performance determined by the driving force. The empirical data prove here the comb as an oscillator (Figs. 4, 5, and 6) and the shimmering activity as the force which drives the comb at half of its basic frequency band. Furthermore, the comb vibrations approached zero in a short exponentially decaying sinusoid function (Fig. 4d), which expresses under-damping of a potentially harmonic oscillator (Nave 2001). In this respect, it is quite remarkable that the diffuse forces produced during mass flight activity (Kastberger et al. 1996) affect the comb at its possibly first harmonic component at 6 Hz. These physical properties of the comb have been exemplified in the experimental nest but may happen in similar performance (possibly at slightly different frequency bands) in other giant honeybee nests of different sizes or ages.

These mechanical features of the vibrations, measured at the comb, and the driving forces of shimmering waves or of mass flight activity (Fig. 5) lead to two different, not exclusive, views of driven oscillator models explaining the compound physical system of an A. dorsata nest in general: it may behave as a physical pendulum or/and as an elastic mechanical plate (Fig. 8a). The question into which model was most fitting to the empirical data would have significance on the extent to which mechanical signals could be transmitted across the bee curtain.

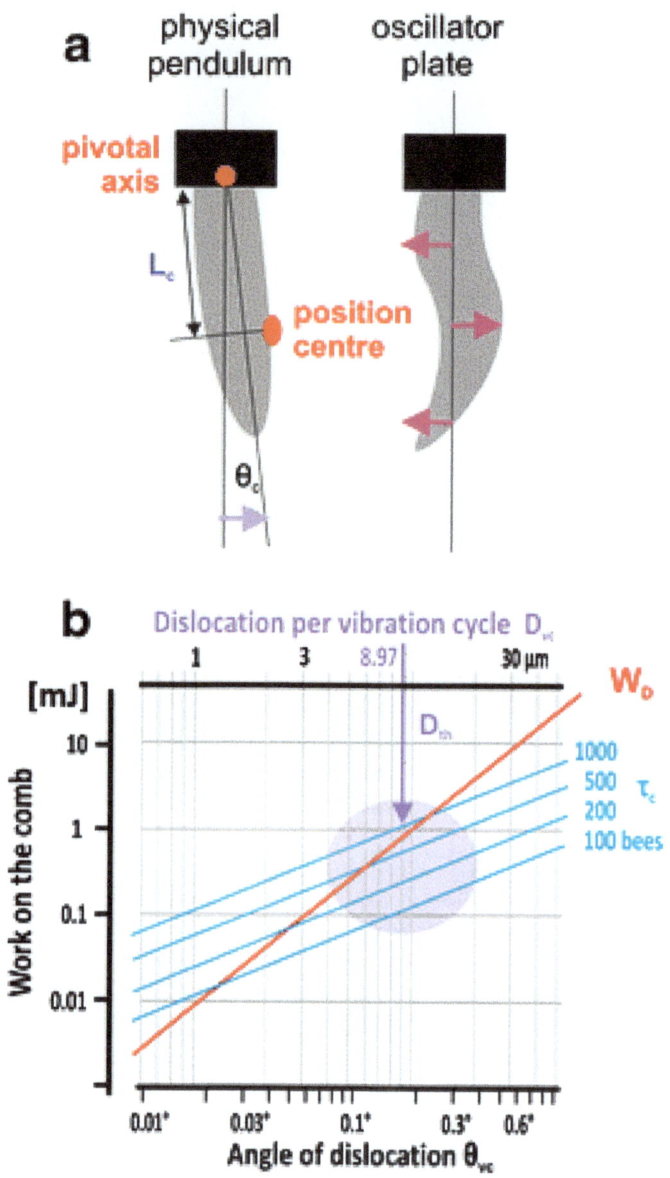

Figure. 8 How empirical data match the pendulum dislocation theory in the experimental A. dorsata nest. **a** Sketches of the comb of an A. dorsata nest as a physical pendulum and as an oscillator plate. Dislocation angle (Θ_c) of the comb, distance between the pivotal axis and the position centre as a substitute for the pendulum length L_c. **b** Lookup table of work on the comb against dislocation length (D_{vc}) and dislocation angle (Θ_{vc}) per cycle of the comb vibration at position centre (L_c=30 cm). The red line gives the work W_D(Eq. 2b) which considers the dislocation and the acceleration in the half cycle of a vibration ($T_{vc}/2$); the blue lines give the torque (Eq. 3) estimated under four conditions ofF_{sh} when 100–1,000 surface bees act synchronized but asymmetrically on one side of the comb during shimmering. The vertical arrow D_{th} gives the threshold dislocation estimated after Sandeman et al. (1996) for a vibration of 2 Hz. It is illustrated that both work values match for the threshold dislocation in the small area denoted by the violet full circle

Matching the empirical data with the physical pendulum theory

A typical giant honeybee (A. dorsata) nest, alike the experimental nest used for this investigation, would display a mass of M_c>30 kg with a vertical extension of L_c>50 cm. If excited as a physical pendulum with a typical torque force of shimmering of F_{sh}<1 N (Eqs. 1and S3b), it may swing around the pivotal axis at its specific frequency ($\Omega_C \approx 2$ Hz) with small angular displacements ($\Theta_c < 0.5°$) and a moment of inertia I_c (Eqs. S1a and S3s) which ranges between $I_c = 0.55$ kg/m² ($M_c = 30$ kg) and $I_c = 0.90$ kg/m² ($M_c = 50$ kg).

In the experimental nest, we proved the coincidence between the empirical data and the physical pendulum model in the work domain twofold (Fig. 8b): under a linear approximation (W_D, Eq. S2) and under the rotational (τ_c, Eq. S3) concept. The work W_D on the mass of the comb during one half of a vibration cycle is defined (Eq. S2b) by the product of the dislocating force F_D in the model nest ($M_c = 50$ kg; $L_c = 30$ cm; measurement position centre) and the empirically determined measures of its basic pendulum cycle (Eq. S2, with $T_c = 0.4638$ s; see Fig. 4b, d) under small dislocations ($D_c = L_c \times \sin \Theta_c < 10$ µm).

The rotational aspect of the pendulum model is quantified by the torque (τ_c) of the comb (Eqs. S3 and S5) and considers the empirically estimated driving force F_{sh} (which is set up by the cohorts of shimmering-active bees at the surface of the nest; Eq. 5), the length of the pendulum rod (with $L_c = 30$ cm, as represented by the distance between the pivotal axis and the measurement position centre) and the empirically determined angle of the natural vibration cycle Θ_c:

$$\tau_c = L_c F_{sh} \sin \Theta_c \; [\text{Joule}] \tag{5}$$

As result, the torque lines (Fig. 8b) of a variable number of shimmering-active bees ($N_{bees} = 100–1{,}000$) show gentler slopes than the W_D curve. We assume that the best correspondence of both work models (W_D, τ_c) is given at the crossing points of the functions which occurred at dislocations of $\Theta_c < 0.2°$, correspondingly $D_c < 12$ µm, per vibration cycle. Remarkably, this crucial dislocation values coincide with that of the sensory threshold of honeybees ($D_{th} = 8.97$ µm) as estimated after Sandeman et al. (1996) for 2 Hz and refer, theoretically, to a driving force produced by a shimmering wave of $N_{bee} = 500–1{,}000$ active bees (Fig. 8b).

Requirements for acceptance of the pendulum hypothesis

This match of theoretical and empirical data in the work domain supports the view that a giant honeybee nest shows the performance of a physical pendulum, with the shimmering activity of surface bees as its driving force. However, for the acceptance of this (shimmering-drives-the-nest-as-a-) pendulum hypothesis, the empirical data of the comb have to meet the further three predictions: First, the natural frequency [ω_c] of the real comb should match with the mathematically expected natural frequency of a physical pendulum of similar shape (M_c, L_c); second, the real comb should show the same natural frequency along its full 'rod' length [L_c]; and third, the amplitudes of the oscillations (Θ_c, D_c) should show a gradient according to their distance from the pivotal axis.

The measurements confirm these expectations: First, the basic ('natural') period of the comb ($T_c = 0.4682 \pm 0.0106$ s; $n_s = 23$ sessions) of the experimental nest ($M_c = 40$ kg, $I_c = 0.7$ kg/m²) under the regime of shimmering waves match the theoretical predictions (see Eq. S1a). Interestingly, under mass flight activity, the vibrations occur at 6 Hz which demonstrates that the comb also functions as an elastic plate. Second, the basic periods of the comb vibration differed only slightly between the main measurement positions (centre: $T_c = 0.4798$ s; att: $T_c = 0.4503$ s; $P = 0.1312$, t test). And third, the data document a gradient in the dislocation amplitude D_c from the attachment zone down to the lower rim in two aspects: the comb displacements $_{rel}D_c$ were larger at position centre than at position att (Fig. 6d; $P = 0.038$, Eqs.2a, 2b, and 2c) and the occurrence of larger displacements was higher at position centre ($P < 0.001$, t test) than at position att. These latter results were achieved in the frame-based (Fig. S1c–e) and cycle-based (Fig. 8b) filtered data of comb vibrations and are also documented in the spectra (Fig. 5).

Matching the empirical data with the alternative elastic mechanical plate model

The second approach to explain the swinging comb of a giant honeybee nest under the regime of shimmering is to compare it with a plane elastic structure oscillating at its natural resonance frequency. This view is obvious because the centrally positioned comb is built of wax stiffed by the watery cell contents of honey or jelly, which may give some local elasticity. This structure can be compared with Apis mellifera hives (Sandeman et al. 1996) where pieces of comb even bounded by frames on all sides still carry a low-frequency signal of about 15 Hz, while higher-frequency displacements of >60 Hz could not be detected four cells away from the source (Sandeman et al. 1996). Low-frequency signals in Western honeybees detected from the comb site are known for producing a grooming invitation dance, when workers stand stationary vibrating their bodies from side to side at a frequency of 4 Hz for 10 s (Land and Seeley 2003). It is also known that vibrations are transmitted in the plane of the comb face (Michelsen et al.1986) when produced by 'beggar' bees which press their thoraces down onto the comb while producing a 320-Hz vibration by the thoracic musculature. Here, displacements of up to 1.5 µm at right angles to the plane of the comb face can be measured even a few centimetres, but not more, away from those bees. Such traits in A. mellifera combs accord with the physical properties of an elastic mechanical plate which acts as a highly damped oscillator.

In contrast to A. mellifera, giant honeybees produce the oscillation of the central comb by the upstrokes of the abdomens of shimmering bees which happens synchronized and cascaded predominantly at the nest surface. These mechanical waves propagate across the nest but are damped down by the multiple layers of the bee curtain before they reach the comb. The elastic mechanical plate hypothesis predicts here that the comb would oscillate at its local natural frequency bands, which may vary regarding the consistence of comb cells (Sandeman et al.1996). Furthermore, the locally provoked vibrations should show displacements at different positions of essentially similar magnitude.

This aspect of an elastic mechanical plate is supported by the spectral properties of the comb of the experimental nest, in particular because of the occurrence of two main frequency bands, one slightly above 2 Hz and the other at 6 Hz (Fig. 6a–e). These two bands happened irrespective whether the nest was investigated under quiescent or arousal conditions (provoked by shimmering waves or by mass flight activity) and whether the vibrations had been detected near the attachment zone (position S1; Fig. 1) or near the lower rim zone (position S3). It also seems as if mass flight activity provoked vibrations at the higher frequency band rather than at the lower one, whereas shimmering waves mainly drove the comb at the lower frequency band. Although the data displayed in Fig. 5 were ephemeral, they document the stable character of both frequency bands, in particular at small vibration amplitudes.

Magnitude of mechanoreceptive signalling

The behavioural traits of the mechanical oscillations at the comb generated by shimmering raise two further questions: First, do the vibrations gain sufficient power to achieve signal value; in other words, are they sufficient to stimulate the curtain bees on the side contra-lateral to the threatened one? If the shimmering process has evolved (in addition to its anti-predatory goals, cf. Kastberger et al. 2008) to enforce colony-intrinsic communication, it should accelerate the comb by the wave-like flashes at intensities which exceed the sensory threshold of honeybees. InA. mellifera, the most sensitive frequencies were found (Sandeman et al. 1996) between 30 and 100 Hz with threshold displacements (D_{th}) of 2 µm and a threshold sensitivity of $D_{th}=6.2$ µm at 10 Hz. Extrapolation of the empirical data (Sandeman et al. 1996) to lower frequencies results in $D_{th}=8.97$ µm for a 2-Hz vibration which corresponds to a threshold acceleration of 0.0355 mm/s^2.

In A. dorsata nests, the vibration spectrum of the central comb, based on a cycle-related analysis during shimmering waves, revealed supra-threshold displacements of more than 9 µm per natural cycle of comb vibrations in 51.08 % of the observation time (Fig. 7), but still 10 % of vibration cycles had the fivefold value of 45 µm and, therefore, accelerate the comb by more than 0.2 mm/s. These vibrations were only present under ongoing shimmering activity but not after its termination and differed significantly in their dislocation spectra between the recording positions (centre, att) in the arousal phase of the experiments.

This means that shimmering flashes do have the potential to provoke comb vibrations as supra-threshold cues for the curtain bees. Due to the pendulum effect, we can expect that factually all curtain bees are able to sense this mechanical signal irrespective of their engagement in shimmering, practically in all layers of the curtain, and on both sides of the comb, but with two restrictions: first, these colony members who are positioned at the lower portions of the bee curtain would perceive the comb vibrations stronger than those positioned near the attachment zone; and second, the loose coupling of the bee curtain with the comb expectedly damps the impact of comb oscillations, in particular for the outer layers of curtain bees. Therefore, the cohorts near the comb could sense such vibrations more strongly than colony members at the outer layers of the bee curtain.

Summarizing, the findings not only support the pendulum's hypothesis for the impact of the pulsed forces of shimmering on the comb of giant honeybee nests but also show that the elastic mechanical plate hypothesis matches the conditions under more stochastic forces such as under the conditions of mass flight activity (Kastberger et al. 1996).

CONCLUSIONS

A giant honeybee (A. dorsata) nest comprises, from a physics viewpoint, a central comb as a stiff plate, attached to a solid substrate and covered by the multilayered bee curtain. Shimmering generates repetitive wave-like flashes in the mechanical domain and provokes vibrations which operate the comb as a driven under-damped oscillator. The repetition rate of shimmering waves ($F_{sh}=0.957$ Hz) is ≈1 Hz lower than the basic frequency of the comb ($f_c=2.112$ Hz), which determines the powering of the comb as a driven oscillator. The oscillation properties of the comb match the model of a mechanical pendulum under the pulses of shimmering waves, whereas under the much weaker and diffuse forces of mass flight activity or even under general quiescence of the nest, the comb shows with two frequency bands, a virtually harmonic property indicative for the performance as an elastic plate. Shimmering waves deliver forces upon the comb above the sensory threshold of honeybees, sufficiently strong for being perceived by worker bees throughout the bee curtain. The pendulum principle may provide mechanical information to those members of the bee curtain who are assembled nearer to the lower nest in the rim zone, while the elastic plate properties mediate shimmering-provoked vibration to those members who are in direct contact with the comb. The findings support the colony-intrinsic communication hypothesis that shimmering serves as a fast way to inform the members of the bee curtain about the momentary defensive state of the nest.

ACKNOWLEDGMENTS

We thank Dr. Madhu Singh, Dr. S.M. Man, Dr. R. Thapa and Dr. M. B. Gewali from Tribhuvan University, Kathmandu, Nepal, for their support regarding logistics, Klaus Maresch for his help with handling of the bees in Chitwan, Horst Bischof and Matthias Ruether from the Technical University of Graz, Austria, for their support with the development of the specific software used for this study and Gerald Eder, University Graz, for the basic evaluation of the primary data in his Bachelor's thesis.

Funding

The study has been supported by the Austrian Science Fund (FWF Project P 20515-B16). The funders had no role in the study design, data collection and analysis, decision to publish or preparation of the manuscript.

Ethics statement

The government office of the Tribhuvan University (Centre for International Relations) of Kathmandu (Nepal) certified the research expedition under the title 'Study on the behaviour of the giant honeybees: observations and recording of behaviours at the nesting site in the Chitwan district of Nepal 2009 and 2010'.

REFERENCES

1. Alcock J (2005) Animal behaviour, 8th edn. Sinauer Associates Inc., Massachusetts
2. Carne PB (1962) The characteristics and behaviour of the sawfly Perga affinis affinis Hymenoptera. Aust J Zool 10:1–34
3. Claridge MF (1985) Acoustic signals in the Homoptera: behavior, taxonomy and evolution. Annu Rev Entomol 30:297–317
4. Cocroft RB (1996) Insect vibrational defence signals. Nature 382:679–680
5. Cocroft RB, Rodriguez RL (2005) The behavioural ecology of insect vibrational communication. Bioscience 55:323–334
6. DeVries PJ, Cocroft RB, Thomas J (1993) Comparison of acoustical signals in Maculinea butterfly caterpillars and their obligate host Myrmica ants. Zool J Linn Soc 49:229–238
7. Evans TA, Inta R, Lai JCS, Lenz M (2007) Foraging vibration signals attract foragers and identify food size in the drywood termite, Cryptotermes secundus. Insect Soc 54:374–382. doi:10.1007/s00040-007-0958-1
8. Fletcher LE, Yack JE, Fitzgerald TD, Hoy RR (2006) Vibrational communication in the cherry leaf roller caterpillar Caloptilia serotinella (Gracillarioidea: Gracillariidae). J Insect Behav 19:1–18
9. Grasse P (1959) La reconstruction du nid et les coordinations inter-individuelles chez Bellicostitermes natalensis et Cubitermes. sp. La theorie de la stigmergie: Essai d'interpretation du comportement des termites constructeurs. Insect Soc 61:41–80
10. Hartbauer M (2010) Collective defense of Aphis nerii and Uroleucon hypochoeridis (Homoptera, Aphididae) against natural enemies. PLoS One 5(4):e10417. doi:10.1371/journal.pone.0010417
11. Hrncir M, Barth FG, Tautz J (2006) Vibratory and airborne sound signals in bee communication Hymenoptera. In: Drosopoulos S, Claridge MF (eds) Insect sounds and communication: physiology, behaviour, ecology and evolution. CRC/Taylor and Francis, Boca Raton, pp 421–436
12. Jones PL, Page RA, Hartbauer M, Siemers BM (2011) Behavioral evidence for eavesdropping on prey song in two Palearctic sibling bat species. Behav Ecol Sociobiol 65:333–340. doi:10.1007/s00265-010-1050-9
13. Kastberger G (1999) The magic trees of Assam—documentary film about the biology of the giant honeybee Apis dorsata. National Geographic, ZDF, ORF & epo-film, Wien
14. Kastberger G, Sharma DK (2000) The predator–prey interaction between blue-bearded bee-eaters (Nyctyonis athertoni) and giant honeybees (Apis dorsata). Apidologie 31:727–736
15. Kastberger G, Winder O, Hoetzl T, Raspotnig G (1996) Behavioural features of a periodic form of massed flight activity in the giant honeybee Apis dorsata. Apidologie 27:381–395
16. Kastberger G, Schmelzer E, Kranner I (2008) Social waves in giant honeybees repel hornets. PLoS One 3(9):e3141. doi:10.1371/journal.pone.0003141
17. Kastberger G, Weihmann F, Hoetzl T (2010) Complex social waves of giant honeybees provoked by a dummy wasp support the special-agent hypothesis. Commun Integr Biol 3(2):179–180
18. Kastberger G, Weihmann F, Hötzl T (2011a) Self-assembly processes in honeybees: the phenomenon of shimmering. In: Hepburn R, Radcliff S (eds) Honeybees of Asia.. ISBN 978-3-642-16421-7 (Print); 978-3-642-16422-4 (Online)
19. Kastberger G, Maurer M, Weihmann F, Ruether M, Hötzl T, Kranner I, Bischof H (2011b) Stereoscopic motion analysis in densely packed clusters: 3D analysis of the shimmering behaviour in giant honey bees. Front Zool 8:3, http://www.frontiersinzoology.com/content/8/1/3
20. Kastberger G, Weihmann F, Hötzl T, Weiss SE, Maurer M (2012) How to join a wave: decision-making processes in shimmering behavior of giant honeybees (Apis dorsata). PLoS One 7(5):e36736. doi:10.1371/journal.pone.0036736
21. Kastberger G, Weihmann F, Hötzl T, Weiss SE, Maurer M (2013) Speeding up social waves. Propagation mechanisms of shimmering in giant honeybees. Front Zool (in press)
22. Kirchner WH (1993) Acoustical communication in honeybees. Apidologie 24:297–307
23. Land BB, Seeley TD (2003) The grooming invitation dance of the honey bee. Ethology 110:1–10. doi:10.1046/j.1439-0310.2003.00947.x
24. Michelsen A, Kirchner WH, Lindauer M (1986) Sound and vibrational signals in the dance language of the honeybee, Apis mellifera. Behav Ecol Sociobiol 18:207–212
25. Nave CR (2001) Hyperphysics. Department of Physics and Astronomy, Georgia State University, Atlanta, http://hyperphysics.phy-astr.gsu.edu/hbase/pendp.html#c2
26. Nieh JC, Tautz J (2000) Behaviour-locked signal analysis reveals weak 200–300 Hz comb vibrations during the honeybee waggle dance. J Exp Biol 203:1573–1579
27. Oldroyd BP, Wongsiri S (2006) Asian honey bees. Harvard University Press, Cambridge
28. Peake TM (2005) Eavesdropping in communication networks. In: McGregor PK (ed) Animal communication networks. Cambridge University Press, Cambridge, pp 13–37
29. Roces F, Tautz J (2001) Ants are deaf. J Acoust Soc Am 109:380–382
30. Roces F, Tautz J, Hoelldobler B (1993) Stridulation in leaf-cutting ants—short-range recruitment through plant-borne vibrations. Naturwissenschaften 80:521–524
31. Ruttner F (1988) Biogeography and taxonomy of honeybees. Springer, Berlin
32. Sandeman DC, Tautz J, Lindauer M (1996) Transmission of vibration across honeycombs and its detection by bee leg receptors. J Exp Biol 199:2585–2594

33. Schmelzer E, Kastberger G (2009) Special agents' trigger social waves in giant honeybees (Apis dorsata). Naturwissenschaften 96:1431–1441. doi:10.1007/s00114-009-0605-y

34. Seeley TD, Seeley RH, Akratanakul P (1982) Colony defence strategies of the honeybees in Thailand. Ecol Monogr 52:43–63

35. Tautz J, Rohrseitz K, Sandeman DC (1996) One-strided waggle dance in bees. Nature 382:32

36. Weihmann F, Hötzl T, Kastberger G (2012) Training for defence? From stochastic traits to synchrony in giant honeybees (Apis dorsata). Insects 3(3):789–820. doi:10.3390/insects3030789

37. Yack JE, Smith ML, Weatherhead PJ (2001) Caterpillar talk: acoustically mediated territoriality in larval Lepidoptera. Proc Natl Acad Sci USA 98:11371–11375

INDEX